Lecture Notes in Mechanical Engineering

Lecture Notes in Mechanical Engineering (LNME) publishes the latest developments in Mechanical Engineering—quickly, informally and with high quality. Original research or contributions reported in proceedings and post-proceedings represents the core of LNME. Volumes published in LNME embrace all aspects, subfields and new challenges of mechanical engineering.

To submit a proposal or request further information, please contact the Springer Editor of your location:

Europe, USA, Africa: Leontina Di Cecco at Leontina.dicecco@springer.com
China: Ella Zhang at ella.zhang@cn.springernature.com
India, Rest of Asia, Australia, New Zealand: Swati Meherishi at swati.meherishi@springer.com

Topics in the series include:

- Engineering Design
- Machinery and Machine Elements
- Mechanical Structures and Stress Analysis
- Automotive Engineering
- Engine Technology
- Aerospace Technology and Astronautics
- Nanotechnology and Microengineering
- Control, Robotics, Mechatronics
- MEMS
- Theoretical and Applied Mechanics
- Dynamical Systems, Control
- Fluid Mechanics
- Engineering Thermodynamics, Heat and Mass Transfer
- Manufacturing Engineering and Smart Manufacturing
- Precision Engineering, Instrumentation, Measurement
- Materials Engineering
- Tribology and Surface Technology

Indexed by SCOPUS, EI Compendex, and INSPEC.

All books published in the series are evaluated by Web of Science for the Conference Proceedings Citation Index (CPCI).

To submit a proposal for a monograph, please check our Springer Tracts in Mechanical Engineering at https://link.springer.com/bookseries/11693.

Tamás Mankovits · Natasa Trisovic
Editors

Mechanical Design and Simulation: Emerging Technologies and Applications

Proceedings of the 6th International Conference on Mechanical Design and Simulation

Springer

Editors
Tamás Mankovits
Department of Mechanical Engineering
University of Debrecen
Debrecen, Hungary

Natasa Trisovic
University of Belgrade
Belgrade, Serbia

ISSN 2195-4356 ISSN 2195-4364 (electronic)
Lecture Notes in Mechanical Engineering
ISBN 978-981-92-2439-5 ISBN 978-981-92-2440-1 (eBook)
https://doi.org/10.1007/978-981-92-2440-1

This Springer imprint is published by the registered company Springer Nature Singapore Pte Ltd.
The registered company address is: 152 Beach Road, #21-01/04 Gateway East, Singapore 189721, Singapore

Preface

The 2026 6th International Conference on Mechanical Design and Simulation (MDS 2026) was successfully held in Debrecen, Hungary, from March 20 to 22, 2026. The conference provides an international platform for researchers, engineers, and practitioners to present and exchange recent advances in mechanical design, simulation, and related interdisciplinary areas, with an emphasis on the integration of theoretical developments and engineering applications.

MDS 2026 attracted over 80 participants from universities, research institutions, and industry, reflecting active international engagement and continued interest in the field. The technical program consisted of keynote speeches and oral presentation sessions, covering a broad range of topics aligned with current developments in modern mechanical engineering.

The keynote presentations addressed several representative and emerging research directions, including engineering system design and component optimization under complex operating conditions, intelligent control strategies for human–machine interaction systems, structural behavior analysis for practical strength-related engineering problems, and advanced tool design and manufacturing technologies. These contributions illustrate the close interplay between mechanical design, computational modeling, and real-world engineering applications.

A total of 127 submissions were received for the conference. Following a peer-review process conducted by the program committee and external reviewers, 39 papers were selected for inclusion in this volume. The accepted papers were evaluated based on their originality, technical quality, and relevance to the scope of the conference.

The contributions collected in this volume reflect a broad spectrum of research in mechanical engineering, including design methodologies, simulation and modeling techniques, as well as studies related to materials and manufacturing processes. The overall organization of the volume follows a coherent thematic structure based on methodological and application-oriented considerations. In particular, a significant proportion of the contributions focuses on simulation and modeling approaches, which is consistent with the current trend in mechanical engineering, where computational methods, multiphysics analysis, and digital engineering technologies play an increasingly central role in both research and industrial practice. Complementary contributions in design-oriented and manufacturing-related topics further ensure the diversity and completeness of the volume.

The organizers acknowledge the efforts of the authors in contributing their work and the reviewers for their careful evaluations. It is expected that this volume will serve as a useful reference for researchers and practitioners and will contribute to further developments in the field of mechanical design and simulation.

MDS 2026 Conference Committee

Committee Members

General Chair

Tamás Mankovits	University of Debrecen, Hungary

Program Committee Chairs

Weidong Li	University of Shanghai for Science and Technology, China
Liang Ge	Southwest Petroleum University, China
Natasa Trisovic	University of Belgrade, Serbia

Program Committee

Ashvin Chaudhari	Turku University of Applied Sciences, Finland
Ádám Kalácska	Ghent University, Belgium
Domonkos Tolnai	Helmholtz-Zentrum Hereon, Germany
Dimitrios Tzetzis	International Hellenic University, Greece
László Könözsy	Cranfield University, UK
Matej Vesenjak	University of Maribor, Slovenia
Ziad Saghir	Toronto Metropolitan University, Canada
Zoltán Major	Johannes Kepler University Linz, Austria
László Zsidai	Hungarian University of Agriculture and Life Sciences, Hungary
Imre Norbert Orbulov	Budapest University of Technology and Economics, Hungary
Gábor Kalácska	Hungarian University of Agriculture and Life Sciences, Hungary
Rúnar Unnthorsson	University of Iceland, Iceland
Gábor Bohács	Budapest University of Technology and Economics, Hungary
Tibor Poós	Budapest University of Technology and Economics, Hungary
János Péter Erdélyi	University of Miskolc, Hungary
Gusztáv Fekete	Széchenyi István University, Hungary
Zoltán Siménfalvi	University of Miskolc, Hungary

Publication Chair

Derui Ding	University of Shanghai for Science and Technology, China

Organizing Committee Chairs

Sándor Bodzás	University of Debrecen, Hungary
Dávid Huri	University of Debrecen, Hungary
Sándor Manó	University of Debrecen, Hungary
István Budai	University of Debrecen, Hungary
Szilvia Barkóczyné Gyöngyösi	University of Debrecen, Hungary
Sándor Pálinkás	University of Debrecen, Hungary
Sheng Lu	Chongqing University of Posts and Telecommunications, China
Jiufei Luo	Chongqing University of Posts and Telecommunications, China

Organizing Committee

Lilla Csonkané Dóró	University of Debrecen, Hungary
Ágnes Ölveti-Szabó	University of Debrecen, Hungary
Mihály Csüllög	University of Debrecen, Hungary
Wei Shen	University of Shanghai for Science and Technology, China
Xiaohui Jiang	University of Shanghai for Science and Technology, China
Yuchen Liang	Jiangsu University, China
Yang Zhao	Chongqing University of Posts and Telecommunications, China
Ying Ma	Chongqing University of Posts and Telecommunications, China

Contents

About the Editors

Tamás Mankovits is a Professor and Head of the Department of Mechanical Engineering at the Department of Mechanical Engineering, Faculty of Engineering, University of Debrecen, Hungary. He holds a Ph.D. in Mechanical Engineering and is pursuing the "dr. habil." qualification in Mechanical Engineering Sciences. His research focuses on solid mechanics, material performance, and structural optimization, with notable contributions to the study of metallic foams and the optimization of automotive rubber components. He has published extensively in engineering optimization, finite element analysis, and numerical modeling, and is actively engaged in industry-oriented engineering education.

Natasa Trisovic is a Full Professor at the Faculty of Mechanical Engineering, University of Belgrade, Serbia, specializing in theoretical and applied mechanics, fatigue, FEM, and vibrations. She has been a guest professor at institutions in Europe and the USA, and participated in major international projects including the EUREKA AWAST program. Author of over 10 ISI journal papers and 60+ conference papers, she is a member of ESIS, DIVK, and secretary of the Serbian Society of Mechanics. She received the 2005 Innovation Award from the Assembly of Belgrade Municipality.

Mechanical Design and System Development

Innovative Design of B-Axis Structure for Double-Table Turntable

Yu Jin, Meiqin Liang(✉), Shuai Liu, Yunhai Chen, and Jichuan Shao

College of Mechatronics Engineering, Beijing Polytechnic University, Beijing 100176, China
liangmeiqin@bpi.edu.cn

Abstract. To meet the demands of high-end equipment manufacturing for five-axis turntables with high precision, high stability, and fast response, this paper focuses on the structural design of the core component B-axis of a five-axis dual-table turntable. The B-axis adopts a "T-shaped layout—outer rotor direct drive—electromechanical-hydraulic integration" design scheme. By optimizing the structural layout, innovating transmission technology, and accurately selecting and verifying components, a performance breakthrough has been achieved. Finite element analysis results show that the maximum stress of the B-axis housing is 19.26 MPa, the maximum displacement is only 0.057 mm, and the strain value is 5.958×10^{-5}, indicating that the structural strength and rigidity meet the design requirements. The selected 442sim-M832 motor enables the B-axis to reach a speed of 42 rpm, with excellent acceleration performance, and inertia and torque ratios that meet operational requirements. This design effectively solves the problems of large transmission gaps, slow response, and frequent maintenance in traditional turntables, providing reliable support for the automation and efficiency of five-axis machining.

Keywords: Double-table five-axis linkage rotary table · B-axis · Outer rotor direct drive · Finite element analysis · Motor selection

1 Introduction

As the core equipment of high-end manufacturing, five-axis machining equipment is the only means enabling the machining of complex parts such as aero-engine impellers and spacecraft structural components (Li 2024; Liu 2023). As a core functional component of high-end CNC machine tools, the performance of a five-axis rotary table directly determines the machining accuracy and production efficiency of complex parts (Liu 2022, 2025). Compared with traditional three-axis rotary tables, five-axis rotary tables achieve "one-time clamping and multi-surface forming" through multi-axis coordination, fundamentally eliminating cumulative errors caused by multiple clampings (which can be controlled within 5 μm). This is crucial for meeting the high-precision requirements in the aerospace sector (e.g., ± 0.05 mm tolerance for aero-engine single crystal blades and ± 0.02 mm hole position tolerance for spacecraft connectors) and the precision mold sector (e.g., cavity machining of automotive panel molds) (Li 2021, ISO 10791. 2020,

T. Mankovits and N. Trisovic (Eds.): MDS 2026, LNME 2026.
https://doi.org/10.1007/978-981-92-2440-1_1

Liu 2022). The five-axis dual-table rotary table adopts a dual-station parallel operation mechanical structure, achieving integrated multi-face machining of workpieces through the coordinated movement of the A1, A2, and B axes. It has broad application prospects in fields such as aerospace and precision molds. The B-axis, as the core support and transmission component of the rotary table, plays a key role in connecting the machine tool body, transmitting torque, and ensuring precise table rotation (Zhang 2025, Zhao 2024). Its structural design must simultaneously meet requirements for rigidity, accuracy, and dynamic response (Zheng 2025, Wang 2024, Li 2025).

The B-axis, as the core supporting and transmission component of the turntable, undertakes the functions of connecting the machine tool body, transmitting torque, and ensuring precise rotation of the worktable (Chen 2023, Wang 2021, Zhang 2022). Its performance directly determines the overall machining accuracy, dynamic response speed, and service life of the turntable. However, traditional B-axis structures mostly adopt "inner rotor motor + worm gear transmission" technology, which has inherent defects: (1) Mechanical backlash in the worm gear pair leads to rotation accuracy errors of more than 10 microns, which is difficult to meet the machining requirements of high-precision parts (Li 2023); (2) Torque loss in mechanical transmission is severe, with energy utilization rate less than 70%, and the maximum rotation speed is limited to 3 rpm, which cannot adapt to high-speed automated production (Rexroth 2022); (3) Wear of transmission components is fast, requiring maintenance every 1–2 months, which increases production costs and downtime (THK 2021).

In response to the above problems, this paper conducts innovative design of the B-axis structure from the aspects of structural layout, transmission technology, integrated design, and performance verification. The research results can not only improve the performance indicators of the five-axis turntable but also provide a reference for the structural optimization of similar core components, which is of great significance for promoting the technological upgrading of high-end CNC machine tools in China and enhancing the core competitiveness of the manufacturing industry.

The research mainly focuses on the following aspects:

In terms of structural layout, a T-shaped symmetric structure is adopted to shorten the force transmission path, balance loads, and improve overall rigidity. In terms of transmission technology, an outer rotor direct-drive motor is introduced; the "outer rotor" design expands the force arm, increasing torque by 30% under the same conditions while eliminating backlash, thus achieving micron-level rotation accuracy. In terms of integrated design, a rotary distributor combining Glyd ring sealing and oil-gas channels is innovated to solve the problem of oil-gas-electricity entanglement and wear, extending the maintenance cycle to more than 2 years. In terms of performance verification, modeling with SolidWorks and analysis with ANSYS are conducted to verify that the maximum stress of the housing is 19.26 MPa and the maximum displacement is 0.057 mm, which meet the requirements for strength and rigidity.

This innovation effectively breaks through traditional bottlenecks, simultaneously improving the accuracy stability, dynamic response speed, and service life of the B-axis. It provides core technical support for the five-axis dual-table rotary table to adapt to the needs of high-end manufacturing, such as high precision, high productivity, and long-term stability.

2 Overall Structural Design of the B-Axis

2.1 Overall Structural Layout

As shown in Fig. 1, the B-axis adopts a symmetrical T-shaped layout with the A1 and A2 axes, and is fixedly connected to the main machine tool via a housing. The A1 and A2 axes are symmetrically distributed on both sides of the housing, forming a dual-station parallel operation structure that supports integrated multi-surface machining of workpieces. The B-axis drive motor is directly mounted on the housing and rigidly connected to the table via a rotary base to drive the table for precise rotational motion. Two sets of control valves correspond one-to-one to the tables of the A1 and A2 axes respectively, ensuring that the dual stations can independently achieve motion control and meet the differentiated needs of parallel operations.

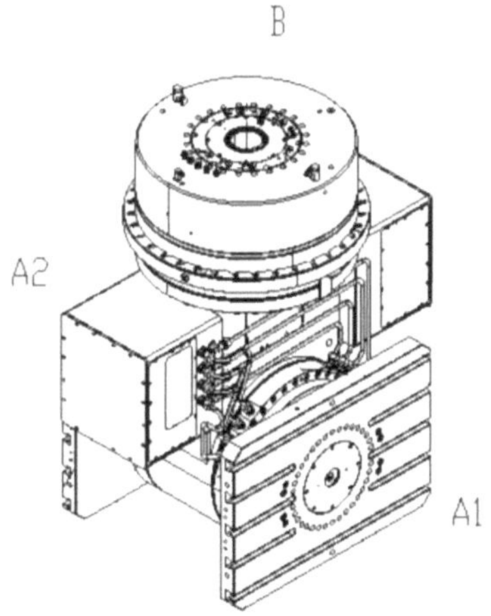

Fig. 1. Double-table five-axis linkage turntable structure drawing.

The core advantages of this layout design are: first, the T-shaped structure shortens the force transmission path, reduces stress concentration, and enhances overall structural rigidity; second, the symmetrical layout balances the load distribution during dual-station operations, preventing accuracy deviation caused by unilateral stress; third, the compact structural design provides spatial accommodation for the installation of the outer rotor direct-drive motor and integrated control system, enabling optimized reduction of the turntable volume.

2.2 Outer Rotor Direct Drive Technology

To break through the performance bottlenecks of traditional transmission schemes, the B-axis transmission system innovatively adopts an outer rotor torque motor design, subverting the layout mode of traditional inner rotor motors. As shown in Fig. 2, the stator of the drive motor forms a fixed connection with the housing, while the rotor of

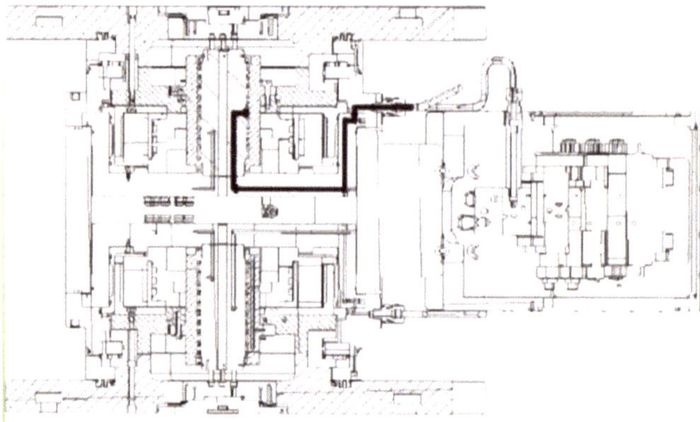

Fig. 2. Full sectional view of the double-table five-axis rotary table

the drive motor is fixedly connected to the tabletop via the rotating base. The distributor comprises a spindle and a rotating sleeve: the spindle is fixedly connected to the fixed base, the rotating sleeve is fixedly connected to the rotating base, and the rotating sleeve has the degree of freedom to rotate relative to the spindle.

With the outer rotor design, the rotor is positioned on the outer side, farther from the axis center. According to the torque formula $T = Fr$, under the same current and magnetic field intensity, the torque strength is increased by 30%. The direct-drive structure of the motor for the rotary table avoids wear in the transmission chain; when combined with YRT bearings, it enhances the equipment rigidity, enabling the table to withstand radial forces and overturning moments. A high-strength special adhesive is added at the magnetic steel sheets to ensure structural stability when bearing large torques. Compared with traditional solutions, the outer rotor direct-drive technology achieves three key breakthroughs:

(A) Accuracy improvement: eliminates mechanical backlash in the drive chain, combined with YRT bearing support, allowing the B-axis rotation accuracy to reach the micron level, providing precise guarantee for five-axis machining;

(B) Volume optimization: eliminates the worm and worm gear mechanism, reducing the overall size of the turntable by 40%, meeting the compact design requirements of CNC machines[Zheng 2025];

(C) Performance upgrade: avoids mechanical transmission wear, extending the maintenance cycle from 1–2 months to over 2 years, while increasing the rotation speed from 10 s per rotation to 2 s per rotation, significantly improving production efficiency.

During the transmission process of the outer rotor direct-drive motor, as shown in Fig. 3, the motor rotor directly drives the rotary table. Unlike the gear transmission of conventional rotary tables, the direct-drive design of the motor does not generate backlash during transmission. While saving space, it can further improve the rotational accuracy of the product. The design of the direct-drive motor is more suitable for five-axis rotary tables to enhance the rotational accuracy during coordinated movement.n addition, magnetic steel of the motor sheets use a high-strength special adhesive for fixation, ensuring structural stability under high-torque conditions and further guaranteeing transmission reliability.

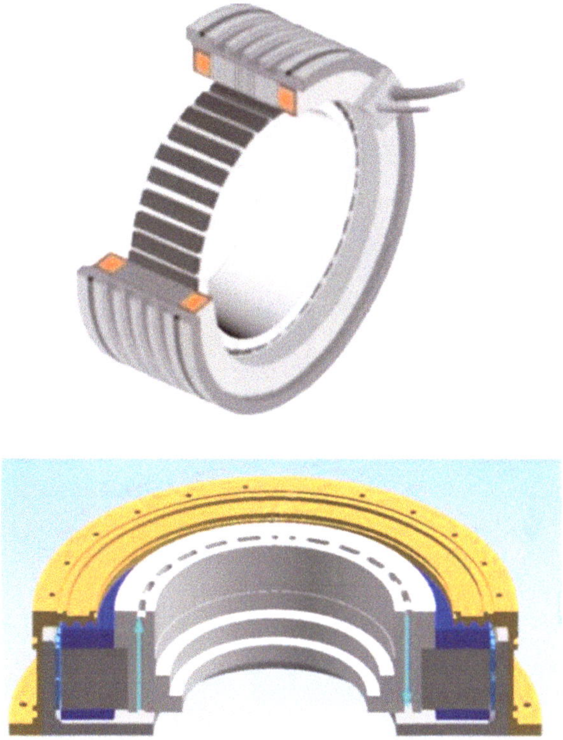

Fig. 3. Comparison diagram of inner and outer rotor motors

2.3 Rotary Distributor Sealing System

To solve the problem of oil, air, and electrical lines being prone to entanglement and wear during B-axis rotation, an innovative Gley ring sealing component oil and gas channel technology was designed, as shown in Fig. 4. The fixed distributor guides the medium into the rotating shaft through gas-liquid passages, and the rotating distributor then transfers the gas and liquid to the transitional shaft through holes and annular grooves, finally connecting to the tabletop gas-liquid output components, achieving interference-free transmission between the fixed end and the rotating end.

The Gleuer ring is made of PTFE material, combining low friction coefficient with high sealing performance. It has a service life three times that of conventional solutions after 2,000 h of continuous operation testing, and maintenance costs are reduced by 60%. The multi-channel design supports independent distribution of different gases and liquids, meeting the medium supply requirements for dual-station synchronous operations.

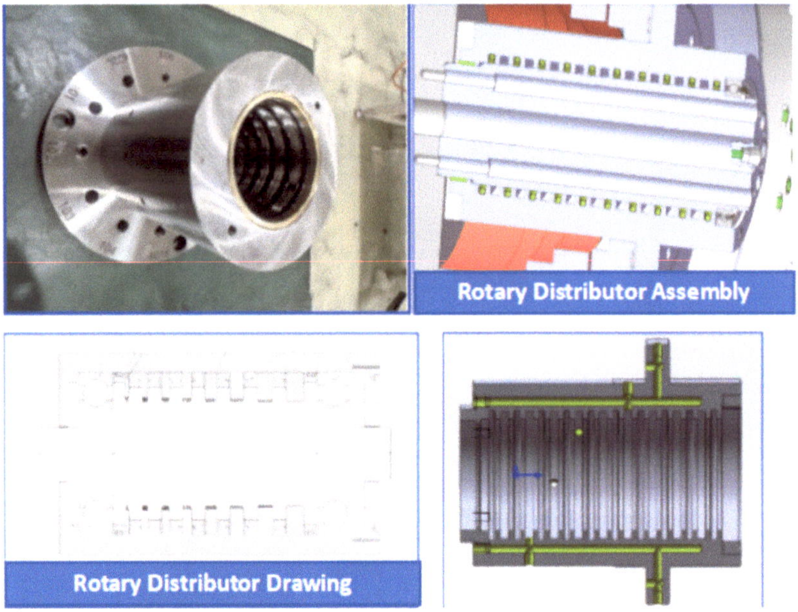

Fig. 4. Rotary distributor seal system diagram.

3 Finite Element Analysis of B-Axis Structure

3.1 Analysis Condition Settings

3.1.1 Geometric Model Simplification and Processing

A three-dimensional solid model of the B-axis housing was created using SolidWorks software, as shown in Fig. 5. To improve computational efficiency, the model was simplified.

3.1.2 Grid Division

Using the automatic mesh division function of ANSYS Workbench, combined with manual control to optimize mesh quality: a hybrid curvature mesh is used overall, with a minimum element size of 0.0128584 m and a maximum element size of 0.136002 m, balancing computational accuracy and efficiency; the final mesh consists of 436,954 elements and 693,101 nodes. The mesh quality check shows that the element aspect ratios are reasonable, meeting the computational requirements.

3.1.3 Load and Boundary Condition Settings

Using the automatic mesh division function of ANSYS Workbench, combined with manual control to optimize mesh quality: a hybrid curvature mesh is used overall, with a minimum element size of 0.0128584 m and a maximum element size of 0.136002 m, balancing computational accuracy and efficiency; the final mesh consists of 436,954 elements and 693,101 nodes. The mesh quality check shows that the element aspect ratios are reasonable, meeting the computational requirements.

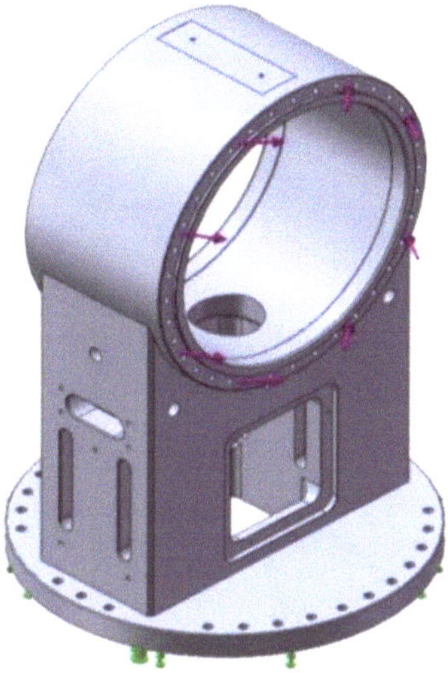

Fig. 5. 3D solid model of B-axis housing.

3.2 Calculation Results and Analysis

As shown in Table 1, Stress Analysis: The high-stress areas of the B-axis housing are concentrated at the connections between the stiffeners and the housing body. The maximum stress value is 19.26 MPa (node 1471), and the minimum stress value is 6.929×10^3 N/m^2 (node 6160). The stress distribution is uniform, with no local stress overload observed.

Displacement Analysis: The maximum total displacement is 0.057mm (node 2619), and the minimum displacement is 0.000 mm (node 17). The displacement is well below the design allowable value, indicating sufficient structural rigidity and no impact on machining accuracy due to deformation.

Strain Analysis: The maximum equivalent strain is 5.958×10^{-5} (element 17843), and the minimum strain is 5.475×10^{-8} (element 41953). The strain levels are low, demonstrating good structural stability.

Finite element analysis results indicate that the B-axis housing structure design meets the strength and stiffness requirements and can withstand the combined loads during dual-station machining.

As shown in Table 2, stress analysis:The high-stress areas of the B-axis housing are mainly concentrated at the connection between the reinforcement ribs and the housing body, with a maximum stress of 19.26 MPa, strain of 5.958×10^{-5}, and the displacement values validate the accuracy of the computational model.

Table 1. Load and boundary condition settings.

Fixture name	Fixture image	Fixture details
Fixed-1	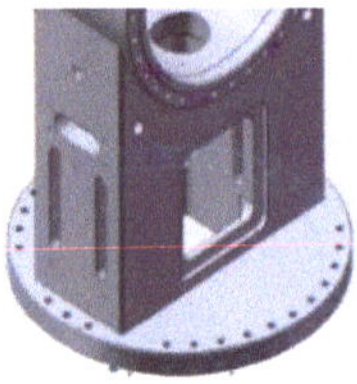	Entity: 1 face Type: Solid geometry
Load name	Load image	Load details
Torque-1	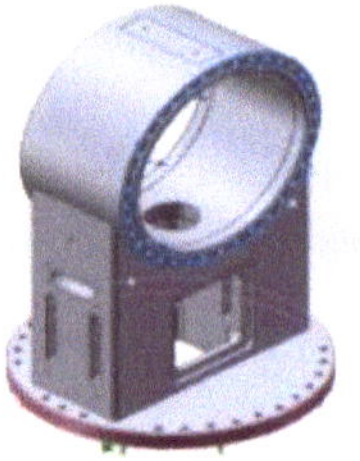	Entity: 1 Face Reference: Face < 1 > Type: Applied Torque Value: 3000 N.m

Table 2. Calculation results.

Analysis items	Values	Minimum	Maximum
Stress	VON: Von Mises Stress	6.929e + 03N/m^2 Node: 6160	1.926e + 07 N/m^2 Node: 1471

(*continued*)

Table 2. (*continued*)

Analysis items	Values	Minimum	Maximum
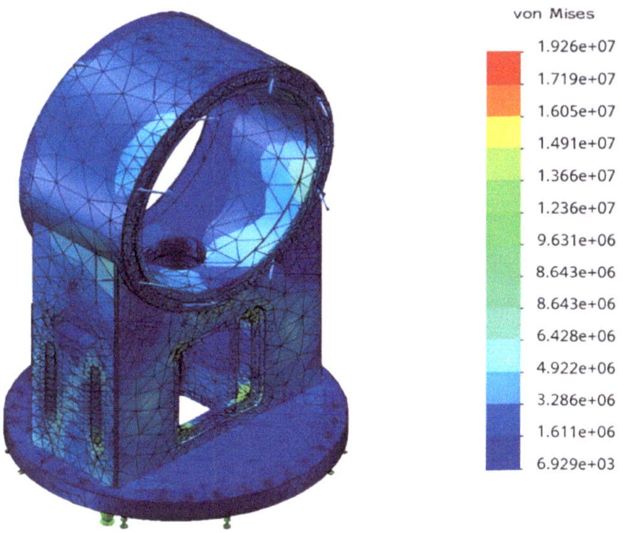 B-axis assembly—static stress analysis—stress-1			
Displacement-1	URES: Total Displacement	0.000 mm Node: 17	0.057 mm Node: 2619
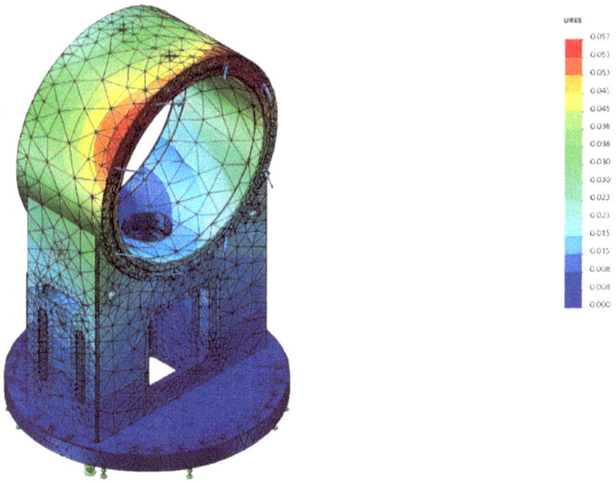 B-axis assembly—static stress analysis—displacement-1			
Strain-1	ESTRN: Equivalent strain	5.475e-08 Element: 41953	5.958e-05 Element: 17843

(*continued*)

Table 2. *(continued)*

Analysis items	Values	Minimum	Maximum

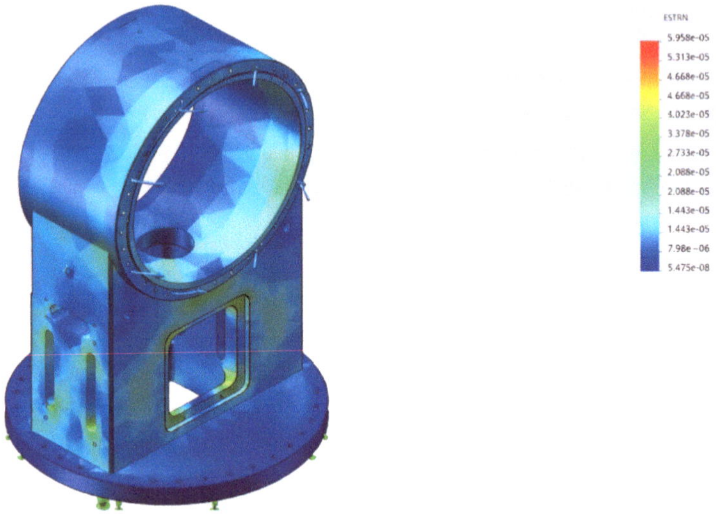

B-axis assembly—static stress analysis—strain-1

3.3 Motor Selection and Verification

The motor selection for the B-axis must meet three core requirements: first, the torque capacity must match the load demand; second, the acceleration performance must ensure dynamic response; third, the moment of inertia matching must optimize energy efficiency.Torque Calculation: Comprehensively consider the B-axis's own moment of inertia and the load moment of inertia, combined with the acceleration requirement, to calculate the required torque value of the B-axis.Moment of Inertia Matching: Take into account the additional load of 150Nm from the Glyd ring and bearing on the C-axis, and calculate the moment of inertia ratio between the motor and the load to ensure the motor's acceleration performance and operational stability.

Table 3. Main dynamic parameters of the B-axis.

Parameter	Value
Rotational speed (n)	42 rpm
Acceleration α	2π rad/s^2
B-axis's own moment of inertia	102.55kg·m^2

As shown in Table 3, for the B-axis, after ignoring the correlation of some parts, the A-axis housing, worktable, and workpiece constitute its full load.

The torque of the B-axis is

$$T_1 = J_B \times \alpha = 102.55 \times 2\pi \approx 644Nm$$

Based on the above results, the 442sim-M832 motor is selected. Its rated torque is 832 *Nm*, maximum torque is 1384 Nm, rated rotational speed is 42 *rpm*, motor moment of inertia is $J_D = 0.892$, and the usage requirement for the motor's moment of inertia ratio is ≤ 200.

According to the above parameters, when the motor drives the B-axis to rotate and workpieces are clamped on both sides of the worktable, the moment of inertia ratio between the load and the motor is

$$\frac{J_B}{J_D} = \frac{102.55}{0.892} \approx 115 \leq 200$$

Due to the additional 150Nm load from the Glyd ring and bearing on the C-axis, the torque ratio is

$$\frac{T_1}{T_D} = \frac{644}{1384 - 150} = 52\%$$

Meeting the design requirements.

4 Conclusions

This paper conducted structural design for the B-axis of five-axis dual-table turntables to meet the requirements of high precision, high stability and fast response, and developed a complete and reliable scheme through layout optimization, technological innovation and performance verification. In the design, a T-shaped symmetrical layout + YRT bearings + outer rotor direct drive was adopted, achieving the compactness and micron-level precision of the B-axis while reducing the turntable volume by 40% compared with traditional designs; meanwhile, the integration of the rotary distributor sealing system and backpack-type solenoid valve solved the problem of entanglement and wear in mechatronics-hydraulics transmission, extending the maintenance cycle to more than 2 years and shortening the clamping response time to 0.3 s.

Performance verification showed that finite element analysis via ANSYS Workbench revealed the B-axis housing had a maximum stress of 19.26MPa and a maximum displacement of 0.057mm, meeting the requirements for strength and stiffness; the selected 442sim-M832 motor (rated torque: 832Nm; speed: 42rpm) ensures drive matching and dynamic response. This design overcomes the shortcomings of traditional turntables (large transmission backlash, slow response, frequent maintenance), improves the turntable's precision stability, response speed and service life, provides support for the efficient automation of dual-station five-axis linkage machining, can be applied to the upgrading of high-end CNC machine tools, and has practical significance for enhancing machining precision and efficiency in fields such as aerospace and precision molds.

Acknowledgments. The research of this paper is supported by the Beijing Polytechnic University College-Level Project (Major Science and Technology) (2025X001-KXD).

Declaration of AI-Assisted Technologies (Only Language Polishing and Translation. AI Rate <40%). During the writing of this manuscript, the author used AI tools during the Chinese-to-English translation process and for checking grammar. The authors take full responsibility for the use of these tools and take full responsibility for the content of the paper.

References

Chen, X., Wang, J., Zhang, L.: Optimization design of double-table turntable B-axis based on response surface methodology. J. Mech. Des. **40**(5), 89–96 (2023). https://doi.org/10.13841/j.cnki.jxsj.2023.05.012

GB/T 3077-2015: Alloy structural steels. Standards Press of China (2015). https://doi.org/10.6052/1000-0952-2015-3-20140687

ISO 10791-7:2020: Test conditions for machining centres—Part 7: Evaluation of positioning accuracy and repeatability of rotary axes. International Organization for Standardization (2020). https://doi.org/10.1515/iso-10791-7-2020

Li, D., Wang, S., Zhao, Y.: Reliability prediction of five-axis turntable B-axis based on fault tree analysis. J. Reliab. Eng. **19**(2), 108–115 (2023). https://doi.org/10.13420/j.cnki.jcre.2023.02.014

Li, J., Chen, W., Liu, H.: Design method of B-axis components of water chamber head turning and milling machining center. Open Mech. Eng. J. **9**, 226–235 (2025). https://doi.org/10.2174/1874155X02509010226

Li, Y.C.: Assembly technology and application of cradle-type direct-drive five-axis linkage CNC turntable. Mech. Res. Appl. **37**(1), 134–137 (2024). https://doi.org/10.16576/j.ISSN.1007-4414.2024.01.035

Li, Y.T.: Research on five-axis CNC machining technology of integral impellers. China Plant Eng. (9), 242–244 (2021). https://doi.org/10.19569/j.cnki.cn11-3303/tb.0198.01.045

Liu, F., Chen, G., Zhang, Z.: Torque calculation and optimization of outer rotor permanent magnet synchronous motor. Small Spec. Electr. Mach. **50**(4), 1–5 (2022). https://doi.org/10.19532/j.cnki.12-1401/tm.2022.04.001

Liu, K.M., Wang, X.B., Liu, J.Y.: Research on axis calibration method of five-axis machining center worktable. Intern. Combust. Engine Parts (7), 66–68 (2025). https://doi.org/10.19475/j.cnki.issn1674-957x.2025.07.022

Liu, T.J.: Research on five-axis linkage CNC machining. China Plant Eng. (15), 84–86 (2023). https://doi.org/10.19569/j.cnki.cn11-3303/tb.0217.01.035

Liu, Z., Zhang, W., Li, C.: Research on dynamic performance of direct-drive five-axis turntable. J. Mech. Eng. **58**(12), 234–242 (2022). https://doi.org/10.3901/JME.2022.12.234

Rexroth: R911308716 Direct drive motor datasheet. Bosch Rexroth AG (2022). https://doi.org/10.1002/9783834198428

THK.: TDX series high precision turntable product manual (Report No. TDX-2021). THK Co., Ltd (2021). https://doi.org/10.1109/IEMDC.2021.9479234

Wang, L.Y., Qin, C.L., Zhao, Q.X.: Development status and trends of five-axis linkage CNC machine tools. South. Agric. Mach. **55**(3), 30–33 (2024). https://doi.org/10.19415/j.cnki.1673-890x.2024.03.009

Wang, Y., Li, Z., Liu, H.: Analysis of transmission backlash of worm gear drive for five-axis turntable. Mech. Sci. Technol. Aerosp. Eng. **40**(8), 1256–1262 (2021). https://doi.org/10.13433/j.cnki.1003-8728.20210245

Zhang, Q., Chen, W., Li, J.: Energy consumption analysis of direct-drive vs. worm gear drive for five-axis turntable. Energy Conserv. Technol. **40**(3), 245–251 (2022). https://doi.org/10.13637/j.issn.1002-6339.2022.03.015

Zhang, S., Wang, L.: CNC machine double rotary table vs. single rotary table comparison. Jiankemach (2025). https://doi.org/10.13140/RG.2.2.36542.58791

Zhao, H., Chen, Y., Yang, Z.: Innovative design and machining verification of a dual axis swivel table for a milling machine. Adv. Sci. Technol. Res. J. **18**(3), 45–56 (2024). https://doi.org/10.12913/22998624/184564

Zheng, J.W., Zhao, L.: Dynamic error detection and correction of five-axis machining center. Mod. Manuf. (3), 28–30 (2025). https://doi.org/10.16567/j.cnki.1000-7008.2025.03.009

Horizontal Stiffness Characteristics of Three-Wire Pendulum in Air Spring Vibration Isolator: Modelling and Simulation

Chang Du, Yongling Fu(✉), and Wanguo Li

School of Mechanical Engineering and Automation, Beihang University, Beijing, China
fuyongling@126.com

Abstract. Air spring vibration isolator is a major approach to attenuate ground/base vibration for precision equipment. Three-wire pendulum can be used as the horizontal isolation mechanism in air spring vibration isolator. Horizontal stiffness variation due to change of equilibrium position, referred to as the horizontal stiffness characteristics, are essential for analysing interaction between different mechanisms in vibration isolator. Analytical model for calculating horizontal stiffness characteristics of three-wire pendulum within a certain range around its initial position is established. Then the three-wire pendulum in an air spring vibration isolator prototype and related variants are analysed using the proposed model. Static and dynamic rigid multibody simulations are also carried out to verify the analytical model and reveal effects of model parameters on the stiffness characteristics. Results show that the difference between simulated and analytical stiffness characteristics is acceptable, and stability can be predicted using the analytical model. Besides, primary parameter affecting the stiffness and stability is identified. The proposed model offers an analytical method to quickly examine the property of three-wire pendulum before further investigation.

Keywords: Horizontal stiffness characteristics · Three-wire pendulum · Rigid multibody simulation

1 Introduction

Air spring vibration isolator (ASVI) plays an important role in protecting precision equipment like scanning electron microscopes, lithography machines from vibrating ground or mounting base (Bai et al. 2024; Mikułowski 2021). It can attenuate ground vibration with frequency higher than specific value and is capable to bear a significant mass of payload. However, ASVI can only cope with vertical vibration, to isolate horizontal ground vibration, other mechanism is needed to assist ASVI, such as three-wire pendulum (TWP) (Liu et al. 2020).

The improvement of precision equipment also places increasingly stringent demands on vibration isolators (Heertjes et al. 2020), which can be fulfilled by quasi-zero-stiffness (QZS) technique. One straightforward and widespread method for achieving QZS is to combine positive stiffness mechanisms with negative stiffness mechanisms (NSMs) in

T. Mankovits and N. Trisovic (Eds.): MDS 2026, LNME 2026.
https://doi.org/10.1007/978-981-92-2440-1_2

parallel configuration (Yan et al. 2022), provided that the stiffness characteristics of both are well understood.

The stiffness characteristic here refers to the stiffness variation with static equilibrium position, which is fundamental for preliminary analysis of QZS design. For example, it can be told by the stiffness characteristic whether a mechanism is stable or not. The stiffness characteristics of NSMs such as X-shape mechanisms (Yan et al. 2022), magnetic springs (Wu et al. 2022), buckling beams (Demir and Yilmaz 2022) and so on are well studied, they are unstable without combination with positive stiffness mechanisms like coil spring, air spring. As for positive stiffness mechanisms, stiffness characteristics of coil spring, air spring (Vo et al. 2021), plate spring (Mizuno et al. 2017) are also well-established.

In our earlier study (Liu et al. 2020), an ASVI with TWP as the horizontal isolation mechanism was designed and manufactured. In that work, the TWP analytical model proposed by Xu et al. (2010) was applied to calculate the natural frequency of the ASVI in the horizontal direction. However, that model is not meant for analysing the horizontal stiffness characteristics, since the variable denoting equilibrium position is eliminated. Therefore, the aim of this paper is to establish and verify the analytical model for analysing the horizontal stiffness characteristics of TWP. This lays the foundation for future improvement of the ASVI, by offering an analytical method to examine the interaction between TWP and other mechanisms for vibration isolator.

The rest part of this paper is organized as follows. In Sect. 2, analytical model of the TWP horizontal stiffness characteristics is established. In Sect. 3, static and dynamic rigid multibody simulation are carried out to verify the model. In Sect. 4, conclusions are drawn.

2 Modelling of the TWP Horizontal Stiffness Characteristics

The schematic of the TWP in our earlier work (Liu et al. 2020) is shown in Fig. 1. It is assumed that all of the components are rigid bodies, so they will not deform under internal or external forces. The payload, the centre column and the swing plate are rigidly connected and move as a single unit. The ball joints in the TWP permits relative rotation between connected components. Consequently, the payload is able to translate and rotate slightly relative to the ground. This guarantees the ability to attenuate horizontal vibration from the ground and therefore protecting the payload.

To model the horizontal stiffness characteristic, we start by applying a quasi-static horizontal external force F_x to the payload centre of gravity denoted by point E. The resulting displacement is denoted by x. Constrained by the mechanism kinematic, a height change denoted by h is also induced, as shown in Fig. 1b. Let point E be the zero of gravitational potential energy, then the gravitational potential energy at point E', denoted by V, is given by (Xu et al. 2010):

$$V = mgh \tag{1}$$

where m is the mass of the payload, g is the acceleration due to gravity.

Since F_x is quasi-static, both the initial and the final kinematic energies are zero, thus V is contributed by the work done by F_x along x:

$$V = \int F_x \mathrm{d}x \tag{2}$$

Then, F_x is given by

$$F_x = \frac{\mathrm{d}V}{\mathrm{d}x} = mg\frac{\mathrm{d}h}{\mathrm{d}x} \tag{3}$$

To figure out the relation between h and x, we project the displaced TWP onto the xz-plane, and focus on the region where $x > 0$, as shown in Fig. 2. To aid the derivation, only necessary points and lines are included.

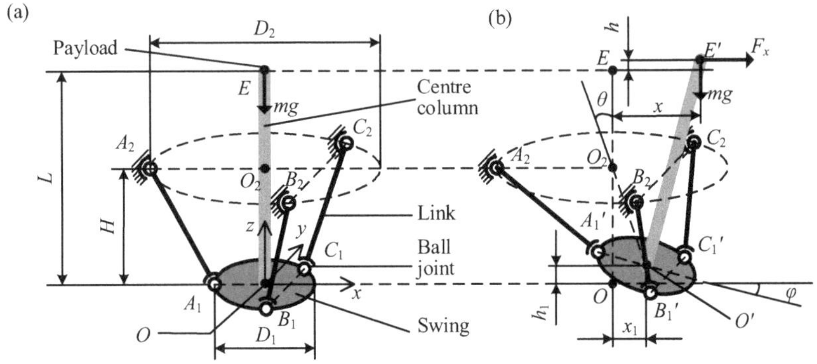

Fig. 1. Schematic of the TWP: **a** at the initial position, **b** at the equilibrium position under static horizontal external force.

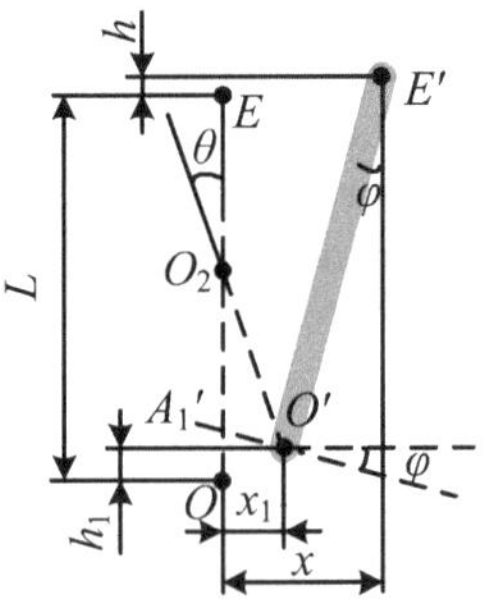

Fig. 2. The projection of the displaced TWP on the xz-plane.

From Fig. 2, we can get

$$\begin{cases} h = h_1 + \left|O'E'\right| \cos\varphi - L = h_1 + L(\cos\varphi - 1) \\ x = x_1 + \left|O'E'\right| \sin\varphi = x_1 + L\sin\varphi \end{cases} \tag{4}$$

and their differentials

$$\begin{cases} \mathrm{d}h = \mathrm{d}h_1 - L\sin\varphi \mathrm{d}\varphi \\ \mathrm{d}x = \mathrm{d}x_1 + L\cos\varphi \mathrm{d}\varphi \end{cases} \quad (5)$$

Assume that θ is small, there are approximations as follows (Xu et al. 2010):

$$\begin{cases} h_1 = H - \left|O'O_2\right| \cos\theta \approx H(1 - \cos\theta) \\ x_1 = \left|O'O_2\right| \sin\theta \approx H \sin\theta \\ \varphi \approx (D_2 - D_1)\theta/D_1 = (D_2/D_1 - 1)\theta = (D_\mathrm{r} - 1)\theta \end{cases} \quad (6)$$

The differentials of the approximations are

$$\begin{cases} \mathrm{d}h_1 \approx H \sin\theta \mathrm{d}\theta \\ \mathrm{d}x_1 \approx H \cos\theta \mathrm{d}\theta \\ \mathrm{d}\varphi \approx (D_\mathrm{r} - 1)\mathrm{d}\theta \end{cases} \quad (7)$$

Combination of Eqs. (3), (5) and (7) gives

$$F_x \approx mg \frac{H \sin\theta - L \cdot (D_\mathrm{r} - 1) \sin\varphi}{H \cos\theta + L \cdot (D_\mathrm{r} - 1) \cos\varphi} \quad (8)$$

Finally, the TWP stiffness characteristic in the x-direction is obtained by taking the derivative of F_x with respect to x:

$$k_x = \frac{\mathrm{d}F_x}{\mathrm{d}x} = \frac{\mathrm{d}F_x}{\mathrm{d}\theta} \cdot \frac{\mathrm{d}\theta}{\mathrm{d}x} \approx \frac{\mathrm{K}_{x1} + K_{x2}}{H \cos\theta + L \cdot (D_\mathrm{r} - 1) \cos\varphi} \quad (9)$$

where K_{x1} and K_{x2} are intermediate results, which are given by

$$\begin{cases} K_{x1} = mg \frac{H \cos\theta - L \cdot (D_\mathrm{r} - 1)^2 \cos\varphi}{H \cos\theta + L \cdot (D_\mathrm{r} - 1) \cos\varphi} \\ K_{x2} = F_x \frac{H \sin\theta + L \cdot (D_\mathrm{r} - 1)^2 \sin\varphi}{H \cos\theta + L \cdot (D_\mathrm{r} - 1) \cos\varphi} \end{cases} \quad (10)$$

The only independent variable in Eq. (9) is θ, so the relation between k_x and x is implicitly defined. The derivation of the TWP stiffness characteristic in the y-direction and the approximations applied are identical to that of the x-direction when θ is small, therefore k_y equals k_x.

Let $\theta = 0$ and we can get the horizontal stiffness of the TWP at the initial position:

$$k_{x0} = k_{y0} = k_0 = mg \frac{H - L(D_\mathrm{r} - 1)^2}{[H + L(D_\mathrm{r} - 1)]^2} \quad (11)$$

The TWP is stable at the initial position if $k_0 > 0$, which yields

$$\begin{cases} H/L > (D_\mathrm{r} - 1)^2, D_\mathrm{r} \neq 1 \\ mg > 0, D_\mathrm{r} = 1 \end{cases} \quad (12)$$

The solution when $D_\mathrm{r} \neq 1$ is the same as the existing model (Xu et al. 2010), so the proposed model is compatible with existing model. The difference is that the proposed model can be applied to analyse horizontal stiffness variation due to equilibrium position change, while the existing model cannot.

3 Verification of the TWP Model

To verify Eqs. (8) and (9), a real TWP in an ASVI, together with related variants, was simulated using COMSOL multibody dynamics module. The reference TWP parameters are given in Table 1 and shared by the analytical calculations and the simulations. The range of θ is set to the operating range of the TWP in vibration isolator.

Table 1. Parameters of the TWP in a real ASVI.

Items	Values	Units	Items	Values	Units
H	112.52	mm	D_1	45.00	mm
L	154.95	mm	D_2	89.07	mm
H / L	0.726	1	$D_r = D_2 / D_1$	1.979	1
m	1000	kg	θ	– 0.02 ~ 0.02	rad

3.1 Static Simulation

These simulations offered numerical force-displacement curves of TWPs, which were converted to stiffness-displacement curves using backward difference and then compared to analytical results.

3.1.1 Simulation Setup

The simulation model and conditions are presented in Fig. 3, all components were treated as rigid bodies. The payload was modelled by a point mass, which was attached to the centre of gravity of the load plate. Then, a static horizontal displacement x was applied to the load plate, and the steady-state solution of the horizontal external force F_x required to maintain/induce the specified x was solved. Batch processing was utilized to automatically vary the static displacement throughout the simulation, such that the static F_x-x relation could be simulated. The same simulation and batch processing were also carried out in the y-direction to get F_y-y relation.

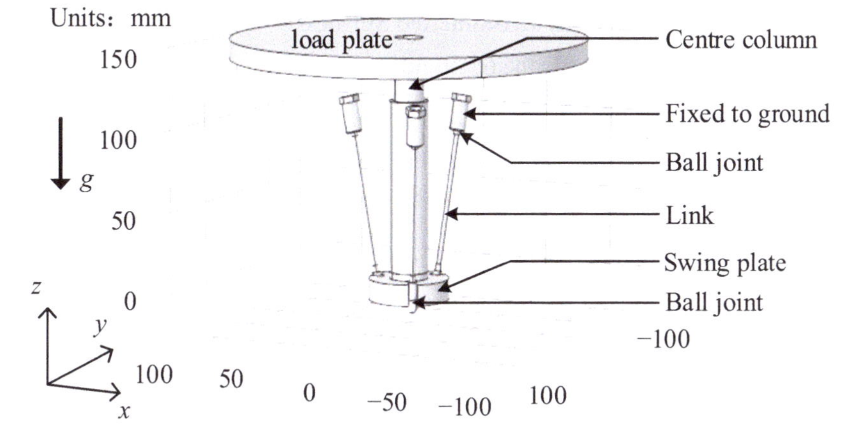

Fig. 3. Simulation model and conditions of the TWP (initial position).

3.1.2 Results

Both simulated and analytical results are shown in Fig. 4. As can be seen, when the horizontal displacement is positive, the corresponding force is negative. This indicates that the external force is maintaining the specified displacement while the TWP itself tends to enlarge the given displacement. Then, the TWP will not return to its initial position if the external force is removed. Consequently, the TWP is unstable. The stiffness is negative, which is another symbol of instability.

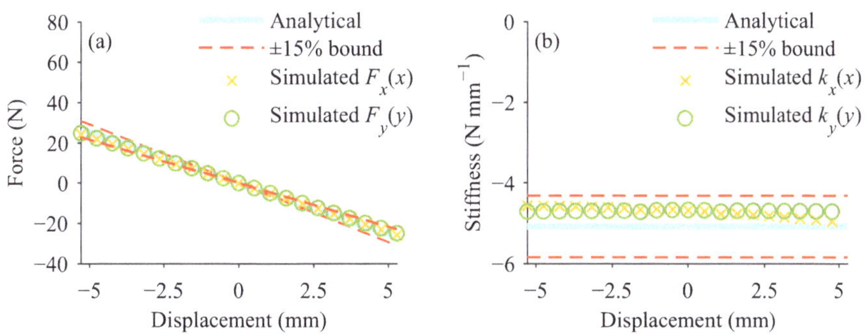

Fig. 4. Comparison between simulated and analytical results: **a** force, **b** stiffness.

The simulated and analytical results match well. Besides, the difference between k_x and k_y is within ±5%, so k_x and k_y can be considered equal when θ is small. Substitute the parameters in Table 1 into Eq. (12), the inequality when $D_r \neq 1$ cannot be satisfied, which also indicates that the TWP is unstable. Thus, Eq. (12) has no conflict with the analytical and simulated results.

3.1.3 Effect of Joint Stiffness on Simulated Stiffness

The deviation in Fig. 4b is due to the non-rigid joint model employed by the simulation. The non-rigid joint model can reduce the computational complexity by lowering the stiffness of the simulation model. The value of joint stiffness is not available, yet it can be adjusted by setting a dimensionless factor p. Figure 5 visualizes the effect of joint stiffness on simulated stiffness by varying p. It shows that the stiffness deviation does relate to joint stiffness and decreases as p decreases. However, p does not change the sign of TWP stiffness, therefore it will not affect the validity of stability prediction based on stiffness.

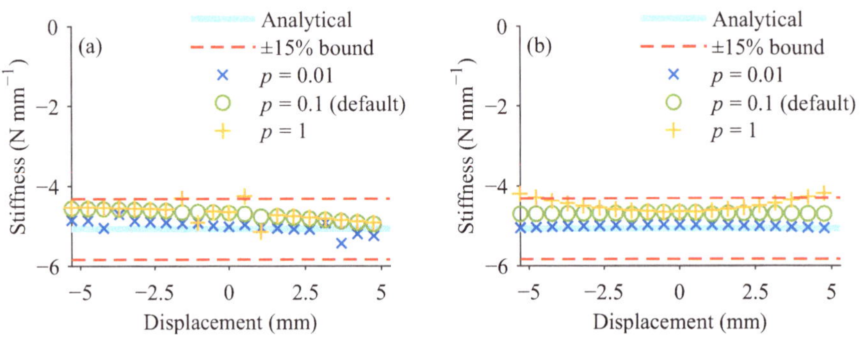

Fig. 5. Effect of joint stiffness factor p on simulated stiffness (scattered points): **a** k_x, **b** k_y.

3.1.4 Parametric Study on the Stiffness Characteristics

Figure 6 displays the effects of D_r and H / L by varying D_l and H, respectively. Unchanged parameters were kept at their values in Table 1. As expected, decrease of D_r and increase of H / L are beneficial for stability of TWP since these operations increase the stiffness. The varying range for D_r and H / L was identical (±25% about the critical value, 1.852 for D_r and 0.959 for H / L), and variation of D_r caused greater change of stiffness. Thus, stiffness is more sensitive to variation of D_r. To raise stiffness level thereby restoring the stability, decreasing D_r is a good choice when redesigning the TWP.

3.2 Dynamic Simulation

To better demonstrate the relationship between stability and stiffness, dynamic simulations were carried out. The TWP defined by Table 1 (TWP1) exhibits negative stiffness and is thus considered unstable. For comparison, a stable TWP from Fig. 6a (TWP2, $D_l = 55$ mm, $D_r = 1.619$) was also simulated. The TWPs were at the initial position in Fig. 3 when time $t = 0$ s. A static force F (magnitude: 10 N, direction: (1, 1, 0)) was applied at $t = 0$ s and removed at $t = 5$ s, and the results are shown in Fig. 7. Isotropic viscous damping (7 N s mm^{-1}) was applied to suppress vibration, allowing for better observation of displacement responses.

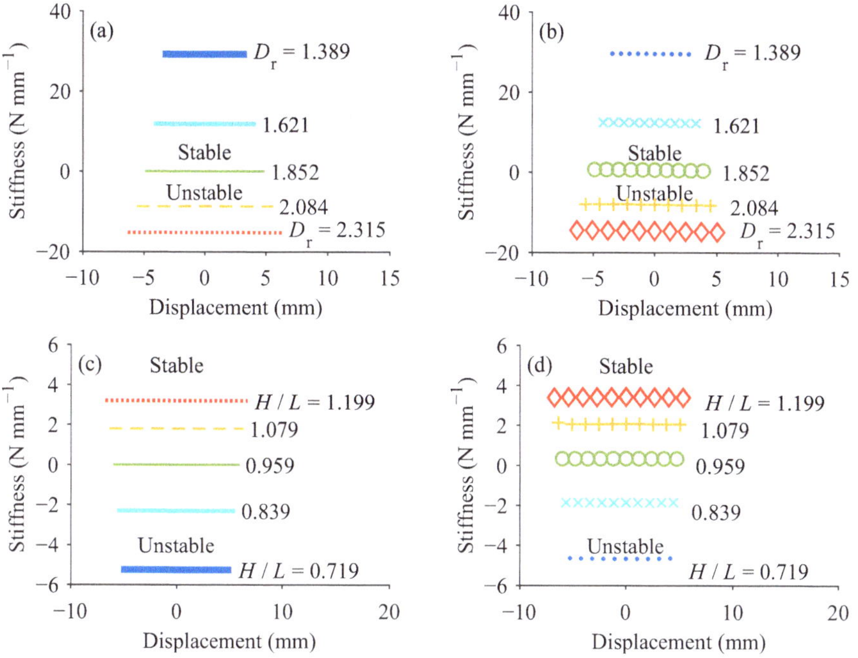

Fig. 6. Effects of D_r and H/L on TWP stiffness: **a** D_r analytical, **b** D_r simulated, **c** H/L analytical, **d** H/L simulated.

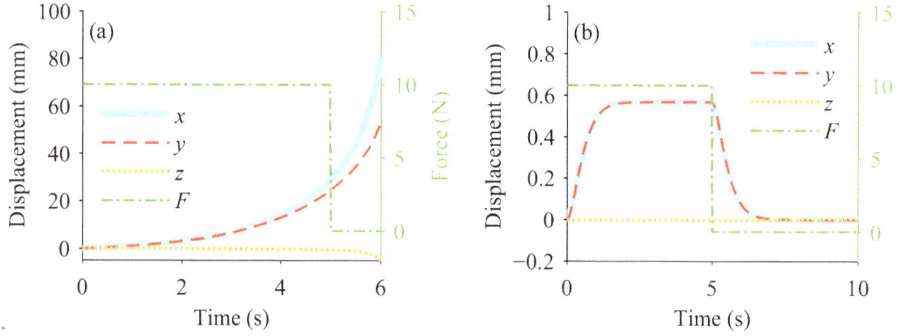

Fig. 7. Simulated dynamic response of TWPs: **a** unstable TWP, **b** stable TWP.

Figure 7 firmly exhibits that TWP1 is unstable and TWP2 is stable, which coincides with predictions based on static stiffness characteristics. Consequently, it is proved that stability can be predicted by static stiffness characteristics from the analytical model. In the simulations, there were no constraints other than the kinematic relationships, hence the TWP1 displacements can exceed operating range. This is the exact dynamic behaviour when a mechanism is unstable, though.

4 Conclusion

This paper established an analytical model of TWP, which can be used to calculate the static stiffness-displacement relation in the horizontal direction and predict the stability. Then, analysis based on the model, as well as static and dynamic rigid multibody simulations, was carried out on a series of TWPs. Finally, analytical and simulated results were compared. It turns out that the proposed analytical model can represent the horizontal stiffness characteristics of the TWP within operating range, and reflect its stability. It is also revealed that the diameter ratio of TWP affects the stiffness and stability most. This paper offers an analytical method to quickly examine the property of TWP before further investigation.

Acknowledgments. This work was supported by the National Science and Technology Major Project of China (No: 2018ZX02201006).

References

Bai, Y., et al.: Stiffness model for pneumatic spring with air-diaphragm coupling effect. Precis. Eng. **91**, 728–738 (2024). https://doi.org/10.1016/j.precisioneng.2024.11.008

Demir, M.U., Yilmaz, C.: Analysis and design of an adjustable stiffness three-axis horizontal vibration isolator using elastic columns and a string in tension. J. Sound Vib. **523**, 116736 (2022). https://doi.org/10.1016/j.jsv.2021.116736

Heertjes, M.F., Butler, H., Dirkx, N.J., Van Der Meulen, S.H., Ahlawat, R., O'Brien, K., Simonelli, J., Teng, K.-T., Zhao, Y.: Control of wafer scanners: methods and developments. In: 2020 American Control Conference (ACC), pp. 3686–3703 (2020). https://doi.org/10.23919/ACC45564.2020.9147464

Liu, C., Li, W., Chen, J.: Key characteristics analysis of vibration isolator used in high precision testing equipment. In: 2020 IEEE/ASME International Conference on Advanced Intelligent Mechatronics (AIM), pp. 810–817 (2020). https://doi.org/10.1109/AIM43001.2020.9159035

Mikułowski, G.: Vibration isolation concept by switchable stiffness on a semi-active pneumatic actuator. Smart Mater. Struct. **30**(7), 075019 (2021). https://doi.org/10.1088/1361-665X/ac008f

Mizuno, T., Namai, M., Takasaki, M., Ishino, Y.: Proposal of magnetic suspension with elastic ferromagnetic substance for vibration isolation system. In: 2017 International Automatic Control Conference (CACS), pp. 1–6 (2017). https://doi.org/10.1109/CACS.2017.8284257

Vo, N.Y.P., Nguyen, M.K., Le, T.D.: Analytical study of a pneumatic vibration isolation platform featuring adjustable stiffness. Commun. Nonlinear Sci. Numer. Simul. **98**, 105775 (2021). https://doi.org/10.1016/j.cnsns.2021.105775

Wu, J., et al.: Analysis and design of a novel arrayed magnetic spring with high negative stiffness for low-frequency vibration isolation. Int. J. Mech. Sci. **216**, 106980 (2022). https://doi.org/10.1016/j.ijmecsci.2021.106980

Xu, D., Gao, Z., Zhu, Y., Zhang, M.: Principle and application of horizontal suspension isolator constituted by trilinear pendulum. J. Beijing Univ. Aeronaut. Astronaut. **36**(8), 1005–1008 (2010). https://doi.org/10.13700/j.bh.1001-5965.2010.08.005

Yan, G., et al.: Nonlinear compensation method for quasi-zero stiffness vibration isolation. J. Sound Vib. **523**, 116743 (2022). https://doi.org/10.1016/j.jsv.2021.116743

Assembly Quality Control and Dynamic Design of High-Speed Rotor System with Curvic Couplings

Shiping Song[1,2], Yanhong Ma[1], Jian Zhang[1], Xueqi Chen[1(✉)], Hailong Guo[2], and Jie Hong[1,3]

[1] Research Institute of Aero-Engine, Beihang University, Beijing, China
chenxueqi@buaa.edu.cn
[2] AECC Hunan Aviation Powerplant Research Institute, Zhuzhou, Hunan, China
[3] School of Energy and Power Engineering, Beihang University, Beijing, China

Abstract. Aiming at solving the engineering problems in the use of rotor system with curvic couplings, including large dispersion of initial unbalance and high-level vibration response at operation state, the mechanical model for the curvic coupling assembly is established to describe the nonlinear relationship between the interface contact state and component nonaxiality, and incorporated into the rotor dynamic model to obtain the numerical solution for the vibration response. Furthermore, the multi-parameter optimization method for assembly quality control and rotor dynamic design is developed, targeting the reduction of initial unbalance and displacement amplitude at working speeds. Case studies exhibit that assembly quality and dynamic characteristics of the high-speed rotor system are both significantly improved through reasonable design of the assembly phase angle and the axial preload, which verifies the effectiveness of the optimal method proposed in this paper.

Keywords: Rotor dynamics · Curvic coupling · Assembly quality · Optimal design

1 Introduction

The rotor system in aero-engine is not a continuous structure, but an assembly of disks, shaft segments and other components through different coupling connections. Proposed by Ernest Wildhaber of Gleason Works in 1940s, the curvic coupling structure relying on end-tooth pairs for torque transmission and a central rod for axial preloading of multi-stage disks, has great advantages of high centering accuracy, strong load-bearing capacity and light weight (Liu et al. 2018). Therefore, the curvic coupling structure has been widely used, and the study of its influence on the dynamic performance of rotor system has become research focus in the field of advanced aero-engine technology. According to the elastic averaging principle, when the tooth pairs are closely fitted under axial pressure, elastic deformation of the tooth surfaces will average out the manufacturing errors, achieving higher accuracy for positioning (Rowe et al. 2015). However, under harsh

T. Mankovits and N. Trisovic (Eds.): MDS 2026, LNME 2026.
https://doi.org/10.1007/978-981-92-2440-1_3

operating conditions, local slip or even separation could occur on the meshing interface of coupling structure, weakening the connection stiffness. Therefore, it is necessary to study the interface contact state and the mechanical characteristics of the connection structure, as well as its influence on the rotor dynamic performance. Yin et al. (1996) considered the tooth pair as beam element, sorting out the beam stiffness according to the deformation mode. The results reveal that the equivalent stiffness of curvic coupling connection is only about 30% of the continuous structure in the same size, and the axial preload loss can be induced under high-speed rotation state due to the centrifugal effect. Richardson et al. (1999) established a finite element model of the fan rotor with curvic coupling and bolt preloads, and the variation of contact stress during the blade off was numerically analyzed, which are in good agreement with the test results. Hong et al. (2019) combined the connection stiffness model with the rotor dynamic model to study the influence of axial force on bending stiffness and dynamic response, and the numerical simulation results reveal that the decrease in joint stiffness leads to a significant concentration of strain energy around the connection structure when the rotor undergoes bending-mode vibrations, which decreases the critical speed and deteriorate the rotor dynamic characteristics.

At this present, there are still many problems in the design and operation stages of curvic-coupled rotor systems. Firstly, the overall unbalance of the rotor system after assembly is often much larger than the sum of the unbalances of individual components, and the unbalance and phase changes significantly after in-situ reassembly. Secondly, even after high-precision dynamic balancing, high level vibration appears frequently when the rotor enters high-speed operating condition, which greatly deviates from the design value (Chen et al. 2025). The main reason for that is the lack of control conaxiality for the multi-stage series rotor system, resulting in high sensitivity of the rotor unbalance to the assembly deformation of the tooth pairs, which further leads to the inclination of the disk inertial principal axis and introduces additional rotational inertia excitation at high speed.

Thus, aiming at solving the engineering problems in the application of high-speed rotor system with curvic couplings such as excessive dispersion of assembly quality and frequent occurrence of dynamic response exceeding the design value, mechanical model and numerical simulation method are established in this paper to describe the bending stiffness and unbalance distribution of curvic coupling structure, and analyze the dynamic characteristics of the rotor system. Base on that, optimization method for the assembly quality control and dynamic design of curvic-coupled rotor system is proposed, which realizes effective control of rotor mass asymmetry under initial assembly conditions and dynamic response within the full rotational speed range.

2 Mechanical Modeling

The gas generator rotor system in the typical turboshaft engine shown in Fig. 1 consists of a four-stage axial compressor, a single-stage centrifugal compressor, and a two-stage gas turbine. Curvic couplings are applied at the front and rear ends of the centrifugal compressor, connected to the axial compressor and the gas turbine respectively, and a central rod is used for axial preloading. The gas generator rotor is assembled in a staged manner, where the front of the central rod is connected to the first stage of the compressor through threads, using the nut 1 to preload the compressor components axially, and then the gas turbine is installed, using the nut 2 at the end of the rear journal to axially preload the turbine component to the compressor part.

In addition to the assembly load of axial preloading, the gas turbine rotor is also subjected to the combined action of centrifugal load, thermal load, aerodynamic load, and other excitations during operation, generating various forms of internal forces inside the rotor components, and causing structural deformation at the curvic couplings position. As shown in Fig. 2, when the rotor undergoes bending deformation, structural deformation occurs between the centrifugal compressor and the gas turbine, resulting in the transition of some tooth pairs from the engaged state to separation. The interface separation of the connection structure not only causes local stiffness loss of the rotor system, but also leads to centroid offset and inertial principal axis inclination of disk and shaft components, which will generate high-level rotational inertia excitation under high-speed rotation and aggravate the vibration response.

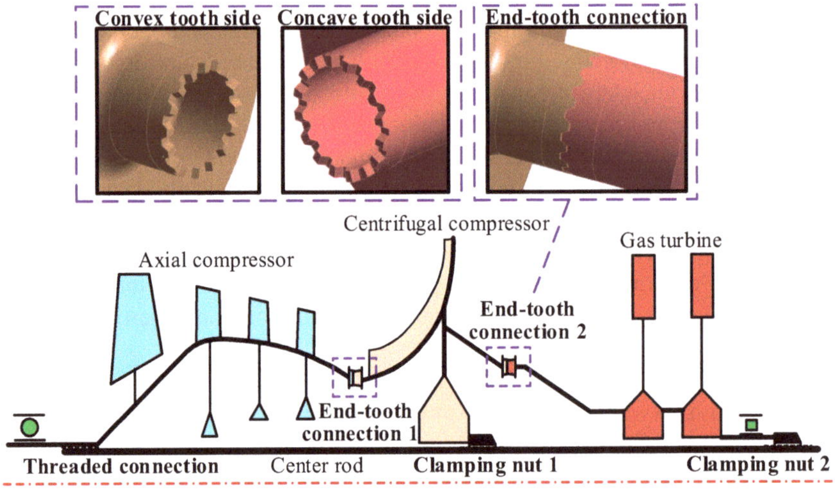

Fig. 1 Schematic diagram of the gas generator rotor system

2.1 Mechanical Model of Curvic Coupling Assembly

The contact state of the curvic coupling interfaces will significantly affect the meshing stiffness of tooth pairs, thereby influencing the overall stiffness characteristics and dynamic performance of the rotor system.

As shown in Fig. 3, the curvic coupling is subjected to axial preloading F_a at the assembled state, where both sides of the teeth are engaged with normal pressures F_{n1}, F_{n2} and tangential friction F_{t1}, F_{at2}. Suppose the compression deformations on tooth

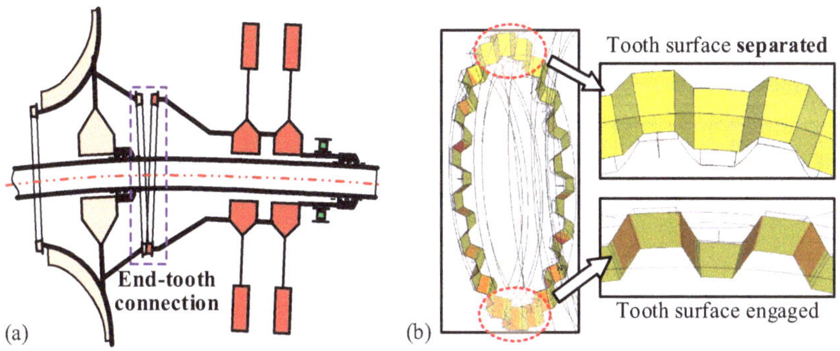

Fig. 2 The contact state on the curvic coupling interfaces, **a** Rotor bending deformation; **b** Local separation at some tooth pairs

sides are the same, then the force balance gives:

$$\begin{cases} F_a = F_{n1}\sin\theta + F_{t1}\cos\theta + F_{n2}\sin\theta + F_{t2}\cos\theta \\ F_{n1} = F_{n2},\ \ F_{t1} = F_{t2} \end{cases} \tag{1}$$

As shown in Fig. 4, the relationship between the normal deformations δn1, δn2 and the axial displacement δ_a at the tooth pair can be expressed as:

$$\delta_{n1} = \delta_{n2} = \delta_a \sin\theta \tag{2}$$

Assuming the meshing stiffness is k_n and the friction coefficient is μ_t, we have:

$$\begin{cases} F_{n1} = k_n \cdot \delta_{n1},\ F_{t1} = \mu_t F_{n1} \\ F_a = (1+\mu_t)k_n(\delta_{n1}+\delta_{n2})\sin\theta = 2(1+\mu_t)k_n\delta_a\sin^2\theta \end{cases} \tag{3}$$

During operating state, the curvic coupling structure transmits the rotor torque, and the tooth pairs are subjected to the tangential force F_t. The compression deformation δ_{n1} on the interface C1 increases while δ_{n2} on the interface C2 decreases, and then the force balance relationship transforms to:

$$\begin{cases} F_a = F_{n1}\sin\theta + F_{t1}\cos\theta + F_{n2}\sin\theta + F_{t2}\cos\theta \\ F_t = F_{n1}\cos\theta - F_{t1}\sin\theta - F_{n2}\cos\theta + F_{t2}\sin\theta \end{cases} \tag{4}$$

The compression deformation δ_{n2} the interface C2 decreases as the tangential force F_t increasing. The interface C2 will be completely separated when the force F_t becomes large enough, as shown in Fig. 3 (c), and the meshing stiffness k_n will show a sudden decrease at the moment of interface separation, which causes the nonlinear stiffness of the curvic coupling structure.

Three typical forms of the curvic coupling deformation are shown in Fig. 5, including lateral offset Δx, Δy, angular deflection $\Delta\beta_y$, $\Delta\beta_x$ and torsion $\Delta\phi_z$. Assuming the left side is fixed and the right side is deformed under the external forces, the relationship between the structure deformation and the compression on the both side of the i-th convex tooth can expressed as:

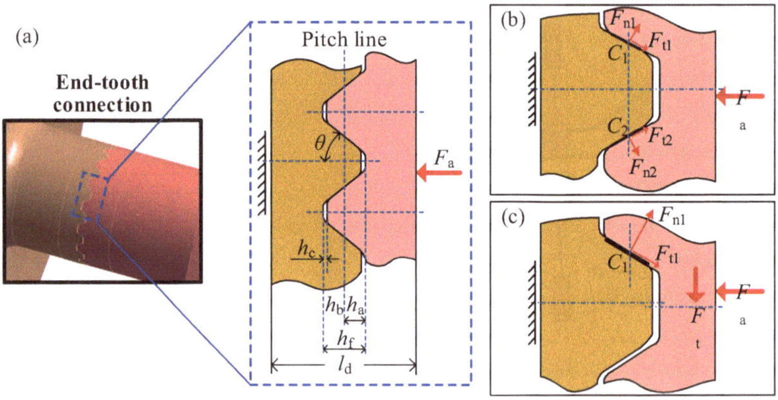

Fig. 3 Forces and compression deformations on the tooth pair interfaces, **a** Curvic coupling structure, **b** Both sides engaged, **c** Single-side engaged

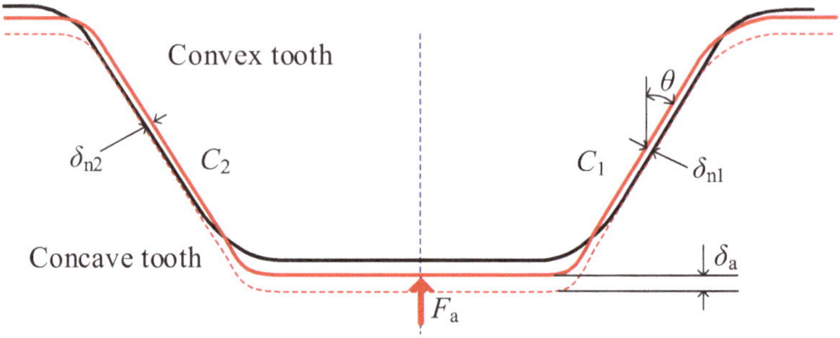

Fig. 4 Compression deformation at the tooth pair interface

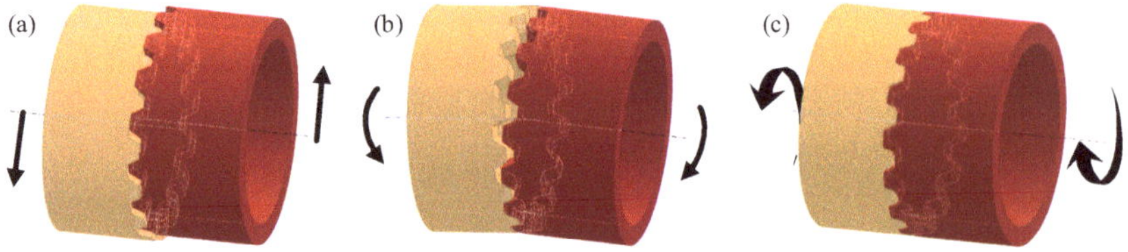

Fig. 5 The typical deformation forms of the curvic coupling structure, **a** Lateral offset, **b** Angular deflection, **c** Torsion

$$\begin{cases} \delta_{n1,i} = \delta_{n0} + \left(\Delta\phi_z r_m - \Delta x \sin\varphi_{1,i} + \Delta y \cos\varphi_{1,i}\right)\cos\theta + r_m\left(\Delta\beta_x \sin\varphi_{1,i} - \Delta\beta_y \cos\varphi_{1,i}\right)\sin\theta \\ \delta_{n2,i} = \delta_{n0} + \left(\Delta x \sin\varphi_{2,i} - \Delta y \cos\varphi_{2,i} - \Delta\phi_z r_m\right)\cos\theta + r_m\left(\Delta\beta_x \sin\varphi_{2,i} - \Delta\beta_y \cos\varphi_{2,i}\right)\sin\theta \end{cases} \quad (5)$$

where r_{m} represents the pitch circle radius, and δ_{n0} represents the normal deformation on the tooth pair interface at the assembly state. $\varphi_{1,i}$, $\varphi_{2,i}$ represent the azimuth angles of the meshing surfaces on the both sides of the i-th convex tooth, which can be calculated by the following expressions:

$$\varphi_{1,i} = 2\pi(i - 1.25)/N, \varphi_{1,i} = 2\pi(i - 0.75)/N \tag{6}$$

where N represents the number of teeth.

The contact pressures and friction $\dot{\delta}_t$ al forces on the both sides of the convex tooth can be expressed as:

$$\begin{cases} F_{n1,i} = k_n\delta_{n1,i}H(\delta_{n1,i}), \ F_{t1,i} = -\mu_t\left|F_{n1,i}\right| \cdot \operatorname{sign}(\dot{\delta}_{t1,i}) \\ F_{n2,i} = k_n\delta_{n2,i}H(\delta_{n2,i}), \ F_{t2,i} = -\mu_t\left|F_{n2,i}\right| \cdot \operatorname{sign}(\dot{\delta}_{t2,i}) \end{cases} \tag{7}$$

where represents the tangential slip velocity of the concave tooth relative to the convex tooth at the meshing positions. $H(\delta)$ and sign($\dot{\delta}$) denote the unit step function and sign function respectively. The relationship between the contact forces on the tooth pair interface and the external loads acting on the curvic coupling structure can be expressed in the following integral form:

$$\begin{cases} F_a = \sum_{i=1}^{N}\left(F_{n1,i}\sin\theta + F_{t1,i}\cos\theta + F_{n2,i}\sin\theta + F_{t2,i}\cos\theta\right) \\ F_x = \sum_{i=1}^{N}\left[\left(F_{n1,i}\cos\theta - F_{t1,i}\sin\theta\right)\cos\varphi_{1,i} - \left(F_{n2,i}\cos\theta - F_{t2,i}\sin\theta\right)\cos\varphi_{2,i}\right] \\ F_y = -\sum_{i=1}^{N}\left[\left(F_{n1,i}\cos\theta - F_{t1,i}\sin\theta\right)\sin\varphi_{1,i} - \left(F_{n2,i}\cos\theta - F_{t2,i}\sin\theta\right)\sin\varphi_{2,i}\right] \\ M_x = r_m\sum_{i=1}^{N}\left[\left(F_{n1,i}\sin\theta + F_{t1,i}\cos\theta\right)\sin\varphi_{1,i} + \left(F_{n2,i}\sin\theta + F_{t2,i}\cos\theta\right)\sin\varphi_{2,i}\right] \\ M_y = -r_m\sum_{i=1}^{N}\left[\left(F_{n1,i}\sin\theta + F_{t1,i}\cos\theta\right)\cos\varphi_{1,i} + \left(F_{n2,i}\sin\theta + F_{t2,i}\cos\theta\right)\cos\varphi_{2,i}\right] \\ T_z = r_m\sum_{i=1}^{N}\left(F_{n1,i}\cos\theta - F_{t1,i}\sin\theta - F_{n2,i}\cos\theta + F_{t2,i}\sin\theta\right) \end{cases} \tag{8}$$

where F_a, F_x, F_y, M_y, M_x, and T_z represent the axial force, lateral forces, bending moments, and torque. Substituting Eqs. (5) and (7) into the above Eq. (8) yields the mechanical model for the curvic coupling assembly. The assembly characteristics of the axial preloading and the non-uniform distribution of the contact state on the meshing interfaces are considered in the model, which can describe the nonlinear relationship between the structural deformation and the external loads.

Finite element method is used to verify the mechanical model, and the FE model with three-dimensional solid elements is developed as shown in Fig. 6. The contact elements are set on the interfaces of the tooth pairs, with the displacement constrains applied to left-side surface of the convex tooth part, and the external loads applied to the right-side surface of the concave tooth part. The contact state of tooth surface is significantly affected by the external loads, especially bending moment. When there is bending deformation, only the bottom side of tooth face stays compressed and tightened, while the upper side is separated, resulting in a reduction in the contact area and structure stiffness.

As shown in Fig. 7, the lateral and angular stiffness of the curvic coupling structure are calculated, showing a nonlinear weakening trend as the external loads increasing, and the stiffness loss drastically increases in a certain range where the number of separated tooth pairs pipes up. It can be seen that the equivalent stiffness obtained by the theoretical analysis model are in good agreement with those of the FE model, which suggests that the proposed model is both effective and accurate, and has potential for application in the dynamic analysis of the rotor system with curvic couplings.

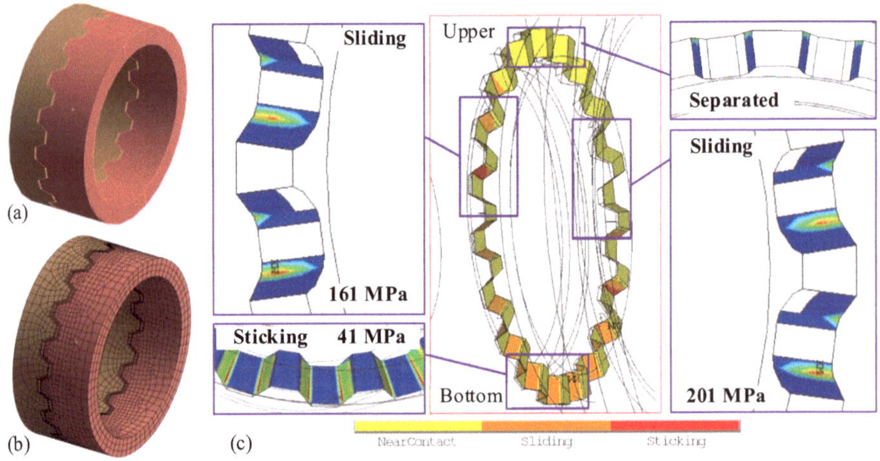

Fig. 6 Finite element model of the curvic coupling structure, **a** Three-dimensional structure, **b** FE model with solid finite elements, **c** Stress distribution under bending moment

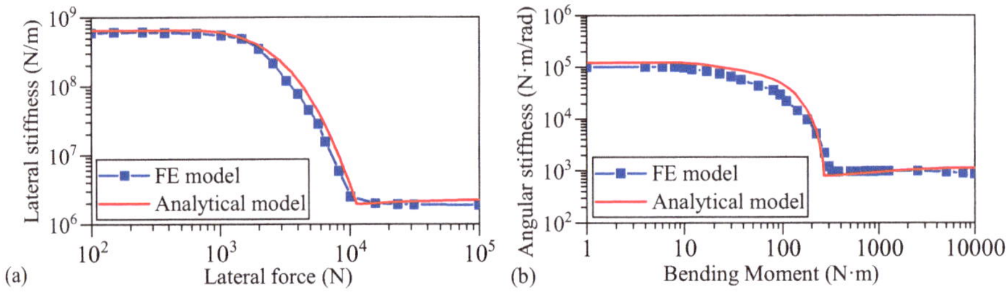

Fig. 7 Curves of the equivalent stiffness varying with external loads **a** Lateral stiffness, **b** Angular stiffness

2.2 Dynamic Model of the Rotor System

The dynamic model of the typical multi-stage rotor system shown in Fig. 1 is developed, using beam elements to discretize the shaft segments and mass elements to simulate the disk components. The dynamic equation of the rotor system is derived as:

$$[\boldsymbol{M}_{\mathrm{s}}]\ddot{q} + (\omega[\boldsymbol{G}_{\mathrm{s}}] + [\boldsymbol{C}_{\mathrm{s}}])\dot{q} + \left([\boldsymbol{K}_{\mathrm{s}}] + \left[\boldsymbol{K}_{\mathrm{et}}(\boldsymbol{q})\right]\right)\boldsymbol{q} = \boldsymbol{F}_{\mathrm{u}} \tag{9}$$

where $[\boldsymbol{M}_{\mathrm{s}}]$, $[\boldsymbol{K}_{\mathrm{s}}]$, $[\boldsymbol{G}_{\mathrm{s}}]$ represent the mass matrix, gyroscopic matrix, and stiffness matrix generated by the shaft segments and the supports, respectively. Rayleigh damping $[\boldsymbol{C}_{\mathrm{s}}] = \alpha[\boldsymbol{M}_{\mathrm{s}}] + \beta[\boldsymbol{K}_{\mathrm{s}}]$ is applied to form the damping matrix, and $[\boldsymbol{K}_{\mathrm{eq}}(\boldsymbol{q})]$ is the equivalent stiffness matrix of curvic coupling structure. $\boldsymbol{F}_{\mathrm{u}}$ represents the load vector unbalance, and $\boldsymbol{q} = [x_1, y_1, \theta_{x1}, \theta_{y1}, \ldots, x_n, y_n, \theta_{xn}, \theta_{yn}]^{\mathrm{T}}$ is displacement vector, where $[x_i, y_i, \theta_{xi}, \theta_{yi}]^{\mathrm{T}}$ denotes the lateral displacements and angular deflections of the i-th node.

3 Numerical Analysis

3.1 Assembly Quality of the Curvic Coupling Structure

The surface quality of each tooth varies in curvic coupling structure due to the manufacture errors, and therefore the compression deformation on different tooth pairs are not consistent, which causes the nonaxiality of connected components and assembly quality deterioration. Usually, radial nonaxiality and angular nonaxiality are used to quantitatively evaluate the assembly quality of the connection structure in the rotation machine, as shown in Fig. 8 (a) and (b). Radial nonaxiality represents the radial misalignment of the connected components' centroids, and angular nonaxiality represents the angular deflection of the centroid lines (Wu et al. 2023). Due to the differences in tooth surface quality, the meshing tightness varies among individual tooth pairs, and further leads to the fluctuations in the component unbalance with the assembly phase angle, as illustrated in Fig. 8 (c).

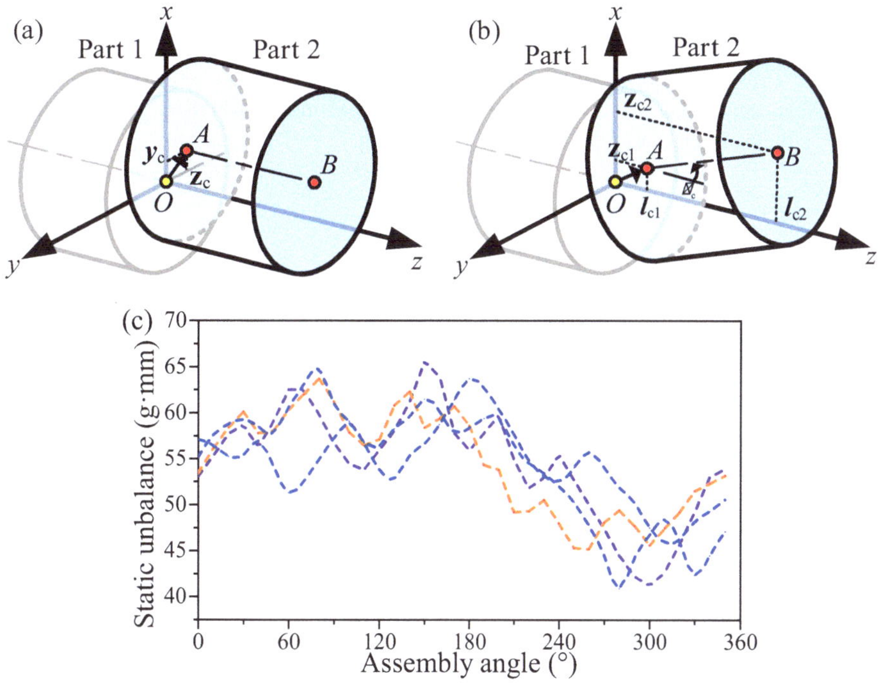

Fig. 8 Schematic diagram of curvic coupling structure nonaxiality, **a** Radial nonaxiality, **b** Angular nonaxiality, **c** Static unbalance fluctuations with assemble angle

The compression inconsistent on the interfaces arising from processing errors is numerically simulated by adjusting the pre-deformation δ_{n0}. According to the measured data, the range of the initial compression on the tooth interfaces is [−0.022, 0] mm, and Monte Carlo method is used to randomly select the values within the range to simulate the uncertainty of the interface compression. The probability density curve fitting is conducted on the frequency of sample results to obtain the angular nonaxiality and angular stiffness of the connection structure, as shown in Figs. 9 and 10. The black and red dashed lines in the figures represent the statistical average and maximum values of the data samples, respectively, and the analysis results verify the dispersion characteristics of the assembly quality. It can be seen that with the increase of axial preload, slight increase in angular nonaxiality occurs due to the intensification of the local deformation on the tooth pair interfaces, and alleviation of the nonaxiality dispersion. On the contrary, increasing axial preload will significantly improve the equivalent angular stiffness but also aggravate the stiffness dispersion.

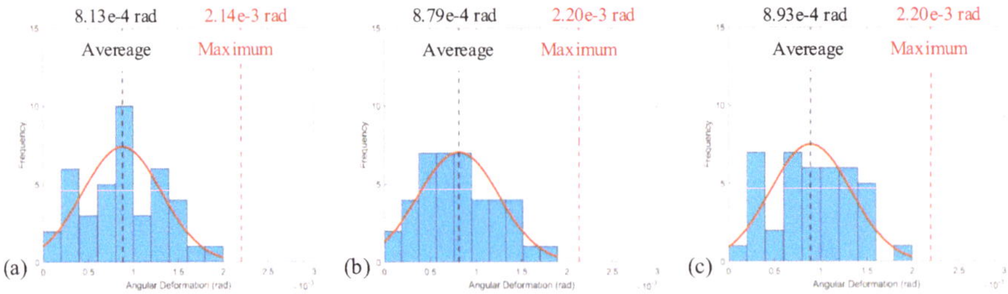

Fig. 9 Angular nonaxiality dispersion of the curvic coupling structure, **a** Preload = 4 kN, **b** Preload = 7 kN, **c** Preload = 10 kN

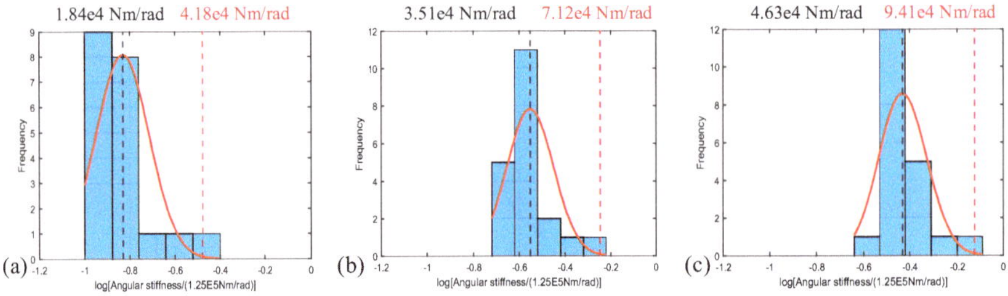

Fig. 10 Angular stiffness of the curvic coupling structure, **a** Preload = 4 kN, **b** Preload = 7 kN, **c** Preload = 10 kN

The inconsistent of the contact state on the tooth pair interfaces not only causes stiffness loss of the connection structure but also increases the unbalance of the assembly, thereby deteriorating the dynamic response of the rotor system. The influence of the curvic coupling structure nonaxiality on the static unbalance and couple unbalance of the rotor system is numerically analyzed, and the probability density of the sample distribution are shown in Fig. 11. The results indicate that the rotor unbalance mainly comes from the accumulation of individual components without regard to the influence of curvic couplings, and the total unbalance is relatively low. However, the assembly nonaxiality would significantly increase the rotor unbalance, where the static unbalance increases from 13.21 g·mm to 1736.2 g·mm, and the couple unbalance increases from

3895 g·mm^2 to 34611 g·mm^2, which will greatly improve the dynamic response during the high-speed operation, especially when passing through the critical speed.

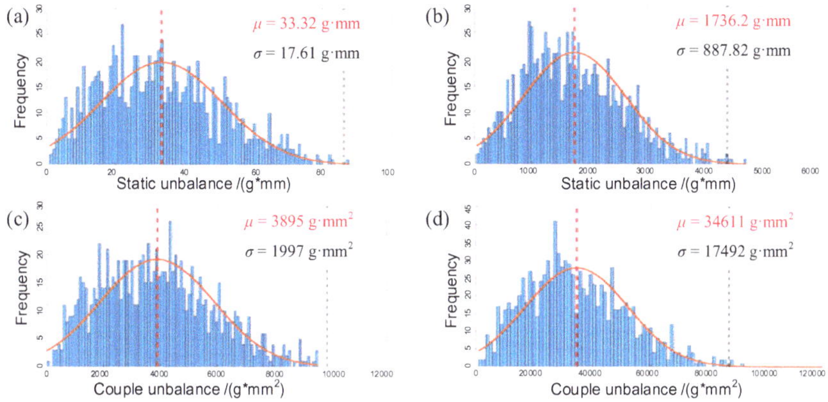

Fig. 11 Unbalance state of the rotor system, **a** Static unbalance, continuous rotor, **b** Static unbalance, curvic-coupled rotor, **c** Couple unbalance, continuous rotor, **d** Couple unbalance, curvic-coupled rotor

In this case, simply adjusting the balancing weights is no longer sufficient to control the rotor unbalance. It is necessary to reduce the angular nonaxiality of connected components to control the deviation of the inertial principal axes from the rotation centerline. Some common methods available here include adjusting the assembly phase angle and optimizing the axial preload.

3.2 Rotor Dynamic Response

Numerical analysis on the dynamic characteristics of the rotor system is conducted, and the influence of the connection stiffness loss on the rotor critical speed is studied first. The natural frequencies and mode shapes are obtained by solving the eigenvalue equations. Modal analysis results with and without considering the influence of the curvic couplings are shown in Figs. 12 and 13. It can be seen in Table 1 that the first two-order modes of the rotor system correspond to rigid-body vibration of translation and pitching, at which the local deformation of the connection structure is rather small, and the decreases of critical speeds are only 1.64 and 0.18%. However, the third-order mode corresponds to the bending deformation, and the deformation shape line at the connection position becomes nonsmooth, which causes the interface separation and stiffness loss, finally leading to a 20.4% decrease in the value of critical speed.

The 4th-order Runge-Kutta method is used to solve the transient response of the rotor system, considering the nonlinear relationship between the curvic coupling meshing state and the rotor whirling state. An unbalance of 20 g·mm is applied at the position of the second-stage turbine disk, and the dynamic response at different rotational speeds are calculated, compared with that of the continuous rotor without curvic couplings. The disk displacement and angular nonaxiality response of the connections are extracted, as

Table 1 Comparison of the critical speed

Order	Critical speed/rpm		Percentage of difference /%
	Continuous rotor	End-tooth connected rotor	
The first order	8973	8826	−1.64
The second order	18498	18465	−0.18
The third order	51921	41329	−20.40

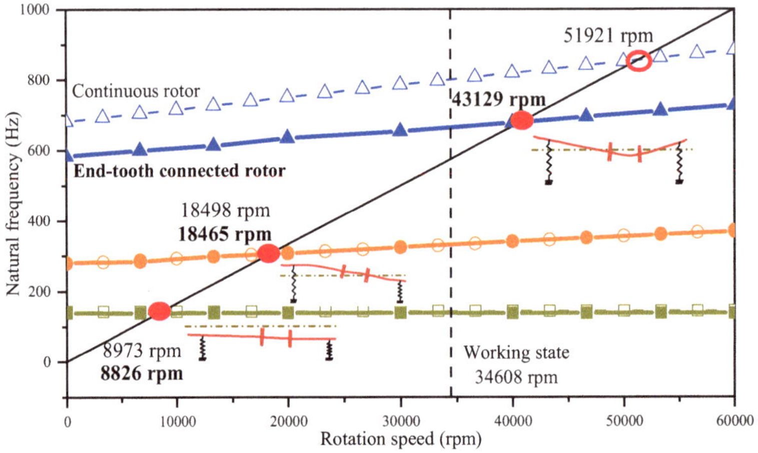

Fig. 12 Campbell diagram

shown in Fig. 13. The results indicate that after passing the rigid-body critical speed, the rotor undergoes bending deformation, and the curvic couplings cause discontinuity of the rotor deformation due to meshing surface separation. Nonaxiality occurs between adjacent disk and shaft components, resulting in local rotational inertia excitation of the disk, and thus the rotor dynamic response keeps increasing with the rotational speed even passing the critical point. This explains why the vibration level at high-speed operating conditions is still much higher than the design value even after well low-speed dynamic balancing for the rotor system with curvic couplings.

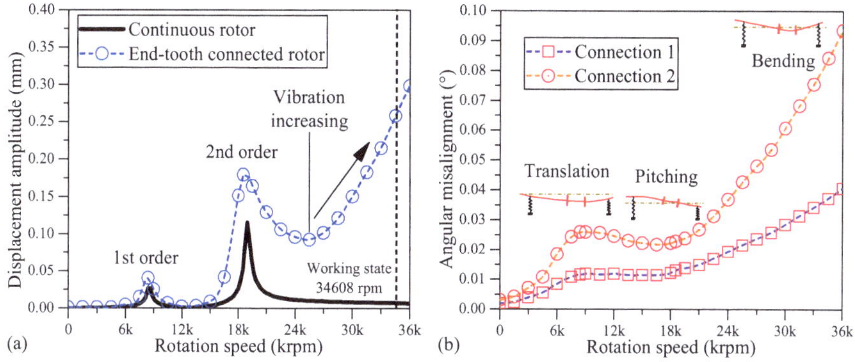

Fig. 13 Amplitude-frequency curves of rotor dynamic response, **a** Turbine disk displacement response, **b** Angular nonaxiality response of the connections

4 Optimal Design

The dynamic response of high-speed rotor in the supercritical state is determined by the rotor mass asymmetry, which is also called operating unbalance. Its amplitude depends on the initial unbalance and the inclination of inertial principal axis under the operation loads, which will be further aggravated by the nonaxiality of the curvic coupling assembly. Therefore, it is crucial to protect the coaxial state from the deflection caused by the rotor bending deformation. This requires improving the assembly quality and the robustness of the connection structure through the assembly parameters optimization.

For the high-speed rotor system with curvic couplings, considering the uncertain compression deformation on the tooth pair interfaces airing from the manufacture error, the optimal design method for the assembly quality and dynamic characteristics is proposed. The assembly phase angles and axial preloads are chosen as the design parameters, with the goals of minimizing the initial rotor unbalance and response amplitude at working state. The basic flow of the optimal design is shown in Fig. 14, which mainly includes the following three steps:

Step 1: Input the structural parameters, the sample of interface compression deformation, and the initial values of the design variables, including the assembly phase angle and axial preload.

Step 2: Establish the analysis models. Quantitatively evaluate the assembly quality and dynamic response based on the numerical solutions.

Step 3: Update the design variables. Complete the optimization if the requirements are satisfied, otherwise reselect the design variables by repeating Step 2 for iteration.

The key of the optimal design lies in the feasible direction method used to search design points (Zhang and Huang 1982). Firstly, decide the search direction according to the opposite gradient of the objective function. Then, obtain new design point by multiplying the direction and the increment step size. Change the direction to its projection on the boundary if the new point does not satisfy the constraint conditions. Finally, input the updated design variable point back to the main program for the optimization.

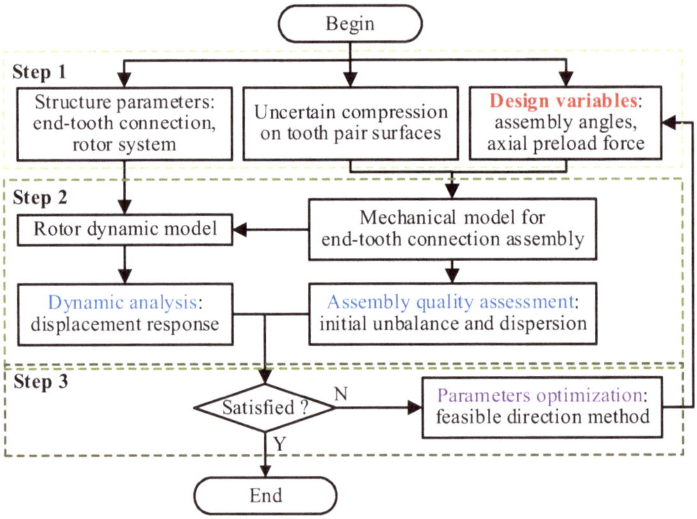

Fig. 14 Optimization path for the assembly process of curvic-coupled rotor system

Based on the optimization method proposed above, the optimal design for the high-speed rotor system shown in Fig. 1 is carried out. The comparison of design variables and target parameters before and after the optimization are shown in Table 2 and 3. The results exhibit obvious improvement in the curvic coupling assembly quality and rotor dynamic characteristics, where the average value and standard deviation of the rotor unbalance decrease by more than 50%, and the rotor dynamic response at operating speeds is also significantly reduced.

Table 2 Comparison of the design variables

	Before the optimization	After the optimization
Assembly angles/(°)	[0, 0]	[40, 150]
Axial preloads/(N)	[5800, 9000]	[5213, 10932]

Table 3 Comparison of the design targets

	Before the optimization	After the optimization
AVE of static unbalance/(g·mm)	33.32	14.77
SD of static unbalance/(g·mm)	17.61	7.33
AVE of couple unbalance/ (g·mm2)	3895	1453
Displacement amplitude/(mm)	0.258	0.067

[a] AVE represents the average value, and SD represents the standard deviation value.

5 Conclusions and Future Work

The mechanical model for curvic coupling assembly to describe the nonlinear relationship between the contact state on the connection interfaces and the rotor motion state is developed in this paper, and numerical analysis as well as quantitative evaluations are carried out for the assembly unbalance dispersion and rotor dynamic characteristics. The optimal design method proposed is verified by the case study. Based on the research results, the main conclusions are drawn as follows:

(1). The contact state on the interfaces can significantly affect the stiffness of the curvic coupling structure, and excessive external will cause large part of the interface separation, which leads to equivalent stiffness loss of the connection structure.
(2). The inconsistent distribution of the compression deformation on the connection interfaces arising from manufacture errors will lead to the radial and angular nonaxiality of the connected components, thereby increasing the rotor unbalance.
(3). Increasing the axial preload will intensify the local deformation on the contact interfaces and increase the angular nonaxiality of connected components, but is helpful for suppressing the unbalance dispersion and improving the angular stiffness of the connection structure.

Furthermore, ongoing efforts will concentrate on enriching the mechanical model by introducing the pre-deformation of the central rod and considering the rod stiffness as design variable. Besides, the tooth wear as well as its influence on the component nonaxiality should be studied theoretically and experimentally. The optimization strategy can be improved by introducing some more efficient algorithms. In summary, the overarching goal is to establish the mechanical model which can describe the nonlinear behavior of the curvic coupling under operation condition with accuracy and efficiency, which help to obtain reasonable assembly process parameters and further improve the dynamic characteristics of rotor system.

Acknowledgments. The authors gratefully acknowledge the financial support provided by the National Natural Science Foundation of China (Nos. 52305088) and National Science and Technology Major Project (Nos. J2022-IV-0004-0021).

References

Chen, X.Q., Li, J., Ma, Y.H., Hong, J.: Influence of assembly on unbalanced state and dynamic response dispersity of high-speed rotors. J. Aerosp. Power. **40**(8), 50–59 (2025). https://doi.org/10.13224/j.cnki.jasp.20240576

Hong, J., Chen, X.Q., Wang, Y.F., Ma, Y.H.: Optimization of dynamics of non-continuous rotor based on model of rotor stiffness. Mech. Syst. Signal Process. **131**, 166–182 (2019). https://doi.org/10.1016/j.ymssp.2019.05.030

Liu, H., Hong, J., Shao, F.Y., Ma, T.L., Li, L.D.: Progress and prospect of structural design and processing technology of CURVIC coupling. J. Propuls. Technol. **39**(4), 721–730 (2018). https://doi.org/10.13675/j.cnki.tjjs.2018.04.001

Richardson, I.J., Hyde, T.H., Becker, A.A., Taylor, J.W.: A comparison of two and three dimensional finite element contact analyses of curvic couplings. WIT Trans. Eng. Sci. **24**, 389–399 (1999). https://doi.org/10.2495/CON990381

Rowe, K.G., Dickrell, D.J., Sawyer, W.G.: Interrupted measurement repositioning using elastic averaging. Tribol. Lett. **59**(1), Article 15 (2015). https://doi.org/10.1007/s11249-015-0527-0

Wu, F. Y., Hong, J., Chen, X. Q., & Ma, Y. H. (2023). Analysis of high-speed rotor vibration failure due to sudden angular deformation of bolt joints. Symmetry, 15(10), article 1937. doi:https://doi.org/10.3390/sym15101937

Yin, Z.Y., Hu, B.A., Wu, J.G., Xu, Y.L., Zheng, Q.X.: Determination of axial preloads of rotors with curvic couplings. J. Aerosp. Power. **11**(4), 405–408 (1996). https://doi.org/10.13224/j.cnki.jasp.1996.04.005

Zhang, Y.S., Huang, J.X.: Method of degree of fullness in structure optimum design. J. Beijing Univ. Aeronaut. Astronaut. **8**(3), 63–73 (1982). https://doi.org/10.13700/j.bh.1001-5965.1982.03.006

Design of Wire Guide Device for Horizontal Cylinder Displacement Gauge of Miter Gate in a Ship Lock

Yezhou Li, Chenglin Hu, and Zhaonong Zhang(✉)

Three Gorges Navigation Authority, Yichang 443000, China
1597848842@qq.com

Abstract. At present, the horizontal cylinder displacement gauge of the miter gate in a ship lock generally adopts the rubber guide wire sleeve protection method, which has the defects of easy wear and tear, and wire rope breakage caused by guide failure. To solve the above problems, a wire guide device with simple structure, light weight, small volume, convenient installation and high reliability is developed in this paper. Firstly, the composition structure and working principle of the device are introduced, and its design characteristics of realizing rolling guidance by using bearing-equipped V-shaped rollers and being fixed at the wire outlet through steel hoops are elaborated. Secondly, the finite element theory is used to carry out simulation analysis on the device. Through material property definition, mesh division and load constraint setting, the strength verification is completed. The results show that the structural stress is much lower than the material yield limit, and the safety factor meets the requirements. On this basis, a bench test is carried out to simulate the actual working conditions, which verifies the effectiveness of the device in supporting the steel wire and improving the friction form. The device operates stably without wire breakage. The device has been successfully applied to this Ship Lock, significantly improving the service life of the steel wire and system reliability, reducing maintenance costs, and is of great value in ensuring the navigation efficiency of the ship lock. It can also be promoted and used in similar shipping hub projects.

Keywords: Ship lock · Miter gate · Horizontal cylinder · Displacement gauge · Wire guide device · Finite element analysis

1 Introduction

At present, an external cable-pulling displacement gauge has been adopted for measuring the opening value of the horizontal cylinder of the miter gate in a ship lock. The external cable-pulling displacement gauge is fixed at the cylinder head of the miter gate oil cylinder, and the steel wire rope of the displacement gauge is drawn out from its interior. The end of the steel wire rope is fixed at the connection between the piston rod of the miter gate oil cylinder and the miter gate to measure the stroke value. The working stroke of the horizontal cylinder of the miter gate exceeds 7 m, and the maximum pressure generated

T. Mankovits and N. Trisovic (Eds.): MDS 2026, LNME 2026.
https://doi.org/10.1007/978-981-92-2440-1_4

by the dead weight of the steel wire rope when the stroke is maximum is 2 N. Figure 1 shows the wire outlet of the displacement gauge. To prevent the steel wire rope from being worn and broken, resulting in the failure of the displacement gauge, the rubber guide wire sleeve is generally used to prevent the friction and breakage of the steel wire rope caused by the direct contact between the steel wire rope and the wire outlet (Bai 2021). The gravity of the aforementioned steel wire rope will gradually wear and fail the rubber guide wire sleeve, leading to the contact between the steel wire rope of the displacement gauge and the edge of the metal right-angle notch at the wire outlet of the displacement gauge, resulting in the breakage of the steel wire rope of the displacement gauge.

Fig. 1. Wire outlet of the displacement gauge

To address the defects existing in the aforementioned prior art, this paper optimizes and designs the mature wire guide devices widely used in fields such as textiles, and develops a wire guide device for the horizontal cylinder displacement gauge of miter gates with simple structure, light weight, small volume, convenient installation and high reliability. Firstly, this paper introduces the structural characteristics of the wire guide device of the displacement gauge, and analyzes its working process and structural features. Secondly, based on the finite element theory and combined with the simulation plug-in in SolidWorks, the finite element analysis is carried out (Rounak 2020; Gómez-López 2013; Yu 2025). Through calculation, it is concluded that the main components of the wire guide device of the displacement gauge can meet the strength requirements. Finally, a test bench is built according to the actual working conditions of the horizontal cylinder displacement gauge of the miter gate in this Ship Lock to carry out experimental verification, which verifies the reliability and feasibility of the wire guide device of the displacement gauge.

2 Structural Introduction

The design of the wire guide device is shown in Fig. 2. When in use, the device needs to be installed at the wire outlet of the external displacement gauge of the horizontal cylinder of the ship lock. It is fastened to the left side of the main body fixing frame 1 of the wire guide device with a steel hoop, which is used to fix the main body fixing frame 1 of the wire guide device and the cylindrical wire outlet of the external displacement gauge. The left side of the main body fixing frame 1 of the wire guide device is designed with a semicircle, so that a large friction force can be generated between the steel hoop and the cylindrical wire outlet of the external displacement gauge, thereby limiting the horizontal movement and rotation of the main body fixing frame 1 of the wire guide device. Both the inner and outer sides of the right end of the main body fixing frame 1 of the wire guide device are designed with wire-passing holes for installing the main shaft 4, and the height of the main shaft 4 can be finely adjusted in the vertical direction. The bearing-equipped V-shaped roller 2 is positioned by the short bushing 5 and the long bushing 6, so that the bottom of its V-shaped groove coincides with the center of the cylindrical wire outlet of the external displacement gauge. The bearing-equipped V-shaped roller 2 is installed together with the main shaft 4, the short bushing 5 and the long bushing 6. After adjusting the height, the inner stop bolt 3 and the outer stop bolt 6 are installed through the stop washer 7. The length of the main shaft 4 is slightly shorter than the distance between the outer ends of the double-sided wire-passing holes, so that the stop bolt 3 can obtain a larger pre-tightening force, and the positions of the main shaft 4 and the bearing-equipped V-shaped roller 2 are limited by tightening the stop bolt. The level meter 4 is installed at the outer end of the left side of the main body fixing frame 1 of the wire guide device, which is convenient for observing and adjusting the main body fixing frame 1 of the wire guide device to be in a horizontal state. During operation, the steel wire at the wire outlet of the external displacement gauge of the horizontal cylinder of the ship lock drives the bearing-equipped V-shaped roller 2 to rotate on the bearing, thereby preventing the wire from sagging and avoiding direct contact with the rigid structure of the wire outlet, which causes wear and tear.

Fig. 2. Wire guide device

The new mechanism has the following characteristics and advantages:

- The roller component adopts a bearing-equipped V-shaped roller. The bearing enables smooth rotation, and the V-shaped roller groove is convenient for the steel wire of the displacement gauge to fall into the bottom of the V-shape, thereby preventing the steel wire of the displacement gauge from sagging.

- The fixing component adopts a steel hoop for fixing the left end of the main component. The main component can be fixed by adjusting its tightness, and at the same time, excessive extrusion that may damage the cylindrical wire outlet of the displacement gauge is avoided; it is also easy to disassemble and install without damaging the original structure of the wire outlet.

Standard installation and maintenance procedures:

- Installation shall be carried out during the operation interval of the ship lock.
- Place the tail end of the main fixing frame 1 at the wire outlet of the displacement gauge, check its levelness with a spirit level, and then fix it with a steel hoop.
- Adjust the installation height of the bearing-equipped V-shaped roller 2 via the guide hole. The wire outlet of the displacement gauge shall be level with the V-shaped bottom of the bearing-equipped V-shaped roller 2, and axial positioning shall be performed by using the main shaft 4, short shaft sleeve 5 and long shaft sleeve 6; if necessary, the length of the long and short shaft sleeves can be changed to adjust their lateral positions. After the position is determined, fix it with the stop bolt 3.
- For maintenance, the main task is to replace the bearing-equipped V-shaped roller 2. Keep the position of the main fixing frame 1 unchanged, remove the old roller, and then repeat Step 2.

3 Finite Element Analysis

3.1 Definition of Material Properties

According to the structural parameters, all components are made of 304 stainless steel to prevent corrosion in the high-humidity environment of the Ship Lock. The elastic modulus $E = 1.9 \times 10^{11}$ Pa, poisson's ratio $\mu = 0.29$, and density $\rho = 8000$ kg/m^3.

3.2 Mesh Division

The intelligent mesh division method in SolidWorks is adopted, with standard mesh parameters and good mesh density. After division, the total number of nodes is 133047, and the total number of elements is 80820.

3.3 Load Application and Constraint Solution

Loading: Under working conditions, the steel wire exerts a vertical pressure of 2 N on the roller. The friction coefficient between the steel wire of the external cable-pulling displacement gauge and the roller is 0.04, the rolling friction force is 0.08 N, and the relative torque to the main shaft is 0.0008 N/m. The friction force provided by the pre-tightening force of the stop bolt is used for radial fixation. The total weight of the main shaft and the roller is 0.2 N, the friction coefficient is 0.2, and under the condition that the safety factor is 2, the theoretical pre-tightening force of the two stop bolts is 2 N, which act on the surfaces in contact with the main component respectively.

Constraints: Full constraint points are set on the inner surface of the main body fixing frame of the wire guide device.

The diagram after adding loads and dividing the mesh is shown in Fig. 3.

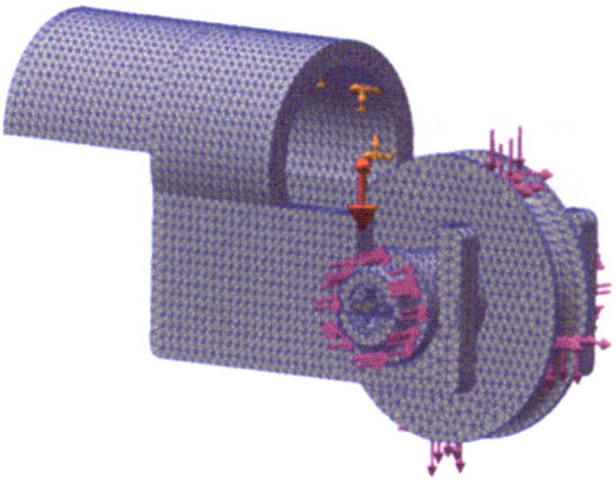

Fig. 3. Mesh division and load

3.4 Simulation Analysis

Figure 4 shows the equivalent stress nephogram of the main body of the wire guide device of the external cable-pulling displacement gauge, and Fig. 5 shows the overall comprehensive displacement nephogram. It can be seen from the figures that the maximum value of the overall equivalent stress is 7.030 MPa, which occurs at the lower end of the wire-passing groove, and the maximum value of the overall comprehensive displacement is 0.0004 mm. The results show that the structure and strain of the main body of the wire guide device of the external cable-pulling displacement gauge meet the design requirements under this working condition.

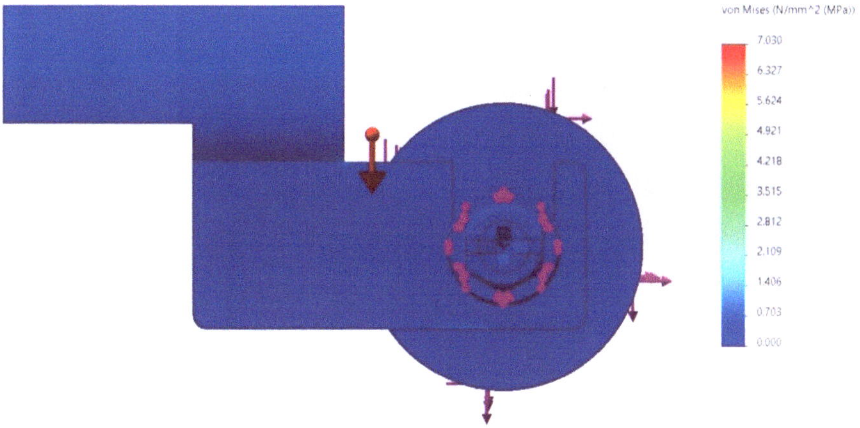

Fig. 4. Overall stress distribution

The maximum equivalent stress of the main body of the wire guide device of the external cable-pulling displacement gauge is 7.030 MPa, which is much lower than the yield strength of 304 stainless steel (206 MPa). The maximum overall comprehensive displacement is 0.0004 mm, which has a negligible impact on the overall structure. The minimum safety factor of the main body of the wire guide device of the external cable-pulling displacement gauge is 29, as shown in Fig. 6, which is much higher than the safety factor standard of general plastic materials. Therefore, from the perspective of finite element analysis, this design meets the requirements of normal operation.

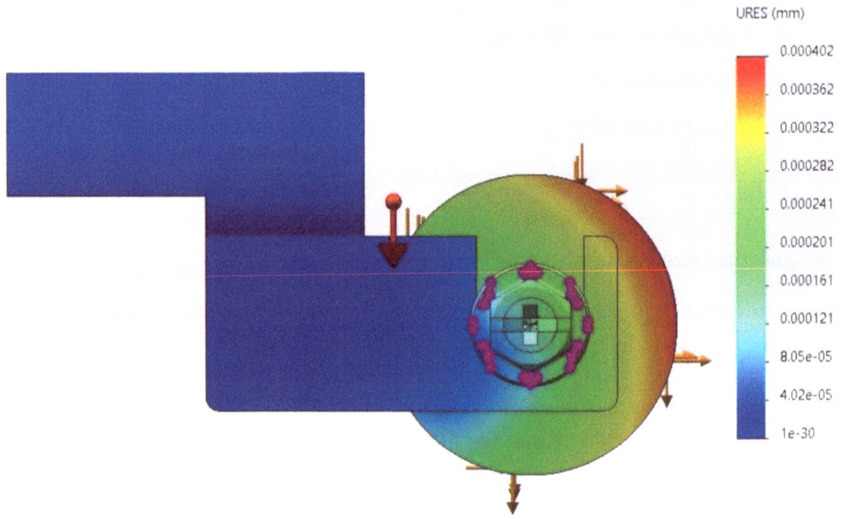

Fig. 5. Overall strain-displacement

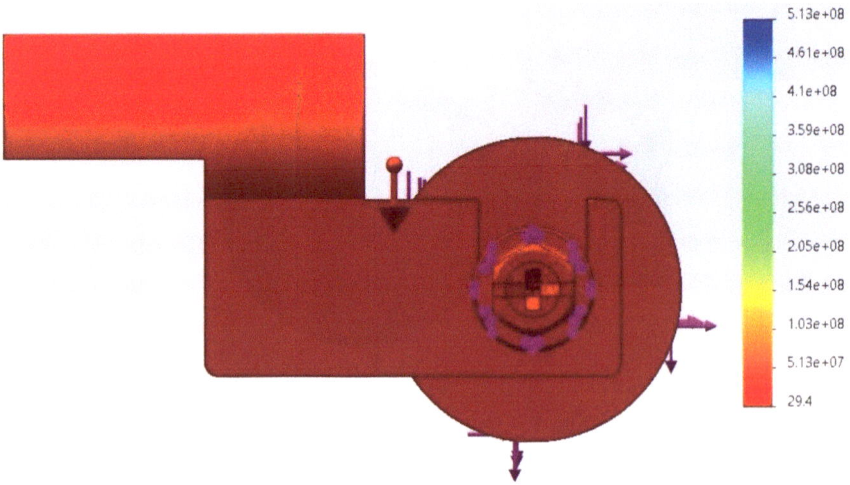

Fig. 6. Safety factor map

4 Experimental Verification and Application

4.1 Experimental Verification Device

To verify the feasibility and reliability of the device, a reciprocating tensile test was conducted on the wire guide device for the horizontal cylinder displacement gauge of the miter gate. The displacement gauge equipped with the device was fixed at the left end of the test bench, and the crank-slider mechanism driven by a motor at the right end was used to simulate the reciprocating tensile working condition of the steel wire rope of the displacement gauge. A digital display counter was installed in the middle to count the number of reciprocations. The experimental verification device is shown in Fig. 7.

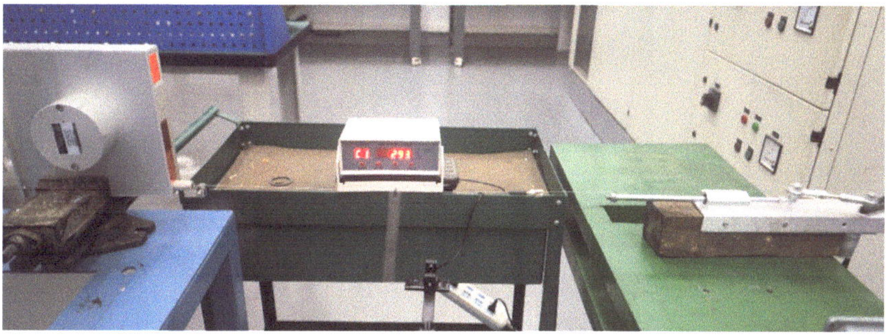

Fig. 7. Verification device

Partial parameters of the experimental verification device are as follows:

- Stroke of the reciprocating motor: 600 mm
- Reciprocating speed of the motor: 4 m/s
- Test environment temperature: 25 ± 5 °C
- Relative humidity: maintained at 60%
- Simulated actual operation times: 18,000
- Pressure on the upper end of the roller: 4 N

4.2 Data and Conclusions

The test lasted for 3 weeks, with counts conducted on a daily basis. The counting results on selected dates are shown in Fig. 8.

A total of 18,000 simulated reciprocating movements were carried out in the experiment. Compared with the wire outlet without the wire guide device of the displacement gauge, the wear condition of the guide wire sleeve was significantly improved, as shown in Fig. 9. Measurements indicated that the wear depth of the guide wire sleeve equipped with the wire guide device of the displacement gauge was less than 0.3 mm, while the wear depth of the guide wire sleeve without the device was approximately 2.5 mm. Meanwhile, the main structure of the wire guide device of the displacement gauge remained in good condition, with no obvious deformation or loosening.

The test results show that the wire guide device of the displacement gauge has high reliability and remarkable effectiveness. It exhibits excellent mechanical properties without deformation or loosening, which is consistent with the simulation results.

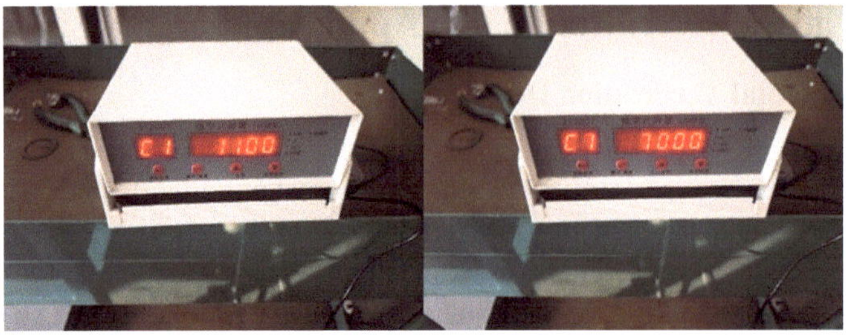

Fig. 8. Counting results on selected dates

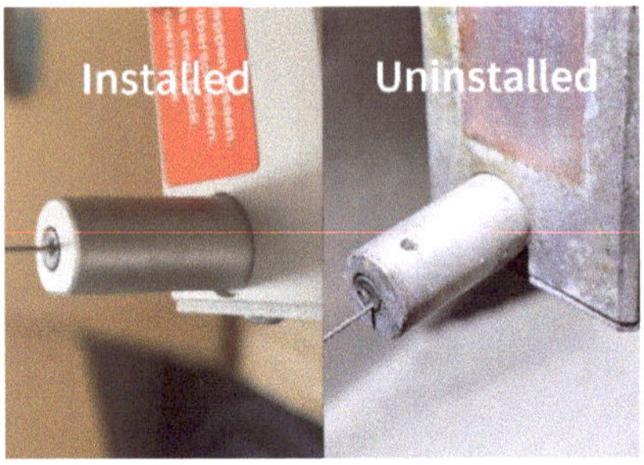

Fig. 9. Comparison diagram of test results

4.3 On-Site Testing

The device was installed at the wire outlet of the horizontal cylinder displacement gauge for the miter gate of a large-scale ship lock, as shown in Fig. 10. For the actually installed device, the main fixing frame, main shaft and shaft sleeve were made of AISI 304 stainless steel. The surface of the main fixing frame was subjected to sandblasting treatment, while the surfaces of the main shaft and shaft sleeve were passivated. The bearing-equipped V-shaped rollers were made of aluminum alloy, and their surfaces were coated with alumina ceramic.

As of the writing of this paper, the device has been operating safely for 150 days, with a cumulative total of more than 2,400 safe reciprocating operations. During the test period, the average temperature was 16 °C, the maximum temperature reached 27 °C, and the minimum temperature dropped to −2 °C, with an average humidity of 70%. The test results after 150 days of on-site operation are shown in Fig. 11: the alumina ceramic coating on the bearing-equipped V-shaped rollers showed no obvious wear, and the main fixing frame had no significant deformation, which was consistent with the laboratory test results. Meanwhile, no noticeable axial or radial deviation was observed in the bearing-equipped V-shaped roller device, indicating its excellent positioning performance.

4.4 Application and Popularization

In view of the different specifications of wire outlets of external displacement gauges in various ship lock fields, Table 1 provides multiple sets of dimensional data for the main

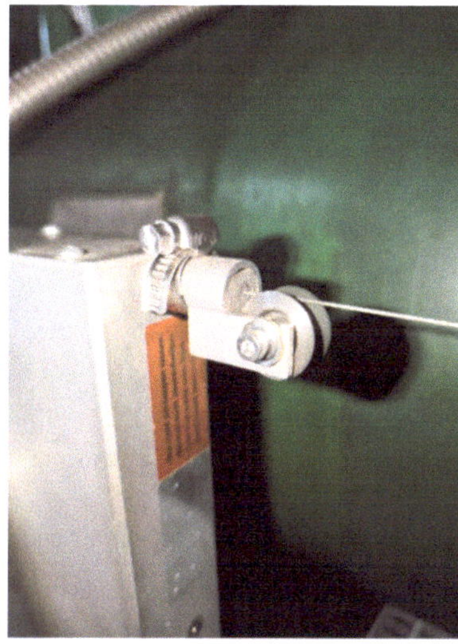

Fig. 10. On-site installation scenario

Fig. 11. On-site test results

fixing frame and main shaft of this device, so as to facilitate selection and popularization. The dimensions of the bearing-equipped V-shaped rollers shall be selected according to different actual working conditions. Table 2 presents the recommended materials and surface treatment schemes for the main components.

This device can effectively solve the wire breakage problem of external cable-pulling displacement gauges. Applied to the guiding field of external displacement gauges for ship locks, it can increase the safe number of reciprocating operations of steel wires to

Table 1. Device dimensions under different working conditions

Wire outlet diameter of displacement gauge/mm	Recommended wall thickness of main fixing frame/mm	Recommended minimum diameter of main shaft/mm
10	1.2	4
12	1.5	6
14	1.5	6
16	2.0	8

Table 2. Recommended materials and surface treatments for main components

Main components of the device	Recommended materials and surface treatments
Main fixing frame	AISI 304 stainless steel (passivation treatment)
Main shaft, shaft sleeve	AISI 304 stainless steel (passivation/polishing treatment)
Bearing-equipped V-shaped Roller	Aluminum alloy (alumina ceramic coating)

more than 6,000 times, significantly reducing the failure rate. It reduces the subjective errors and workload of manual inspection and judgment, cuts down the time and cost of equipment return for maintenance, and improves the efficiency of on-site operation and maintenance. By ensuring the stable operation of the miter gate stroke detection system, it avoids the impact of displacement gauge failures on the normal navigation of ship locks. It is of great significance for giving full play to the shipping benefits of water conservancy hubs and ensuring the safe and efficient operation of ship locks, with remarkable economic and indirect benefits (Yang 2024).

5 Conclusion

The research results of this project can be directly applied to the upgrading and transformation of all external displacement gauges for the miter gate cylinders of this ship lock, which can meet the actual on-site operation requirements. Its core technical ideas and device design concepts can also be extended to the planning and construction of other ship lock-related equipment, and are applicable to other equipment scenarios with similar cable-pulling detection devices, narrow installation spaces or long-stroke operating conditions, especially for the same type of stroke detection equipment in fields such as shipping hubs, demonstrating broad adaptability. As this project was developed to meet the requirements of high adaptability and high reliability for the aforementioned ship lock, the results can also be flexibly adapted to similar scenarios of other equipment scales, featuring high promotion and application value as well as promising application prospects.

References

Bai, X.: Fault analysis and handling of the miter gate displacement gauge for ship locks. J. Equip. Manag. Maintenance. **16**, 50–51 (2021). https://doi.org/10.16621/j.cnki.issn1001-0599.2021.08D.27

Gómez-López, L.M., Miguel, V., Martínez, A., Coello, J., Calatayud, A.: Simulation and modeling of single point incremental forming processes within a solidworks environment. J. Procedia Engineering. **63**, 632–641 (2013). https://doi.org/10.1016/j.proeng.2013.08.253

Rounak, M., Dhirendra, N., Sasanka, C., Pragyan, P.: Design and numerical analysis of spur gear using solid works simulation technique. J. Mater. Today: Proceedings. **41**, 340–346 (2020). https://doi.org/10.1016/j.matpr.2020.09.554

Yang, B.J., Yang, X., Li, Z., Deng, J.P.: Statistical analysis of navigation situation in the three gorges-Gezhouba hub. J. Pearl River Water Transport. **11**, 126–128 (2024). https://doi.org/10.14125/j.cnki.zjsy.2024.11.030

Yu, Z.W.: Simulation design and finite element analysis of bearing block based on SolidWorks. J. Mech. Manag. Development. **40**(11), 271-272+275 (2025). https://doi.org/10.16525/j.cnki.cn14-1134/th.2025.11.091

Structure Design of Helical Dual-Channel Nozzles for 3D Printers and Investigation of Part Mechanical Properties of Printed Parts

Shiyuan Li(✉)

College of Mechanical Science and Engineering, Northeast Petroleum University, Daqing, Heilongjiang, China
lsy.research@outlook.com

Abstract. This paper overcomes the shortcomings of low mechanical properties for Fused Deposition Modeling (FDM). A dual-channel, single-nozzle print head with a helical mixing channel is designed to realize co-extrusion and in-situ mixing between pure ABS and fiber reinforced ABS. The finite element analysis characterizes the nozzle thermal field distribution and compares the tensile and compression stress strain behavior between pure ABS and composite materials. The thermal simulation validated the design feasibility to produce a steady thermal gradient. Mechanics Analysis according to materials' characteristics showed the composite theoretical ultimate tensile strength is greater than pure ABS, furthermore, FEA results confirmed higher elastic modulus and stiffness for the composite with same loading. This result confirms that the coextrusion architecture is capable of supporting loads and provides an engineering solution to improve FDM's strength properties.

Keywords: Integrated dual-channel nozzle · Co-extrusion process · Helical channel · Mechanical properties

1 Introduction

Conventional manufacturing is mainly based on the subtractive and formative processing modes. However, the small-batch production in those paradigms suffer from low cost-efficiency and large amount of materials waste, while the process planning in case of complex components is complicated (Zhao 2020). Thus, the idea of 3D printing was developed to solve these drawbacks. Using the concept of additive manufacturing technology, three dimensional printing overcomes some of this shortcomings and can be considered as complimentary to conventional techniques. At present, China's AM Innovation System. The Chinese AM innovation system covers all the areas of AM: materials, processes, equipment technology, and important engineering applications (Wang and Lu 2022).

Recent work on printer designs has aimed to improve efficiency and stability. Zhang et al. (2020a) designed a 3D printer using a chamber thermal dual circulation heating

T. Mankovits and N. Trisovic (Eds.): MDS 2026, LNME 2026.
https://doi.org/10.1007/978-981-92-2440-1_5

and insulation, improving the system efficiency. Pan et al. (2022) designed a parallel multi-nozzles 3D printer, which improved the printing efficiency and accuracy than that of single nozzle structure. Zhou et al. (2023) used synergistic optimization approach of feeding mechanism and nozzle to improve extruding continuity and stability of filaments. Moreover, Zhang et al. (2020b) designed a 5-DOF hybrid 3D printer with a large positional and orientational workspace that is nonsingular, enabling bidirectional and tri-directional printing by rotating the position of the printhead.

Although these works have been successful within certain domains, we are yet to achieve great mechanical performances on printed parts. The tensile strength, flexural strength, hardness, wear resistance and dimensionally stability of the normal printed part needs to be improved (Ahn et al. 2002; Dey and Yodo 2019).Thus, we address process issues related with achieving in-situ compounding efficiency and microstructure control on solid fiber filaments and molten ABS matrix inside the helical channel before extrusion by employing a process flow with double inlet-single nozzle architecture, independent dual-path feeding, helical chaotic mixing and synergistic formation. This enables a combined initial structure optimization and printed component mechanical property improvement. As for the problem of material residue in the single nozzle multiple materials system proposed by Tian et al. (2024), this device uses static material pathways that prevent crosscontamination.

2 Mechanical Structure Design

2.1 Overall Mechanical Structure

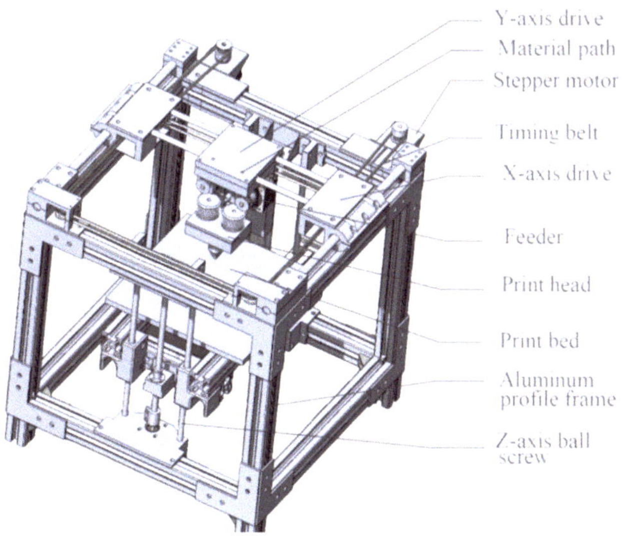

Fig. 1 3D model of the 3D printer

The proposed printhead model fulfills the main criteria of mechanical rigidity and robustness which is necessary for multiple use cases. In Fig. 1 it can be seen that the framework uses an aluminium profile structure as its main frame, the moving platform adopts an XYZ-three degree-of-freedom drive mechanism, and the nozzle has a two-way, single outlet device with an integrated double nozzle including a spiral channel within the nozzle.

The aluminium profile frame provides a structure rigidity, modularization and scalability to ensure the stability of high speed running. An XYZ compound transmission mechanism, which is driven by independent stepping motor to achieve smooth work and fast response. Ball screw of the Z axis can eliminate interlayer dislocation, while XY-axis timing belts achieve the required printing accuracy in controlling the position of the print head; and Heat bed made from heat resistant polymer, safety and reliability assurance. Strong bonding force and low heat expansion coefficient avoid distortion and delamination.

2.2 Print Head Design

The extrusion type print head is used for the deposition of materials. The working principle consists in melting solid filament into liquid form, followed by nozzle extrusion and layer-by-layer deposition forming a solid part. We chose a direct drive configuration in order to feed. The direct drive system provides traction force and responsiveness of materials, eliminating the hysteresis and clog hazards of remotely fed feeders. To overcome the shortcomings of ordinary printed parts, two channels and one nozzle hybrid extrusion system were developed (Fig. 2).

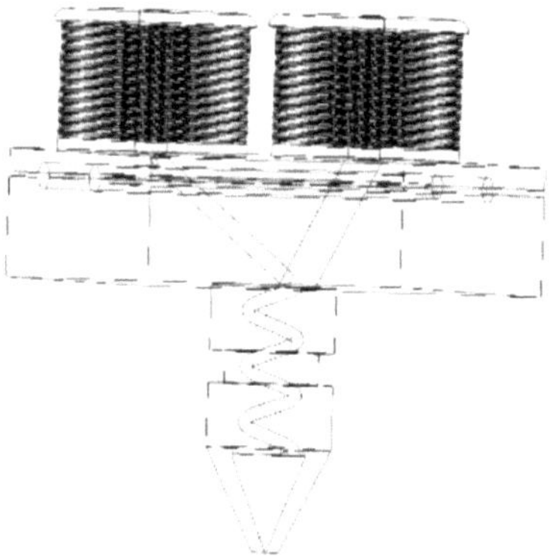

Fig. 2 Helical channel and triangular outlet geometry

Fibers filaments reduce the shrinkage and warp in ABS printing by reducing the thermal deformations. Compared with pure ABS, the composite has improved layer bonding performance. The tensile strength, bending strength and heat deformation temperature are greatly increased (Du et al. 2023; Pang et al. 2024; Han et al. 2017). The print head cold end uses a stepper motor driven feed gear to move the filament with two inlets. The hot end contains inner dual channels as well as an outlet. The heat source is the heater cartridge and the heat sink is the thermistor. A helix-like inner channel design was adopted as shown in Fig. 2 for the nozzle. This choice has been made because of following reasons; the ability of inducing Dean secondary flows, creating alternate directions of shearing forces at laminar flow with high viscosity which are characteristics for FDM process (Li 2021; Guo et al. 2011).This allows for efficient 3D chaotic mixing between solid fibres and the melt phase in the vicinity of the exit plane, where the flow exits through a truncated tetrahedron shaped channel having a 0.4 mm diameter orifice at its base.

2.3 Print Head Design Verification

Thermal simulation of the print head was conducted via 3D simulation software. Boundary conditions included a convection coefficient of 100 W/(m^2·K), a heating temperature of 573.15 K, and ambient room temperature (Chen et al. 2022). Figure 3 illustrates the resulting thermal distribution.

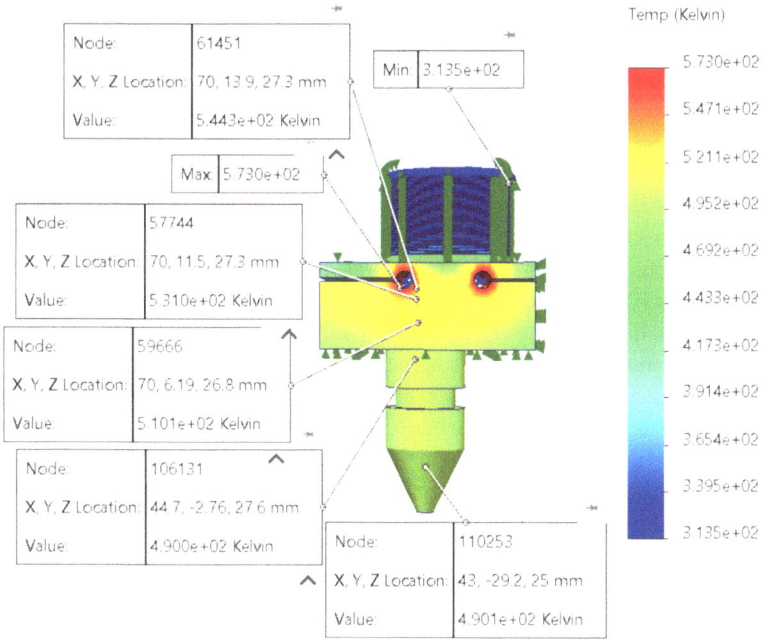

Fig. 3 Thermal analysis simulation of the print head

One of the most important aspects on printability and part integrity is the heating/cooling behaviour of the extruder nozzle. A fast acting, accurate thermal control system has to be capable of maintaining a constant heat profile inside the extruder. Models have been imported for the purpose of simulation into software and discretised, discretising the continuum of a solid model into FE's. The computation by using 3D software resulted in the thermal distribution gradient. The thermal field distribution contour of Fig. 3 shows that thermal gradients are stable: heater block temperature is stabilized within a required value (~510–544 K), while keeping the temperature in the heat sink area under that of the ABS glass transition temperature (~378 K) to prevent any kind of feed clogging due to heat creep theoretically.

3 Simulation Analysis of Printed Specimen Mechanical Performance

3.1 Simulation Model Establishment

The boundary conditions were applied by fixing one end (the surface with dimensions 20 mm × 15 mm) and applying the pressure force as loading condition over the entire area equal to 12, Tensile Load Application: 000 N/m^2 on the 20 mm × 50 mm face (load vector is normal). Tensile Load Application: 12,000 N/m^2 on the two opposite 15 mm × 20 mm faces. Where is parallel to the tension axis. The fibre reinforced composite model required for precise solution to fibre matrix interfacial micromechanics behaviour explicit modelling coupled with very non-uniform progressive meshing scheme was used. The key concept in this case was application of high density local mesh refinement in the area of highest stresses (fibre and neighbouring interface). A minimum size of elements equal 0. The mesh size of 367,742 mm was sufficient for capturing the interfacial stress gradient and transferring loads. The mesh size was gradually increased from the above-mentioned mesh size value towards outside (pure matrix area) controlling the computational scale. The location of the largest element size (7.35485 mm) was at the far field boundary of the matrix, a domain of little effect in the macroscopic mechanical response. This approach resulted in a high-fidelity model generation of 1,200,347 elements and 1,712,929 nodes.

On the other hand, pure ABS characterization as a homogeneous continuum used global uniform meshing. The characteristic element size was in range of 0.674118–2.02258 mm, leading to a total number of 19,345 elements and 28,608 nodes. The order of magnitude difference in mesh size between the composite and pure ABS model is due to the local refinement required for explicitly representing massive fibers interfaces. This difference illustrates the computing complexity of multiscale modeling. The simulation model parameter is listed as Table 1.

Table 1 Parameter comparison: pure ABS versus composite material

Parameter	Pure ABS	Composite material
Dimensions	20 × 15 × 50(mm)	20 × 15 × 50(mm)
Tensile strength	40 MPa	60 MPa
Elastic modulus	2.0 GPa	5.0 GPa
Poisson's ratio	0.35	0.30

3.2 Simulation Results

Figure 4 shows that there is no stress concentration on the pure ABS sample. It can be seen from this figure that the maximum von mises stress was found to be 7.22×10^5 Pa with a corresponding maximum strain value of 2.616×10^{-4}. As mentioned earlier, the maximum stress observed here is still much lower than the theoretical compressive strength of ABS which usually ranges between 34 and 55 MPa. Thus, the material is in an elastic deformation state under loading and there is no plastic yield.

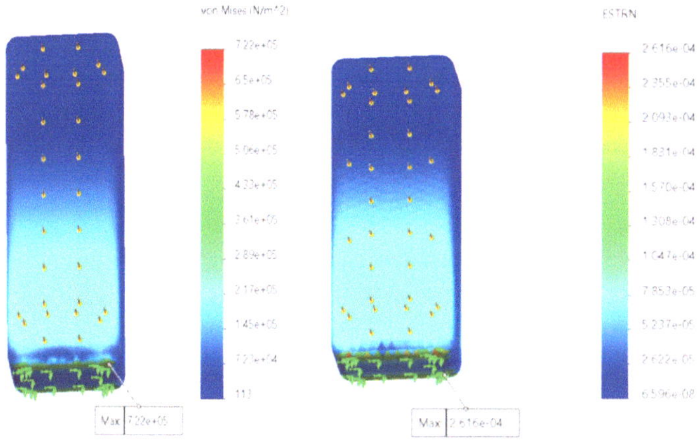

(a) Compressive Stress State in Pure ABS (b) Compressive Strain State in Pure ABS

Fig. 4. Simulation analysis of compressive stress and strain in pure ABS

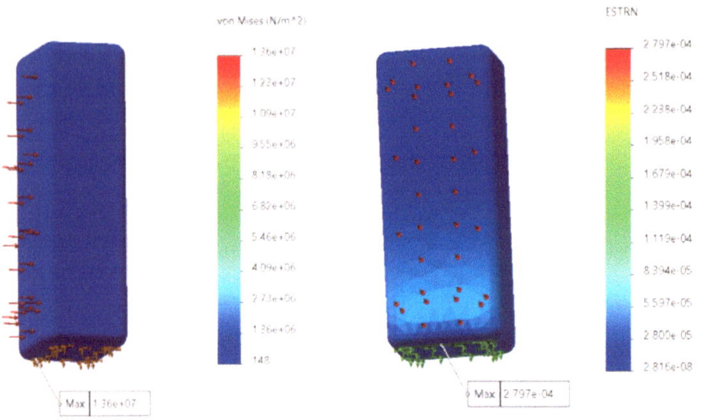

(a) Compressive Stress State in Composite Material (b) Compressive Strain State in Composite Material

Fig. 5. Simulation analysis of compressive stress and strain in ABS-fiber composite

According to Fig. 5, the maximum stress and maximum strain of the composite were respectively 1.36×10^7 Pa and 2.797×10^{-4} at the same load condition. In addition, the maximum stress obtained by simulating the composite was mainly caused by the high elastic modulus (5.0 GPa), which results in higher internal force production to overcome the resistance of deformation. The reason is mainly due to the large modulus of elasticity and fine division of the mesh size, but it does not exceed the maximum allowable stress of the FRP. According to the results above can be found that when changing the parameters in Table 1, because the stiffness of the FRP is larger than, due to the increased elastic modulus and strength limit.

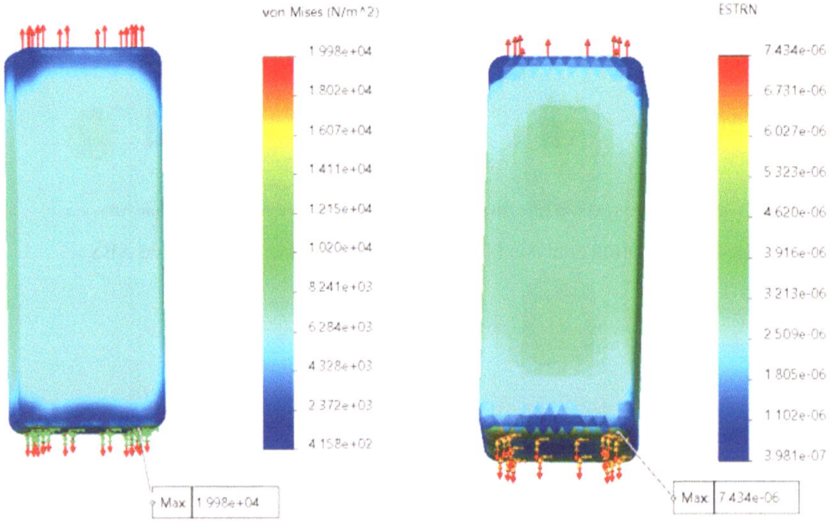

(a) Tensile Stress State in Pure ABS (b) Tensile Strain State in Pure ABS

Fig. 6. Simulation analysis of tensile stress and strain in pure ABS

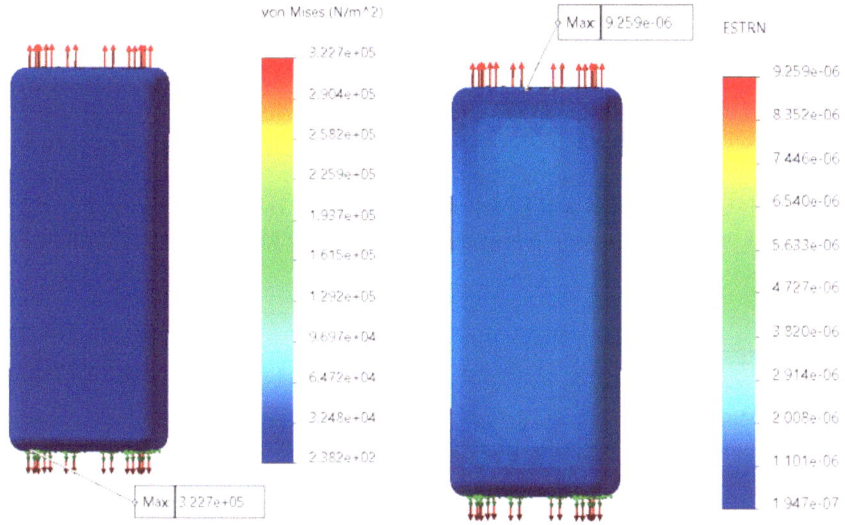

(a) Tensile Stress State in Composite Material (b) Tensile Strain State in Composite Material

Fig. 7. Simulation analysis of tensile stress and strain in ABS-fiber composite

As indicated in Figs. 6 and 7, following tensile loading, the pure ABS specimen registered a maximum stress value of 1.998×10^4 N/m^2 and a maximum strain of 7.434×10^{-6}. In contrast, the composite material exhibited a maximum stress of 3.227×10^5 N/m^2 and a maximum strain of 9.259×10^{-6}.

The simulation shows, with the same load on the outside, lower stresses than the strength of both materials, proving that structure of the designs is safe. In particular, it was shown that the composite can withstand higher stresses, which is consistent with a higher stiffness of the underlying polymer. Therefore, one can expect that under realistic conditions, this bi-layered film would be able to bear heavier loads than pure PMMA films.

4 Conclusion

In this paper, we have designed and verified (via simulation) a two-channel single nozzle print head with a helical mixing channel. Simulation results on the thermal analysis proved that our design for thermal management is valid since it establishes a steady-state thermal gradient. Our mechanical simulations show that there will be a great advantage in terms of stiffness in such a composite structure. We see an increase of 150% in the value of the elastic modulus compared to pure ABS, well as to minimize deformation under the action of loads. At the same time, we also have an estimation using the physical parameters of materials which indicates that the UTS should be improved by about half. The above result proves from both aspects (material composition and rigidity) that our proposed dual nozzle extrusion method is effective to improve the mechanical characteristics of FDM objects.

The current work is simulated only, it has not yet been fabricated or tested experimentally. Any conclusions must be verified by experimental tests. The constitutive model used for the composite is very simple, excluding fiber orientation/deposition path effects on the resulting macroscopic properties. In addition, we have not yet evaluated the effect that large aperture nozzles will have in terms of surface print quality. Prototype fabrication and function test through actual print are future works, Standard tensile specimen and flexure specimen will be fabricated, followed by the acquisition of empirical mechanical properties through a universal test machine to correlate and calibrate with simulations.

References

Ahn, S.H., Montero, M., Odell, D., Roundy, S., Wright, P.K.: Anisotropic material properties of fused deposition modeling ABS. Rapid Prototyping J. **8**(4), 248–257 (2002). https://doi.org/10.1108/13552540210441166

Chen, W.P., Huang, J., Yu, J., Chen, W.W.: Analysis and optimal design of temperature field of nozzle system based on FDM-3D printer. J. Qilu Univ. Technol. **36**(3), 46–52 (2022). https://doi.org/10.16442/j.cnki.jlgydxxb.2022.03.007

Dey, A., Yodo, N.: A systematic survey of FDM process parameter optimization and their influence on part characteristics. J. Manuf. Mater. Process. **3**(3), 64 (2019). https://doi.org/10.3390/jmmp3030064

Du, J., Xu, Y., Li, T., Li, X.J., Wang, B.: Preparation and mechanical properties of fiber reinforced ABS composites. Plastics Sci. Technol. **51**(3), 29–32 (2023). https://doi.org/10.15925/j.cnki.issn1005-3360.2023.03.006

Guo, X.Y., Zhao, C.Y., Wang, L.Y., Lin, Z.M.: Numerical study on secondary flow intensity in helical pipe. J. Gansu Sci. **23**(3), 87–92 (2011). https://doi.org/10.16468/j.cnki.issn1004-0366.2011.03.015

Han, S., Xiao, Y., Qi, T., Li, Z., Zeng, Q.: Design and analysis of fused deposition modeling 3D printer nozzle for color mixing. Adv. Mater. Sci. Eng. **2017**, 2095137 (2017). https://doi.org/10.1155/2017/2095137

Li, X.: Numerical study on evolution of secondary flow and boiling heat transfer characteristics in helical tube. Master's thesis, Zhengzhou University (2021). https://doi.org/10.27466/d.cnki.gzzdu.2021.004974

Pan, S.C., Liu, Y.Q., Hu, H., Zhang, X., Feng, Y.: Research on control system of 3D printer based on multi-nozzle parallel. J. Eng. Des. **29**(1), 100–106 (2022). https://doi.org/10.3785/j.issn.1006-754X.2022.00.003

Pang, Y.N., Li, J.H., Qi, H.J.: Study on mechanical and thermal properties of CF reinforced PC/ABS composites. Plastics Sci. Technol. **52**(11), 68–72 (2024). https://doi.org/10.15925/j.cnki.issn1005-3360.2024.11.013

Tian, Z.Y., Chen, J.F.: Analysis of the influence of feeding mechanism on the properties of multi-material 3D printing by fused deposition modeling. Shandong Chem. Indus. **53**(16), 103–106 (2024). https://doi.org/10.19319/j.cnki.issn.1008-021x.2024.16.005

Wang, L., Lu, B.H.: Research on the development of additive manufacturing technology and industry in China. Strategic Study CAE **24**(4), 202–211 (2022). https://doi.org/10.15302/J-SSCAE-2022.04.018

Zhang, H.Q., Du, J.B., Zhao, J.H., Chen, X.J.:. Design and research on molding space heating and insulation system of FDM 3D printer. Mach. Des. Manuf. **3**, 141–145 (2020a). https://doi.org/10.19356/j.cnki.1001-3997.2020.03.034

Zhang, J., Chi, C.C., Tang, T.F., Fang, H.L.: Design and kinematic analysis of a 5-DOF hybrid 3D printer. J. Central South Univ. (Sci. Technol.) **51**(10), 2822–2833 (2020b)

Zhao, J.N.: Development and application trend of 3D printers. Paper Equip. Mater. **49**(3), 114 (2020). https://doi.org/10.11817/j.issn.1672-7207.2020.10.014

Zhou, S.L., Zhang, X.F.: Research on collaborative optimization of wire feeding mechanism and nozzle of 3D printer. Mach. Tool Hydraulics **51**(13), 27–33 (2023). https://doi.org/10.3969/j.issn.1001-3881.2023.13.005

Research on the Die Structure Development of a Mounting Cover

Binghua Jiang[1,2] and Deping Chen[1,2](✉)

[1] School of Material and Environmental Engineering, Chengdu Technological University, Chengdu, Sichuan 611730, China
chendeping1030@qq.com

[2] Mold Technology Laboratory, Chengdu Technological University, Chengdu, Sichuan 611730, China

Abstract. This paper focuses on improving the processing route of a mounting cover, transforming the original manufacturing process from drawing and wire cutting to a one-set mold forming process. To this end, CAE tools are introduced in the development of the stamping plan to ensure the smooth progress of mold production and reduce trial-and-error costs. A key structural feature of this mold is the double-sided piercing at the bottom of the product; the fine blanking process is adopted to guarantee the precision and flatness of the bottom bridge. After the improvement, production efficiency is increased by 4 times, the number of operators is reduced by 8, and production cost is decreased by 80%.

Keywords: Manufacturing process route · Stamping scheme · Cost reduction

1 Introduction

With the continuous introduction and iteration of automation and information technology in the manufacturing industry, conventional processing routes for traditional mechanical products have encountered significant challenges (Kumar et al. 2026; Oztemel and Gursev 2020; Sabri and Gaceb 2026). In addition to the ever-increasing demands on machining processes and die equipment, severe market competition has also led to a substantial reduction in the production costs of mechanical components compared to traditional methods. This project focuses on a critical component of general-purpose equipment—a drawn part featuring a bridge base, referred to as the "Mounting Cover." The original mature process involved multiple drawing and blanking operations followed by wire-cut electrical discharge machining (WEDM) (Olguner et al. 2025). However, the significant disparity in production rhythm between stamping and specialized machining processes, compounded by intermediate steps such as part sorting, logistics, dedicated

T. Mankovits and N. Trisovic (Eds.): MDS 2026, LNME 2026.
https://doi.org/10.1007/978-981-92-2440-1_6

fixture clamping, and WEDM, resulted in extremely low production line efficiency, rendering it insufficient to meet current market demands. Through further analysis of the material, product precision, and assembly requirements, a fully integrated stamping die process was adopted to enhance efficiency and quality stability.

2 Product Analysis

The two-dimensional drawing and three-dimensional geometry of the Mounting Cover are shown in Fig. 1 and Fig. 2, respectively. The required monthly production volume is approximately 20,000 parts.

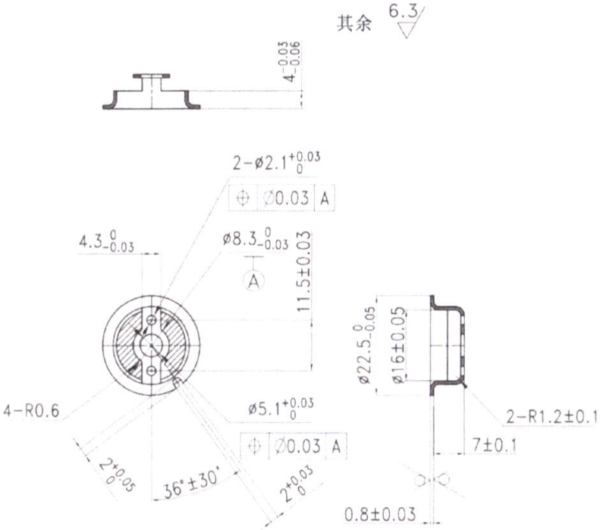

Fig. 1. 2D drawing

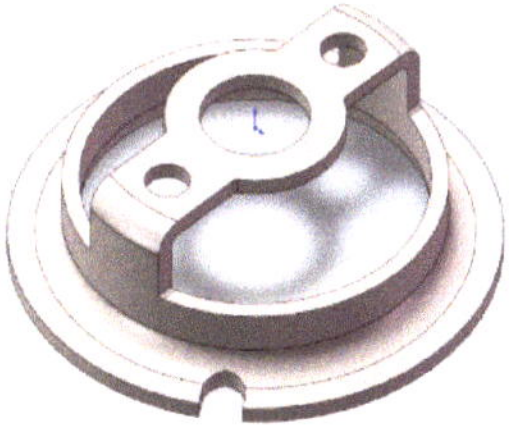

Fig. 2. 3D geometry

2.1 Technical Requirements

1. Primary requirements are detailed in the drawings. Notably: the flange positioning holes require an angle of $36^\circ \pm 30'$ (critical parameter); the center distance between the two holes is 11.5 ± 0.03 mm, with associated position tolerances of 0.03 mm relative to datum A.
2. The bridge width dimension is mm.
3. The flatness of the mounting cover flange shall be controlled within 0.02 mm.
4. The product shall exhibit no significant burrs (critical parameter).

2.2 Material and Performance Specifications

The material selected in this study is 30CrMnSiA alloy steel. This material demonstrates a favorable combination of high plasticity and excellent mechanical strength. Owing to its outstanding load-bearing capacity, 30CrMnSiA alloy steel has been widely adopted in the manufacture of aerospace structural components and critical mechanical parts. In addition, its superior corrosion resistance renders it a preferred candidate for high-value-added equipment, particularly for key connecting and mounting components. The chemical composition of the material is listed in Table 1, and its key mechanical properties are provided in Table 2(Lei et al. 2017; Wang et al. 2018; Wang et al. 2023; Xiao et al. 2024).

Table 1. Material composition (%)

C	Si	Mn	S	P	Cr	Ni	Cu
0.28 ~ 0.34	0.90 ~ 1.20	0.80 ~ 1.10	Allowable residual amount $\leq$ 0.025	Allowable residual amount $\leq$ 0.025	0.80 ~ 1.10	Allowable residual amount $\leq$ 0.030	Allowable residual amount $\leq$ 0.025

Table 2. Material mechanical properties

Tensile strength	Tensile strength	Elongation	Reduction of area	Impact energy	Impact toughness value	Hardness
$\geq$1080 MPa	$\geq$835 MPa	$\geq$10%	$\geq$45%	$\geq$39 J	$\geq$49 J/cm^2	$\leq$229HB

3 Stamping Process Planning

3.1 Strip Layout Design

According to the product specifications and die design requirements provided by the customer, together with accumulated industrial manufacturing experience, two forming strip layouts were designed, as illustrated in Figs. 3 and 4. The initial stamping process (Fig. 3) adopted scrap cutting at the first station. Nevertheless, the forming quality of the bridge base was seriously unsatisfactory, especially in terms of its flatness and dimensional accuracy along the width direction. Extensive adjustments to process and dimensional parameters did not eliminate these defects. Therefore, process optimization was carried out, and a revised stamping process (Fig. 4) was proposed. To reduce die tryout costs and improve process reliability, CAE simulation analysis was conducted on the optimized Scheme 2 (Fig. 4).

3.2 Determination of Blank Holder Force and Die Clearance

Figures. 5, 6 and 7 presents the analysis results obtained under different blank holder forces (BHF) and various punch-die clearances (expressed as multiples of material thickness). Given that the analysis pertains primarily to formability, simulation of the first drawing station was deemed sufficient. Figure 5 indicates that using a clearance of 1.1 times the material thickness facilitates excessive material flow, leading to insufficient deformation. Figure 7 shows that excessive BHF restricts material flow, resulting in insufficient safety margins at the punch radius, potentially causing cracking during physical tryout(Wang et al. 2026).

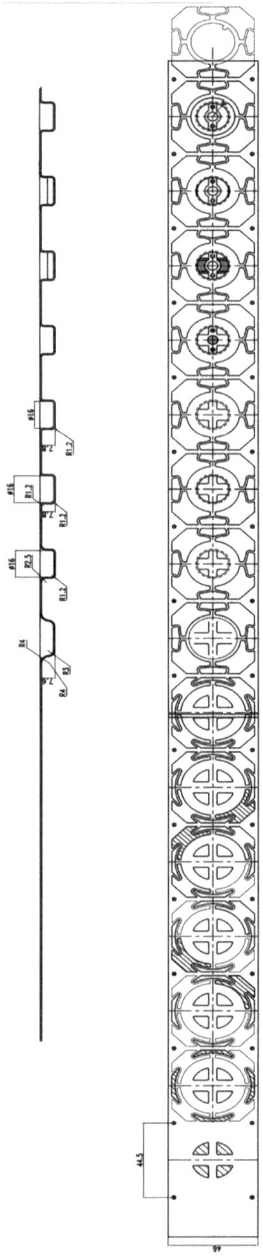

Fig. 3. Layout plan 1

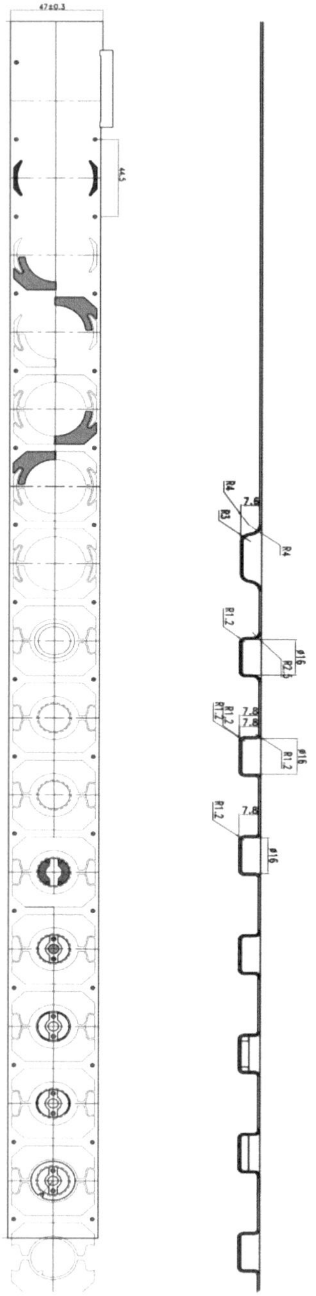

Fig. 4. Layout plan 2

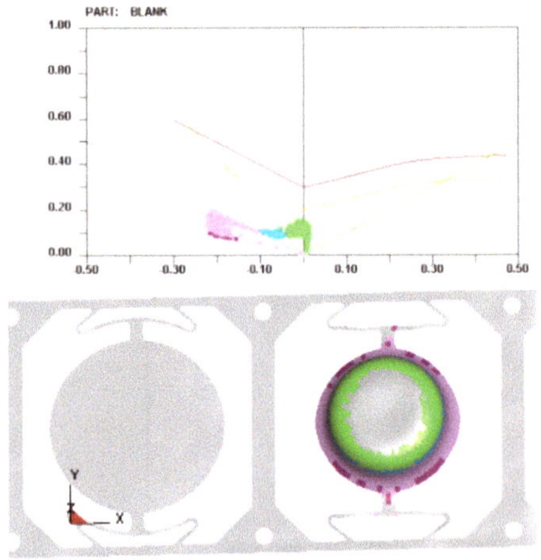

Fig. 5. BHF 5Mpa, gap 1.1 times blank thickness

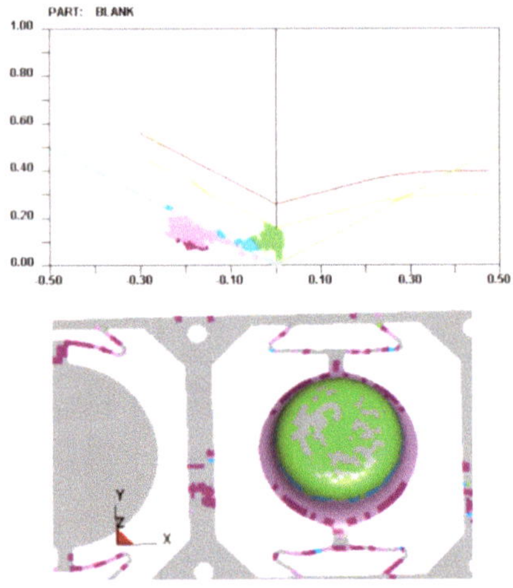

Fig. 6. BHF 5Mpa, gap 1.05 times blank thickness

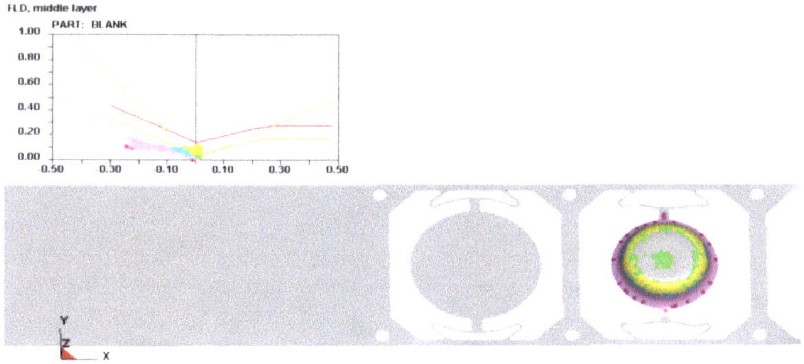

Fig. 7. BHF 6Mpa, gap 1.05 times blank thickness

Through iterative adjustment of the BHF (surface pressure) during the drawing simulation, Fig. 6 demonstrates that neither cracking nor wrinkling occurred at the punch radius or edge regions. Figure 8 displays the results after the first and second drawing operations, followed by the restriking operation (third step), indicating favorable outcomes. Therefore, the BHF was converted based on a surface pressure range of 5–6 MPa, and clearance with 1.05 blank thickness. The total BHF was calculated using the die binder surface area, serving as the basis for selecting and arranging springs.

As the two most critical parameters in die design, the blank holder force (BHF)and die clearance will be determined based on the above forming analysis method, using the criterion that a thinning rate exceeding 15% in the drawn part indicates a risk of fracture, while a thickening rate exceeding 10% (i.e., −10%) indicates a risk of wrinkling. By simulating die clearances ranging from 1.0 times the material thickness, 1.05 times the material thickness, to 1.1 times the material thickness, and blank holder forces varying from 5 MPa, 5.5 MPa, to 6.0 MPa, the following material thickness thinning rates are obtained in Table 3.

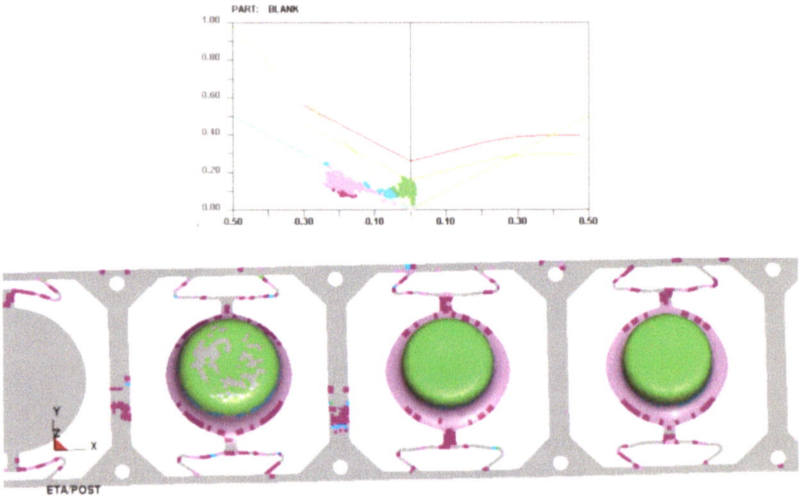

Fig. 8 restriking operation of the 3rd step

Table 3. Material thinning (%)

Binder force clearance	5.0 Mpa		5.5Mpa		6.0Mpa	
	Punch radius	Flange	Punch radius	Flange	Punch radius	Flange
1(material thickness)	15.9 Crack tenancy	−12.4 wrinkle	16.7 Crack tenancy	−10.6 wrinkle tenancy	17.8 Crack	−10.2 wrinkle tenancy
1.05(material thickness)	13.4	−9.82	14.9	−9.57	15.6 Crack tenancy	−9.51
1.1(material thickness)	11.6	−10.63 wrinkle tenancy	14.1	−9.81	15.2 Crack tenancy	−9.62

The regions prone to cracking are the punch radius zones, and those prone to wrinkling are the flange areas. Accordingly, only these two regions are selected as evaluation areas in Table 3 to assess the formability of the part (Jemal et al. 2024; Liao et al. 2015; Sheng et al. 2006).

As can be observed from Table 3, three sets of parameters exhibit relatively feasible forming performance: a die clearance of 1.05 times the material thickness combined with a blank holder force of 5 MPa; a die clearance of 1.05 times the material thickness with a blank holder force of 5.5 MPa; and a die clearance of 1.1 times the material thickness with a blank holder force of 5.5 MPa. For parts formed with the material thickness clearance, cracking or a tendency to crack appeared in the punch radius region, and wrinkling or a tendency to wrinkle occurred in the flange region. Given that the tensile strength of the material reaches 1000 MPa, springback deformation after drawing may significantly impair dimensional accuracy. Therefore, sufficient plastic deformation must be ensured during the drawing process. The optimal condition is that the thinning rate at the punch radius is close to 15%, while the thickening at the flange region is minimized. Accordingly, based on the comprehensive analysis results, the most suitable process parameters are determined to be a die clearance of 1.05 times the material thickness and a blank holder force of 5.5 MPa.

3.3 Analysis of Carrier Deformation at Strip Edge

The significant shape changes observed in the drawing results necessitated an examination of the forming operation's influence on the feeding carrier and pilot hole deformation (Xie et al. 2023). Undetected deformation would compromise feeding piloting accuracy in subsequent stations, thereby jeopardizing product precision. As subsequent operations are primarily blanking processes with negligible impact on contour change, analysis was confined to the influence of the initial forming stations on the carrier.

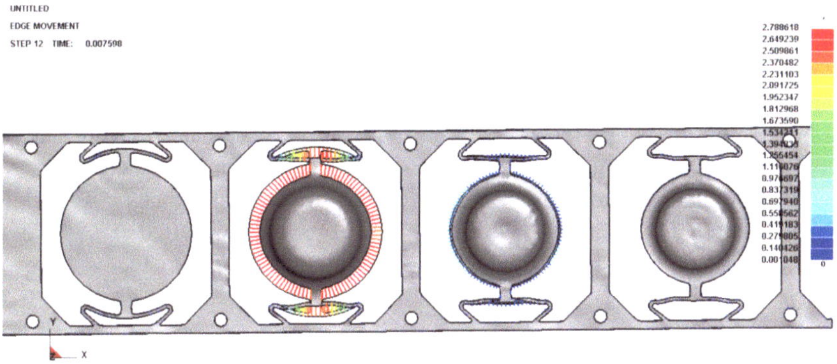

Fig. 9. Edge movement

Figure 9 reveals that the pre-defined carrier geometry and width are appropriate. The boundary contour change is substantial after the first drawing operation, diminishes after the second drawing operation, and material flow is virtually absent during and after the restriking step. The deformation-induced displacement, as shown in Fig. 9, is primarily concentrated on the upper and lower carrier edges and the product's peripheral profile. Crucially, the upper and lower edges of the strip carrier remain unchanged. Consequently, deformation of the part's edge carrier has no impact on the strip carrier itself. The edge of the remaining portions of the strip remains unaltered, thus posing no interference with preceding or subsequent steps. Material adjacent to the pilot holes exhibits no displacement, ensuring piloting accuracy. Therefore, the carrier design for this drawing process is deemed reasonable and feasible.

3.4 Verification

Based on the above analysis, experiments were conducted using BHF(blank holder force) and clearance between punch and die as parameters. The mold process parameters were determined based on the final product quality. Table 4 shows the experimental data and the corresponding product.

Table 4. BHF of 5.5 with clearance verification

One material thickness	1.05 times material thickness	1.1 times material thickness
		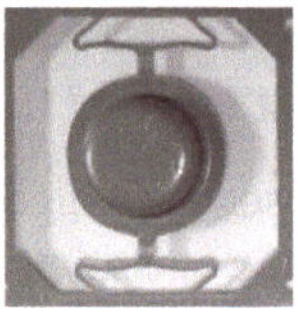
crack tendency	appropriate deformation	Insufficient deformation
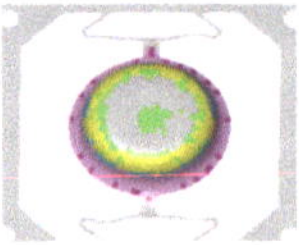	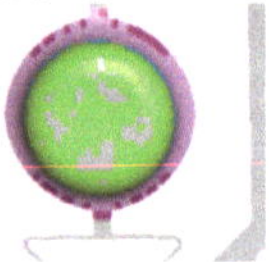	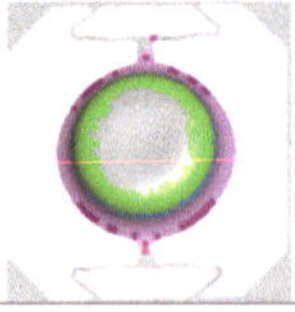

In summary, Scheme 1 (Fig. 3)—comprising flange blanking, two-stage drawing, restriking, semi-circular piercing, small-hole piercing, two-stage side notching, and final cut-off—is a feasible forming process plan. It shows good correlation between physical tryout and simulation prediction from Figures 5 to 8 and the parameters of BHF listed in Table 3 and Table 4; therefore, the blank holder force was converted to 5.5 MPa, with a clearance of 1.05 times blank thickness, which should be appropriate.

4 Die Structure Design

As shown in the process flow (flange blanking, two-stage drawing, restriking, semi-circular piercing, small-hole piercing, two-stage cam piercing, and final cut-off), special emphasis in die design was placed on the two-stage side notching of the small bottom holes. During piercing, cracking was prone to occur at the junction between the bridge and the base. Meanwhile, the technical requirements specified that these two holes must be burr-free. Therefore, a side-piercing scheme with minimal clearance was employed, combined with an increased force applied to the side-piercing pad to improve material plasticity and suppress cracking. The two-stage side notching mechanisms for these holes are presented in Figures 10 and 11. During die manufacturing, critical components were fabricated using ASP2030 high-speed steel to ensure service life and process stability.

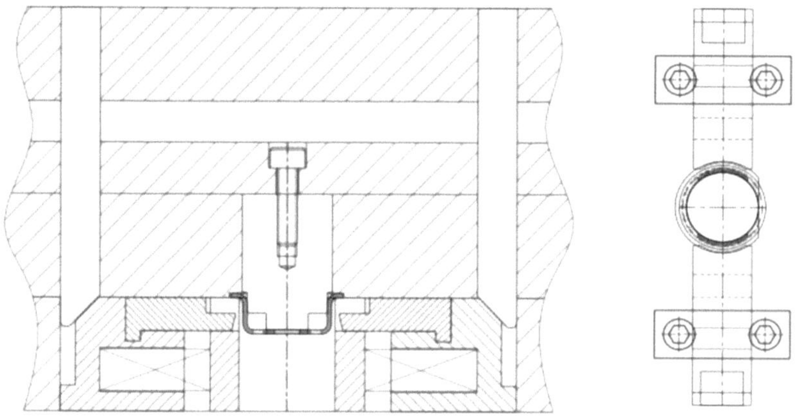

Fig. 10. 1st side notching (cam piercing)

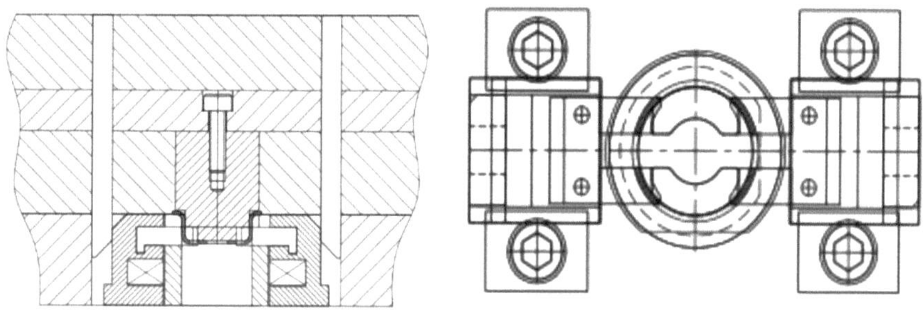

Fig. 11. 2nd side notching (cam piercing)

Figure 12 shows the die assembly drawing. The overall dimensions are length 840 mm and height 237 mm. Due to the considerable die length, the die insert was split into two sections to mitigate risks such as heat treatment distortion. Each section is guided by four independent guide pins. The front section primarily involves blanking operations and was designed as a single die block to facilitate tryout and sharpening. The rear section encompasses drawing, side piercing, and cut-off operations, designed as a separate die block. Key forming components utilize insert structures.

Figure 13 provides close-up photographs of the physical die structure for several critical rear-section processes. Figure 14 depicts the die during production tryout. After a period of production testing and adjustment, the junction of the two-stage cam piercing operations is clear and well-defined. Parts are free of burrs, and cracking at the bridge base junction is absent. The die operates stably, efficiency is effectively realized, and the production rhythm meets expectations.

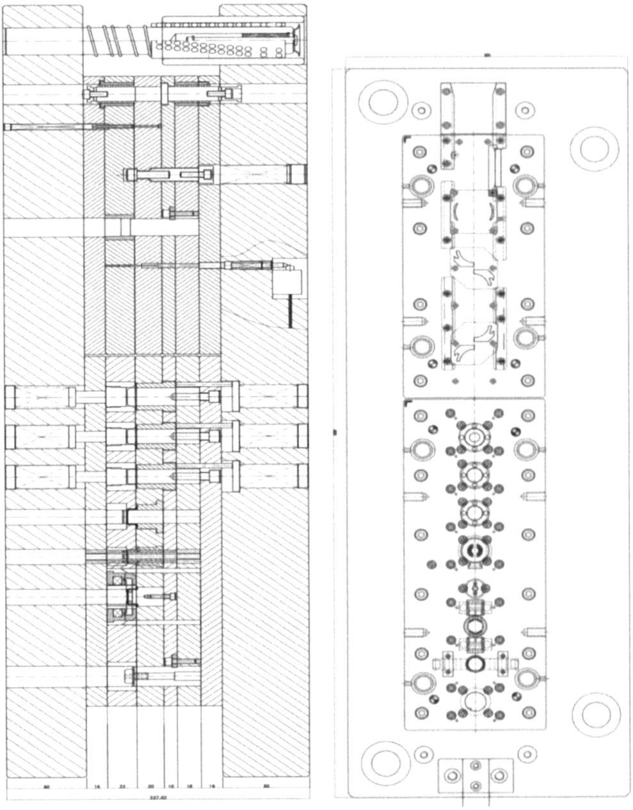

Fig. 12. Die structure.

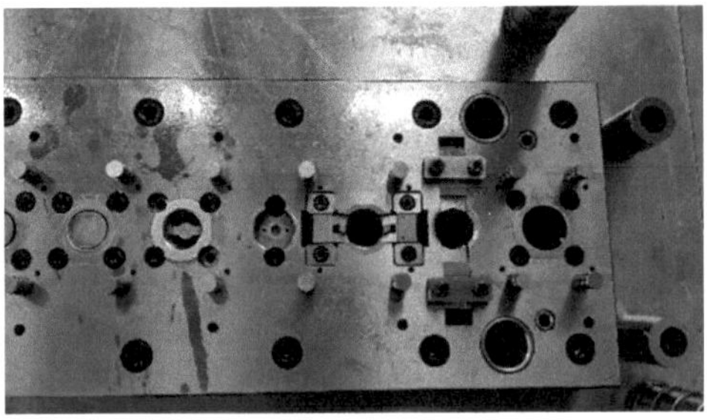

Fig. 13. Die structure of the 2-stage cam piercing

Fig. 14. Die in producing

5 Conclusion

Through the improvement of the processing route for this product, the application scope of stamping technology has been expanded to replace certain machining or specialized processing methods. Against a backdrop of increasingly intense market competition, the successful implementation and delivery of this project has accumulated valuable experience, providing strong support for future market development.

1. **Replacement of conventional stamping + WEDM process with a single progressive stamping die.** The original process required 9 operators per shift, producing 800–1000 parts per day. Following the process route improvement and die implementation, a single die operated by one person produces 4000 parts per shift. This significantly increases production capacity and reduces manufacturing costs. The hardware of 2 die sets and 2 presses (Three sets of single-process molds are used on the original production line.), and 8 operators at different stations, listed in Table 5, can be eliminated in conventional lines.

Table 5. Dispensable hardware or operators in conventional lines

Punch Operator	Die set	Press	Grinding deburring	Process Quality Inspector	Wire-cut operator	Parts Transfer	Remove Recast Layer (from EDM)
2	2	2	1	2	1	1	1

2. **Utilization of CAE analysis enabling one-step forming without any idle (or reserve) stations** Following the initial stamping route failure, CAE simulation was employed to pre-analyze feasibility and reliability. Based on CAE results, the second stamping route was adopted and optimized to meet technical quality requirements, and eliminated all reserve stations that during tryout can be modified into production-ready stations in subsequent steps.

3. **Achievement of bridge width tolerance and minimal burrs via two-stage side piercing** This approach incorporated principles of fine blanking with minimal clearance, successfully resolving the issue of cracking at the bridge-base junction.
4. **Foundation for future die development** The successful development of this product establishes a foundation for transitioning similar components from traditional stamping + specialized machining (or completely machining) to one set mold stamping forming. This enriches technological capabilities, substantially improves production efficiency, and directly reduces production costs by an estimated 80%.

References

Jemal, A., Salau, A.O., Wondimu, A.: Finite element method–based multi-objective optimization of press-brake bending of sheet metal. Int. J. Adv. Manuf. Technol. **130**(9–10), 4263–4275 (2024). https://doi.org/10.1007/s00170-023-12942-5

Kumar, D., Addula, S.R., Lind, M., Brown, S., Odion, S.: AI-driven hybrid deep learning and swarm intelligence for predictive maintenance of smart manufacturing robots in industry 4.0. Electronics. **15**(3), 715 (2026). https://doi.org/10.3390/electronics15030715

Lei, Z., Li, B., Ni, L., Yang, Y., Yang, S., Hu, P.: Mechanism of the crack formation and suppression in laser-MAG hybrid welded 30CrMnSiA joints. J. Mater. Process. Technol. **239**, 187–194 (2017). https://doi.org/10.1016/j.jmatprotec.2016.08.033

Liao, M., Liu, J., Liu, Y., Wang, S.: Optimal stamping direction for an automotive part. Int. J. Adv. Manuf. Technol. **79**(1–4), 285–297 (2015). https://doi.org/10.1007/s00170-015-6826-7

Olguner, S., Bozdana, A.T., Kütük, M.E.: Experimental investigation on the combined effects of ultrasonic vibrations and servo press motion scenarios on drawability of low carbon steels. Int. J. Adv. Manuf. Technol. **137**(11–12), 5699–5717 (2025). https://doi.org/10.1007/s00170-025-15456-4

Oztemel, E., Gursev, S.: Literature review of industry 4.0 and related technologies. J. Intell. Manuf. **31**(1), 127–182 (2020). https://doi.org/10.1007/s10845-018-1433-8

Sabri, K., Gaceb, M.: Real-time hybrid AI for strip steel surface monitoring in industry 4.0. Int. J. Adv. Manuf. Technol. **142**(3–4), 1467–1493 (2026). https://doi.org/10.1007/s00170-025-17113-2

Sheng, Z.Q., Taylor, R., Strazzanti, M.: FEM-based progressive drawing process design. Int. J. Adv. Manuf. Technol. **36**(3–4), 226–236 (2006). https://doi.org/10.1007/s00170-006-0829-3

Wang, L.Z., Wu, J.J., Huang, X.F., Hong, X.F.: SLM-manufactured 30CrMnSiA alloy: mechanical properties and microstructural effects of designed heat treatment. Opt. Laser Technol. **107**, 89–98 (2018). https://doi.org/10.1016/j.optlastec.2018.05.020

Wang, L. et al.: Achieving high tensile properties and impact toughness in ultrahigh strength lean alloy steel by quenching and partitioning treatment. Archives of Civil and Mechanical Engineering. **24**(1), 32 (2023). https://doi.org/10.1007/s43452-023-00841-1

Wang, S., Wang, H., Zeng, R., Wan, L., Che, G.: Mudguard stamping springback control for agricultural tractors: collaborative multi-strategy approach and experimental verification. CIRP J. Manuf. Sci. Technol. **64**, 187–202 (2026). https://doi.org/10.1016/j.cirpj.2025.12.008

Xiao, H., Geng, Y., Zheng, H., Xiao, Y., Cao, X., Wu, R.: Microstructure evolution mechanisms and mechanical properties of 30CrMnSiA steel repaired by laser cladding. Mater. Today Commun. **40**, 109845 (2024). https://doi.org/10.1016/j.mtcomm.2024.109845

Xie, Y., Li, W., Liu, C., Du, M., Feng, K.: Optimization of stamping process parameters based on improved GA-BP neural network model. Int. J. Precis. Eng. Manuf. **24**(7), 1129–1145 (2023). https://doi.org/10.1007/s12541-023-00811-w

Design of an Integrated Three-Axis Micro-Mechanical Quartz Tuning Fork Gyroscope

Maoyan Fan(✉), Yaqiu Zhang, Shengping Zhao, Hairu Zhao, Ye Tu, and Wenzhen Yang

Faculty of Engineering, Yuxi Normal University, Yuxi, China
fmy@yxnu.edu.cn

Abstract. This paper presents a dual-mass H-type experimental tuning fork structure. Structural analysis and optimization were carried out using ANSYS software, demonstrating its advantages of high sensitivity and high resonant frequency. Mass fine-tuning is applied to the mass blocks at the ends of the tuning fork to improve overall accuracy, thereby achieving consistency and interchangeability. The design integrates a three-axis micro-mechanical quartz tuning fork gyroscope. A limit stop structure is introduced to effectively prevent finger collision, making it more suitable for extreme environments. A scheme for reducing cross-coupling errors in redundant micro-mechanical gyroscopes is also discussed, which helps broaden its application fields.

Keywords: Three-axis · Quartz tuning fork gyroscope · Redundant configuration · Impact resistance · Structure design

1 Introduction

The Quartz Resonator Suspended (QRS) gyroscope is a small solid-state inertial device characterized by small size, simple structure, high reliability, and low cost. It has been widely used in fields such as autonomous vehicles, consumer electronics, and military inertial navigation technology. Precision-guided munitions have become indispensable weapons on the battlefield. However, their high research and development cost and limited cost-effectiveness restrict large-scale deployment in modern warfare. Shock resistance testing, especially in terminal guidance fuses, requires detection of accelerations exceeding 10,000 g. This requires the micro-mechanical gyroscope to possess not only high sensitivity but also high resonant frequency and wide bandwidth, so that carrier attitude parameters can be measured rapidly and accurately. In practical use, it is often necessary to obtain angular velocity parameters of pitch, yaw, and roll along three axes. The demand for three-axis gyroscopes continues to grow. In particular, the application of high-overload micro-mechanical quartz resonator suspended gyroscopes in inertial guidance has become a research focus. (Li and Zhao 2021; Chen et al. 2025).

T. Mankovits and N. Trisovic (Eds.): MDS 2026, LNME 2026.
https://doi.org/10.1007/978-981-92-2440-1_7

2 Research and Application at Home and Abroad

Research on quartz tuning fork gyroscopes abroad originated in the 1980s. In 1987, the Draper Laboratory in the United States developed the first quartz tuning fork gyroscope capable of outputting precise angular velocity. After years of optimization, tuning fork gyroscopes have been widely deployed in both military and civilian applications. Currently, the primary company capable of mass-producing high-performance quartz tuning fork gyroscopes globally is the American BEI company. It stands as the most widely applied and superior quartz tuning fork gyroscope product supplier, with the vast majority of its developed products being double-ended quartz tuning fork gyroscopes. Among them, the QRS 11 model gyroscope has been extensively used in various inertial control systems since 1990 due to its excellent performance.

The use of three-axis gyroscopes in navigation and positioning shows that monolithic integrated gyroscopes capable of measuring three-axis angular velocity represent a clear development trend. Three-axis testing in inertial navigation can be realized in two ways. One method installs three independent micro-mechanical gyroscopes along three mutually perpendicular directions. The other method uses an integrated micro-mechanical gyroscope composed of three vacuum-encapsulated quartz tuning forks, whose sensitive axes X, Y, and Z are arranged orthogonally. In the first method, it is difficult to guarantee installation accuracy among the three axes, and problems such as large size and high cost remain. In contrast, the second method benefits from high machining precision, resulting in better inter-axis positional accuracy. The orthogonal installation of three vacuum-encapsulated quartz tuning forks also offers advantages including smaller size, lower cost, convenient vibration isolation, and strong impact resistance.

In 1989, Rockwell and BEI jointly developed the QRS inertial sensor and inertial measurement unit, which entered commercialization in 2000. Ninety sets of quartz tuning fork gyroscopes produced by Rockwell in the United States accumulated 1.2 million hours of operation without malfunction. In 1998, the Center for Sensors and Actuators (BSAC) at the University of California, Berkeley, developed the first monolithic integrated three-axis gyroscope using the surface micromachining technology of Sandia National Laboratories (SNL) MFMS. The wire-vibrating gyroscope proposed by Honeywell can operate in high-overload environments up to 20,000 g. An internal isolation device is designed to reduce the influence of shock and vibration. (Yang and Lu 2018)

Many domestic universities and research institutes have also carried out research on high-overload gyroscopes. Southeast University developed a new three-axis MEMS gyroscope with a peak impact response of 10,000 g and a pulse width of 0.2 ms. After impact, the maximum displacement was 2.3 μm along the X-axis, 3.02 μm along the Y-axis, and 2.3 μm along the Z-axis. The structural stresses under impact in the three directions were 236.8 MPa, 333.7 MPa, and 264.8 MPa, respectively. These values are far below the ultimate stress of silicon, which is 790 MPa. (Liu et al. 2025, Wang 2024)

3 Operating Principle of Integrated Three-Axis Micro-Mechanical Gyroscope

Three-axis angular velocity detection is realized using a single-axis quartz tuning fork gyroscope based on the piezoelectric effect of quartz and the Coriolis force principle. The three quartz tuning forks are vacuum-encapsulated separately, and mass correction ensures consistency and interchangeability. The sensitive axes X, Y, and Z are installed orthogonally to measure angular velocities in three directions. Installation accuracy helps reduce cross-coupling errors. Maintaining consistency, interchangeability, and compatibility among the three vacuum-encapsulated quartz tuning forks is essential. The integrated three-axis micro-mechanical gyroscope uses a unified drive circuit composed of one drive module and three detection modules. Compared with combining three independent single-axis gyroscopes, this design increases circuit commonality and reduces size and power consumption.

The quartz tuning fork vibrates at a constant frequency with equal amplitude. When rotation occurs, the Coriolis force excites vibrations perpendicular to the original vibration plane. The amplitude is proportional to the angular velocity. Electrical signals are generated through the piezoelectric effect, allowing angular velocity sensing and decoupling of drive and detection signals. The motion equation of the detection mode of the quartz tuning fork gyroscope is:

$$J\dot{\theta} + C\ddot{\theta} + K\theta = -F_k - F_\tau \quad (1)$$

Where J is the moment of inertia, C represents the damping coefficient, K is the stiffness coefficient, F_k represents the Coriolis force, and F_τ is the tangential inertia force. When driven at a frequency P, the vibration of the mass point on the quartz tuning fork is $R = R_0 + R_a \sin(pt)$, where R_a is the amplitude; the tangential inertia force is: $F_\tau = mR\dot{\omega}$. Solving the motion equation shows that the rotation angle reaches its maximum only when the natural detection frequency $w_k = \sqrt{K/J}$ equals the driving frequency P.

3.1 Mass Block Adjustment and Vibration Mode

To suppress mechanical coupling in quartz tuning forks, the frequency difference between the drive mode and the detection mode should remain within a certain range. Frequency tuning aims to keep adequate bandwidth and sensitivity for both the drive and detection modes. This can be achieved by adjusting the mass blocks on the drive and detection fingers. The key point in tuning a quartz tuning fork is to maintain structural symmetry while modifying the mass distribution, so that the relationship between mechanical coupling and frequency difference can be improved. The influence of balancing mass blocks on mechanical coupling and frequency difference is as follows: (1) The balancing mass block of the drive finger has a significant effect on mechanical coupling and is the main way to reduce it; (2) Adjusting the balancing mass block of the drive finger can decrease the frequency difference; (3) Adjusting the balancing mass block of the detection finger is used to increase the frequency difference. (Feng et al. 2021, HODJAT 2020)

Using a picosecond laser with a wavelength of 535 nm, the laser power was set to 0.3 W, the frequency to 1,000 kHz, and the stepping motor speed to 10 μm/s. The driving fork was etched at different positions 10 times each. Compared to the initial characteristic frequency, the driving mode frequency decreased, and it decreased linearly with the number of elastic beam etchings. Automatic tuning connects testing and tuning equipment through control software to read test data in real time. The system then controls the motion platform and uses laser or direct writing methods to remove or add mass in specific areas. Through repeated testing and adjustment, mechanical coupling and frequency deviation are kept within acceptable limits. The core of this process lies in the tuning algorithm embedded in the control software.

3.2 Reducing the Mechanical Coupling Mechanism

Structural asymmetry of the quartz tuning fork causes mechanical coupling errors. If the mass center of the driving tine does not align with the driving axis, vibration produces not only radial signals but also a component along the output axis. This leads to output signals even without angular velocity input, forming a mechanical coupling error. The mechanical coupling output is:

$$\Delta z_s = l_s \cdot A\left(k_{xz} - \beta m_d \cdot \omega^2\right) \cdot \Delta x_d \cdot l_d \tag{2}$$

where Δz_s is mechanical coupling, l_s represents distance from the proof mass to the tuning fork's rotational center, and A represents constant determined by the quartz tuning fork structure. k_{xz} is combined stiffness, β denotes coupling coefficient, m_d is common mode component of the mass of the two driving blocks of the quartz tuning fork, ω represents resonant frequency at the driving end, and Δx_d denotes differential mode component of the displacement of the driving mass of the quartz tuning fork. l_d is distance from the driving mass to the tuning fork's rotational center.

From Eq. (1), when the centroid of the driving mass is adjusted so that the z-direction force caused by stiffness coupling offsets the inertial force in the same direction, mechanical coupling can be eliminated. Ideally, the driving axis and detection axis are perpendicular, and vibration does not couple. However, due to quartz anisotropy and manufacturing errors, a deviation angle H exists, leading to coupling errors. Without angular velocity input, vibration at the driving end transfers to the detection end, producing orthogonal error signals. Eq. (2) introduces a scheme in which two sets of integrated three-axis micro-mechanical gyroscopes are mounted inversely. Six single-axis gyroscopes are arranged orthogonally, and their sensitive axes differ by 180°, forming the orthogonal matrix shown by:

$$H_{\text{orthogonal}} = \begin{bmatrix} 1 & 0 & 0 \\ 0 & 1 & 0 \\ 0 & 0 & 1 \\ -1 & 0 & 0 \\ 0 & -1 & 0 \\ 0 & 0 & -1 \end{bmatrix} \tag{3}$$

Under this configuration, the matrix is simple, computational load is low, and measurement error is small. Therefore, the inverse matrix and its transpose can be obtained by:

$$H_{\text{orthogonal}}^T H_{\text{orthogonal}} = \begin{bmatrix} 2 & 0 & 0 \\ 0 & 2 & 0 \\ 0 & 0 & 2 \end{bmatrix} = \frac{6}{3} I_3 \tag{4}$$

In Eq. (4), I denotes the identity matrix, which physically represents the identity transformation in linear transformations. Its role is similar to "1" in scalar multiplication, preserving the original state of a vector or space. Orthogonal matrices satisfy the condition that their inverse equals their transpose, which ensures that the transformation is invertible and preserves geometric properties. The orthogonal configuration scheme meets the requirements of optimal navigation performance. In addition, since the integrated three-axis micro-mechanical gyroscopes are arranged orthogonally, installation and calibration of the whole system become more convenient.

The inertial measurement system must operate in the harsh missile-borne environment, which places strict demands on the reliability and accuracy of its sensors. Due to the limitations of single-axis gyroscope measurement technology, one integrated three-axis micro-mechanical gyroscope alone cannot satisfy the required accuracy. To solve this problem, redundancy technology is introduced to improve system reliability and precision. The optimal number of sensors is determined through system reliability analysis, and the configuration scheme is selected according to navigation performance requirements. Considering factors such as quantity, configuration, cost, and available space inside the missile, installing two integrated three-axis micro-mechanical gyroscopes with their sensitive axes oriented 180° apart is regarded as the optimal scheme. (Bu 2020, Mao et al. 2024) Multi-sensor data fusion is essentially multi-source information fusion, and its purpose is to improve the accuracy of state and feature estimation. The use of artificial intelligence methods to enhance gyroscope accuracy is also important. By establishing mathematical models to identify errors and applying corresponding compensation, deterministic drift can be largely removed. Random drift is mainly caused by the random component of disturbance torque. Through improved feature extraction and statistical modeling algorithms for random error analysis and compensation, the drift can be reduced, which further improves gyroscope accuracy.

4 Structural Design of Integrated Three-Axis Micro-Mechanical Quartz Tuning Fork Gyroscope

The design of a quartz tuning fork gyroscope begins with selecting the cut type of the quartz wafer. The selected cut type should provide a large piezoelectric constant to obtain high sensitivity. At the same time, the frequency-temperature characteristics must ensure stable output within the operating temperature range. After determining the cut type and cut angle, the dimensions of the tuning fork, including length, width, and thickness, define its frequency characteristics. Both the drive vibration and detection vibration operate near their respective resonant frequencies, and the amplitude is closely related to frequency.

According to the dynamic equation of the quartz tuning fork gyroscope, maximum angular vibration output is obtained only when the drive frequency matches the detection frequency, which leads to maximum sensitivity. The structure of the integrated three-axis micromechanical quartz tuning fork gyroscope is shown in Fig. 1. The device adopts a three-layer structure corresponding to the X-sensitive axis, Y-sensitive axis, and Z-sensitive axis. Each layer of vacuum-packaged devices, as well as the drive and detection circuit boards, can be interchanged, provided that installation accuracy is maintained and appropriate vibration dampers and isolators are selected. By choosing suitable adhesives and adjusting their formulation to modify stiffness and damping, the structural form can be optimized to achieve the required vibration damping transfer rate under specific deformation modes and frequency ranges, thus meeting high overload impact resistance requirements. (Zeng 2020, Du 2023)

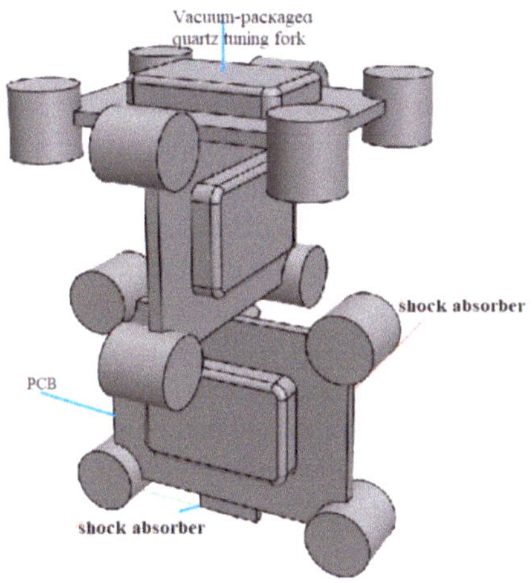

Fig. 1. Overall structure diagram of a three-axis micromechanical gyroscope

4.1 ANSYS Structural Dynamics Simulation

ANSYS is a widely used finite element analysis software. Numerical simulation of the micro-mechanical gyroscope during the design stage helps identify possible structural weaknesses, optimize the structure, shorten the design cycle, and reduce manufacturing cost. Modal analysis mainly serves to ensure that the drive mode and the sensitive mode correspond to the first and second principal modes, to ensure close matching between their natural frequencies, and to separate these frequencies from other modes, especially those that may cause structural damage. The analyzed crystal is cut at a specific angle by rotating around an axis within the plane. The electrodes are arranged so that the two arms generate opposite induced electric fields, which excite planar vibration. Based on finite element analysis, the physical parameters of quartz are listed in Table 1.

Definition: The base length is denoted by (h). For a symmetric crystal, the two arms have equal length (L) and equal width (d). Free vibration (F_{cf}) refers to the vibration of

Table 1. Physical parameters of Quartz

Young's modulus	$E\left(c_{22}^{'}\right) = 8.9455 \times 10^{-10}\left(g \times mm/s^2 \times mm^2\right)$
density	$\rho = 2.6487 \times 10^{-3}(g/mm^3)$
Poisson's ratio	$\delta\left(\frac{s_{12}^{'}}{s_{11}^{'}}\right) = 0.122894$

the crystal without support. The resonant frequency under free vibration is denoted by (F_{oc}). To evaluate the influence of support conditions, simulations were carried out to calculate vibration frequency, displacement, and stress distribution under both free and clamped states. The selected element type was the hexahedral 20-node three-dimensional solid element C3D20, and the analysis step was set as Static, General. The time step was 1, the initial step length was 0.1, the minimum step length was $1e^{-5}$, and the maximum step length was set to the default value of 1. Fig. 2. shows the two-dimensional geometric model and finite element mesh model of the sample tuning fork. The stress distribution at the driving end displacement of $1e^{-3}$ is shown in Fig. 3., and the relationship between the maximum stress and the driving end displacement is shown in Fig. 4.. From the simulation results, it can be seen that as the displacement of the driving end increases, the maximum stress in the model basically increases linearly, and the location of the maximum stress is at the intersection of the two sides of the driving end.

The double tuning fork structure was adopted in the design, and by optimizing the structural parameters, the excitation mode and detection mode of the tuning fork reached resonance, providing a reliable theoretical analysis basis for the design of the gyroscope structure. The application of finite element analysis method in the optimization of quartz tuning fork gyroscope structure was discussed. The first step in the design of a quartz tuning fork gyroscope is to consider the selection of the quartz chip cutting type. The selected cutting type should have a large piezoelectric constant to achieve high sensitivity, and the frequency temperature characteristics of the cutting type should also be considered to ensure stable output within the operating temperature range. The driving vibration and detection vibration of a tuning fork both operate near the resonant frequency, and the amplitude is directly related to the frequency. According to the dynamic equation of a quartz tuning fork gyroscope, only when the driving frequency and detection frequency reach resonance can the maximum output angular vibration be obtained, thereby achieving maximum sensitivity.

4.2 Modal Simulation Analysis of Quartz Tuning Fork

Modal simulation is used to determine the first-order resonant frequency and vibration direction of a structure by analyzing its undamped resonant state. The relationship between the natural resonant frequency and structural stiffness in the detection direction of a micro-mechanical quartz tuning fork gyroscope is given by $f = 1/2\pi\sqrt{k/m}$. The resonant frequency increases as stiffness increases. When stiffness becomes larger, the

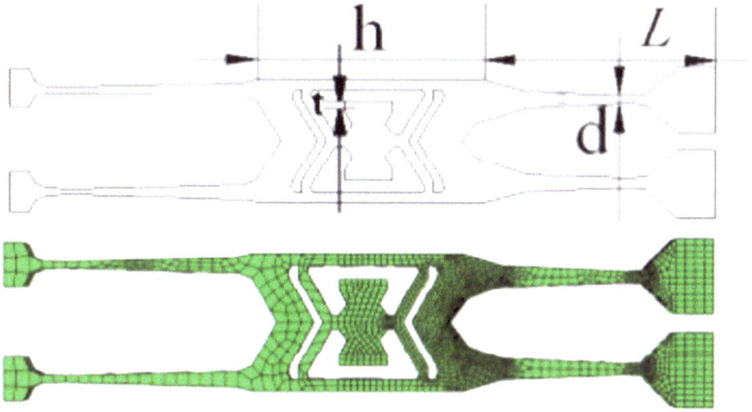

Fig. 2. Two-dimensional tuning fork and finite element mesh model ($L = 3.5$ mm; $d = 0.17$ mm)

($L = 3.5$ mm; $d = 0.17$ mm)

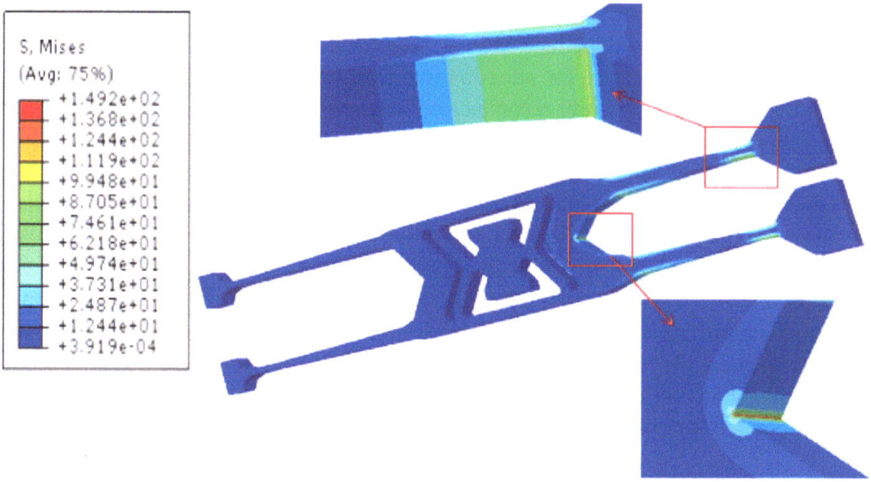

Fig. 3. Stress distribution when the displacement of the drive end is $10e^{-3}$ mm

resonant frequency rises accordingly. However, stiffness is inversely related to mechanical sensitivity. Therefore, the balance between resonant frequency and sensitivity must be considered during design. It is also necessary to avoid the situation in which the excitation signal is close to the first-order natural frequency, since this may cause forced resonance and lead to bending or even fracture. A finite element modal analysis was carried out on the micro-mechanical quartz tuning fork gyroscope. The boundary conditions restricted all degrees of freedom at the central bonding position. The selected analysis step was Frequency, and the modal shapes were obtained. The seventh-order

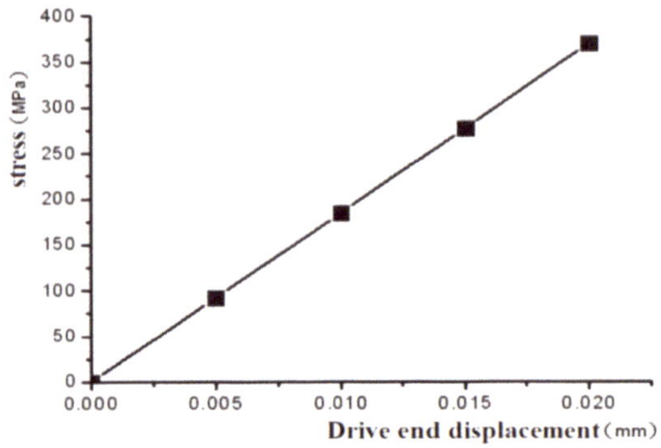

Fig. 4. Relationship between maximum stress and displacement of the driving end

mode has a frequency of f = 11060 Hz, and the eighth-order mode has a frequency of *f* = 13525 Hz, as shown in Fig. 5..

The resonant frequency interval between two types of vibrations is the main factor affecting performance (such as sensitivity and bandwidth). The smaller the difference between the two, the greater the sensitivity of the gyroscope. However, it cannot be too close, otherwise it will reduce the working bandwidth and cause significant zero drift. This is a basic principle for selecting two resonant frequencies. By optimizing the structural parameters to achieve resonance between the tuning fork excitation mode and the detection mode, the maximum output angular vibration can be obtained, thereby achieving maximum sensitivity. Based on the operating characteristics of the device, the seventh-order mode is selected as the drive mode and the eighth-order mode as the sensing mode. In both modes, the two prongs vibrate in opposite directions. When rotation occurs, the Coriolis inertial forces acting on the two prongs are also opposite, which produces differential sensitive vibration. The output of the quartz gyroscope is obtained by summing the charges collected from the sensing electrodes on both prongs. This approach doubles the electrical signal and reduces interference from nearby modes. Additionally, general external disturbances tend to act equally on both prongs, allowing the gyroscope to function reliably in harsh environments.

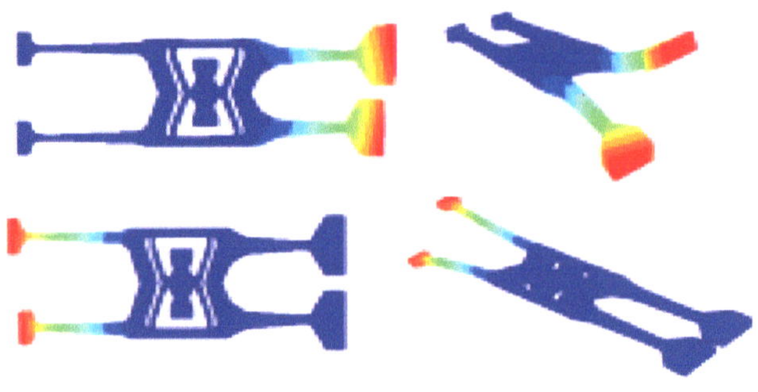

Fig. 5. Vibration modes of driving and sensing modes of micro-mechanical quartz tuning fork

The shock response spectrum corresponds to the shock characteristics of the actual operating environment, indicating that the product has been evaluated under relevant shock conditions. The shock spectrum provides a practical method in engineering to analyze and quantify shock environments and to estimate possible damage. In the maximum stress analysis during rotation, boundary conditions restrict node movement along the rotation axis. The load simulation shows how the maximum stress in the tuning fork increases with rotational speed. When the rotational speed reaches 20,000 rad/s, the maximum stress approaches 600 MPa, which exceeds the overload resistance limit. Quartz will yield and fracture once the stress surpasses its allowable value. Therefore, in structural design, sharp corners at the intersection of the two surfaces at the driving end should be avoided, and smooth circular arc transitions should be adopted. (Bu et al. 2021)

5 Conclusion

This paper discusses the theoretical basis and structural design approach for reducing cross-coupling error in integrated three-axis micro-mechanical quartz tuning fork gyroscopes. The causes of orthogonal error in micro-mechanical gyroscopes and their influence on the output signals are analyzed. By comparing different orthogonal error compensation methods, a mass balancing approach based on laser and direct writing technology is proposed to improve consistency and interchangeability of vacuum-packaged three-axis micro-mechanical quartz tuning forks. A redundant configuration scheme for reducing cross-coupling error is also presented, taking into account the number of redundant sensors, single-axis configuration, and fault detection and isolation capability. When six redundant inertial sensors are combined with accurate orthogonal installation and combined theoretical and experimental verification, the feasibility and effectiveness of the proposed method are confirmed.

Acknowledgments. FundingThis research was supported by the Key Laboratory of Complex System Dynamics and Applied Analysis, Yunnan Provincial Department of Education.

References

Bu, F.: Noise model considering electrical feed-through under force rebalance closed-loop detection of MEMS gyroscope. J. Micromech. Microeng. **30**(5), 055007 (2020). https://doi.org/10.1088/1361-6439/ab7c34

Bu, F., Fan, B., Xu, D., Guo, S., Zhao, H.: Bandwidth and noise analysis of high-Q MEMS gyroscope under force rebalance closed-loop control. J. Micromech. Microeng. **31**(6), 065002 (2021). https://doi.org/10.1088/1361-6439/abf32e

Chen, G.W., Zhang, Y.F., Zhao, Q.C., Qu, F.Y.: Design and simulation of piezoelectric resonant tuning fork gyroscope structure. Instrum. Technol. Sens. **4**, 345–349 (2025) https://read.cnki.net/web/Journal/Article/YBJS202504009.html

Du, Y.Z.: Research on the mechanism and suppression methods of quadrature error in quartz tuning fork gyroscopes (Master's thesis) Southeast University. (2023). https://doi.org/10.27014/d.cnki.gdnau.2023.004938

Feng, R. et al.: Real-time built-in self-test of MEMS gyroscope based on quadrature error signal. Micromachines. **12**(9), 1115 (2021). https://doi.org/10.3390/mi12091115

Hodjat-Shamami, M., Ayazi, F.: Eigenmode operation of piezoelectric resonant gyroscopes. Microsyst.S Nanoeng. **6**(1), 1–11 (2020). https://doi.org/10.1038/s41378-020-00204-3

Li, Y., Zhao, Z.P.: Research status and progress in MEMS gyroscopes. Micro Nano Electron. Technol. **58**(9), 757–768 (2021). https://doi.org/10.13250/j.cnki.wndz.2021.09.001

Liu, J., Meng, Z.J., Cao, H.L., Tang, J., Li, J., Shi, Y.B.: Overview of anti-high-overload MEMS inertial devices and test equipment. Navig. Control. **24**(3/4), 13–39 (2025). https://doi.org/10.3969/j.issn.1674-5558.2025.h3.002

Mao, R.Y. et al.: Research on error separation of quartz tuning fork gyroscope. Autom. Instrum. **11**, 307–311 (2024). https://doi.org/10.14016/j.cnki.1001-9227.2024.11.307

Wang, J. Q. (2024). Research on reliability enhancement and high-precision measurement techniques for MEMS gyroscopes in airborne environments (doctoral thesis). North University of China. https://xueshu.baidu.com/ndscholar/browse/detail?paperid=1u0u08f0pt5p0v00a57k0p60gh424723

Yang, B., Lu, C.F.: Design and analysis of a new three-axis micro-gyroscope. Microsyst. Technol. **24**, 1213–1222 (2018). https://doi.org/10.1007/s00542-017-3488-4

Zeng, P.J.: A thesis submitted in partial fulfillment of the requirements for the degree of master of engineering (Master's thesis). Huazhong Univ. Sci. Technol. (2020). https://doi.org/10.27157/d.cnki.ghzku.2020.004151

Design and Analysis of a Shrimp Preprocessing Machine

Yaxuan Song, Qifan Zhang, Zaitian Tang, Yushan Cai, Jianlu Wang, and Keying Liu(✉)

Dalian Polytechnic University, Dalian 116034, China
1021993666@qq.com

Abstract. Because of the demand of the market and the low integration of existing shrimp processing equipment, this lecture introduce a new modular shrimp machine. The machine is made by six functional module and can be adjusted for different shrimp sizes. The lecture will use mechanical analysis, motor torque calculation, and finite element analysis to verify structural strength and optional reliability. Experimental results demonstrate stable operation and acceptable processing accuracy, indicating the feasibility of the proposed design for automated shrimp preprocessing.

Keywords: Shrimp preprocessing · Modular design · Mechanical analysis · Finite element analysis · Automation equipment

1 Introduction

Shrimp are an important aquatic product, and their preprocessing stage directly affects product quality, safety. With growing market demand, higher requirements are imposed on the performance of shrimp preprocessing like efficiency and consistency. Recent studies have indicated that automation technologies are increasingly being explored in aquatic product processing, which can improve optional efficiency (Senthilkumar et al. 2025). Now, most of the company still rely on labors to process shrimp, which result in low efficiency and unstable quality. In recent years, there have some studies demonstrated that aquatic product processing tend to use automation and integrated processing lines to reduce labor intensity and elevate production efficiency (Mainali and Li, 2025).

For agricultural modernization, preprocessing mechanization has seen a key component in it. In 2023, the Ministry of Agriculture and Rural Affairs of China issued *Guidelines on Accelerating the High-Quality Development of Agricultural Product Preprocessing Mechanization*, which emphasizing the promotion of mechanized preprocess for agricultural and aquatic products. In addition, *the Guiding Opinions on Accelerating Agricultural Mechanization and the Transformation and Upgrading of the Agricultural Machinery Industry,* which released by the State Council, encouraging technological upgrading in aquaculture and agriculture product preprocessing. These policies provide clear guidance for the development of specialized shrimp preprocessing equipment.

T. Mankovits and N. Trisovic (Eds.): MDS 2026, LNME 2026.
https://doi.org/10.1007/978-981-92-2440-1_8

Most existing systems still suffer from limited integration and poor adaptability to vary shrimp size and often lack systematic mechanical analysis. To address these issues, this study develops a modular shrimp preprocessing machine. This paper through mechanical modeling and finite element analysis to verify the structural strength and operational reliability of key mechanisms. The experimental results also demonstrate the feasibility of the proposed system. Fig. 1 shows the three-dimensional model of the entire machine.

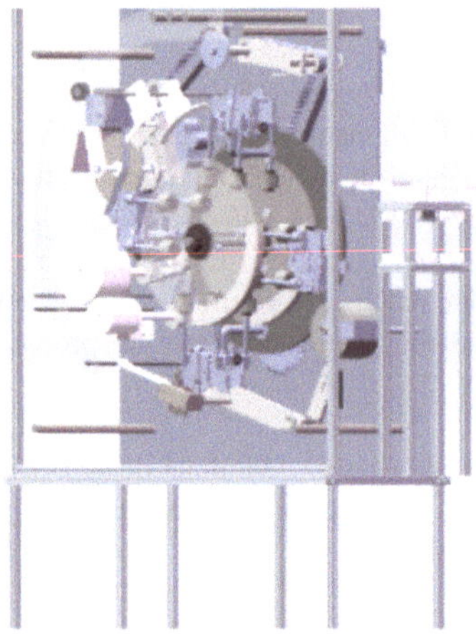

Fig. 1. Overall three-dimensional view.

2 Functional Design of key Modules

2.1 Gripper Positioning and Conveying Module

The gripper positioning and conveying module is the core unit of the proposed machine in order to implement stable clamping, continuous conveying, and accurate positioning of shrimp during processing. In automated food processing systems, grippers are regarded as critical end-effectors, and various mechanical gripping solutions have been developed to address object variability and handling stability (Abdullayev et al. 2025). This research adopt a boss-controlled track to guide the passive self-adaptive gripper, which can deal with the common problems like variations in shrimp size and slippery surfaces.

The gripper system consists of a cam plate, symmetrically arranged grippers, bosses, and tension springs. The bosses mounted on the cam plate act as motion constraints, while the springs provide the required clamping force. The grippers are rigidly connected to the transmission shaft and wheel, enabling continuous conveying driven by the transmission system.

During operation, the drive roller and the cam followers are settled by the both sides of the grippers, giving it force and make it open. Then carry it to enter the clamping area. Once the grippers move away from the cam followers region, the grippers will close automatically under spring action to achieve stable clamping. After that, the clamped shrimp is conveyed with a nearly constant posture, which can provide a consistent reference for subsequent processing.

This module use purely mechanical and passive control design to result in a compact structure with high reliability without additional actuators or complex control systems. The gripper improves force balance and adaptability for different size of shrimp, which made the machine multi-functional. Subsequently, this paper is going to analysis structural strength and mechanical modeling to verify the feasibility of module under actual working conditions. Fig. 2 is a simplified two-dimensional diagram of the module.

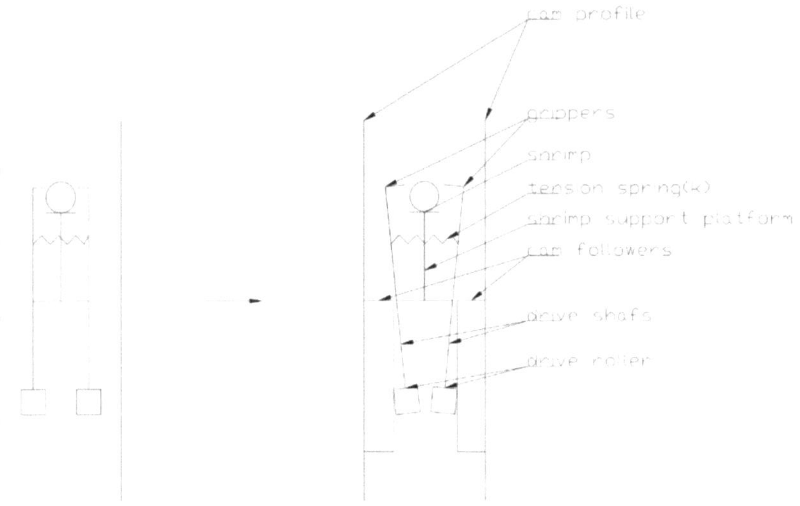

Fig. 2. Two-dimensional simplified schematic.

2.2 Gear Transmission and Shrimp Pressing Module

As the first operating stage of the proposed machine, the gear transmission and shrimp press module provides initial positioning. This module also provides the posture adjustment and pressing conveyance to ensure stable processing for the subsequent processing. A hybrid gear–synchronous belt transmission system is adopted to achieve compact structure and stable power transfer. The multi-stage transmission structure achieves optimized speed matching and torque distribution through coordinated operation so that enhancing the overall system synchronization.

The transmission system consists of a three-stage gear train that adjusts rotational speed and torque. It also transfers power to the synchro4nous belt pulley to drive the pressing mechanism. The gear ratios are arranged as 2:4:1 from bottom to top. The main gear, which located on the center of the transmission system, also driving the gripper positioning and conveying module. A 1:1 synchronous belt ensures stable rotation of the pressing wheel, which applies a steady constraint force to limit vertical displacement and correct shrimp posture. The main gear also improving stability before further processing. The processing wheel which combined with synchronous belt form a continuous constraint and maintain the shrimp posture when the shrimp are processing. Fig. 3 and Fig. 4 are the three-dimensional diagram and two-dimensional simplified diagram of this section.

Fig. 3. Three-dimensional diagram

2.3 Reciprocating Shrimp Back-Cutting Module and its Transmission Mechanism

This module integrates a reciprocating back-cutting mechanism and a blade transmission mechanism to realize shrimp back cutting. The blade is connected to the transmission shaft and crank. The vertical of the blade motion is controlled by an elliptical cam. There is a rebound plate located on the behind of the module, which can give this module a force forward.

During operation, when the shrimp reaches the cutting position, the cam rotates to a non-supporting state. After that, the restoring force of the rebound plate can drive the blade downward for cutting. When the shrimp leaves the cutting region, the cam lifts the blade to prevent unintended contact. Blade rotation is driven by a stepper motor through a 1:1 synchronous belt transmission. Figs. 5 and 6 are the 3D diagram and simplified 2D diagram of this section.

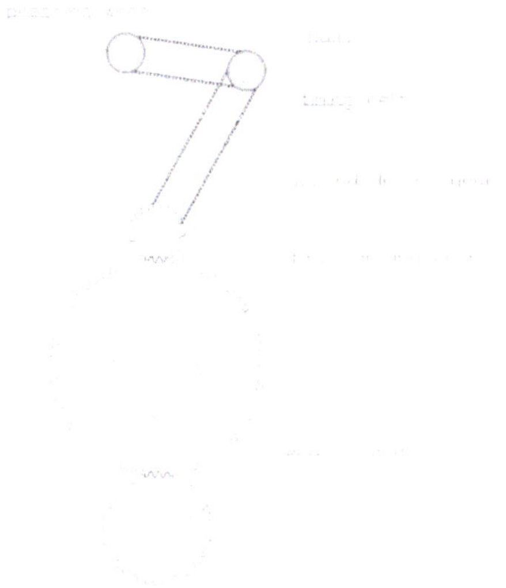

Fig. 4. Two-dimensional simplified diagram

Fig. 5. Three-dimensional model diagram.

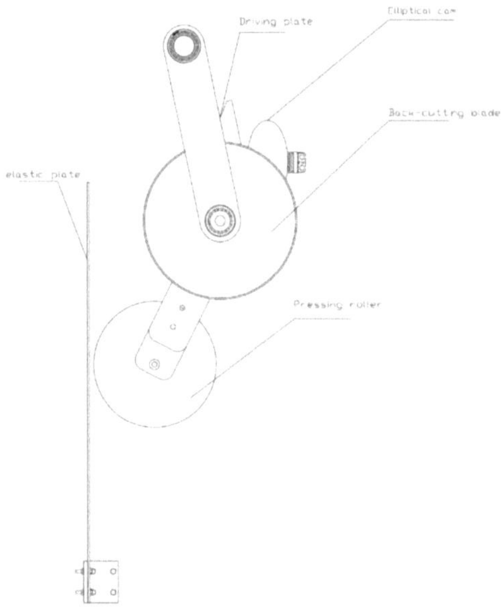

Fig. 6. Two-dimensional schematic diagram

3 Design and Calculation

3.1 Motor Selection and Calculation

The machine adopts a 2:4:1 gear transmission, and the total mass is approximately 40 kg. The torque is determined by the F and r:

$$Mout = F \cdot r \tag{1}$$

Let the total transmission ratio be i and the overall efficiency be μ. The required motor shaft torque is expressed as:

$$M_m = \frac{M_{out}}{i \times \mu} \tag{2}$$

Therefore, the rated output torque of the selected motor should be no less than 1.08 kg·cm to meet the operational requirements of the machine (Abdullayev et al. 2025).

3.2 Clamping Force Calculation

Based on the principle of frictional force equilibrium, the required clamping force should satisfy:

$$F_c \geq \frac{F_s}{2\mu} \tag{3}$$

Let $\mu = 0.65$, representing the friction coefficient between the wet shrimp shell and the aluminum alloy surface. And Fs = 12 N, which denotes the shell–meat separation force.

Calculation procedure:

$$F_c = \frac{12}{2 \times 0.65} \times 2.5 = 23.2N \tag{4}$$

($\times 2.5$ is adopted as the safety factor to compensate for dynamic load fluctuations.)

3.3 Conclusions and Optimization Recommendations

Strength verification shows that the calculated stress is 8.7 MPa, which is lower than the allowable stress $[\sigma] = 137.5$ MPa (defined as half of the yield strength, $\sigma_S/2$). Therefore, the strength requirement is satisfied. The following section presents the simulated stress analysis of the gripper. Fig. 7 shows the stress distribution of the gripper,

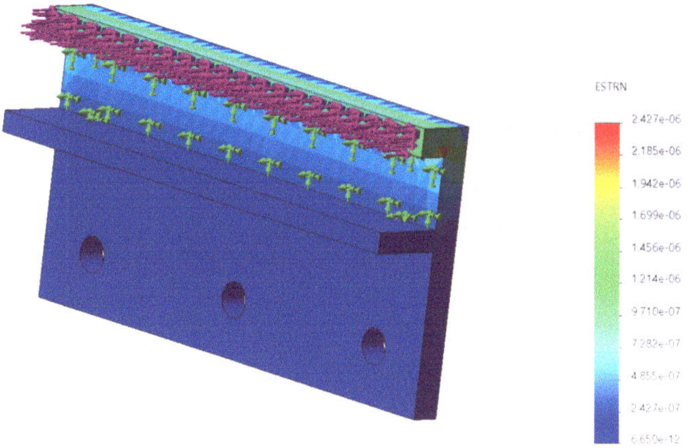

Fig. 7. Gripper stress distribution diagram.

4 Experimental Results

All experiments were conducted using Pacific white shrimp (*Litopenaeus vannamei*), with body lengths of 120–180 mm, weights of 15–35 g, and body widths of 12–22 mm. Because of the inherent structure of shrimp, it will lead to 18% unpredictable and inevitable loss in the processing progress. Below is the experimental data Table 1.

A processing success rate of 47.8% and incomplete rate of 21% reflect this machine have the feasibility as a first-generation machine under the limitation of the hardware. Previous studies have also indicated that high-speed handling of fragile food materials inevitably involves trade-offs between processing efficiency and damage control (Asif

Table 1. Experimental Data Record Sheet

Shrimp size(mm)	Sample size(n)	Successful processing(n)	Success rate (%)	Incomplete peeling(n)	Incomplete rate (%)
120–140	30	16	53.3	6	20
141–160	30	14	46.6	5	16
161–180	30	13	43.3	8	26
Total/average	90	43	47.8	19	21

et al. 2025). Therefore, the performance of the proposed prototype is considered reasonable at the validation stage. In the future iterations, the parameter and control strategy are expected to enhance processing accuracy.

In terms of operational speed, the machine is equipped with four grippers arranged circumferentially to perform periodic processing. Each gripper operates with a stable cycle time of approximately 12 s per shrimp. Under continuous o2peration, the multi-gripper configuration enables parallel processing within a single rotation cycle. This strategy ensure stable throughput.

5 Cost Accounting

Table 2. Cost accounting statement

Drive system	3000RMB
Gripper module	2000RMB
Pressing module	1000RMB
Cutting module	2000RMB
Control system	2000RMB
Total	10000RMB

The overall estimation error is controlled within ±500 RMB, which is considered acceptable for prototype-level economic analysis. Table 2 is the cost accounting statement.

6 Program Framework

Due to the reliability and the ease to integration of the control logic, microcontroller-based control systems have widely adopted for automatic agricultural and process control applications. In particular, microcontroller-based designs have been demonstrated to effectively manage real-time control tasks such as automated water supply using feedback mechanisms in plant experiments (Sugiura et al. 2024). In this study, the control system is implemented using an STM32 microcontroller programmed with the standard peripheral library. This program employed a close loop control system. Motor speed is regulated by PWM signals to ensure continuous rotation.

There is an infrared sensor located on the rear plate of the machine, and a shading plate is fixed to the gripper assembly on the aluminum frame. When the gripper rotates, the shading plate move and block the infrared sensor, giving the feedback signals to the position recognition.

The back-cutting module and the shell-meat separation module are arranged at ninety degrees along the rotational path. When the gripper reaches the particular processing position, the plate will be shade and the signal will be trigger. Through this way, this machine don't need complex rotation sensor to achieve rotate continuously.

7 Conclusion

A shrimp primary processing machine which integrating clamping, transmission, and back-cutting functions is designed to adderss the low efficiency and insufficient automation in shrimp primary processing. The structural design and kinematic analysis of the clamping mechanism, transmission system, and cam mechanism verify the feasibility of the proposed scheme.

In addition, the strength and stability of key components are evaluated through driving torque calculation and finite element static analysis. The results show that the proposed machine has a reasonable structural design, and the key components meet strength requirements with stable and reliable operation. Prototype experiments further demonstrate that the machine can effectively complete shrimp primary processing tasks and improve processing efficiency. This study provides a useful reference for the automation of shrimp processing equipment.

References

Abdullayev, H., Huseynzade, E., Karimov, A., Mammadov, R.: Robotic grippers in food industry: a short review. J. Eng. Manag. Inf. Technol. **5**(1), 45–56 (2025). https://doi.org/10.61552/JEMIT.2025.04.007

Asif, M.R., Thompson, J., Rossi, L., Webb, P.: Robotic grippers for high-speed handling of fragile food products: a review of design trade-offs and challenges. Int. J. Food Eng. **21**(2), 101–118 (2025). https://doi.org/10.1080/01691864.2025.2508785

Mainali, S., Li, C.Q.: A robotic fish processing line enhanced by machine learning. Aquac. Eng. **108**, 102481 (2025). https://doi.org/10.1016/j.aquaeng.2024.102481

Senthilkumar, T., Kumar, A., Patel, R., Singh, V., Sharma, P.: Automation in the shellfish aquaculture sector to ensure sustainability and food security. AgriEngineering. **7**(11), 387 (2025). https://doi.org/10.3390/agriengineering7110387

Sugiura, D., Mitsuya, S., Takahashi, H., Yamamoto, R., Miyazawa, Y.: Microcontroller-based water control system for evaluating crop water use characteristics. Plant Methods. **20**, 179 (2024). https://doi.org/10.1186/s13007-024-01305-0

Dynamic Characteristics and Parametric Study of Vibration Dampers Based on Modal Simulation

Kangkang Lu, Yang Qiao, and Yan Li(✉)

National Research Center of Fluid Machinery Engineering and Technology, Jiangsu University, Zhenjiang, Jiangsu 212013, China
554707891@qq.com

Abstract. Dampers are important components for suppressing wind-induced vibrations in transmission lines, and their dynamic characteristics directly affect vibration damping performance. This paper studies FD-type, FR-type, and FDNJ-type dampers. By establishing three-dimensional finite element models, their natural frequencies, mode shapes, and the effects of structural parameters were systematically investigated using methods including equivalent Young's modulus identification, modal analysis, and transient response simulation. First, based on cantilever beam theory, the formula for the equivalent Young's modulus of stranded wires was derived, addressing the accuracy issue in modeling the stranded wire structure. Subsequently, modal analysis was conducted on the three types of dampers to obtain their multi-order natural frequencies and mode shapes. Then, the effects of key parameters such as stranded wire length and hammer diameter on natural frequency and energy dissipation were systematically studied. In the study, the block Lanczos method was used for modal analysis to ensure the accuracy and convergence of the simulation results, and transient dynamic analysis was conducted based on the Newmark-β method, with a time step set to 1e-4 s to capture the subtle dynamic responses of the damper. The results show that FD-type, FDNJ-type, and FR-type dampers have 2, 3, and 4 main resonance frequencies, respectively. The length of the stranded wire is negatively correlated with the resonance frequency and positively correlated with energy dissipation. Increasing the hammer head diameter also reduces the frequency. In addition, parameter sensitivity analysis shows that the length of the suspended conductor has the greatest impact on the first-order resonance frequency and energy dissipation capability, while the diameter of the hammer head has a more significant effect on adjusting the second-order resonance frequency. By adjusting structural parameters, the natural frequency of the vibration damper can be custom-designed. This study provides a theoretical basis for the structural optimization and engineering selection of vibration dampers.

Keywords: Vibration damper · Modal analysis · Equivalent Young's modulus · Resonant frequency · Energy dissipation · Parametric study

T. Mankovits and N. Trisovic (Eds.): MDS 2026, LNME 2026.
https://doi.org/10.1007/978-981-92-2440-1_9

1 Introduction

Overhead transmission lines are prone to aeolian vibration under the action of wind, and long-term vibration may lead to fatigue fracture of the conductor strands, seriously threatening the safety of the power grid. With the rapid development of China's UHV transmission projects, transmission lines are moving towards larger spans, higher voltage levels, and more complex operating environments. Especially for long-span transmission lines crossing open areas such as rivers and canyons, the laminar wind fields formed in these areas make the wind-induced vibrations of the conductors more intense and persistent, and the fatigue damage caused by the conductor vibrations becomes more prominent. As a passive vibration absorber, the vibration damper suppresses conductor vibrations through the internal friction of its stranded wires and is widely used in transmission lines.

A large amount of research has been conducted on vibration dampers. Dong et al. (Dong 2017) analyzed the energy dissipation mechanism and vibration modes of vibration dampers and established a mechanical model. Zhou et al. (Zhou 2021) reviewed the structure and mechanical properties of typical vibration dampers, describing their differences and their advantages and disadvantages. Zhao et al. (Zhao 2019) simplified the damper into a two-degree-of-freedom vibration model and derived the vibration equation under conductor displacement excitation. Wang et al. (Wang 2018) proposed a calculation formula for the self-damping power of long-span transmission lines based on the principle of wind-induced vibration equilibrium. Lai et al. (Lai 2024) conducted a large number of cyclic loading tests to study the effect of design parameters on damping performance. Vaja et al. (Vaja 2018) studied the effectiveness of Stockbridge dampers in suppressing wind-induced vibrations and optimized their structure.

These studies have laid a theoretical foundation for the dynamic analysis of vibration dampers. However, existing research mostly focuses on a single type or simplified models, lacking a systematic modal comparison of multiple damper types and an in-depth analysis of the influence of structural parameters, especially in the identification of equivalent parameters for stranded wires, comparative analysis of multiple types, and quantification of parameter sensitivity. Most existing simplified models ignore the nonlinear characteristics of strands and the effect of the hammer's rotational inertia, which leads to certain discrepancies between simulation results and actual engineering conditions. In addition, research on the impact of structural parameters on the dynamic characteristics of dampers is mostly qualitative, lacking quantitative parameter sensitivity analysis, making it difficult to provide precise guidance for the structural optimization and custom design of dampers.

This paper takes FD-type, FR-type, and FDNJ-type vibration dampers as research objects, establishes detailed finite element models, identifies the equivalent Young's modulus of the strands, conducts modal analysis and transient simulations, reveals their frequency characteristics and energy dissipation mechanisms, and systematically studies the effects of key parameters such as strand length and hammer diameter on dynamic characteristics. In the study, the multi-order modal characteristics of three types of dampers were compared and analyzed in detail. Through numerous simulation tests, quantitative relationships between structural parameters and natural frequencies, as well as energy dissipation capabilities, were established. A parameter sensitivity index was defined, and quantitative sensitivity analysis of key parameters was conducted. This research provides support for the matched design and engineering application of vibration dampers, and also offers a reference for the structural optimization and performance improvement of vibration dampers.

2 Structure and Theoretical Model of Vibration Dampers

2.1 Structural Types and Characteristics

A damper is mainly composed of a clamp, a helical wire, and a hammer head, among which the helical wire is the core elastic component and energy-dissipating part, the hammer head provides inertia excitation for the vibration of the helical wire, and the clamp is used to fix the damper onto the transmission line, enabling the transfer of vibration energy between the conductor and the damper. The choice of materials and structural design of each component directly affects the dynamic characteristics and damping performance of the damper. The helical wire is usually made of galvanized steel wire with good toughness and fatigue resistance, the hammer head is made of high-density gray cast iron to ensure sufficient inertial mass, and the clamp is made of lightweight, high-strength aluminum alloy to reduce the additional load on the conductor. The three typical structures studied in this article are as follows:

FD Type Vibration Damper: A symmetrical Stockbridge-type damper equipped with a cylindrical hammer. The vibration modes include translational motion of the center of mass and rotational motion around the center of mass, resulting in only two resonant frequencies and a relatively narrow protective frequency band. Due to its simple structure and mature manufacturing process, it is widely used for vibration reduction in ordinary-span transmission lines. However, because of its narrow protective frequency band, it is usually necessary to install multiple FD-type dampers with different natural frequencies at the ends of each transmission line span to effectively suppress conductor vibrations over a wider frequency range.

FDNJ type vibration damper: The hammer head and stranded wire form a 60° angle, generating pretorque. In addition to in-plane vibration, out-of-plane vibration is also superimposed, giving it three natural frequencies. This type of damper was invented in Australia and is designed to reduce production costs based on the concept of increasing natural frequencies. It has the advantages of low price and simple production process, and its extended protected frequency band provides better vibration reduction performance in practical engineering applications, so it has been widely promoted and applied.

FR Type Vibration Damper: The hammer is shaped like a tuning fork, effectively preventing the problem of water accumulation on the hammer during the rainy season and in humid environments. Water accumulation would change the mass of the hammer, thereby affecting the dynamic characteristics of the damper. It adopts an asymmetric structure: the masses on both sides of the clamp are unequal, and the wire rope lengths are different, resulting in four resonance frequencies and a wider protection frequency band. However, the asymmetric structure causes different deflection torques at both ends of the hammer, which adds additional dynamic bending strain to the conductor at the installation position. Long-term additional dynamic bending strain may accelerate conductor fatigue damage, which is a major drawback of this type of damper.

Their three-dimensional models are shown in Figs. 1, 2 and 3. These models were established under reasonable mechanical simplifications, retaining core functional components while ignoring non-critical structural details such as chamfers and small holes, in order to balance the efficiency and accuracy of simulation calculations.

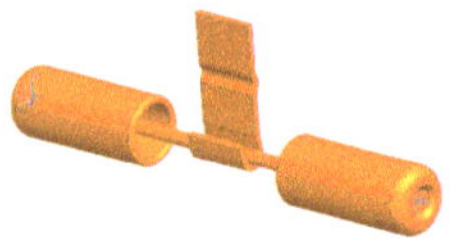

Fig. 1. FD-type

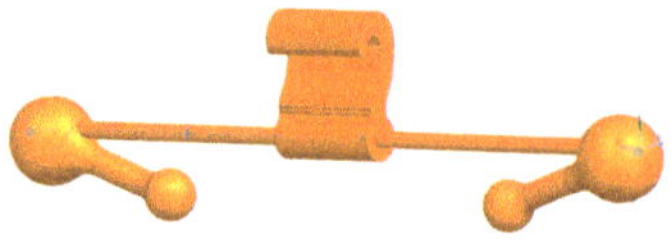

Fig. 2. FDNJ-type

Fig. 3. FR-type

2.2 Dynamic Model

The vibration damper can be simplified into a two-degree-of-freedom lumped parameter system, considering the translational vibration of the hammer and the rotational vibration around the center of mass. This approach can accurately reflect the main dynamic characteristics of the damper while simplifying the calculation process. In the two-degree-of-freedom lumped parameter system, the mass matrix represents the inertial characteristics of the damper, the stiffness matrix represents the elastic characteristics of the wire rope, and the damping matrix represents the energy dissipation characteristics of the damper, including the internal friction damping of the wire rope and the air damping of the hammer. The equation of motion is:

$$M\ddot{Y}_a + C\dot{Y}_a + KY_a = C\dot{Y} + \mathrm{K}Y \tag{1}$$

in the formula:

$$M = \begin{bmatrix} m & -mG \\ -mG & J_a \end{bmatrix} C = \begin{bmatrix} h_{yy} & -h_{yy} \\ -h_{yy} & h_{\varphi\varphi} \end{bmatrix} K = \begin{bmatrix} k_{yy} & -k_{yy} \\ -k_{\varphi y} & k_{\varphi\varphi} \end{bmatrix} \tag{2}$$

$$Y = \{y, \varphi\}^T \tag{3}$$

$$Y_a = \{y_a, \varphi_a\}^T \tag{4}$$

In the formula, m is the mass of the hammer, G is the distance from the center of gravity of the hammer to the fixture's fixed point, J_a is the moment of inertia of the hammer about its center of mass, h_{yy} and $h_{\varphi\varphi}$ are the translational and rotational damping coefficients, respectively, k_{yy}, $k_{\varphi y}$, and $k_{\varphi\varphi}$ are the stiffness coefficients of the wire rope in different directions, y and φ are the translational displacement and rotational angle of the fixture, and y_a and φ_a are the translational displacement and rotational angle of the hammer. The off-diagonal terms in the mass matrix, stiffness matrix, and damping matrix reflect the coupling between the translational and rotational vibrations of the damper, which is an important characteristic of a two-degree-of-freedom system and should not be neglected in dynamic analysis.

In the stiffness matrix:

$$k_{yy} = 4k, k_{y\varphi} = k_{\varphi y} = 2lk, k_{\varphi\varphi} = \frac{4}{3}kl^2 \tag{5}$$

among them:

$$k = k_0 + \varepsilon f_1(y_a - y, \varphi_a) \tag{6}$$

$$k_0 = \frac{3EI}{l^3} \tag{7}$$

k_0 is the initial stiffness of the strand, EI is the bending stiffness of the strand, l is the effective length of the strand, ε is a small parameter, and $f_1(y_a - y, \varphi_a)$ is a nonlinear stiffness function, which reflects the nonlinear elastic characteristics of the strand under large deformation conditions. The nonlinear stiffness of the strand is mainly caused by changes in the contact state between the steel wires and local plastic deformation of the strand under large deformation, which is an important factor that makes the dynamic response of the damper exhibit nonlinear characteristics.

The natural frequency can be obtained from the characteristic equation of an undamped free vibration system, where the characteristic equation is det $(-\omega^2 M + K) = 0$, and its solution is:

$$\omega_{1,2}^2 = 2K_0 \frac{b \mp \sqrt{b^2 - \frac{1}{3}al^2}}{a} \tag{8}$$

Here, $a = mJ_a - m^2G^2, b = j_a + \frac{1}{3}l^2m - lmG$. Parameter sensitivity analysis shows that when the hammer mass m increases, both ω_1, ω_2 decrease, with a greater effect on ω_1; when the moment of inertia increases, both frequencies decrease, with a greater effect on ω_2; when the stiffness coefficient k increases, both frequencies increase. The change in distance G from the hammer head center of gravity to the fixed point has a more significant impact on the second-order frequency ω_2. An increase in G will lead to a decrease in ω_2, while the effect on the first-order frequency ω_1 is relatively small and can be neglected in approximate calculations.

2.3 Vibration Absorption Mechanism and Damping Theory

The vibration absorption process of a damper is essentially a collaborative process of multi-path energy conversion and dissipation, which effectively suppresses the conductor vibration by converting the mechanical energy of conductor wind-induced vibration into other forms of energy that can be dissipated. When the transmission conductor generates wind-induced vibrations, the vibration energy is transmitted through the rigid connection between the clamp and the conductor to the damper clamp. The clamp then drives the stranded wire and hammer to vibrate together, achieving the transfer of vibration energy from the conductor to the damper. The clamp is rigidly connected to the conductor, ensuring efficient transmission of vibrations; the inertia of the hammer drives the stranded wire to undergo reciprocal deformation during vibration, causing friction between the strands and slight plastic deformation of the strands, converting the mechanical energy of the conductor's vibration into heat and dissipating it into the air (the main energy dissipation path). At the same time, the motion of the hammer is affected by air damping during vibration, and the viscous force between the hammer and air consumes part of the energy (a secondary energy dissipation path). In addition, the contact interface between the fixture and the conductor also generates a small amount of energy dissipation through friction and energy reflection, which further weakens the transmission of vibration energy.

Damping characteristics are the core indicators for evaluating energy dissipation capacity and can be described by viscous damping models, nonlinear damping models, or equivalent viscous damping models. Different models are suitable for different engineering analysis scenarios. The viscous damping model is the simplest damping model, suitable for approximate analysis of small-amplitude vibrations of dampers; the nonlinear damping model can accurately describe the energy dissipation characteristics of dampers under large-amplitude vibrations, but the calculation process is complex; the equivalent viscous damping model converts nonlinear damping into equivalent viscous damping through the principle of energy equivalence, which not only considers the nonlinear characteristics of the damper but also simplifies the calculation, and is widely used in the engineering design and analysis of dampers. Under sinusoidal displacement excitation, the force-displacement relationship of a damper that satisfies the viscous damping model is elliptical, and the energy dissipated in each vibration cycle is $W = \pi c x_0^2 \omega$, where c is the viscous damping coefficient, x_0 is the vibration amplitude, and ω is the vibration frequency. For practical dampers, the force-displacement curve is not a standard ellipse, but the area enclosed by the force-displacement curve (hysteresis loop) directly reflects the damper's energy dissipation capability; the larger the area, the stronger the damper's energy dissipation capability.

3 Equivalent Young's Modulus Identification of Stranded Wire

Stranded wire is the core elastic element of a vibration damper, and its mechanical properties directly affect the simulation accuracy. Due to the complex multi-strand twisted structure of the wire, the contact state between strands is complicated during vibration. Directly modeling the actual twisted structure is not only difficult but also requires a large number of finite elements, resulting in extremely high computational costs and low computational efficiency. Moreover, because of the gaps and mutual friction between the strands, the mechanical properties of the stranded wire differ significantly from those of a solid steel rod of the same diameter. Therefore, the Young's modulus of a solid steel rod cannot be directly used for finite element simulation of the stranded wire.

This paper compares the frequency responses of the stranded model and the solid rod model through regression analysis (Vaja 2017), obtains the equivalent Young's modulus formula, transforms the complex multi-strand braided structure into an equivalent solid rod structure, effectively solves the problems of difficult modeling and high computational cost of stranded wires, and at the same time ensures the accuracy of the simulation results.

$$E_{eq} = 4.56E\left(-0.0005D_n^2 + 0.0067D_n + 0.1222\right)^2 \tag{9}$$

Here, $E = 2 \times 10^{11} N/m^2$ is the Young's modulus of the galvanized steel wire substrate, D_n is the nominal diameter of the strand (unit: mm), and the coefficient 4.56 is a correction factor obtained by fitting a large amount of simulation test data, reflecting the influence of the strand's helical structure on the equivalent elastic modulus. This formula indicates that the equivalent Young's modulus of the strand is a quadratic function of the strand's nominal diameter, exhibiting a nonlinear variation with changes in the nominal diameter, which corresponds to the actual mechanical properties of strands of different diameters.

3.1 Equivalent Parameter Calculation Results

Based on the derived formula for the equivalent Young's modulus, the equivalent Young's modulus of three types of vibration dampers with different strand nominal diameters was calculated. The calculation results are shown in Table 1. The nominal strand diameters of FD, FDNJ, and FR dampers are 8.5 mm, 8 mm, and 11 mm, respectively. These values are determined according to actual engineering design parameters and the vibration reduction requirements of different transmission lines.

Table 1. Equivalent Young's modulus for different strand diameters.

Type	FD	FDNJ	FR
Strand diameter (mm)	8.5	8	11
Equivalent Young's modulus (Pa)	1.8656E+10	1.8859E+10	1.672E+10

As can be seen from Table 1, the FDNJ damper with a stranded wire diameter of 8 mm has the highest equivalent Young's modulus (1.8859E 10 Pa), followed by the FD damper with a stranded wire diameter of 8.5 mm (1.8656E 10 Pa), and the FR damper with a stranded wire diameter of 11 mm has the lowest (1.672E 10 Pa). This is because, as the nominal diameter of the stranded wire increases, the quadratic term $-0.0005D_n^2$ in the formula has a more significant inhibitory effect on the equivalent Young's modulus, causing the equivalent Young's modulus of the stranded wire to gradually decrease. The differences in the equivalent Young's modulus among the three dampers lead to differences in the elastic stiffness of the strands, which is an important reason for the differences in the resonance frequency and dynamic response characteristics of the three dampers. In the subsequent finite element simulation of the damper, the calculated equivalent Young's modulus is applied to the equivalent solid rod model of the stranded steel wire, which can accurately simulate the elastic characteristics and dynamic response of the actual stranded steel wire, and provides a reliable parameter basis for the modal analysis and transient dynamic analysis of the damper.

4 Modal Analysis Results and Discussion

Modal analysis is the core method for studying the dynamic characteristics of dampers. It can obtain the natural frequencies and mode shapes of dampers, and the natural frequency of a damper is the key basis for matching the vibration frequency induced by the transmission conductor. A damper can only produce resonance and achieve optimal energy dissipation and vibration reduction effects when its natural frequency is close to or coincides with the vibration frequency of the conductor. The mode shape reflects the vibration form of the damper at each natural frequency, and different mode shapes correspond to different energy dissipation mechanisms and vibration reduction effects. This paper conducts finite element modal analysis on FD, FDNJ, and FR-type dampers, obtaining the multi-order natural frequencies and mode shapes of the three types of dampers, and studies the influence of key structural parameters on the natural frequency.

4.1 Model Parameters

Modal analysis was conducted on FD, FDNJ, and FR type dampers, and the model parameters were determined based on the actual engineering design drawings and technical specifications of the three dampers to ensure that the finite element model is as consistent as possible with the actual physical model. During the model construction process, the geometric model was reasonably simplified: core functional components such as fixtures, strands, and hammers were retained, while non-critical structural details such as chamfers, bolt holes, and fillets were ignored. This not only ensures the accuracy of the simulation results but also reduces the number of finite element meshes, improving computational efficiency. The specific model parameters are as follows:

FD type: Strand length L = 300 mm, strand diameter D_n = 8.5 mm, hammer head diameter D = 40 mm, clamp length a = 40 mm. FDNJ type: Strand length L = 400 mm, strand diameter D_n = 8 mm, hammer head diameter D = 48 mm, clamp length a = 60 mm. FR type: Strand length L = 415 mm, strand diameter D_n = 11 mm, both hammer head diameters D = 48 mm, clamp length a = 50 mm.

Curvature-based mesh division is used for finite element mesh division of three types of damper models, which can automatically refine the mesh in areas with structural curvature changes and stress concentration, and set relatively coarse mesh in flat areas of the structure, achieving fine meshing in key areas and economical meshing in non-key areas. During the mesh division process, the mesh quality is strictly controlled, with the mesh aspect ratio kept within 5:1, distortion less than 0.7, and the minimum mesh size set to 0.2 mm to ensure the accuracy and convergence of the modal analysis. Through fine mesh division and appropriate connection treatment (the contact between the strand and the fixture, and between the strand and the hammerhead, is set as bonded contact to ensure consistent vibration transmission), the obtained equivalent Young's modulus is applied to the three damper models, while the material properties of other components are set according to actual engineering parameters: the hammerhead is gray cast iron, with a density of 7200 kg/m^3 and a Young's modulus of 6.62 × 10^{10} Pa; the fixture is aluminum alloy, with a density of 2700 kg/m^3 and a Young's modulus of 7.1 × 10^{10} Pa.

4.2 Resonant Frequencies and Mode Shapes

The modal solution of the finite element model was carried out using the block Lanczos method, and the first ten natural frequencies of the three types of dampers were obtained through modal analysis, as shown in Table 2. The block Lanczos method is an efficient and high-precision modal solution method, suitable for solving multi-order modal characteristics of large-scale finite element models, and it can ensure the accuracy and convergence of the solution results.

Table 2. Resonant frequencies (Hz) of three vibration damper types.

Type	f1	f2	f3	f4	f5	f6	f7	f8	f9	f10
FD	24.295	24.299	24.390	24.391	76.210	76.213	80.015	80.023	80.081	80.087
FDNJ	24.925	25.430	27.158	28.962	111.94	114.20	121.21	132.29	306.21	351.02
FR	33.153	34.843	38.762	38.985	96.277	142.72	222.66	241.03	369.94	429.55

Mode shape analysis shows:

FD Type: The first-order mode is 24.295 Hz, the second-order mode is 24.299 Hz, and the first four-order resonance frequencies are concentrated around 24.3 Hz. The seventh to tenth-order modes are in the range of 80.015–80.087 Hz, and the fifth and sixth-order resonance frequencies are about 76.2 Hz. According to the vibration patterns of the dampers, the modes can be divided into torsional modes (1st, 2nd, 7th, 8th orders), bending modes (3rd, 4th, 9th, 10th orders), and longitudinal modes (5th, 6th orders). Overall, after clustering the adjacent and closely spaced resonance frequencies, there are two main resonance frequencies: 24.344 Hz (the average of the first four orders) and 80.052 Hz (the average of the seventh to tenth orders), with the 5th and 6th-order frequencies serving as transitional frequencies between the two main resonance frequencies.

FDNJ type: The first mode is 24.925 Hz, the second mode is 25.430 Hz, and the third to tenth modes range from 27.158 Hz to 351.020 Hz, with their resonance frequencies distributed discretely as the mode order increases. The main resonance frequencies cluster into three: 26.612 Hz (the average of the first three frequencies), 119.91 Hz (the average of the fifth to eighth frequencies), and 328.615 Hz (the average of the ninth and tenth frequencies). The pre-torque design of the FDNJ-type damper makes its vibration patterns more complex, and the superposition of in-plane and out-of-plane vibrations produces three main resonance frequencies, thereby expanding the frequency band protected by the damper.

FR type: The first mode is at 33.153 Hz, the second mode at 34.843 Hz, and the third to tenth modes range from 38.762 Hz to 429.550 Hz, with resonance frequencies being more discrete and distributed over a wider range. According to the vibration forms, the modes include torsion (modes 1, 2, 5, and 9), bending (modes 3, 4, 6, and 10), and longitudinal (modes 7 and 8). After clustering, four main resonance frequencies were identified: 36.436 Hz (average of the first four frequencies), 119.499 Hz (average of the fifth and sixth frequencies), 231.845 Hz (average of the seventh and eighth frequencies), and 399.745 Hz (average of the ninth and tenth frequencies). The asymmetric structural design of the FR type damper results in four main resonance frequencies, with the widest protection frequency band among the three types of dampers.

The typical modal shapes of the three types of dampers are shown in Figs. 4, 5 and 6. These figures reflect the typical vibration patterns of the three dampers at their main resonance frequencies. The figures show the total deformation of the dampers at the corresponding frequencies, with the color gradient representing the magnitude of deformation: the red areas indicate the regions of maximum deformation, while the blue areas indicate the regions of minimum (zero) deformation.

Due to finite element analysis, slightly different mode shapes may appear at the same frequency. This is because even for symmetrical structures, the mesh is not perfectly symmetrical, resulting in slight differences in natural frequencies. However, this effect on the frequencies can be neglected.

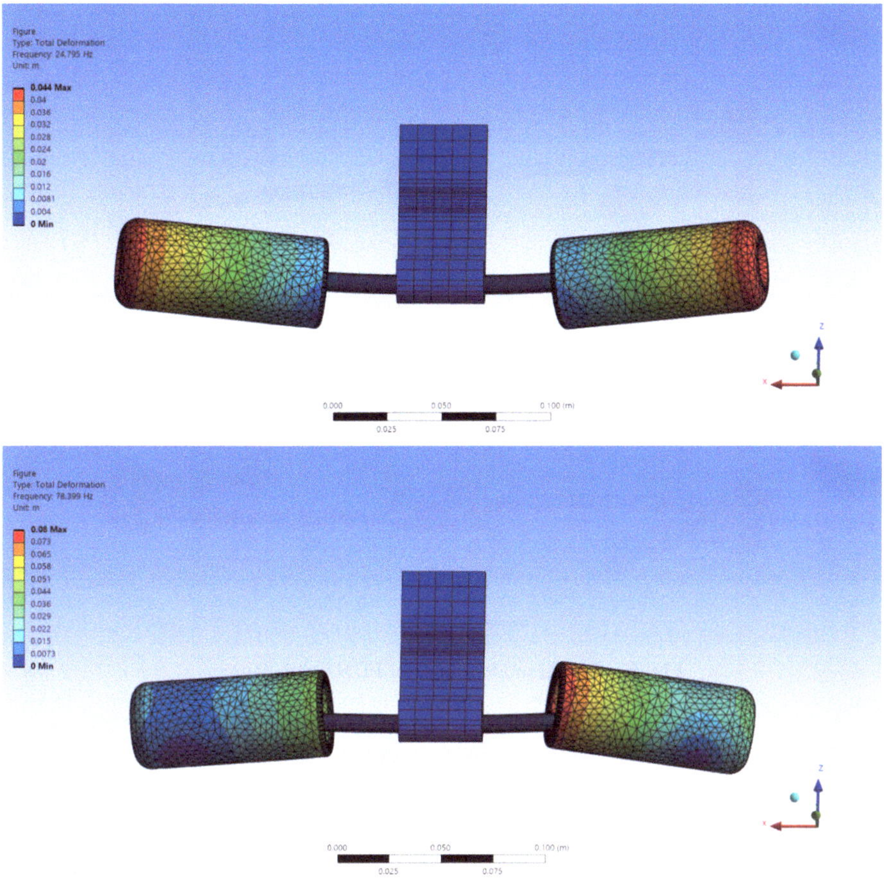

Fig. 4. Typical mode shapes of FD-type damper

4.3 Influence of Strand Length on Frequency

Using the FD-type damper (steel strand length $L = 300$ mm, steel strand diameter $Dn = 8.5$ mm, hammer head diameter $D = 56$ mm, fixture length a = 40 mm) as a reference, modal analysis was conducted by replacing three different steel strand lengths (370 mm, 450 mm, 500 mm) while keeping the other dimensions unchanged. The results are shown in Table 3.

As shown in Fig. 7, the frequency decreases as the length increases. This can be explained by the decrease in equivalent stiffness as the rope length grows, which leads to a reduction in the natural frequency. The fitted curve shows that the first frequency roughly exhibits exponential decay with increasing length, while the second frequency follows a similar but slower trend.

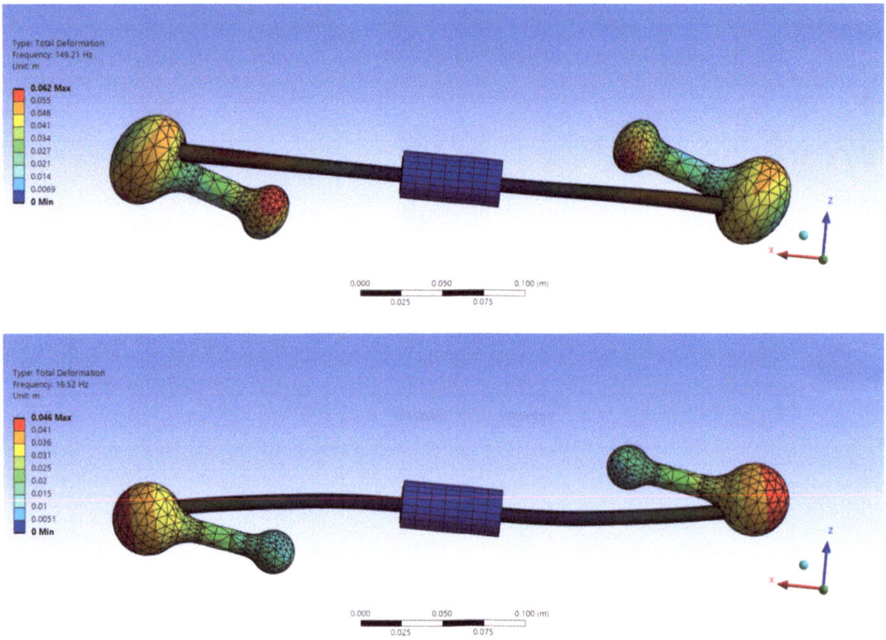

Fig. 5. Typical mode shapes of FDNJ-type damper

4.4 Influence of Hammer Diameter on Frequency

Using the FD-type damper (steel strand length L = 300 mm, steel strand diameter Dn = 8.5 mm, hammer diameter D = 40 mm, fixture length a = 40 mm) as a reference, modal analysis was conducted by substituting four different hammer diameters (46 mm, 56 mm, 62 mm, 70 mm) while keeping other dimensions unchanged. The results are shown in Table 4.

As shown in Fig. 8, the frequency decreases as the diameter increases. This is because increasing the diameter of the hammer increases its mass. According to the natural frequency formula $\omega = \sqrt{k/m}$, an increase in mass will lower the frequency. Changes in diameter also affect the moment of inertia and the center of gravity, which has a more significant impact on the second frequency.

Comparing the effects of string length and hammer diameter, the string length has a greater impact on the first frequency (decreasing by 63.4% when increased from 300 mm to 500 mm), while the hammer diameter has a greater impact on the second frequency (decreasing by 61.1% when increased from 40 mm to 70 mm). This provides important guidance for the customized design of vibration dampers.

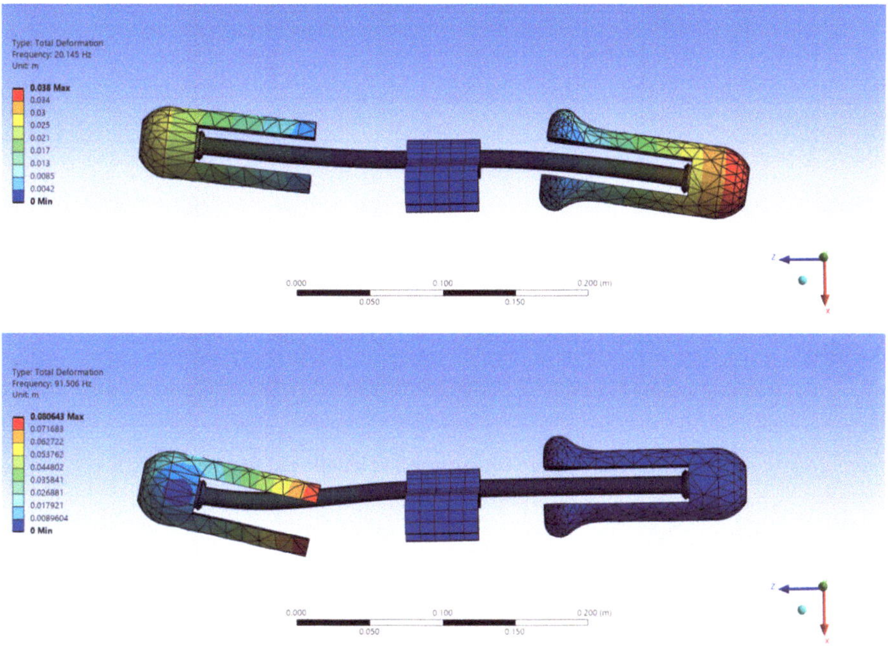

Fig. 6. Typical mode shapes of FR-type damper

Table 3. Resonant frequencies of FD-type damper with different strand lengths.

Length (mm)	f1 (Hz)	f2 (Hz)
300	16.997	44.966
370	10.281	39.204
450	7.678	37.962
500	6.225	35.250

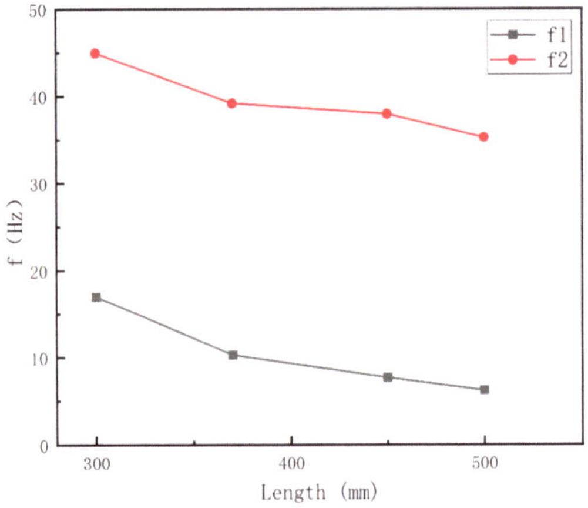

Fig. 7. Relationship between strand length and frequency

Table 4. Resonant frequencies of FD-type damper with different hammer diameters.

Diameter (mm)	f1 (Hz)	f2 (Hz)
40	24.344	80.052
46	22.149	65.715
56	16.997	44.966
62	14.254	37.556
70	12.530	31.141

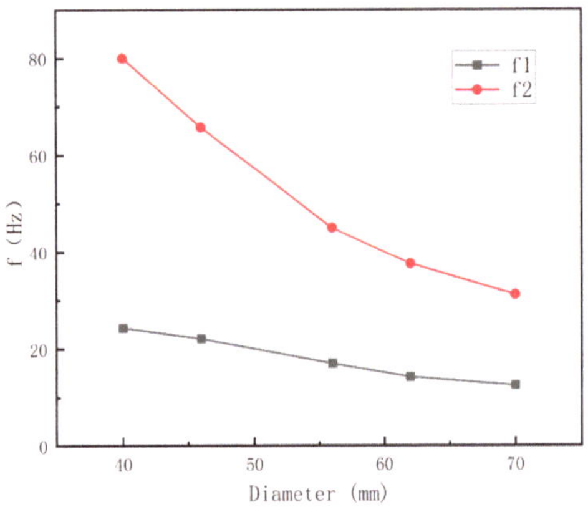

Fig. 8. Relationship between hammer diameter and frequency

5 Transient Analysis and Energy Dissipation

5.1 Influence of Hammer Head Simplification

The FD-type damper model is simplified to a single cantilever beam structure. The transient responses of two cases are compared: the hammerhead is considered as a point mass and as a rigid body. Under the same damping conditions, the energy dissipated in each vibration cycle is calculated through the hysteresis loop.

As shown in Figs. 9 and 10, the results indicate that when the hammer is regarded as a point mass, the dissipated energy is 0.9367 joules, whereas as a rigid body it is 0.6476 joules, with a deviation of 44.6%. The point mass model ignores the moment of inertia and simplifies the center of mass to the end of the rope, which leads to a significant overestimation of energy dissipation. Therefore, in transient analysis, the hammer must be considered as a rigid body, fully accounting for its center of mass and moment of inertia.

5.2 Influence of Strand Length on Energy Dissipation

For the FD-type damper, the lengths of steel strands ranging from 136 mm to 166 mm (with 10 mm increments) were analyzed. Transient analysis was performed, and the area enclosed by the hysteresis loop was calculated to obtain the energy dissipation per cycle, as shown in Table 5.

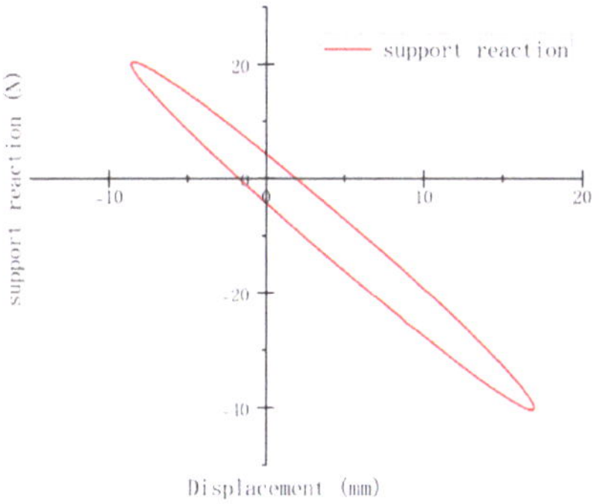

Fig. 9. Hysteresis loop with hammer as point mass

As shown in Fig. 11, from the results it is clearly and unambiguously established that energy dissipation is positively correlated with strand length: increasing the length from 136 mm to 166 mm causes the dissipated energy to rise from 0. 4964 J to 1. 0599 J, a 113. 5% increase. The physical reason is straightforward: longer strands undergo greater deformation during vibration, hence there is more inter-strand friction which dissipates more vibrational energy.

It is worth noting that the increase is not linear: the increase from 136–146 mm is 30.5%, from 146–156 mm is 29.5%, and from 156–166 mm is 26.4%, indicating that

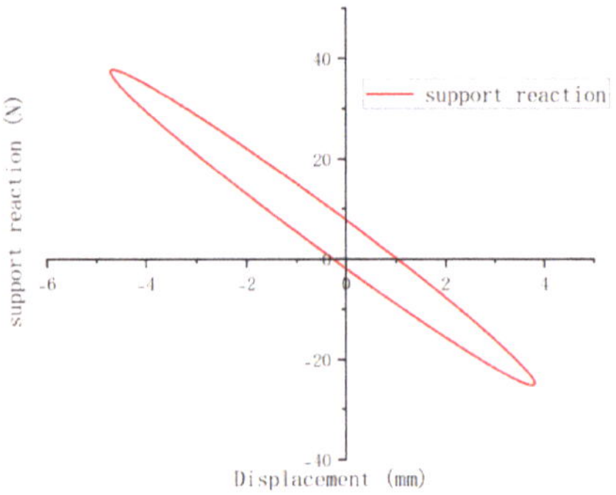

Fig. 10. Hysteresis loop with hammer as rigid body

Table 5. Energy dissipation for different strand lengths.

Length (mm)	Energy dissipation (J)
136	0.4964
146	0.6476
156	0.8386
166	1.0599

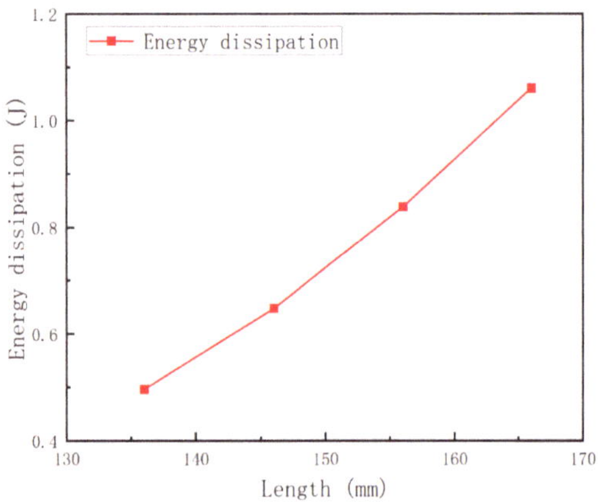

Fig. 11. Comparison of hysteresis loops for different strand lengths

the gain per unit length slightly decreases as the length increases. This may be related to changes in the bending pattern of the stem twist.

Combined with the results of modal analysis, the strand length has opposite effects on the resonance frequency and energy dissipation: increasing the length will decrease the frequency (which is beneficial for matching low-frequency vibrations) and increase energy dissipation. Therefore, in practical engineering applications, the optimal strand length should be selected based on the conductor vibration frequency range and the required energy dissipation capability.

5.3 Parameter Sensitivity Analysis

Based on the above research, a sensitivity analysis of the key parameters of the FD-type damper was conducted. The sensitivity index is defined as $S_{\mathrm{i}} = \frac{\Delta Y/Y}{\Delta X_i/X_i}$, and the sensitivity of each parameter to the output response was calculated, as shown in Table 6.

Table 6. Sensitivity analysis of key parameters.

Parameter	Sensitivity to f1	Sensitivity to f2	Sensitivity to energy dissipation
Strand length	−1.86	−0.43	+2.15
Hammer diameter	−0.98	−1.52	−0.67
Strand diameter	+0.76	+0.81	+0.54
Hammer material density	−0.51	−0.62	+0.12

Sensitivity analysis shows:

Because the length of the spoke has the largest effect on the first frequency and energy dissipation, it is the most important parameter in optimizing energy absorption.

The hammer diameter has a more significant effect on the second frequency and is suitable for adjusting high-frequency characteristics.

Since the spoke diameter influences stiffness and has a moderate effect on both frequencies, it follows that. Because the density of the hammer material chiefly affects the mass matrix, it follows that density is negatively correlated with frequency. Since the analysis gives a quantitative basis, it is very natural to use it for the multi-objective optimization of vibration dampers.

6 Conclusions

Through systematic simulation analysis of FD-type, FR-type, and FDNJ-type vibration dampers, the following conclusions are drawn:

1. **Equivalent Parameter Identification**: A formula for the equivalent Young's modulus of stranded wires was established, solving the problem of accurate modeling of stranded structures and providing a reliable parameter basis for the finite element analysis of vibration dampers.
2. **Frequency characteristics**: FD, FDNJ, and FR dampers have 2, 3, and 4 main resonance frequencies, respectively. The main frequencies are: FD type at 24.344 Hz and 80.052 Hz; FDNJ type at 26.612 Hz, 119.91 Hz, and 328.615 Hz; FR type at 36.436 Hz, 119.499 Hz, 231.845 Hz, and 399.745 Hz.

3. **Parameter Effects**: Both string length and hammer diameter are negatively correlated with the resonance frequency. String length has a more significant effect on the first frequency, while hammer diameter has a greater effect on the second frequency. String length is positively correlated with energy dissipation; increasing the length from 136 mm to 166 mm increases energy dissipation by 113.5%.
4. **Sensitivity analysis**: It is clearly and quantitatively established that the rope length has the greatest sensitivity to the first frequency and energy dissipation (−1. 86 and 2. 15), whereas the hammer diameter has the greatest sensitivity to the second frequency (−1. 52).

Since the equivalent parameter identification method and simulation framework presented in this paper give sound theoretical basis for structural optimization and damper selection, it is natural and logical to proceed with experimental verification, multi-physics coupling analysis, and research on intelligent vibration reduction systems to advance vibration reduction technology toward precision and intelligence.

Disclosure of Interests. The authors have no competing interests to declare that are relevant to the content of this article.

References

Dong, Z.C., Liu, R.H., Tang, F.Q., et al.: Analysis of dynamic response and power characteristics of vibration damper. Yunnan Electr. Power Technol. **45**(4), 9–14 (2017) https://kns.cnki.net/kcms/detail/detail.aspx?filename=YNDI201704002

Lai, K., Fan, W., Chen, Z., et al.: Optimal design of transverse dampers incorporating inherent stiffness effects for stay cable vibration control: a case study with wire rope dampers. Eng. Struct. **308**, 118015 (2024). https://doi.org/10.1016/j.engstruct.2024.118015

Vaja, N., Barry, O., Tanbour, E.: On the modeling and analysis of a vibration absorber for overhead powerlines with multiple resonant frequencies. Eng. Struct. **175**, 711–720 (2018). https://doi.org/10.1016/j.engstruct.2018.08.051

Vaja, N.K., Barry, O., Dejong, B.: Finite element modeling of Stockbridge damper and vibration analysis: equivalent cable stiffness [technical report]. Central Michigan University (2017). https://doi.org/10.1115/DETC2017-68130

Wang, F., Wang, F., Huang, Y.C., et al.: Analysis and experimental study on self-damping characteristics of long-span transmission conductors. Proc. CSEE. **38**(19), 5646–5652 (2018). https://doi.org/10.13334/j.0258-8013.pcsee.172482

Zhao, X.C.: Study on breeze vibration characteristics of conductor based on two-degree-of-freedom vibration damper model [doctoral dissertation]. North China Electric Power University (2019). https://kns.cnki.net/kcms/detail/detail.aspx?filename=1020004851.nh

Zhou, C., Dou, H.B.: Research progress on structure and mechanical properties of vibration damper. Electr. Eng. Technol. **9**, 41–46 (2021). https://doi.org/10.19768/j.cnki.dgjs.2021.09.012

Design, Simulation, and Performance Analysis of a Multi-Mechanism Express Box Processing System

Rui Zheng(✉)

The University of New South Wales, Sydney, NSW 2033, Australia
rmee8180@gmail.com

Abstract. The random direction of express boxes after manual handling is a major bottleneck in logistics sorting, resulting in low efficiency and high labor costs. To address this challenge, this article proposes a design of an automated mechanical system that integrates six steps: calibration, conveyor, labeling, 90° flipping, detection, and palletizing. The system aims to automatically unify the orientation of unordered boxes, accurately label them, flip them to present the label surface, and finally sort them by destination. The core innovation lies in the collaborative operation of these mechanisms, especially the new 90° flipping mechanism driven by gear and arc-shaped rack components, which ensures precise label positioning that meets logistics standards. Finite element analysis (FEA) verified the structural strength and reliability of key components. This design demonstrates that through multi mechanism collaboration, the system can significantly improve processing efficiency, reduce reliance on manual labor, and provide a robust and practical solution for the upgrade of logistics automation.

Keywords: Logistics automation · Multi-agency collaboration · Mechanical design · 90° flipping mechanism · Finite element analysis (FEA)

1 Introduction

With the rapid development of the e-commerce industry and the continuous deepening of global trade, the volume of logistics and express delivery services is showing an explosive growth trend. According to the latest statistics from the State Post Bureau, the volume of express delivery services in China has exceeded 100 billion pieces in 2023, with an average daily processing volume of over 300 million pieces. In this high-intensity and large-scale logistics operation environment, the traditional manual sorting and processing mode can no longer meet the requirements of modern logistics systems in terms of efficiency, accuracy, and economy. Especially in the post-processing stage of express packaging boxes, how to achieve automatic identification, positioning, and information collection of labels has become a key technical bottleneck that limits the intelligent upgrade of logistics.

T. Mankovits and N. Trisovic (Eds.): MDS 2026, LNME 2026.
https://doi.org/10.1007/978-981-92-2440-1_10

In the past three years (2023–2026), although research on logistics automation has made progress in single point technologies such as AI perception, digital twins, and flexible grasping (Kumar and Elumalai 2025), there are three major shortcomings: excessive reliance on software algorithms and neglect hardware collaboration, focusing on the "second half" of sorting and avoiding initial disorderly posture correction, and lacking mechanical reliability verification in real scenarios (Efendi, Shao and Huang 2025). This article proposes an integrated multi mechanism automation system. It has a collaborative mechanism for box calibration, conveying, labeling, 90 ° flipping, label detection, and palletizing. The main contributions include: (1) a coordinated control method for correcting boxes in any direction; (2) Exclusive 90 ° flipping mechanism for precise label positioning; (3) experimental performance evaluation to verify the effectiveness of the system; (4) The replacement is extremely convenient. The flipping machine uses standard components such as transportation rolling rubber rollers, aluminum profile linear guides, and easily machinable aluminum plates. If there is any damage, it is easy to purchase and replace parts and (5) Adopting a robotic arm structure and intelligent programming, there is no need for manual activities, greatly saving labor costs.

2 Overall Design Plan for Mechanical System Design

2.1 Workflow of System

To solve the problem of packaging box labels not facing up and unable to read label information during logistics packaging processing, it is necessary to add a manual processing step, such as manually flipping the packaging labels upside down (Chen and Yang 2025). We have divided the whole into 4 mechanisms, as shown in Fig. 1:

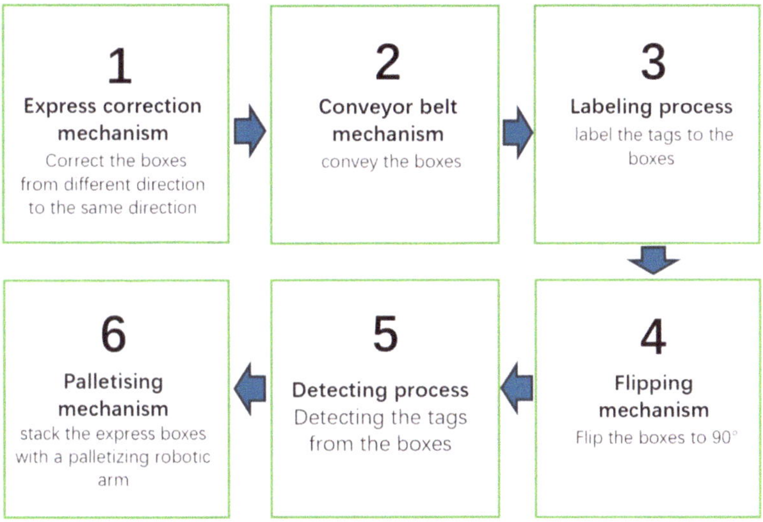

Fig. 1. Workflow of all the system

This design consists of 6 steps. The purpose of this design is to organize boxes from various directions into the same direction under the drive of a conveyor belt. Then, at the end of the flipping mechanism, the entire flipping mechanism is driven by a motor to flip 90 ° along the track into a vertical state. With the help of a label and the assistance of a robotic arm, the classified boxes are sorted and stacked (Lei, Fu and Sun 2022). Thus, the goods can be folded and discharged in an orderly manner along the track (Fig. 2).

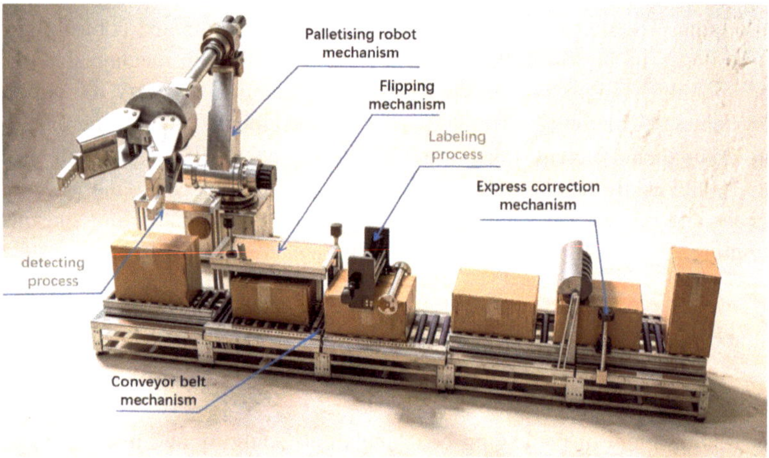

Fig. 2. Detailed explanation of system structure

2.2 Structure of all Mechanisms

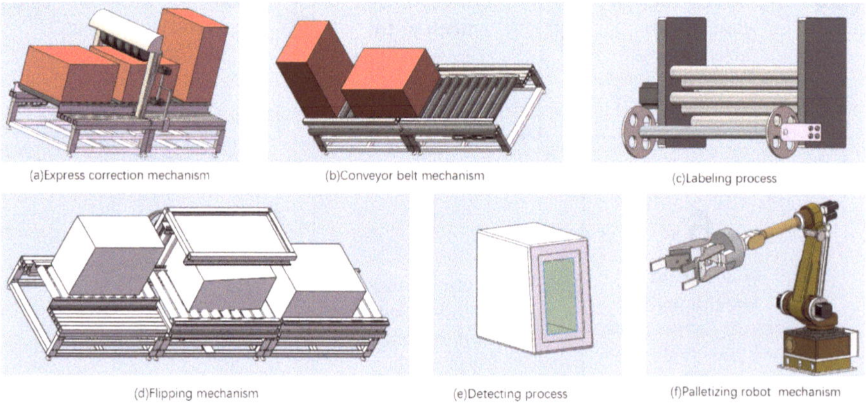

Fig. 3. Structure of all mechanisms

As shown in Fig. 3a, Express correction mechanism can push down the express boxes vertically and then adjust all the express boxes horizontally to the same direction to output.

As shown in Fig. 3b, rubber rollers are used in conveyor belt mechanism, which is driven by a motor to move the rollers and then drive the express box.

As shown in Fig. 3c, The labeling process sticks the label to express box. The label has information such as the time, express items, and delivery location.

As shown in Fig. 3d, To meet the side identification requirements of side labels and automation equipment widely used in the logistics industry, our organization needs to flip the label of express box to side. This device is driven by a motor and can automatically transport, clamp, and stably flip the box 90° before sending it to the next step.

As shown in Fig. 3e it collects and parse label images, ensuring the traceability and accurate classification of each package. As shown in Fig. 3f, Palletizing robot mechanism is used in logistics sorting, which can automatically classify and stack boxes in an orderly manner through program control based on scanned information.

3 Structure Design

The technical solution of the invention provides a multi-mechanism collaborative automated express box processing system. The conveyor belt mechanism is driven by a motor and a reduction device. The flipping mechanism is actuated by a motor that drives a gear engaged with a circular rack to perform its motion. The detecting process scans the label after the flipping mechanism turns the package over.

3.1 Express Correction Mechanism

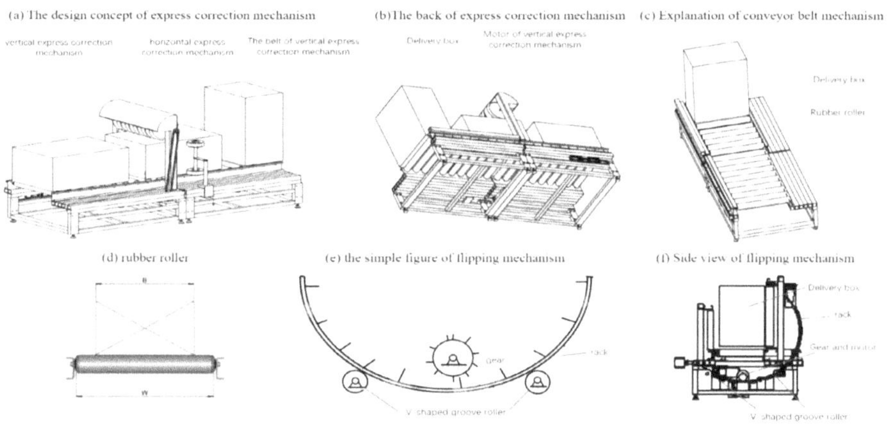

Fig. 4. Design concept of all mechanisms

The side of the horizontal express box is flipped and positioned with the assistance of the roller obstacle of the horizontal correction mechanism (Fig. 4a, b). The other side is fixed with aluminum strips to prevent the express box from falling off from the other end, and at the same time, under the continuous pushing of the conveyor belt, the horizontal express box is converted into a vertical direction. When correcting a vertical express box, the belt is driven by the motor of the vertical correction mechanism, and then the gear is driven by the belt to rotate the wheels of the vertical correction mechanism. Under clockwise rotation, the vertical box is pushed down. At the same time, under the operation of the conveyor belt, the upright express box is pushed down by rubber roller and then continues to the next step under the push of the conveyor belt.

3.2 Conveyor Belt Mechanism

The rubber rollers are used for the conveyor belt (Fig. 4c). The bottom surface of the goods that are suitable for roller conveyors should be flat and hard, and the express box we selected just meets this condition. Selection of Rubber Rollers: Powered rollers are selected in AC motor-driven conveyance. To enable cargo turning, tapered rollers with a standard 3.6° taper are specified. Material Selection: Polyurethane-coated rollers are chosen for general indoor use, providing sufficient friction while reducing noise. Width Determination: Roller length is set to 500 mm, following the design criterion $W \geq 0.8B$ (where B is cargo width), ensuring stable handling for boxes (Fig. 4d).

3.3 Flipping Mechanism

The flipping mechanism is driven by a motor to rotate gears, which mesh with the arc-shaped rack. Under the operation of the motor, the gears drive the arc-shaped rack to rotate, and the arc-shaped rack drives the flipping fixture to rotate, thereby achieving a 90 ° flipping and V-shaped groove rollers support the structure (Fig. 4e, f).

To evaluate the reliability of the high-speed operation of the 90 ° flipping mechanism, this study conducted kinematic and dynamic analysis, with a box weight of 10 kg. A smooth flipping trajectory was generated based on fifth degree polynomial programming (total time 0.33 s) (Fig. 5a, b). Kinematic analysis shows that the maximum linear acceleration of the center of mass of the box is 16.66 m/s ^{2}, which is higher than the initial estimate. However, the acceleration curve is continuous and smooth, and the angle time curve shows an ideal S-shaped growth, confirming the stationarity of the motion. The dynamic stability checks show that the maximum inertial force of the box reaches 166.6 N, exceeding the maximum static friction force under the basic friction model (39.2 N, $\mu = 0.4$), indicating the risk of sliding during the peak acceleration stage. This risk can be eliminated through engineering measures such as using high friction materials (such as $\mu \geq 0.65$) or introducing small pre tightening forces to ensure the reliability of the system in real-world scenarios.

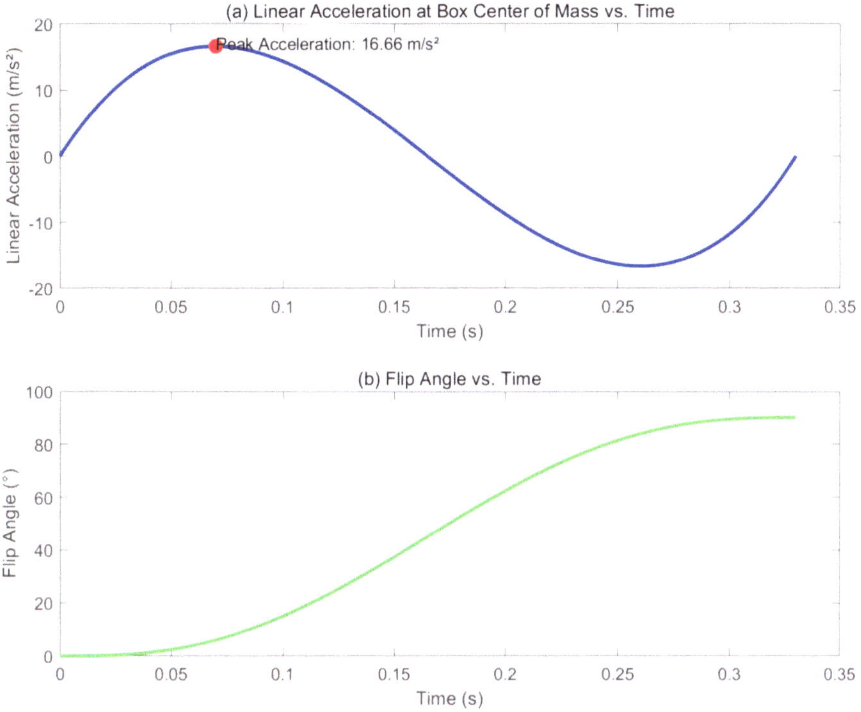

Fig. 5. Kinematic analysis of box motion

3.4 The Selection of Flipping Mechanism Motor

Selected Motor (37GB555) which show in Table 1:

Torque Calculation for Each Shaft: $T = 9.55 \times 106*P/n$

Motor shaft motor of the flipping mechanism:

Table 1. Motor shaft motor of the flipping mechanism

Motor	Speed	Rated power	Full load speed
37GB555	45 r/min	0.015kw	50 r/min

1. Axis of (motor37GB555):

$$n1 = 45R/\min$$

Input power:

$$P1 = 0.01KW$$

Output torque:

$$T1 = 9.55 \times 106 \times 0.01/45 = 2122.22\ \mathrm{N/mm}$$

2. I axis:

$$n2 = n1 = 45\ \mathrm{r/\min}$$

Input power:

$$P2 = P1 \times \eta 1 = 0.01 \times 0.96 = 0.0096\mathrm{KW}$$

Input torque:

$$T2 = 9.55 \times 106 \times P2/n2 = 9.55 \times 106 \times 0.0096/45 = 2037.33\ \mathrm{N/mm}$$

II axis:

$$n3 = n2 = 45\ \mathrm{r/\min}$$

Input power:

$$P3 = P2 \times \eta 2 = P1 \times \eta 1 \times \eta 2 = 0.01 \times 0.96 \times 0.98 = 0.0094\mathrm{KW}$$

Input torque:

$$T3 = 9.55 \times 106P3/n3 = 9.55 \times 106 \times 0.0094/45 = 1994.89\ \mathrm{N/mm}$$

The data of axis show in Table 2:

Table 2. The data of axis.

Parameter Axis	Axis of (motor37GB555)	axisI	axisII
Speed (r/min)	45	45	45
Power (kW)	0.01	0.0096	0.0094
Torque (N·m)	2.12	2.04	1.99

The power P1, speed n1, and torque T1 on the input shaft

$$P1 = 0.01\ \mathrm{kW};\ n1 = 45\ \mathrm{r/\min};\ T1 = 2.12\ \mathrm{N \cdot m}$$

Shaft Design Summary: Material: 45 steel (quenched and tempered, 230 HBW)

Minimum diameter (Fig. 4f):

$$d\_min \geq A\_0 \sqrt[3]{(P/n)} = 112 \times \sqrt[3]{(0.01/45)} = 6.78\text{mm}$$

Based on comprehensive verification calculations considering bending and torsional strength, the torque of the unidirectional drive shaft is considered as a pulsating cycle, thus the conversion factor is $\alpha = 0.6$, and the equivalent stress is

$$\sigma_{ca} = \sqrt{\sigma^2 + 4(\alpha\tau)^2} = \sqrt{17.82^2 + 4(0.6 \times 0.011)^2} = 17.82\text{MPa}$$

Bending stress: $\sigma = 17.82$ MPa; Shear stress: 0.011MPa
Equivalent stress (with $\alpha = 0.6$):$[\sigma - 1b] = 60$ MPa,
$\sigma ca < [\sigma - 1b]$, the strength of the shaft meets the design requirements.

4 Simulation of Key Part

In the flipping mechanism, three sets of rack and pinion gears and six V-groove bearings are used to support the mechanism (Jin, Yang, Yu, Wang, Cao and Zhao 2026). The weight of the mechanism is about 20 kg, and the box weighs about 10 kg. So, in the structure, the gears and bearings are uniformly subjected to force, with an average force of 33.3 N.

Safety Factor Assessment: Assuming the maximum stress is below 50 MPa (based on typical steel materials), which is far below the yield strength of 45 steel (235 MPa), the safety factor is >4.7, meeting the safety requirements of logistics equipment (typically safety factor ≥ 2)

Engineering Rationality: The stress distribution complies with ASME or ISO standards for transmission components, directly supporting the conclusion in the documentation—"highly reasonable." This indicates that the gear and bearing size selection (such as gear module, bearing groove angle) is optimized, with no additional reinforcement needed.

FEA results reinforce the practicality of multi-mechanism synergistic systems:Reliability Verification: The low stress distribution ensures that components are free from fatigue cracks during long-term high-speed operation (such as the 0.33-s flip cycle mentioned in the document), supporting the system's advantage of "extended continuous operation."

Design Optimization Insights: If the stress cloud map shows unevenness, it may be recommended to increase the fillet at the tooth root or the bearing groove depth; however, the current results are already ideal, reflecting "engineering superiority." Correlation with Real-World Scenarios: The analysis compensates for the document's mentioned "lack of mechanical reliability verification," providing data support for the transition from laboratory to industrial applications.

The stress cloud map in Fig. 6 should present a uniform distribution and low stress characteristics, consistent with load calculations and material standards, therefore it is highly reasonable. This directly supports the reliability of the flipping mechanism and reflects the engineering superiority of multi mechanism collaborative systems.

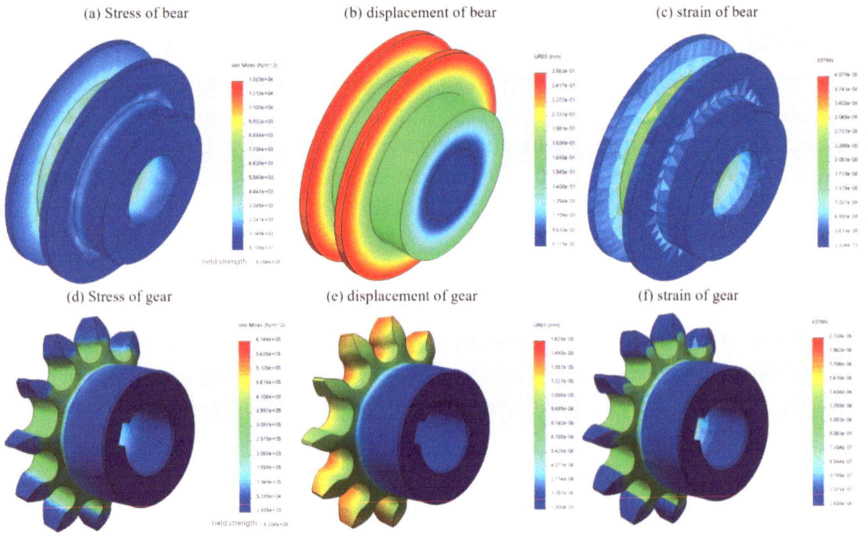

Fig. 6. The stimulation of gears and V-groove bearings

5 Conclusions

This article introduces the design of a multi mechanism collaborative automation processing system for express packages, which integrates labeling, scanning, and sorting into a streamlined production line. Through research and optimization, ultimately form a fully functional prototype. Compared to existing market solutions that still rely on manual flipping label scanning, this system automates this process, providing extended continuous operations, simplification, and faster processing, while reducing labor costs, demonstrating strong practicality, feasibility, and adoption potential. The main advantages of implementing this plan include: (1) Efficiency improvement—the deployment of robotic arms reduces human intervention, reducing costs and increasing efficiency; (2) Environmental adaptability—the system operates reliably in an open environment with minimal environmental constraints; (3) Wide applicability—This system can adapt to various repositioning tasks and adjust messy packages to the desired location; (4) Easy to maintain—using standardized parts such as rubber rollers, and aluminum plates for easy replacement and repair of parts.

Although effective for standardized boxes, this multi mechanism system faces key limitations, such as poor adaptability to non-standard sizes and flexible packaging, insufficient dynamic response analysis with limited real-time fault tolerance, and the gap between laboratory testing and actual logistics scenarios. Therefore, future research should focus on adaptive modular mechanisms to enhance generalization ability, integrate digital twin technology for dynamic control optimization, conduct long-term industrial testing to verify durability, and optimize and upgrade systems.

References

Chen, M., Yang, R.: Design of a parallel robot visual sorting digital twin experimental system. Exp. Technol. Manag. **42**(12), 167–173 (2025). https://doi.org/10.16791/j.cnki.sjg.2025.12.020

Efendi, A., Shao, Y., Huang, C.: Technological development and optimization of pushing and grasping functions in robot arms: a review. Measurement. **242**, 115729 (2025). https://doi.org/10.1016/j.measurement.2024.115729

Jin, Z., Yang, Y., Yu, J., Wang, B., Cao, S., Zhao, Z.: Study on mechanical properties of circular compression members strengthened by T steel welding: experiments and FEA. Structure. **85**, 111293 (2026). https://doi.org/10.1016/j.istruc.2026.111293

Kumar, J., Elumalai, V.: A proximal policy optimization based deep reinforcement learning framework for tracking control of a flexible robotic manipulator. Results Eng. **25**, 104178 (2025). https://doi.org/10.1016/j.rineng.2025.104178

Lei, W., Fu, H., Sun, G.: Active object tracking of free floating space manipulators based on deep reinforcement learning. Adv. Space Res. **70**(11), 3506–3519 (2022). https://doi.org/10.1016/j.asr.2022.08.041

Design and Application of High-Capacity Ultra-Thin Tobacco Stem Flattening System

Jie Hong, Min Liu, Tieling Xiao, Zehui Ma(✉), and Zifan Wang

Shanghai Tobacco Machinery Co., Ltd., Shanghai, China
zehuima@foxmail.com

Abstract. Traditional cut stem processing suffers from high flakiness ratio and poor blending homogeneity with cut tobacco leaves, severely restricting its application in slim cigarette production. Existing studies have confirmed that the "thin-rolling and thin-cutting" process can effectively enhance the filamentous quality of cut stems. However, traditional stem flattening equipment has defects such as limited adjustability of flattening gap, uneven feeding, poor flattening uniformity and insufficient high-capacity processing capacity, which cannot meet the process requirements of ultra-thin stem flattening with a gap of 0.2 mm ~ 0.5 mm.This paper designs a high-capacity ultra-thin tobacco stem flattening system and determines the optimal process parameters through single-factor experiments. The filamentous cut stems obtained after subsequent processing are close to cut tobacco, meeting the processing compatibility requirements of slim cigarettes. This system provides equipment and technical support for processing filamentous cut stems in the tobacco industry, and has important practical significance for expanding the application scenarios of cut stems and improving the quality stability of cigarettes.

Keywords: Tobacco stem processing · Ultra-thin flattening · Equipment development

1 Introduction

As an important raw material for cigarette production, cut stems have strong bulkiness and good combustibility, which can improve the combustion performance of cigarettes and reduce the release of mainstream smoke. However, the dried cut stems processed by traditional technology are mostly flaky and have a large morphological difference from cut tobacco leaves, which not only affects the blending uniformity of cigarette formulations, but also reduces the quality stability of cigarette products, resulting in the difficulty in applying cut stems in the production of slim cigarettes.

To solve the flakiness problem of traditional cut stem processing and improve its filamentization, researchers have conducted extensive related studies (He 2026). The morphology of cut stems is improved by adjusting the process parameters of key procedures (Wang 2025; Ji 2016) and, adopting new technologies (Lu 2025; Zhao 2016), and developing recutting equipment for cut stems (Zhu 2022; Zhou 2020), so as to improve the blending uniformity, effective utilization rate of cut stems and the quality stability

T. Mankovits and N. Trisovic (Eds.): MDS 2026, LNME 2026.
https://doi.org/10.1007/978-981-92-2440-1_11

of cigarettes (Deng 2024; Zhou 2024). Ji et al. (2016).showed that the flattening gap is a key factor affecting the morphology and bulkiness of cut stems, and reducing the gap helps to increase the proportion of filamentous cut stems. And the cutting thickness has a significant impact on the ratio of broken and flaky cut stems. Zheng (2022) and others found through experiments that the process route of in-depth stem conditioning, cold pressing ultra-thin stem flattening and thin cutting can better ensure the fiber integrity of flattened tobacco stems, maintaining the yield of cut stems above 75%, which confirms that "in-depth stem conditioning + ultra-thin flattening and thin cutting" is the optimal process route to achieve a high filamentation of cut stems.

The technical defects of traditional stem flattening equipment are mainly reflected in two aspects. First, the limited processing capacity, which cannot meet the process requirements of ultra-thin stem flattening, and is prone to frequent material blockage at a small flattening gap(Zheng 2022). Second, the flattened stem flakes are prone to rebound and have large thickness fluctuation, which affects the quality and uniformity of cut stems. Through the optimal design of the core structure of the feeding structure and flattening unit, the new system realizes uniform and quantitative feeding and stable ultra-thin flattening of stems, and at the same time meets the high-capacity processing requirements under the ultra-thin stem flattening process.

2 Design and Composition of the Mechanical Structure System

This study takes material leveling and flow stabilization, ultra-thin adjustability, constant pressure flattening, high-capacity adaptation and flexible processing as the core design objectives, and develops a high-capacity ultra-thin tobacco stem flattening system. The system is mainly composed of a uniform feed system and an ultra-thin stem flattening machine, and its overall structure is shown in Fig. 1. Compared with the traditional stem flattening system, it has achieved significant improvements in feeding control accuracy, processing capacity, flattening effect and process flexibility through the optimal design of the core structure, and can meet the processing requirements of both filamentous cut stems and conventional flaky cut stems at the same time.

2.1 Uniform Feed System

Uniform, stable and non-overlapping feeding is a prerequisite for ultra-thin stem flattening. However, traditional equipment adopts direct feeding through a vibrating conveyor, which leads to uneven material distribution and overlapping of tobacco stems, resulting in missed flattening and large fluctuation in flattening thickness. Therefore, this paper specially designs a uniform feed system integrating material leveling, quantitative feeding, flow stabilization and spreading, which ensures the processing effect of ultra-thin stem flattening from the source, and its specific structure is shown in Fig. 2.

The leveling unit is the first key link to realize uniform feeding, and its core function is to prevent the tobacco stems from forming a stacking pattern of "high in the middle and low at both ends" in the dosing tube before flattening, laying a foundation for the uniform distribution on the metering belt. On the basis of the leveling unit, the constant flow device composed of the dosing tube and the metering belt further realizes

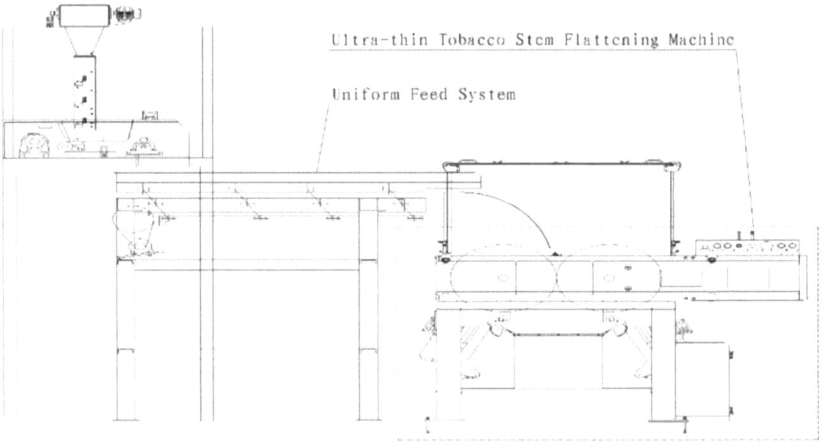

Fig. 1. High-capacity ultra-thin tobacco stem flattening system.

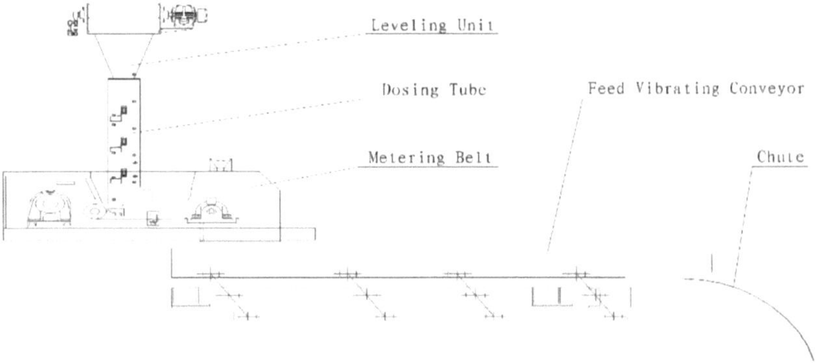

Fig. 2. Uniform feed system.

the precise flow control. Among them, the dosing tube has the functions of quantitative feeding, flow stabilization, material level monitoring and anti-blocking, and the metering belt adopts a volume flow control mode based on the gap parameters of the ultra-thin stem flattening machine. The stems are secondarily spread by a high-frequency and low-amplitude feed vibrating conveyor, and then transported to the flattening unit through a specially designed chute. The chute has a gradually changing curvature, which can effectively reduce the friction, avoid overlapping and form a uniform material curtain. Through the coordinated cooperation of the leveling unit, constant flow device, feed vibrating conveyor and chute, the uniform feed system can realize stable flow, uniform distribution and non-overlapping feeding of tobacco stems, providing a reliable material guarantee for the stable operation and good flattening effect of the subsequent ultra-thin stem flattening unit.

2.2 Ultra-Thin Stem Flattening Unit

The ultra-thin stem flattening unit, as the core executive unit of the whole system, has the functions of fine gap adjustment, constant pressure control, blockage monitoring and foreign body protection, and its specific structure is shown in Fig. 3.

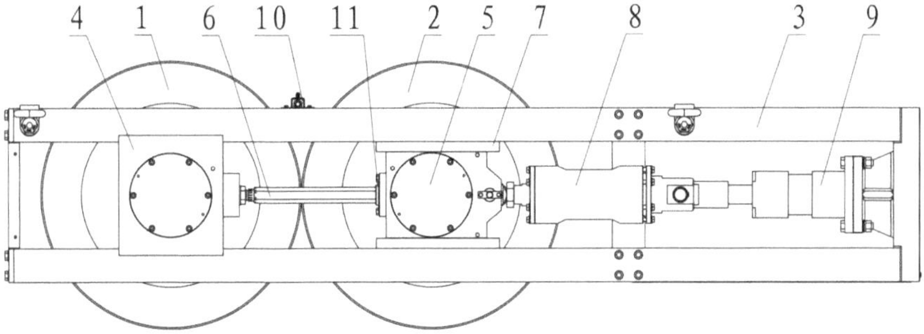

Fig. 3. Ultra-thin Tobacco Stem Flattening Unit. ***Note*** 1-Fixed Roller, 2-Adjustable Roller, 3-Frame, 4-Fixed Roller Bearing Housing, 5-Adjustable Roller Bearing Housing, 6-Gap Adjustment Unit, 7-Guide Rail, 8-Belleville Spring Assembly, 9-hydraulic cylinder, 10-Photoelectric Sensor, 11-Guard.

The ultra-thin stem flattening unit is composed of core components including a fixed roller (1), an adjustable roller (2), a frame (3), a fixed roller bearing housing (4), an adjustable roller bearing housing (5), a gap adjustment unit (6), a guide rail (7), a Belleville spring assembly (8), a hydraulic cylinder (9) and a photoelectric sensor (10). All components cooperate with each other to jointly complete the ultra-thin flattening operation of tobacco stems. The core process parameters of the unit can be precisely regulated by exclusive components: The ultra-thin flattening gap between the two rollers is determined by the thickness of the gasket in the guard (11) and fine screw adjustment, meeting the dual process requirements of ultra-thin stem flattening and conventional stem flattening; The flattening pressure of the pressure rollers is realized by adjusting the pre-compression amount of the Belleville spring in the Belleville spring assembly, and the elastic deformation of the Belleville spring is used to offset the pressure difference caused by uneven feeding to achieve constant pressure flattening; The photoelectric sensor can real-time monitor the material state between the two rollers. When blockage is detected, the hydraulic system will drive the adjustable roller to retreat quickly, effectively preventing material blockage. At the same time, the unit has a foreign body protection function. When metal foreign bodies enter between the two rollers, the hydraulic cylinder automatically releases oil to drive the adjustable roller to retreat, protecting the roller surface from damage. The overall structural design ensures the stable and noiseless operation of the flattening process, and realizes the flexible regulation of the ultra-thin flattening gap and flattening pressure, meeting the process requirements of ultra-thin stem flattening. The unit adopts large-specification pressure rollers, which increase the stress area of tobacco stems compared with traditional pressure rollers, ensure that the flattened tobacco stems undergo sufficient plastic deformation, and effectively avoid the rebound of stem flakes.

3 Experimental Verification and Results

To verify the actual processing effect of the high-capacity ultra-thin tobacco stem flattening system and determine the optimal process parameters adapted to the system, a series of experimental studies were carried out in this paper. The tobacco stems used in the experiments are conventional materials in the tobacco industry, and the in-depth stem conditioning technology is adopted in the stem conditioning process to minimize the hard core phenomenon of tobacco stems. The detailed experimental process and results are as follows.

3.1 Study on Influencing Factors of Stem Flattening Quality

The stem flattening quality is affected by a variety of process parameters, among which the moisture content before flattening, flattening gap and flattening flow rate are the core factors that directly affect the deformation effect, fiber integrity of tobacco stems and the uniformity of flattened stem flakes. In this section, single-factor experiments are carried out to explore the influence laws of each parameter on stem flattening quality, providing a basis for determining the optimal process parameters of the system.

Table 1. Effect of stem moisture content on flattening quality.

Moisture Content/%	>31	30 ± 0.5	<29
Presentation		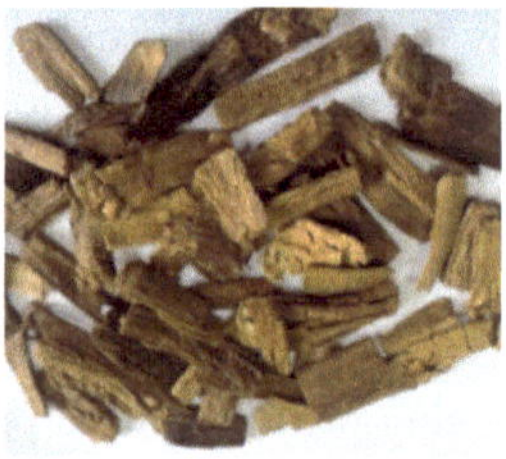	

Effect of Moisture Content before Flattening on Stem Flattening Quality. The moisture content of tobacco stems before flattening is a basic process parameter affecting the flattening effect, which directly determines the swelling degree of tobacco stem cellulose and the overall ductility of tobacco stems, and then affects the deformation effect in the flattening process. Experimental results are presented in Table 1, when the moisture content of tobacco stems before flattening is 30 ± 0.5%, the flattening deformation effect of tobacco stems is the best, with a complete and uniformly flaky morphology and intact fiber structure, which is the optimal moisture content condition for realizing ultra-thin flattening of tobacco stems.

Effect of Flattening Gap on Stem Flattening Quality. The flattening gap is a core structural parameter for realizing the "ultra-thin flattening" process, and its size directly determines the thickness and uniformity of flattened stem flakes. Under the fixed condition of tobacco stem moisture content of 30 ± 0.5%, this experiment explores its influence on stem flattening quality by setting different ultra-thin flattening gaps. Experimental results are presented in Table 2, when the flattening gap is adjusted to 0.3 mm, the epidermis and fiber structure of the flattened tobacco stems remain intact, presenting a uniform flaky shape with a thickness of 1.0 ± 0.1 mm, no missed flattening or crushing phenomenon, and a good flattening deformation effect, which can meet the processing requirements of filamentous cut stems.

Table 2. Effect of flattening gap on stem quality.

Flattening Gap/mm	0.4	0.3	0.2
Presentation			

Effect of Flattening Flow on Stem Flattening Quality. As a core index of high-capacity design, the flattening flow rate directly affects the production efficiency of the system. On the premise of ensuring the stem flattening quality, under the fixed conditions of tobacco stem moisture content of 30 ± 0.5% and flattening gap of 0.3 mm, this experiment changes the flattening flow rate by adjusting the frequency of the metering belt to explore the flattening effect under different flow rates. Results show that when the frequency of the metering belt before flattening is 6HZ, corresponding to a flattening flow rate of about 1250 kg/h, the feeding amount of tobacco stems matches the processing capacity of the stem flattening machine. The tobacco stems can fully contact with the pressure rollers, forming a flaky shape after flattening with an intact epidermis, no missed flattening and a good deformation effect. When the flow rate exceeds 2000 kg/h, overlapping feeding of tobacco stems is prone to occur, leading to an increase in missed flattening and a decrease in flattening uniformity.

3.2 Comparison of Stem Flattening Effects

To compare the processing effects of this system and traditional stem flattening equipment, under their respective optimal process parameters, the thickness uniformity of flattened stem flakes is taken as the core evaluation index, the thickness of flattened stem flakes in 50 batches is measured, and the statistical results are shown in Fig. 4. The characteristics of the two devices are shown in Table 3.

Test results show that the average thickness of flattened stem flakes by the conventional stem flattening machine is 1.35 mm, with a maximum of 1.81 mm and a minimum

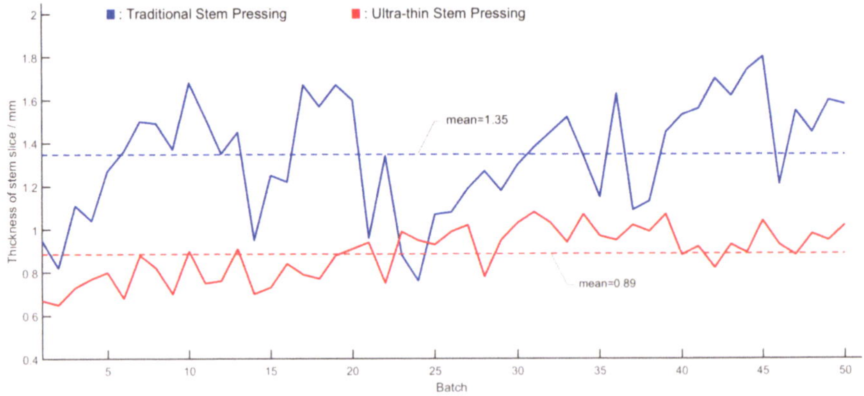

Fig. 4. Comparison of stem flattening results.

of 0.77 mm and a large standard deviation; The average thickness of flattened stem flakes by the ultra-thin stem flattening machine is 0.89 mm, with a maximum of 1.08 mm and a minimum of 0.65 mm and a standard deviation of only 0.014, with small thickness fluctuation and good uniformity, which fully meets the technical process requirements of filamentous cut stem processing.

Table 3. Equipment specification comparison.

Index	Conventional stem flattening machine	Ultra-thin stem flattening machine
Roll specification/(mm × mm)	1000 × 600	1400 × 800
Minimum pressing gap/mm	0.6	0.3
Rated capacity/(kg · h^{-1})	1500 ~ 2000	1200 ~ 3000
Feeding control mode	Vibrating conveyor and deflector	Uniform feed system

3.3 Effect of Different Stem Cutting Thickness on Cut Stem Quality

Stem cutting is the main process for the morphological transformation of tobacco stems into cut stems, and the matching between stem cutting thickness and flattening thickness directly affects the filamentation ratio, integrity rate and breakage rate of cut stems. To determine the optimal stem cutting process parameters for filamentous cut stems, on the basis of the above optimal stem flattening process parameters (moisture content of 30 ± 0.5%, flattening gap of 0.3 mm, flattening flow rate of 1250 kg/h), multiple batches of tobacco stems are organized for the same preprocessing, and three stem cutting thickness gradients of 0.10 mm, 0.09 mm and 0.08 mm are set to explore their influence on the morphology and physical quality of cut stems. The experiment takes the morphology, integrity rate and breakage rate of dried cut stems as the core evaluation indexes, and the results are shown in Table 4.

Table 4. Effect of cutting thickness on stem silk quality.

Cutting thickness/mm	0.10	0.09	0.08
Post-drying morphology	flaky majority	fine and even	severely damaged
Integrated/%	67.09	63.6	61.07
Broken/%	4.36	5.95	7.31
Bulkiness/($cm^3 \cdot g^{-1}$)	6.9	6.78	6.66

Experimental results show that when the stem cutting thickness is 0.10 mm, the width of cut stems is too large with existing flaky cut stems and a poor filamentation effect; When the stem cutting thickness is 0.09 mm, the dried cut stems are uniformly slender and filamentous without flaky cut stems. Although the integrated filamentation rate decreases slightly, the broken filamentation rate is still within a reasonable range and the bulk value meets the cigarette production requirements, which is the optimal stem cutting thickness; When the stem cutting thickness is reduced to 0.08 mm, the cut stems suffer from serious breakage and a sharp drop in integrated filamentation rate, losing practical application value. Therefore, the optimal stem cutting thickness adapted to this system is determined to be 0.09 mm.

3.4 Evaluation of Finished Strand-like Cut Stems

To comprehensively verify the large-scale practical application effect of this system, a comprehensive quality verification of the finished filamentous cut stems is carried out from three dimensions: compliance of process parameters in the processing process, stability of physical indexes of finished products and compatibility of morphological indexes. Meanwhile, its morphological indexes are compared with those of conventional cut stems processed by traditional technology and cut tobacco leaves.

Process Standard Evaluation. The key process parameters in the whole processing process of filamentous cut stems are recorded and counted in real time, including hot air temperature, moisture after addition, post-drying moisture, etc. The process standard values are compared with the actual average values to verify the stability of the processing process, and the results are shown in Table 5.

Table 5. Comparison of compliance with key process parameters.

Process	Thin plate temperature/°C	Hot air temperature /°C	Moisture after addition/%	Post-drying moisture /%	Flavored moisture /%
Standard	140 ± 10	140 ± 10	32.5 ± 1.0	12.9 ± 0.5	12.7 ± 0.5
Actual	133.2	140.1	32.76	12.89	13.16

It can be seen from Table 5 that in the processing process of filamentous cut stems, the actual average values of the key process parameters in each key procedure all meet the requirements of process standard values without obvious deviation; The flow rate is stable in the production process, and the key equipment such as stem flattening, stem cutting and cut stem drying operate normally without material blockage, caking, equipment failure and other problems, which verifies the operational stability and process controllability of the system and its suitability for large-scale production.

Physical Quality Evaluation. The physical indexes of 10 batches of finished filamentous cut stems produced on a large scale are detected, including integrity rate, breakage rate, bulk value, moisture content and filamentation rate. The results indicate that all physical indicators have good stability and meet the requirements of the tobacco industry process standards for filamentous cut stems: The average integrity rate reaches 75.66%, while the average breakage rate is only 2.45%; The average bulk value is 6.81cm^3/g, maintaining the good bulkiness of cut stems; The average filamentation rate is as high as 90.6%, realizing a high filamentation ratio of cut stems; At the same time, the average moisture content is 12.83%, meeting the raw material moisture content requirements for cigarette production. The results verify that the finished filamentous cut stems processed by this system have stable physical indexes and a high filamentation ratio, with good application value.

Evaluation of Morphological Indicators. The morphological indexes of the finished filamentous cut stems processed by this system are compared with those of cut stems processed by traditional technology and cut tobacco leaves, including average width, standard deviation, coefficient of variation and uniformity coefficient. The results are shown in Table 6, and the morphological comparison is shown in Fig. 5.

Table 6. Parameter comparison among finished filamentous cut stem, conventional cut stem, and cut tobacco leaves.

Category	Average width/mm	Standard deviation/mm	Coefficient of variation	Uniformity coefficient
Conventional cut stems	1.9	0.443	23.31	5.08
Filamentous cut stems	1.05	0.25	23.26	5.28
Cut tobacco leaves	0.93	0.094	10.14	11.91

A. conventional cut stems B. filamentous cut stems C. cut tobacco leaves

Fig. 5. Morphology comparison among finished strand-like cut stem, regular cut stem, and cut tobacco.

The following conclusions can be drawn from Table 5 and Fig. 5:

The filamentous cut stems produced by the ultra-thin stem flattening process have an average width of 1.05 mm, which is much smaller than 1.9 mm of the conventional, with a smaller standard deviation, indicating that the width of filamentous cut stems is more uniform and the flakiness problem is effectively solved;

The average width of filamentous cut stems is close to 0.93 mm of cut tobacco leaves, and their morphology is more similar to that of cut tobacco leaves, which improves the compatibility between cut stems and cut tobacco leaves;

The uniformity coefficient of filamentous cut stems is 5.28, which is higher than 5.08 of conventional cut stems, indicating that the ultra-thin stem flattening process helps to improve the morphological uniformity of cut stems, laying a foundation for the uniform blending of cut stems and cut tobacco leaves in the subsequent process.

The multi-dimensional evaluation results show that the finished filamentous cut stems processed by this system have significant improvements in morphological uniformity and filamentation ratio, with lower average width and standard deviation than conventional cut stems, and their morphology is more similar to that of cut tobacco leaves, fully meeting the compatibility requirements of slim cigarette production. All the above experimental results fully verify the design rationality and application feasibility of the high-capacity ultra-thin tobacco stem flattening system, providing a solid experimental basis for the summary of the research achievements in this paper.

4 Conclusion

Aiming at the technical problem that traditional stem flattening equipment cannot meet the process requirements of ultra-thin stem flattening and restricts the processing of filamentous cut stems, the high-capacity ultra-thin tobacco stem flattening system presented in this paper, through systematic mechanical structure design and a series of experimental verifications, has achieved the following core research achievements and industrial application values:

A tobacco stem uniform feed system is innovatively developed, which effectively solves the problems of uneven feeding and tobacco stem overlapping of traditional stem flattening equipment, ensures the uniformity and quantitative control of incoming materials from the source, and lays a solid material foundation for stabilizing the flattening thickness and realizing the high-capacity ultra-thin flattening technology.

A large-specification ultra-thin stem flattening unit with constant pressure flattening is developed, which breaks through the ultra-thin stem flattening bottleneck of traditional equipment. It significantly improves the flattening uniformity, reducing the defective tobacco stems after flattening by 86.3%. And it provides core equipment support for the realization of the "ultra-thin flattening and thin cutting" process.

The optimal process parameters for the system have been determined and verified through experiments. Under these parameters, the system realizes high-capacity ultra-thin stem flattening, and its processing effect is significantly better than that of traditional stem flattening machines, with high uniformity of flattened stem flakes and intact fiber structure.

The finished filamentous cut stems processed by this system have excellent morphology and uniformity, and are more similar to cut tobacco leaves in appearance, with a good effect in practical production and application. Not only the quality of cut stems and the filament yield of tobacco stems have been improved, but the application proportion of cut stems has increased to 7%. At the same time, the stable sensory quality of cigarettes can also be guaranteed.

In order to address the issue of material sticking on the roller and enhance the continuous and stable operation capability of the equipment, further research will be conducted in the following three aspects: modification of the roller surface, optimization of the cleaning device, and coordinated regulation of process parameters.

References

Deng, J.Q., Yao, Z., Pang, T.H.: Research on the influence of different cutting parameters on the morphology of cut tobacco stems. J. Light. Ind. Technol. **40**(6), 6–8 (2024) https://kns.cnki.net/kcms2/article/abstract

He, C.: Research progress on the influence of cut tobacco shape on the comprehensive quality of cigarettes. Process. Agric. Prod. **1**, 265–268 (2026) https://doi.org/10.19754/j.nyyjs.20260130046

Ji, X., Li, X., Yao, E.M.: Effects of filamentation process parameters on morphological distribution and effective utilization rate of cut stem. Hubei Agric. Sci. **55**(1), 96–99 (2016) https://doi.org/10.14088/j.cnki.issn0439-8114.2016.01.027

Lu, Y.X., Lei, Z., Li, J.Y.: A pre-pressing process for tobacco stems in threshing and redrying line and quality analysis of cut stems produced. Tob. Sci. Technol. **58**(8), 77–84 (2025) https://doi.org/10.16135/j.issn1002-0861.2024.0904

Wang, W., Li, X., Cao, X.Y.: Study on the influence of different process parameters on the quality of cut tobacco stems. Process. Agric. Prod. **6**, 13–18 (2025) https://doi.org/10.16693/j.cnki.1671-9646(X).2025.06.003

Zhao, Y.C., Zou, Q., Liao, X.X.: Selection and optimization of processing technologies for microwave expansion of cut stem. Tob. Sci. Technol. **49**(12), 60–70 (2016) https://doi.org/10.16135/j.issn1002-0861.2016.0023

Zheng, X., Guan, A.Z., Liu, W.F.: A low-temperature pretreatment process for tobacco stems and quality characteristics of the prepared cut stems. Tob. Sci. Technol. **55**(8), 65–71 (2022) https://doi.org/10.16135/j.issn1002-0861.2022.0160

Zhou, L.J., Wang, X., Zhu, B.: The analysis of physical quality of multi specification cigarettes blended with different forms of cut stem. Farm Prod. Process. **10**, 76–81 (2024) https://doi.org/10.16693/j.cnki.1671-9646(X).2024.20.016

Zhou, L.J., Zheng, L.W., Li, H.T.: Effects of stem rolling process and stem cutting process on filamentation characteristics of flake cut stem. Acta Tabacaria Sinica. **26**(5), 47–53 (2020) https://doi.org/10.16472/j.chinatobacco.2020.067

Zhu, B., Shan, K., Guan, X.: Study on processing procedure applicable to produce cut stem with roller cutting. Tob. Sci. Technol. **55**(1), 84–90 (2022) https://doi.org/10.16135/j.issn1002-0861.2021.0463

Design Method of Sphere Parameters for Passive Activation Device of Independent Heat Exchanger in Sodium-Cooled Fast Reactor

Zhenjia Chen[1], Jiong Zhang[1], Yichen Li[1], Xiangyang Liu[2](✉), Tudi Zhang[2], and Yaquan Xu[2]

[1] China Institute of Atomic Energy, Beijing, China
[2] School of Energy and Power Engineering, Xi'an Jiaotong University, Xi'an, China
lxyyang@mail.xjtu.edu.cn

Abstract. As the switch for the residual heat removal system of the independent heat exchanger in an integrated sodium-cooled fast reactor, the passive start-up device (PSD) is a critical component ensuring reactor safety in the event of a malfunction. The weight and dimensions of the sphere inside the PSD determine its ability to open and close. By analyzing the relationship between the forces acting on the sphere, the pressure loss of the device, and the inlet velocity, a design method for the PSD sphere parameters was established. To verify the reliability of this design method, a flow field simulation model of the PSD was built. The forces acting on the sphere and the pressure loss of the device under different inlet velocities were simulated and the variation patterns were analyzed. The simulation results show that the sphere parameters determined using this design method can meet the opening and closing requirements of the passive start-up device.

Keywords: Sodium-cooled fast reactor · Passive start-up device · Design methodology · Force analysis · Numerical simulation

1 Introduction

Nuclear energy, as a clean and efficient energy source, plays a significant role in addressing global climate change and ensuring energy security (Wang et al. 2023). However, the Fukushima nuclear accident has re-emphasized the importance of nuclear safety, shifting the focus of Generation IV nuclear energy system development towards passive safety technologies capable of operating without external power during extreme accident scenarios, such as Station Blackout (Cahalan et al. 2004). Among various Generation IV reactor concepts, the Sodium-cooled Fast Reactor (SFR) has garnered attention due to its superior breeding performance and inherent safety (Futagami et al. 2025). As the final barrier in the defense-in-depth strategy for SFRs, the reliability of the Passive Decay Heat Removal System (PDHRS) is directly linked to the capability of removing decay heat and preventing core meltdown post-accident (Xue 2025). Consequently, it is imperative to design a decay heat removal loop with robust passive opening/closing mechanisms (Peide et al. 2025) to ensure reactor safety.

T. Mankovits and N. Trisovic (Eds.): MDS 2026, LNME 2026.
https://doi.org/10.1007/978-981-92-2440-1_12

Scholars have carried out extensive research on PDHRS: Meng et al. (2015) optimized the tube bundle structure of passive heat exchangers to improve their heat transfer capacity; Xiao-ming et al. (2012) identified the main failure mode of AP1000 PDHRS via fault tree analysis. Meanwhile, the U.S. and France are advancing relevant reactor projects based on their respective technical experience (Azih 2025; Wang et al. 2021), and research on passive activation components is still critical—Baluyev et al. (2015) developed a hollow sphere check valve for BN-1200's emergency heat removal system and completed scaled hydraulic tests, and Pakholkov et al. (2017) verified the passive safety performance of the BN-1200 system through full-scale and scaled model experiments on check valves.

However, existing design research on PSD is largely based on simplified assumptions of steady-state flow fields and constant physical properties. In reality, the temperature difference between the cold and hot pools in SFR fluctuates drastically between 350 and 550 °C before and after an accident (Yoo et al. 2016). Although the above researches have made progress in system-level design, there is little research on the micromechanical properties and parametric design methods of key actuators—passive start-up device.

The opening and closing of the PSD depend on the forces acting on the sphere, which are intrinsically determined by the sphere's mass and dimensions. Therefore, the determination of sphere parameters is the key to structural design. To date, there is a lack of reported methodologies specifically for the parameter design of spheres in such devices under these complex conditions. Addressing this gap, this paper proposes a parameter design method for the sphere of a PSD in a pool-type integrated SFR, incorporating force analysis that accounts for variable properties and transient effects. The reliability of this method is subsequently validated through numerical simulation.

2 Structure and Parameter Design Model of the Passive Start-Up Device

2.1 The PSD Structure

The PSD is mounted at the inlet of the independent heat exchanger (Fig. 12.1). It stays closed during normal reactor operation: the main pump runs to drive high sodium flow in the primary loop, and the combined upward dynamic pressure and buoyancy of the fluid push the sphere to the upper base, sealing the exchanger channel. In case of reactor malfunction or emergency, the main pump shuts down, leading to a sharp drop in flow rate and upward fluid force; when their sum with buoyancy is less than the sphere's weight, the sphere falls to the lower base, opening the channel and enabling the establishment of natural circulation (Zinkle and Was 2013).

2.2 Force Analysis

In the design of the PSD, a rigorous force analysis of the sphere is essential. Assuming the sphere is at a specific position with instantaneous velocity $v = 0$, the primary forces include gravity F_g, buoyancy F_b, and drag force F_d, F_g acts vertically downward, while buoyancy F_b and drag F_d act vertically upward, consistent with the upward flow direction of the working fluid during both closing and opening phases.

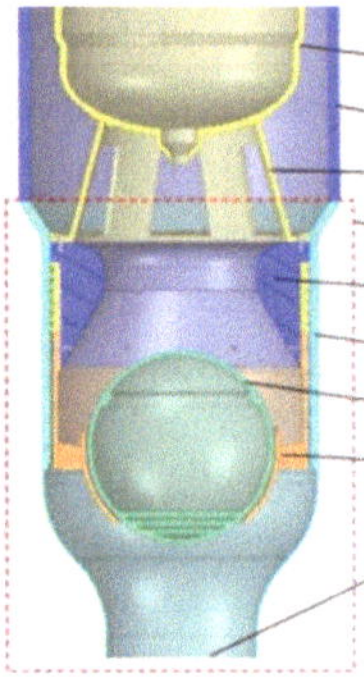

Fig. 12.1. Schematic diagram of PSD

Considering the temperature fluctuations in SFR, the physical properties of liquid sodiumare functions of temperature. The force analyses are modified as follows:

(1) Closing Criterion: When $F_d + F_b > F_g$, the ball moves upward under the combined action of various forces, and the device closes;
(2) Opening Criterion: When $F_d + F_b > F_g$, the position of the sphere remains unchanged under the combined action of various forces, and the device remains in the closed state; when $F_d + F_b < F_g$,the sphere begins to fall, and the device opens.

2.3 Design Process

The process involves: The stroke l divided into 5 characteristic positions y_i (0,80,160,240,290 mm). For each position, different inlet velocities are applied at the inlet. Based on the calculation results, the pressure loss $\Delta p_{i,n} = f(v_n, y_i)$ and forces $F_{\mathrm{i,n}} = (F_{\mathrm{b}} + F_{\mathrm{d}})_{\mathrm{i,n}}$ on the sphere at different positions and inlet velocities can be obtained. The relationship between the pressure loss of PSD and the force on the sphere at that location and the inlet flow velocity can be obtained: $\Delta p_{y_{\mathrm{i}}} = f(v)$, $F_{y_{\mathrm{i}}} = g(v)$.

The sum of the above two equations is the total pressure loss of the independent heat exchange system, which can be expressed as Δp_s, and its expression is

$$\Delta p_s = \Delta p_{1,1} + f(v) = f_s(v) \tag{12.1}$$

Substituting the opening pressure difference $\Delta p_1 = \Delta p_s$ from the design basis into the above formula, we can obtain the inlet velocity of the device at different positions of the sphere during the opening or closing process $v_{y_{\mathrm{i}}}(\Delta p_1)$.Then, substituting $v_{y_{\mathrm{i}}}(\Delta p_1)$ into Eq. (12.3) yields the force $F_{y_{\mathrm{i}}}(\Delta p_1)$ acting on the sphere at different positions. Similarly, by applying the same treatment to the closed pressure difference Δp_2, we can obtain $v_{y_{\mathrm{i}}}(\Delta p_2)$ and $F_{y_{\mathrm{i}}}(\Delta p_2)$.

The design goal of the PSD is to achieve a suitable sphere weight within a given structural configuration. Combining the requirements for opening and closing, the sphere weight must satisfy:

$$\max\left[F_{y_{\mathrm{i}}}(\Delta p_1)\right] < F_g < \min\left[F_{y_{\mathrm{i}}}(\Delta p_2)\right] \tag{12.2}$$

2.4 Governing Equations and Boundary Conditions

The simulation analysis process is based on Ansys Fluent to calculate and analyze the flow process of water and sodium fluids within the PSD (Matsson 2025). The main governing equations are as follows, Mass Conservation Equation:

$$\frac{\partial \rho}{\partial t} + \nabla(\rho U) = 0 \tag{12.3}$$

Momentum Conservation Equation:

$$\frac{\partial(\rho U)}{\partial t} + \nabla(\rho UU) = -\nabla p + \mu \nabla^2 U + F + \rho g \tag{12.4}$$

where ρ is the fluid density (866 kg/m^3); t is time; U is fluid velocity; p is fluid pressure; F is body force; g is gravitational acceleration; μ is fluid viscosity (0.0002966 Pa·s).

The standard k-ε turbulence model was selected to calculate the flow characteristics. The inlet boundary condition is velocity inlet, and the outlet boundary condition is pressure outlet, consistent with the actual working state (Blazek 2005). Second-order upwind scheme is used to discretize the governing equations. The SIMPLE algorithm is used to solve the pressure-velocity coupling, and the residuals of the continuity and momentum equations are controlled within 10^{-5} to ensure convergence.

3 Design Model Verification

To verify the reliability of the proposed design model, simulation calculations were performed on a specific case. The inlet diameter of the device $d_1 = 576$ mm, the sphere diameter $d_2 = 780$ mm, and the sphere stroke $l = 293$ mm. The design basis is the cold-hot pool pressure difference Δp_i under different reactor operating conditions. When $\Delta p_1 = 31250$ Pa, the device close from the open state; when $\Delta p_1 = 2812.5$ Pa, the device open from the closed state. Since the opening and closing times are short, the pressure difference is assumed constant during the movement. The relationship between the device pressure loss and the inlet velocity v is: $\Delta p_{1,1} = 1240.3 \times v^2 + 4.25$.

$\Delta p_{1,1}$ is the relationship between the flow resistance of the independent heat exchanger and the inlet velocity of the device; v is the inlet velocity of the device (m/s).

3.1 Grid Independence Verification

Grid independence verification is performed at position $y_i = 0$ mm with an inlet velocity of 3 m/s. The total pressure loss and sphere force (sum of drag and buoyancy) are used as criteria. By calculating four scenarios with mesh sizes of 10, 15, 20, and 25 mm, the total pressure loss and spherical force deviation obtained at 20 mm fall within 3%. So the mesh size of 20 mm is selected for subsequent calculation.

3.2 Analysis of Simulation Results

Taking the sphere center position $y_i = 0$ mm as an example to explain the data processing method, the calculation results are summarized in Table 12.1, where the data for 0 m/s are theoretical values. Buoyancy is calculated independently based on $F_b = \rho g V$, therefore remains constant at different velocities. The drag force is caused by fluid motion (comprising pressure drop resistance and frictional resistance), as determined by CFD simulation. The total force is obtained by adding the drag force and buoyancy.

Table 12.1. The force on the sphere and the pressure loss of the device ($y_i = 0$ mm)

Import speed m/s	Drag force/N	Buoyancy/N	Total force/N	Pressure loss/Pa
0	0.00	2110.85	2110.85	0.00
1	136.62	2110.85	2247.47	588.41
2	462.38	2110.85	2573.23	2383.48
3	985.55	2110.85	3096.40	5249.04
4	1664.21	2110.85	3775.06	9473.41
6	3751.78	2110.85	5862.63	21258.04
8	6646.98	2110.85	8757.83	37795.96

The force exerted by the fluid upon the sphere during their relative motion, which is the drag force F_d mentioned above. Reynolds number is defined as followed:

$$\mathrm{Re} = \frac{\rho_l u_l d_s}{\eta_l} \tag{12.15}$$

where: ρ_l is the fluid density (997.05 kg/m^3); u_l is the velocity based on the lower outlet area; d_s is the sphere diameter (0.576 m); η_l is the fluid dynamic viscosity (taken here as the viscosity of water at ambient temperature, 0.00089 Pa·s).

Drag coefficient can be calculated using the following equation:

$$C_d = \frac{2F_d}{A \times \rho_l u_l^2} \tag{12.6}$$

where A is the projected area of the sphere. C_d is the sphere's drag coefficient, which is a function of the Reynolds number, namely $C_d = f(\mathrm{Re})$.

The relationship between the calculated drag coefficient and Reynolds number is shown in Fig. 12.2, it is readily apparent that as the Reynolds number increases, the drag coefficient eventually stabilises. The flow then enters the self-modelling region.

Based on Table 12.1, the relationship between sphere total force/device pressure loss and velocity can be plotted, as shown in Fig. 12.3. The pressure loss of PSD are as follows:

$$\Delta p_s = \Delta p_{1,1} + f(v) = 1830.94 \times v^2 - 0.72 \tag{12.7}$$

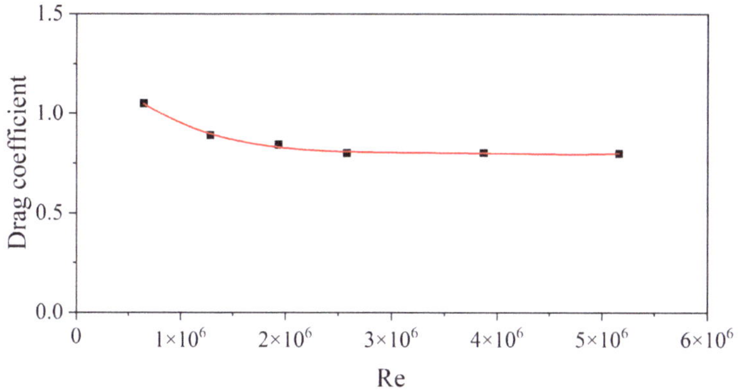

Fig. 12.2. Variation of drag coefficient with Reynolds number

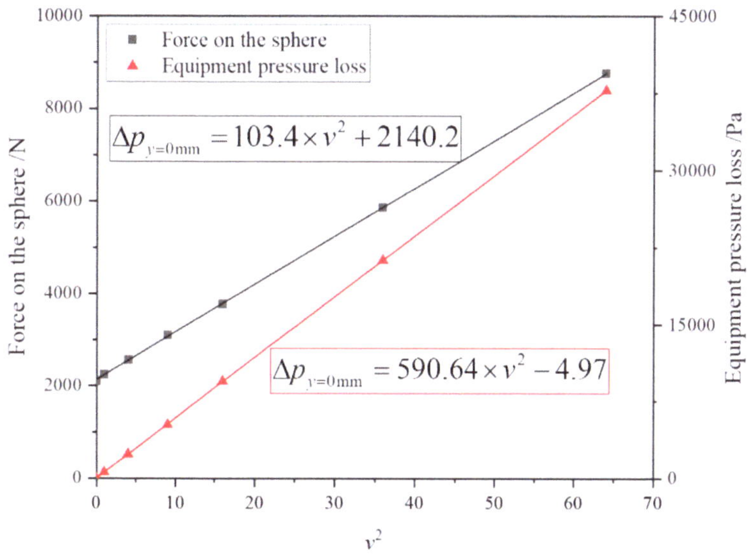

Fig. 12.3. The relationship between sphere total force/device pressure loss and velocity (y = 0 mm)

When $\Delta P_s = \Delta P_1$(2812.5Pa) or ΔP_2(31250Pa), substituting into the derived equations yields the inlet velocities at different locations. Substituting these velocities into the force equation yields total sphere forces. The results are summarized in Table 12.2.

According to Table 12.2, during the opening process, the inlet velocity increases due to the increasing flow area between the sphere and the upper seat, while the sphere force decreases, ensuring smooth opening; For closing, the inlet velocity decreases due to the decreasing flow area, while the sphere force increases, ensuring smooth closing.

The sphere weight should satisfy:

$$3641.11\text{N} < F_g < 3905.05\text{N} \tag{12.8}$$

Table 12.2 The inlet velocity and force on the sphere

Center of the ball/mm	$\Delta P_S = 2812.5$ Pa		$\Delta P_S = 31250$ Pa	
	Import speed/m·s^{-1}	Force on sphere/N	Import speed/m·s^{-1}	Force on sphere/N
0	1.24	2299.07	4.13	3905.05
80	1.25	2417.79	4.17	5440.54
160	1.14	2550.39	3.92	6806.99
240	0.91	3021.77	3.05	12251.27
290	0.51	3641.11	1.68	18742.59

Figure 12.4 shows the non-monotonic variation in inlet velocity with sphere position: inlet velocity rises slightly when the centre of the sphere moves between 0 and 80 mm (as the enlarged flow area weakens throttling), but drops sharply after 80 mm (as the sphere enters the constricted region of the channel and the flow area shrinks drastically, intensifying throttling). The 80 mm position marks the transition point of the channel's constriction, where the suction effect of the constriction and the obstruction effect of the sphere reach equilibrium. This leads to a transient inlet velocity peak at this position.

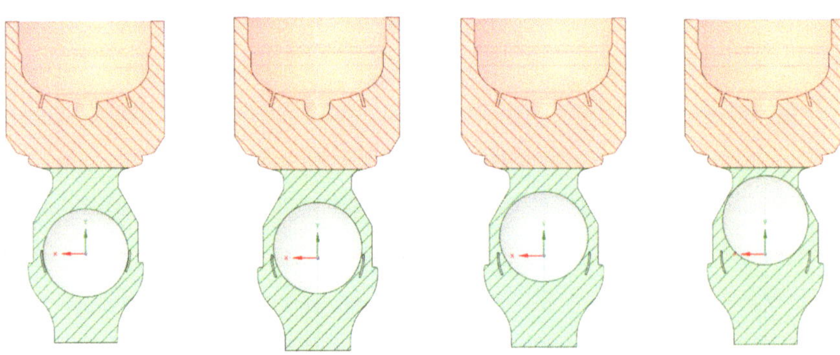

(a) $y_i = 0$ mm (b) $y_i = 80$ mm (c) $y_i = 160$ mm (d) $y_i = 290$ mm

Fig. 12.4. Cross-sectional views of the sphere at different positions

It can be seen that when the weight of the sphere conforms to formula (12.8), the expected goal can be achieved under the condition of meeting the design basis. In order to achieve the smooth opening of the device, the weight of the sphere can be taken as 1.05 times $\max\left[F_{y_i}(\Delta p_1)\right]$,calculated as 3823.2 N (approx. 389.7 kg), it meets the requirement. Overall, the proposed design method allows for the analysis of forces at different positions and the determination of sphere mass satisfying the design basis.

4 Conclusion

This paper proposes a design model for the sphere parameters of a PSD suitable for the decay heat removal system of a sodium-cooled fast reactor. The method determines sphere parameters based on the force conditions required for opening and closing. A numerical simulation model of the device under specific conditions was established. Based on the simulated force and pressure loss results, the sphere weight and dimensions required for the device to open and close were obtained through the design model, validating its feasibility. The results show that the device pressure loss and sphere force are linearly related to the square of the inlet velocity. For a device with an inlet diameter of 576 mm and a stroke of 293 mm, under cold-hot pool pressure differences of 31250 Pa (closing) and 2812.5 Pa (opening), a sphere with a mass of 389.7 kg and a diameter of 780 mm can satisfy the opening and closing requirements.

Of course, this study has limitations as the established pressure loss relationship is based on the assumption of incompressible fluid with constant physical properties, which may deviate under extreme thermal conditions. Specifically, high temperatures will induce fluid compressibility effects that increase actual pressure loss and alter material properties (e.g., viscosity) to change flow Reynolds number and friction factor, even modifying turbulent flow structures. Thus, the findings should not be blindly extrapolated to extreme thermal conditions, and subsequent research will adopt fluid-thermal-solid coupling methodologies to consider variable material properties and refine the pressure loss equation.

References

Azih, C.: Development of experimental and modelling capabilities and tools at Canadian nuclear laboratories for investigations on inherent and passive safety designs of advanced reactor concepts (STI/PUB/2108) (2025). https://inis.iaea.org/records/3p0ea-yr746

Baluyev, D.E., Gusev, D.V., Meshkov, S.I., Nikanorov, O.L., Osipov, S.L., Rogozhkin, S.A., Rukhlin, S.V., Shepelev, S.F.: Study of functional characteristics for safety system check valve using scaled model. Nucl. Energy Technol. **1**(2) (2015). https://doi.org/10.1016/j.nucet.2015.11.020

Blazek, J.: Computational fluid dynamics: principles and applications, p. 491. Elsevier (2005). https://doi.org/10.1016/C2013-0-19038-1

Cahalan, J., Hahn, D., Chang, W., Kwon, Y.M.: Passive safety optimization in liquid-sodium cooled reactors. OSTI OAI (U.S. Department of Energy Office of Scientific and Technical Information) (2004). http://www.osti.gov/scitech/biblio/925169

Futagami, S., Kubo, S., Sofu, T., Gauthe, P.: Development of safety design criteria and safety design guidelines for generation IV sodium cooled fast reactors (STI/PUB/2108) (2025). https://inis.iaea.org/records/h2tsa-q3x98

Matsson, J.E.: An Introduction to Ansys Fluent 2025, p. 953. SDC Publications (2025)

Meng, W., Wei, C., Yan-shen, L.V.: Research on optimal design of passive residual heat removal heat exchanger. At.Ic Energy Sci. Technol. **49**(3) (2015). https://doi.org/10.7538/yzk.2015.49.03.0455

Pakholkov, V.V., Anfimov, A.M., Baluev, D.E., Nikanorov, O.L., Osipov, S.L., Rogozhkin, S.A., Rukhlin, S.V., Shepelev, S.F.: Integrated R&D to validate innovative emergency heat removal system for BN-1200 reactor. In: International conference on fast reactors and related fuel cycles: Next generation nuclear systems for sustainable development (FR17). Programme and papers (2017). https://inis.iaea.org/records/k2mc6-nkf43

Peide, Z., Xisi Z., Yun H.U., Weiwei F., Lin L.I.U., Han Y.A.N., Fenglong W., Shixi W., Qiang Z., Xinyu L.I., Yingyun S., Fangyuan X.U.E.: Research on general design requirements of integrated fast reactor. At.Ic Energy Sci. Technol. **59**(S1) (2025). https://doi.org/10.7538/yzk.2025.youxian.0122

Wang, Q., Guo, J., Li, R., Jiang, X.: Exploring the role of nuclear energy in the energy transition: a comparative perspective of the effects of coal, oil, natural gas, renewable energy, and nuclear power on economic growth and carbon emissions. Environ. Res. **221**, 115290 (2023). https://doi.org/10.1016/j.envres.2023.115290

Wang, X., Kuang, B., Liu, P., Hou, J., Hu, W., Li, Y.: Performance test of conceptually designed hydraulically suspended passive shut down subassembly for CFR600. Prog. Nucl. Energy **140**, 103906 (2021). https://doi.org/10.1016/j.pnucene.2021.103906

Xue, S.: Review of conceptual design and fundamental research related to the passive residual heat removal system in molten salt reactors (2025). https://doi.org/10.1016/j.nucengdes.2025.114275

Xiao-Ming, Q., Dao-Gang, L.U., Yu, Y.U.: Design and optimization of AP1000 passive residual heat removal system based on fault tree analysis. At.Ic Energy Sci. Technol. **46**(8) (2012). https://doi.org/10.7538/yzk.2012.46.08.0927

Yoo, J., et al.: Overall system description and safety characteristics of prototype gen IV sodium cooled fast reactor in Korea. Nucl. Eng. Technol. **48**(5), 1059–1070 (2016). https://doi.org/10.1016/j.net.2016.08.004

Zinkle, S.J., Was, G.S.: Materials challenges in nuclear energy. At.Ic Energy Sci. Technol. **61**(3), 735~758 (2013). https://doi.org/10.1016/j.actamat.2012.11.004

Research on in-Situ Size Measurement Method of Filament Winding Workpiece

Xinyi Zhang(✉) and Dongliang Xu

School of Mechanical and Electrical Engineering, Wuhan University of Technology, Wuhan 430070, China
19907909222@163.com

Abstract. The geometric dimension accuracy of filament wound composite workpiece is the key factor to determine its performance. Aiming at the problems of low efficiency of traditional detection methods and difficulty in taking into account the measurement range and accuracy of a single sensor measurement scheme, this paper proposes a high-precision in-situ size measurement method based on the fusion of depth camera and laser displacement sensor. Firstly, the Intel RealSense D435 depth camera is used to quickly scan the machine tool shaft and the workpiece clamped on the machine tool shaft to obtain the initial point cloud model. Then, through principal component analysis (PCA) and point cloud symmetry analysis algorithm, the workpiece is intelligently judged whether it is a rotary structure. If it is determined to be a rotary body, the BLG-250 N-485 laser sensor of BOE is controlled to perform precise scanning along the direction of the workpiece rotation axis. By synchronously collecting the laser ranging value and the sensor displacement, the workpiece contour point cloud with sub-millimeter accuracy is reconstructed. The experimental results show that this method can effectively identify the type of workpiece and realize high-precision size parameter refinement of the rotary workpiece. Compared with the single visual reconstruction method, the size measurement accuracy is improved to 0.01 mm level, which provides an effective technical solution for the in-size measurement of filament-wound products.

Keywords: Point cloud · Dimensional measurement · Fiber winding · Depth camera

1 Introduction

Filament winding is a processing method of wrapping the fiber on the mandrel. Usually, a specific shell structure is formed by winding high-strength fibers, such as glass fiber, Kevlar fiber, carbon fiber, etc. on a certain shape of mandrel. Winding products can become axisymmetric parts and non-axisymmetric parts according to the geometric structure. The surface of the axisymmetric workpiece is a rotating surface, and its common components are pressure vessels, pipes, cones, gas storage tanks, etc. Non-axisymmetric workpieces are mainly shaped pressure vessels, pipelines, ring shells, etc. Filament winding products are widely used in aerospace, petrochemical, energy storage

T. Mankovits and N. Trisovic (Eds.): MDS 2026, LNME 2026.
https://doi.org/10.1007/978-981-92-2440-1_13

and transportation, sports and other fields, especially in the weight-to-strength ratio sensitive sectors, such as rocket engine shells, missiles, space shuttles and satellite delivery systems, showing its unparalleled superiority (Zu 2007). The quality of the workpiece covers multiple dimensions, of which dimensional accuracy is one of the basic indicators (Zhong 2025). With the rapid development of China 's manufacturing industry, the traditional contact measurement of workpiece scale such as manual measurement, using profilometer, laser measuring instrument, three coordinate measuring machine and other measuring tools, although the measurement accuracy can reach micron level, but there is a large reading error, low efficiency, complex operation, high instrument cost (Lin 2024; Bao 2023), so real-time online non-contact measurement technology has become an important demand of modern production line (Zhang 2023). Machine vision measurement technology based on optics, image processing and artificial intelligence has been widely used in the field of industrial measurement because of its advantages of non-contact, high efficiency and high precision (Shi 2023; Dai 2025). In the current mainstream non-contact measurement methods, binocular vision is characterized by the ability to identify, locate, and three-dimensionally reconstruct the target object, and its application to production can effectively reduce costs and improve work efficiency (Yang 2025). However, the measurement accuracy of binocular vision is limited by its physical principles and hardware. There are usually problems such as low accuracy, vulnerability to ambient light interference, and poor effects on reflective or light-absorbing surfaces. Point laser measurement technology is a method of measuring objects or scenes by laser beam, and its essence is laser ranging. The laser beam emitted by the laser is irradiated on the surface of the object, and after reflection, it converges on the receiver. The position of the imaging spot will change with the change of the distance of the measured object surface (Li 2023). Point laser measurement can achieve relatively high accuracy. However, there are also problems that the measurement is easily affected by the roughness, temperature and vibration of the workpiece to be measured, and the point cloud data processing is difficult (Liu 2021). In this paper, an in-situ measurement method combining depth camera and point laser displacement sensor is proposed for the dimension measurement of filament winding workpiece. Firstly, the depth map of the measured workpiece is taken by the depth camera to obtain the point cloud data, and the three-dimensional reconstruction and geometric structure determination are carried out according to the point cloud data. Then the point laser is used to scan to obtain more accurate point cloud data to calculate the parameters such as the size of the workpiece.

2 System Hardware Composition and Working Principle

The experimental hardware selected in this paper includes Intel RealSense D435 depth camera and BLG-250 N-485 laser distance sensor. In terms of imaging principle, the depth camera mainly includes binocular vision, time-of-flight (TOF) technology and structured light. The Intel RealSense D435 depth camera used in this paper adopts the principle of structured light, which mainly includes an RGB camera, two infrared cameras and an infrared generator. In the traditional binocular stereo vision, the distance between a point in space and the camera is obtained by using the principle of triangular ranging. However, if there is no texture in the field of view, it is difficult to match the pixels, so that the stereo vision fails. For the scene lacking texture, the D435 depth camera uses an infrared light projector to project the spot onto the object to add texture to the object, and then the left and right infrared cameras use the binocular vision principle to calculate the depth. In addition, the D435 depth camera has automatic calibration function, which is only used in software debugging, and does not need to use the calibration plate for calibration. Through the depth data captured by the left and right infrared cameras and the auxiliary light source of the infrared dot matrix transmitter, D435 can generate point cloud data representing the three-dimensional spatial coordinates of the object. Camera Component as shown in Fig. 1.

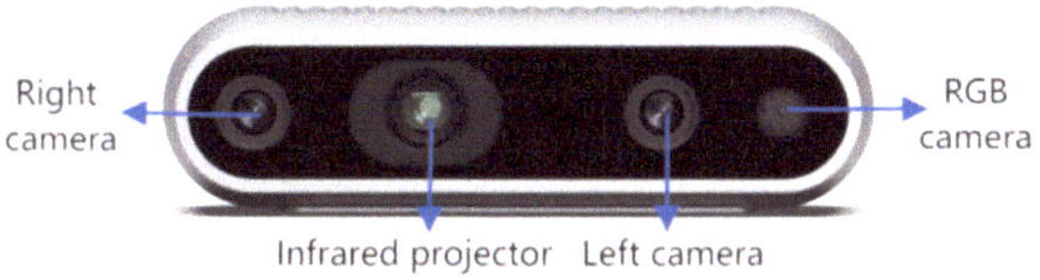

Fig. 1. Diagram of camera component structure (D435)

There are two main working principles of point laser distance sensor: triangulation method and time-of-flight method. The working principle of the BLG-250 N-485 point laser distance sensor used in this paper is the triangulation method. The laser emitter emits a concentrated laser beam to the surface of the measured object to form a spot, and then the optical receiving lens receives the light scattered from the laser spot of the measured object. The position sensitive detector (PSD) is used to accurately detect the position of the light spot on its light surface, which plays a key role. The fixed distance B between the optical axis of the laser and the main optical axis of the receiving lens and the focal length f of the receiving lens are known manufacturing parameters. The position x of the laser point imaging on the PSD is a directly measured value. According to the lens imaging formula and the triangular relationship, the derivation formula of the measured distance D is obtained: The processor inside the sensor detects the change of x in real time and calculates the accurate distance value D according to the above formula.

$$D \approx \frac{B^{*}f}{x} \tag{1}$$

3 Adaptive Reconstruction Method Based on Multi-sensor

3.1 Coordinate Transformation Between Depth Camera and Sensor

Since the depth camera and the point laser distance sensor are not in the same position and do not work at the same time, the points measured by the point laser distance sensor should be converted to the coordinates of the depth camera. In order to obtain the transformation matrix of the point laser sensor relative to the depth camera, the depth camera and the sensor are usually calibrated to obtain their respective coordinate systems. However, since the point laser distance sensor will move with the machine tool car during the working process, if the position of the sensor itself is regarded as the sensor coordinate system, the coordinate system position will also change. In order to facilitate the calculation of the coordinate transformation matrix between the point laser sensor and the depth camera, the initial position of the sensor is set as the origin of the sensor coordinate system L. The sensor only moves horizontally along the X direction of the rotation axis. The distance value d in the Z axis direction is measured, and it does not move in the Y axis direction. The output point can be expressed as (x_l, 0, d), and the value of x varies with the movement distance of the car. The depth camera coordinate system C is a fixed coordinate system, and the measured points are expressed as (x_c, y_c, z_c). There is a point P in the real coordinate system, PL = [x_l, 0, d] T is the point coordinate in the sensor coordinate system, PC = [x_c, y_c, z_c] T is the point coordinate in the camera coordinate system, R is the transformation matrix of the two coordinate systems, T = [t_x, t_y, t_z] T is the translation vector. The coordinate transformation formula is as follows:

$$P_C = R \times P_L + T \tag{2}$$

3.2 Coarse Scanning and Initial Reconstruction Based on Depth Camera

When the workpiece is not clamped, the depth camera is used to photograph the rotation axis and analyze the position of the rotation axis. After the workpiece is mounted, a picture of the workpiece is taken at a certain angle according to the rotation frequency of the rotating shaft of the machine tool. The camera captures multiple frames of depth maps and color maps from a fixed perspective. Due to the messy environment of the measurement site, the point cloud obtained by the depth camera will also represent the point cloud unrelated to the workpiece, so the collected point cloud should be preprocessed. In this paper, the invalid points are removed by depth map filtering and coordinate transformation to realize the original data purification. The depth range and ROI region clipping are used to clip the space to leave an effective part; statistical filtering and radius filtering are used to remove the noise of point cloud. The simplification of point cloud data is realized by using voxel downsampling while maintaining the shape of point cloud and greatly reducing the number of points. The ICP method is used to join point cloud data from many angles, and then the TSDF volume merging is used to make a closed mesh for the 3D model of the workpiece.

3.3 Workpiece Geometry Type Intelligent Judgment

In order to clarify the scanning route of the point laser range finder, it is necessary to judge the geometric type of the workpiece. If the workpiece is a rotating body, the laser range finder only needs to follow the car used for spinning on the machine tool to move along the axis direction. Principal component analysis (PCA) is performed on the coarse point cloud processed in the previous step. The longest direction of point cloud distribution is the first principal component direction, which is the axis of the workpiece. The axis of the workpiece is compared with the rotation axis of the machine tool. If it coincides, the point cloud is cut into multiple slices along this direction. The point cloud on each slice is projected onto a plane perpendicular to the axis, and the minimum circumcircle or fitting circle of these projection points is calculated. If the variance of the fitting circle radius of all slice projection points is less than a threshold and the center of the circle basically coincides on the rotation axis, the workpiece is determined to be a rotating body. Otherwise, it is judged to be a non-rotating body or a complex symmetric body. The current implementation is mainly for rotary workpieces. For ordinary non-rotating workpieces, the geometric feature recognition algorithm can be superimposed on the basis of the existing D435 + TSDF reconstruction. Firstly, the rough three-dimensional model realized by D435 is analyzed. According to the model, the laser scanning path of the specific point is analyzed. For the plane, the RANSAC fitting plane is used to measure the distance and thickness of the plane. For the cylinder, the fitting cylinder is used to extract the radius and axial length. For the reinforcement of complex workpieces such as fiber winding parts, curvature analysis or local feature point extraction is used to give the local radius. To achieve a more general size measurement framework.

3.4 Fine Scanning and Size Refinement Based on Laser Sensor

In the case that the measured workpiece is a rotating body, the machine tool is controlled to drive the laser sensor to move. The moving direction is parallel to the rotation axis and the height of the laser emitted by the sensor is consistent with the axis of the rotation axis. The rotary axis of the machine tool drives the workpiece to rotate at a constant speed. The laser sensor collects distance data at a fixed frequency, and records the linear displacement position of the laser sensor corresponding to each data point to achieve high-precision measurement. Each laser ranging value d is written in the sensor coordinate system (x_l, 0, d), and it is converted into a point corresponding to the depth camera coordinate system by the conversion matrix. The X-axis is the rotation axis, so that the laser sensor can scan the object with the car to obtain a point cloud band. If the workpiece is a rotating body, the point cloud band is both the contour line of the workpiece on the two-dimensional plane. Rotating the point cloud band around the rotation axis is the three-dimensional model of the workpiece. The high-precision point cloud obtained by the laser scanning is registered with the coarse point cloud obtained by the depth camera, and the key dimensions of the workpiece are corrected with the high-precision point cloud.

4 Experimental Data Processing and Analysis

4.1 Experimental Environment Construction

The hardware of the measurement system consists of three parts: Intel RealSense D435 depth camera, Boyi Jingke's point laser distance sensor BLG-250 N-485 and fiber winding machine. The workpiece is clamped on the filament winding machine tool to follow the rotation axis of the machine tool. The depth camera is placed on the platform directly in front of the workpiece to obtain the approximate point cloud data of the workpiece. The point laser distance sensor is fixed at the yarn outlet part of the machine tool. A laser line is emitted from the axis of the rotating shaft and follows the yarn car for reciprocating motion to obtain the accurate point cloud data of the workpiece. In the software part, the development language is Python3.8, and the core libraries are Open3D0.19, pyrealsense2, Numpy and so on. The obtained point cloud data is preprocessed on Pycharm, multi-angle matching and three-dimensional reconstruction of the workpiece. The measurement site is shown in the Fig. 2.

Fig. 2. Survey site

4.2 Rough Scan Reconstruction Effect Display

Due to the limited fixed shooting angle of the camera, some workpieces cannot be photographed and the point cloud cannot be obtained, but it can still be analyzed and judged whether the workpiece is a rotating body according to the main part of the workpiece. The radius variance threshold of the point cloud is set to 1.5 mm2, and the center deviation threshold is 2 mm. PCA analysis was performed on the initial point cloud model, the spindle was extracted, and the point cloud was sliced and circle fitted to verify the rotational symmetry analysis. The model is identified as a rotating body because the deviation values stay within the allowed limit. The slice thickness is 5 mm, and there are 90 effective slices, so the radius variance is 1.4688 mm^2. The average center deviation is 0.4412 mm, and the maximum deviation is 1.1826 mm. Both are below the threshold. The axis of the rotating body does not move away from the rotation axis, and the laser range finder can scan the rotation axis for fine scanning. As shown in Figs. 3, 4, and 5.

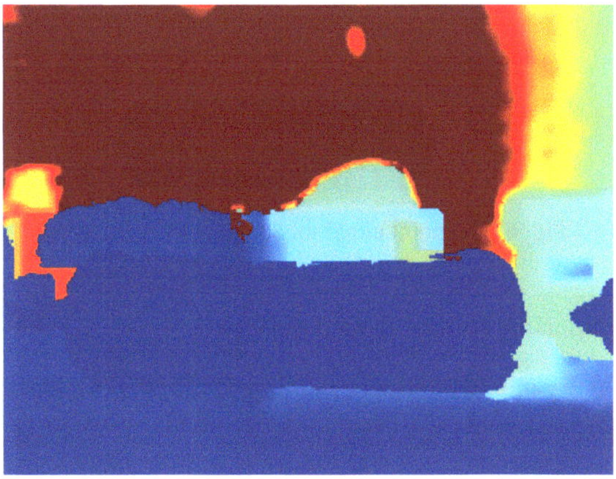

Fig. 3. Raw color images captured by D435

Fig. 4. Rough point cloud after preprocessing (denoising, down-sampling)

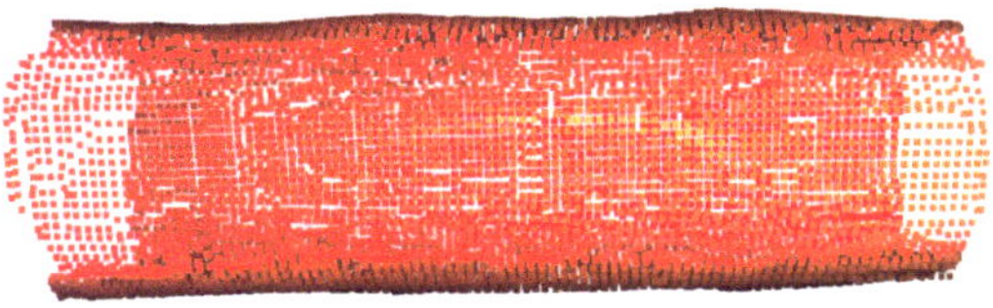

Fig. 5. Point cloud model after initial 3D reconstruction

4.3 Fine Scan Results

After analysis, it is found that when the workpiece is a rotating body, the laser range finder follows the movement of the yarn car for fine scanning. The starting speed of the car is 0 acceleration of 96 mm/s^2, and it accelerates to 48 mm/s. The range finder will generate a data table with time and the distance value d obtained by scanning. The x value of the point from the laser range finder can be found from time and speed. The y value is 0, and the z value is the distance d. The measured point cloud coordinates are converted to the coordinate data in the depth camera coordinate system through the conversion matrix, and then the point cloud is rotated around the rotation axis to obtain the map. As Fig. 6. shows.

Fig. 6. 3D point cloud model after laser scanning

4.4 Repetitive Experiments and Data Analysis

In order to evaluate the stability and repeatability of the proposed measurement method, this section conducts repeated measurement experiments on the same workpiece. Under the same environmental conditions and the same system parameter settings, the same workpiece is continuously and independently measured multiple times, and the whole process is performed from data acquisition to point laser precision measurement modeling and size extraction. The data acquisition results of the point laser sensor are shown in the following Fig. 7.

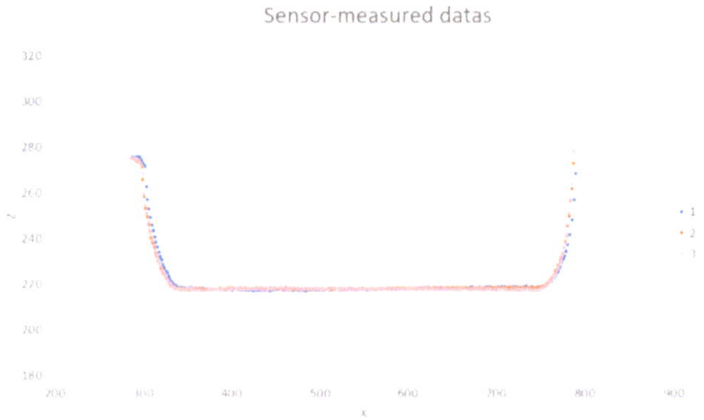

Fig. 7. Laser scanning repeatability test data

The diameter d of the cylindrical part, the length D of the straight section of the cylinder, the radius of curvature R1, R2 of the left and right ends, the length L1, L2 of the non-straight section of the left and right ends and the total length L are selected as the key dimensions of the rotary workpiece. PCA gives the first main part of the point cloud as the axial direction. The distance between the two farthest points along this direction is the total length of the workpiece. Sections are taken at equal steps along the axial direction, and the diameter of each section is used to get the diameter distribution curve. The diameter of the straight section of the cylinder should be stable. The continuous area where the diameter tends to be stable is found by statistics. The diameter of the cylindrical part and the length of the straight section of the cylinder can be obtained by calculating the average diameter and axial range of the area. According to the boundary of the straight section of the cylinder, the point cloud is divided into three parts: the circular arc point cloud at the left and right ends and the straight section of the cylinder.

The length of the non-straight section at the left and right ends is the distance from the left and right ends of the axial coordinate to the boundary of the straight section of the cylinder. The left and right ends are fitted with spherical surfaces respectively, and the radius of curvature of the left and right ends is fitted. The workpiece size is shown in the Fig. 8.

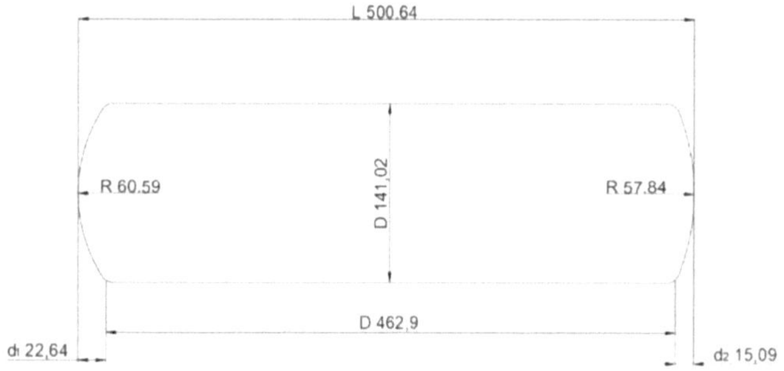

Fig. 8. Workpiece dimension diagram

Table 1. Repeat measurement results (Unit: mm)

	L	D	d	l1	l2	R1	R2
1	500.64	462.90	141.02	22.64	15.09	60.59	57.84
2	500.06	462.41	140.92	22.62	15.03	61.45	57.13
3	499.87	462.37	140.76	22.56	14.94	60.74	57.52
4	499.94	364.23	140.81	115.55	20.16	73.46	63.25

The workpiece size calculated according to the repeated experimental data can be seen in Table 1: The fourth set of data is abnormal, which is not used for the following table statistics.

Table 2. shows the standard deviation for each size is less than 0.4, and the coefficient of variation is less than 0.5%. For the total length L, the standard deviation is 0.328, and the coefficient of variation is 0.0656%. This parameter does not change much in repeated tests and is very stable. For the diameter D and the length of the cylindrical section, the coefficients of variation are 0.0521% and 0.0762%.In contrast, the coefficient of variation of the non-straight segment length at both ends and the standard deviation of the radius of curvature are slightly larger. The main reason is that the point cloud density at the end area is low, the depth noise and the truncated boundary have a more significant effect on the fitting results. Even so, the coefficient of variation is still controlled within a reasonable range to meet the stability requirements of engineering measurement. It can be observed that the length of the straight section of the fourth group of cylinders in Table 1. is 364.23 mm, and the length of the left non-straight section is 115.55 mm, which has a large deviation from the nominal values of 462.5 mm and 22.5 mm, while the total length and diameter remain within a reasonable range. The analysis shows that this is due to the uneven distribution of point cloud density in the end area or the improper selection of slice threshold, which leads to the obvious shortening of the recognition range of the cylindrical area, thus misallocating some cylindrical areas to the non-straight section of the left end. Other parameters that may cause data errors in the experiment include voxel size, ICP distance threshold, etc. The larger the voxel is, the more rough but stable the point cloud is, which is biased towards low-frequency large-scale size. The smaller the voxel is, the more detailed the point cloud is, but the noise is large, which is more sensitive to local size. The influence of ICP on the data is that the overall rigid body error will be transmitted to TSDF fusion, axis estimation, and further affect all axis vectors such as total length, straight length, and end length. In summary, after excluding the abnormal results caused by the failure of the cylindrical region identification in the fourth group, the errors of the key size parameters in the other two measurements are controlled within 1 mm, indicating that this method has good stability and repeatability when repeatedly measuring the same workpiece.

Table 2. Repeatability statistics results

Parameter	Reference value	Average	Standard deviation	Maximum absolute error	Coefficient of variation CV
L	500	500.19	0.328	0.64	0.0656%
D	462.5	462.56	0.241	0.4	0.0521%
d	140.5	140.9	0.107	0.52	0.0762%
l1	22.5	22.61	0.034	0.14	0.151%
l2	15	15.02	0.0616	0.09	0.411%
R1	–	60.93	0.375	–	–
R2	–	57.50	0.290	–	–

5 Summary

In summary, the current proposed multi-sensor fusion method shows potential in the realization of workpiece geometric type recognition and key point detection, but there are still limitations: the algorithm is mainly optimized for rotary workpieces, and the recognition ability is limited in the face of other symmetrical types, and the robustness needs to be improved; the system integrates depth camera, laser sensor and motion control unit. Its calibration and maintenance are complex, and there are some difficulties in actual deployment. In the future, we can continue to promote research work from the following aspects: further introduce artificial intelligence methods, use large-scale point cloud data to train deep learning models, and achieve more powerful and adaptive work types and key point intelligent recognition to improve real-time performance; the application scenarios of the system in industrial inspection, such as weld detection and casting burr recognition, are expanded to verify its generalization ability and practical value. At this identical time point, the hardware integration method is optimized, for example through using a more condensed sensor construction or carrying out the combination of the robot mechanical arm for the flexible scanning work, therefore to promote the whole automation degree and engineering the suitable degree of this system.

References

Bao, H.J., Liu, S.Y., Ren, Z., Zhang, Y.H., Hu, Z.Y., Ge, Y.P.: A machine vision-based method for measuring sprocket dimensions. J. Jilin Univ. (Eng. Ed.). **53**(10), 2795–2806 (2023). https://doi.org/10.13229/j.cnki.jdxbgxb.20211384

Dai, Y.Q.: Machine vision-based detection technology for motor vehicle external dimensions. Automot. Maint. Repair. **13**, 56–59 (2025). https://doi.org/10.16613/j.cnki.1006-6489.2025.13.051

Li, C.G.: Research on parameter measurement technology for synchronous pulley using a point laser [Master's thesis, Beijing University of Technology]. (2023). https://doi.org/10.26935/d.cnki.gbjgu.2023.000222

Lin, H., Liu, X.Y.: Precision measurement methods for irregularly shaped parts based on machine vision. Electromechanical Eng. Technol. **53**(8), 183–185, 214 (2024)

Liu, P., Liu, J.T.: A study on the evaluation of flatness measurement errors using a point laser. J. Shenyang Inst. Eng. (Nat. Sci. Ed.). **17**(2), 92–96 (2021) https://doi.org/10.13888/j.cnki.jsie(ns).2021.02.017

Shi, L.: A machine vision-based measurement system for screw dimensions. Instrum. Technol. Sens.S. **7**, 71–74, 87 (2023)

Yang, X.F., Xing, S.H., Li, S.W., Li, Z.Q.: A method for measuring the dimensions of small and medium-sized objects based on stereo vision. Sens.S Microsyst.S. **44**(9), 162–164, 168 (2025). https://doi.org/10.13873/J.1000-9787(2025)09-0162-03

Zhang, B.X., Gao, J.W., Wang, J.C., Liu, J.H.: Design of a machine vision-based measurement system for circular washers. Tool Technol. **57**(7), 141–145 (2023)

Zhong, Z.Z.: Design of a machine vision-based method for workpiece dimension measurement. Comput. Times. **8**, 61–65 (2025) https://doi.org/10.16644/j.cnki.cn33-1094/tp.2025.08.012

Zu, L. (2007). Structural design and damage detection of fibre-wound annular containers [Master's thesis, Xi'an University of Technology]. https://kns.cnki.net/kcms2/article/abstract?v=OGEEvuKkrhZCutcL1ng0lnliV4bcUKuysJ0YdTMfLkJPzBU3m38Ht4rrTtkheGAL3ELe1HdpxeiTg5n_GDYemqV5Ollue8W4oKmBvLYfUla06jNwIbBFUGHsGaLi4JMn58KY_6u4EuW6gT8JpP8QYh55PiNI7oWoMhMj8MOmyDAk7bqw5bSCpg==&uniplatform=NZKPT&language=CHS

Design and Simulation of Stem Separation Device Based on CFD-DEM Coupling

Ganfu Wei, Ping Liang, Ziyu Zhang, and Qian Liu(✉)

China Tobacco Guangxi Industrial Co., Ltd., Nanning, Guangxi 530000, China
100568861@qq.com

Abstract. During cigarette production, tobacco shreds are often wasted when stems are removed. To reduce shred loss and stem contamination, this study uses CFD–DEM simulation with realistic particle models in FLUENT–EDEM. The negative-pressure separation process in a recovery device was analyzed under different conditions. Results show that the optimal negative pressure is −12,000 Pa, achieving a stem escape rate of 16.20%, shred escape rate of 10.67%, and 3.68% shred content in stems. This provides a theoretical basis for optimizing tobacco shred recovery devices.

Keywords: Tobacco shreds recovery · Stem-shreds separation · Numerical simulation · CFD-DEM

1 Introduction

In production, the ZJ116/ZJ17 unit with a secondary separation device still underperforms compared to PROTOS-M5. The separated stems retain over 20–30% tobacco shreds, indicating continued shred loss and significant long-term waste.

To address the issue of tobacco shred waste in cigarette manufacturing machines. Dai optimized the Z-shaped winnowing chamber, achieving a 14.7% improvement in separation efficiency compared to the initial design and enhancing tobacco shred utilization (Dai 2025). Tan applied CFD-DEM to analyze the motion characteristics of tobacco, establishing the relationship between tobacco density and distribution, and found that at an airflow velocity of 24 m/s, the residence time of tobacco can be reduced (Tan 2025). Yang employed machine vision for tobacco shred separation, reducing the tobacco shred content in the rejected material to 11.24% (Yang 2025). Yang designed an online sorting and recycling device for tobacco shreds, improving their purity and preventing waste (Yang 2022).

However, CFD-DEM coupling studies on tobacco shred and stem particles are still insufficient, often neglecting the interactions among tobacco shred particles. Therefore, an external agglomerated tobacco shred recovery device for cigarette production units was developed. A physical model of tobacco shred particles was established, fully considering inter-particle interactions to improve simulation accuracy and determine the optimal negative pressure for tobacco shred separation (Gao 2022).

T. Mankovits and N. Trisovic (Eds.): MDS 2026, LNME 2026.
https://doi.org/10.1007/978-981-92-2440-1_14

2 Model Construction

2.1 Construction of the Tobacco Shred Recovery Device

As shown in Fig. 1, the agglomerated tobacco shred recovery device uses a dispersing mechanism to break up the tobacco shreds, which are then conveyed by a vibrating plate to the recovery chamber for separation. A monitoring system detects the separation status of the tobacco shreds in real time and adjusts the separation negative pressure and motor speed accordingly.

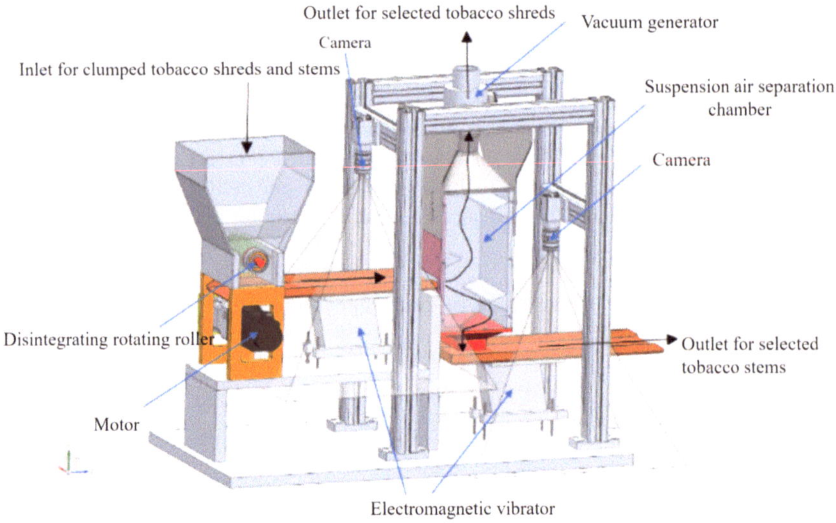

Fig. 1. Agglomerated tobacco recovery device

2.2 Construction of the Winnowing Chamber Model

Figure 2 shows the model parameters of the suspension winnowing chamber. Three guide vanes are arranged inside the chamber. Tobacco shreds and stems enter the chamber through the inlet, where separation is achieved under the negative pressure generated by a vacuum generator. A Cartesian meshing method is adopted, with a cell size of 3.5 mm (Zhao 2025).

2.3 Structural Optimization of the Winnowing Chamber

To improve the winnowing performance of the tobacco shred winnowing chamber, an orthogonal experimental design was conducted to optimize the positions of the guide vanes. A three-factor, three-level orthogonal test was adopted. The levels were as follows: the right guide vane (A) at 109.5 mm, 114 mm, and 119 mm; the upper guide vane (B) at 23.5 mm, 28.5 mm, and 33.5 mm; and the lower guide vane (C) at 45 mm, 48.5 mm, and 53.5 mm. Simulations were carried out using FLUENT software, with the outlet pressure set to −20,000 Pa and the inlet pressure set to atmospheric pressure. The Realizable k–ε turbulence model was employed. The experimental results are shown in Table 1.

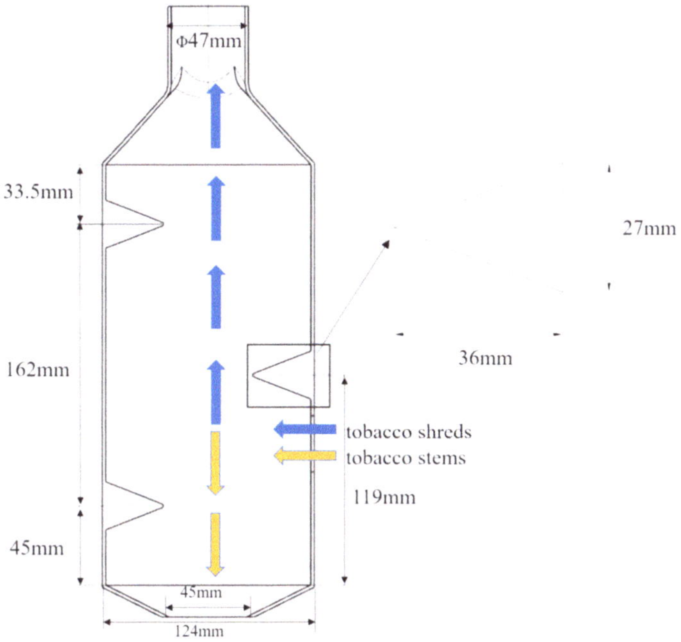

Fig. 2. Winnowing chamber model parameters

From the table, A: $k_3 > k_2 > k_1$; B: $k_3 > k_1 > k_2$; C: $k_1 > k_2 > k_3$, with influence order C > A > B. Higher A and B, and lower C, increase airflow velocity. The optimal combination is $A_3B_3C_1$, corresponding to 119 mm (single deflector), 33.5 mm (upper), and 45 mm (lower).

Table 1. Orthogonal experiment table

	A	B	C	Velocity(m/s)
1	109.5	33.5	53.5	38.6
2	109.5	28.5	45	38.6
3	109.5	23.5	48.5	38.8
4	114	33.5	45	39.1
5	114	28.5	48.5	38.9
6	114	23.5	53.5	38.5
7	119	33.5	48.5	38.9
8	119	28.5	53.5	38.6
9	119	23.5	45	39.1
K1	116	116.4	116.8	
K2	116.5	116.1	116.6	
K3	116.6	116.6	115.7	
k1	38.67	38.8	38.93	
k2	38.83	38.7	38.87	
k3	38.87	38.87	38.57	
R	0.6	0.5	1.1	

2.4 Winnowing Chamber Model Parameters

Tobacco shreds (0.4*2 mm, 3–20 mm long) and stems (2*2 mm, 5–15 mm long) were measured and modeled in UG, then imported into EDEM as non-spherical particles. As shown in Tables 2 and 3, physical properties and collision parameters were obtained from literature. Particles were generated randomly with varying sizes, yielding a range factor of 0.5–1.2 for shreds and 0.5–1.4 for stems (Zhang 2019).

Table 2. Physical property table

	Density (kg/m^3)	Poisson's ratio	Shear modulus(MPa)	Equivalent diameter(mm)	Length (mm)
Tobacco shreds	248	0.4	1.5	1	3–20
Tobacco stems	536	0.3	4	2	5–15

Table 3. Crash characteristics table

	Shreds -shreds	Shreds-stems	Stems-stems	Shreds-wall	Stems-wall
Compensation coefficient	0.5	0.5	0.5	0.5	0.5
Static friction coefficient	0.5	0.6	0.5	0.3	0.3
Static friction coefficien	0.03	0.03	0.03	0.01	0.01
Contact type	Hertz	Hertz	Hertz	Hertz	Hertz

3 Coupled Simulation

3.1 Coupling Calculation Method

The fluid–solid coupling used the Eulerian–Eulerian model with Realizable k–ε turbulence (Zhao 2024), Hertz–Mindlin contact (Shi 2023), and Ergun and Wen–Yu drag (Brandt 2025) to account for particle volume effects. As shown in Fig. 3. The mesh was imported into EDEM: inlet1 releases particles, outlet1 discharges shreds, and outlet2 discharges stems. In FLUENT, inlet1 and inlet2 are pressure inlets, and outlet1 is a pressure outlet.

In FLUENT, inlet pressures were set to 0 Pa with air at room temperature. In EDEM, a particle factory generated tobacco shreds and stems, simulating 50% of actual production (500 stems and 150 shreds) while maintaining the real shred–stem ratio. Particles were released from t = 1 s to t = 6 s with initial velocity − 2 m/s (x and z) and gravity − 9.81 m/s^2 (z). The simulation lasted 7 s, using time steps of 5×10^{-3} s in FLUENT and 5×10^{-7} s in EDEM.

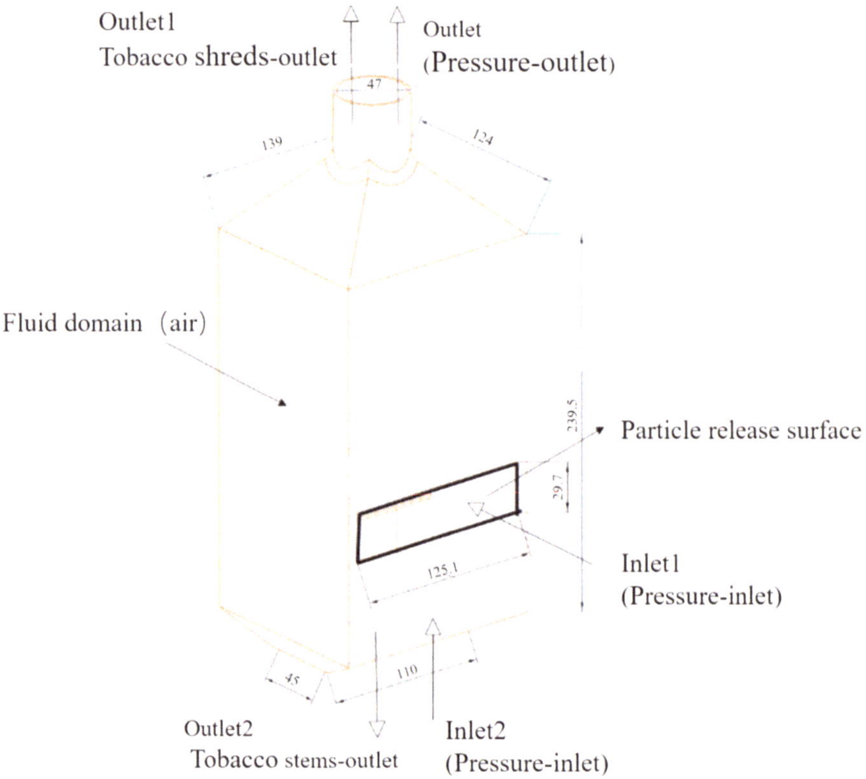

Fig. 3. Boundary condition

3.2 Simulation Results

Separation was simulated at 10 outlet negative pressures. EDEM tracked captured and escaped tobacco shreds and stems, calculating stem shred content, shred escape rate, and stem escape rate. Results are shown in Table 4.

Results show (Figs. 4 and 5) that between −10,000 Pa and − 20,000 Pa, increasing negative pressure raises stem escape, lowers tobacco shred escape, and reduces stem shred content. Below −12,000 Pa, stem escape fluctuates and remains high; above −12,000 Pa, it stabilizes, with further pressure reduction having little effect.

To determine the optimal separation negative pressure, the tobacco shred purity (proportion of shreds at outlet1) was analyzed. It can be seen that the highest tobacco shred purity occurs at −12,000 Pa, while further decreasing the negative pressure results in excessive shred content in the stems. In summary, a negative pressure of −12,000 Pa is the theoretically optimal parameter, at which the stem shred content is 3.68%.

Table 4. Simulation results

Pressure-outlet(pa)	Escaped tobacco shreds	Separated tobacco stems	Separated tobacco shreds	Escaped tobacco stems	Shreds content in stems (%)	Tobacco shred purity (%)
−20000	1	289	149	211	0.34	41.39
−19000	2	290	148	210	0.68	41.34
−18000	3	306	147	194	0.97	43.10
−17000	4	312	141	188	1.27	42.85
−16000	4	325	146	175	1.22	45.48
−15000	7	327	142	171	2.37	45.36
−14000	13	334	141	166	3.75	45.93
−13000	16	382	134	118	4.02	53.17
−12000	16	419	134	81	3.68	62.33
−10000	20	421	130	79	4.62	62.20

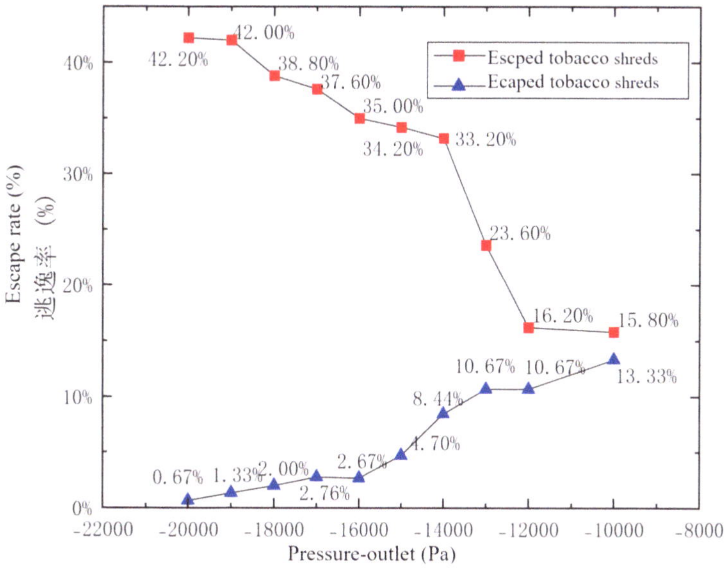

Fig. 4. Escape rate of tobacco material

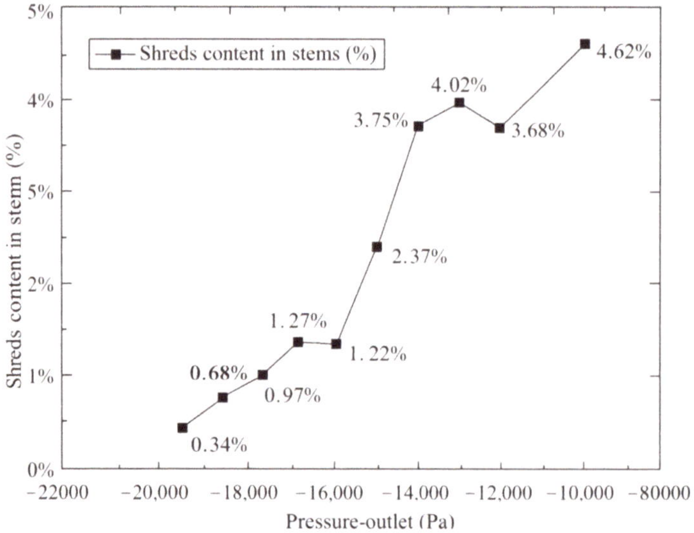

Fig. 5. Shreds content in stems

As shown in Figs. 6 and 7, the particle trajectories at a negative pressure of −12,000 Pa indicate that tobacco shreds can pass through the suspension winnowing chamber and be successfully recovered at outlet1, while most stems are removed through outlet2. At the same time, the tobacco shred and stem particles do not remain suspended in the chamber for a long time, ensuring rapid separation of the feed material and maintaining a high separation speed. This also demonstrates that at a negative pressure of −12,000 Pa, the airflow in the suspension winnowing chamber achieves optimal performance.

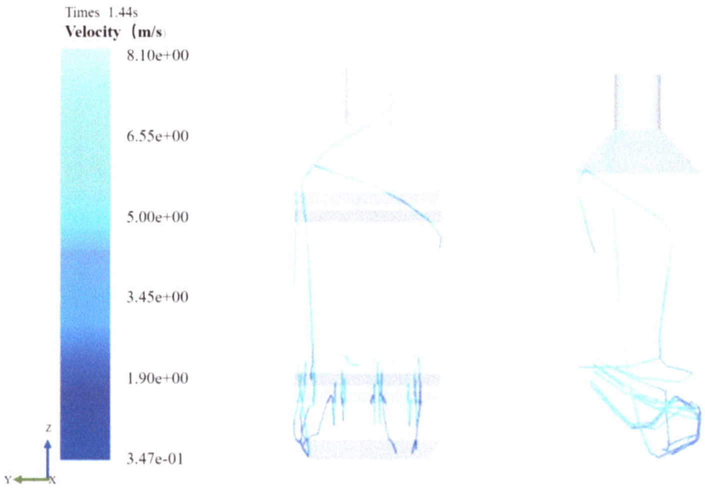

Fig. 6. Tobacco stem trajectory diagram

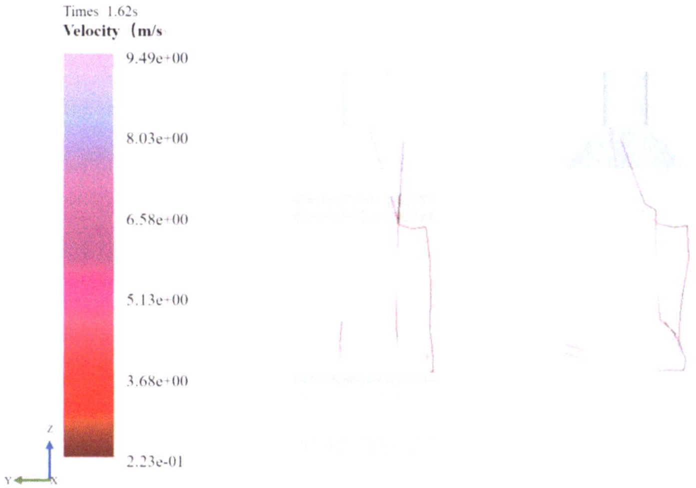

Fig. 7. Tobacco shreds trajectory diagram

4 Conclusion

By constructing and measuring the particle models, the most realistic tobacco shred and stem models were obtained, reducing the discrepancy between simulation data and actual production data. Through CFD–DEM coupled simulation, at a vacuum generator negative pressure of −12,000 Pa, the stem shred content was reduced to 3.68%, and the tobacco shred escape rate was reduced to 10.67%.

References

Brandt, V., Grabowski, J., Jurtz, N., Kraume, M., Emden, K.: CFD-DEM simulation of a circulating fluidized bed using a particle location based multi-level coarse-graining approach. Powder Technol. **460**, 121075 (2025). https://doi.org/10.1016/j.powtec.2025.121075

Dai, Y.S. et al.: Structural optimization and experimental research on tobacco shreds air separation based on Z-field jet diffusion effect. Agric. Equip. Veh. Eng. **63**(10), 68–78 (2025). https://doi.org/10.3969/j.issn.1673-3142.2025.10.011

Diego, S.P., Luís, F.N.S., Fereshteh, R., José, L.P., Marcelo, M.S., Emílio, C.N.S.: Topology optimization of time dependent gas–solid flow using Euler–Euler model with gravity. Adv. Eng. Softw. **216**, 104125 (2026). https://doi.org/10.1016/j.advengsoft.2026.104125

Liu, M., Jiang, C., Khoo, C.B., Zhu, H.F., Gao, G.J.: A cell-based smoothed finite element model for the analysis of turbulent flow using realizable k-ε model and mixed meshes. J. Comput. Phys. **501**, 112783 (2024). https://doi.org/10.1016/j.jcp.2024.112783

Shi, J.P., Shan, Z.D., Yan, H.Q.: Research on the macro-and meso-mechanical properties of frozen sand mold based on hertz-Mindlin with bonding model. Particuology. **88**, 176–191 (2024). https://doi.org/10.1016/j.partic.2023.08.019

Tan, G., Huang, D., Wang, G., Wang, H.Y., Li, C.G., Zhu, W.K.: Numerical simulation study on the conveying characteristics of a multi-component flexible particle system in industrial pneumatic conveying pipeline. Powder Technol. **466**, 121450 (2025). https://doi.org/10.1016/j.powtec.2025.121450

Yang, L., Tian, D.X., Liu, B., Pu, Y.W., Ma, J.P.: Optimization study of cigarette leaf shred recovery device based on fine processing. Manuf. Upgrad. Today. **7**, 142–144 (2022) https://kns.cnki.net

Yang, Y., Tian, C.F., Liu, D.K., Zhao, W.K., Xie, Z.Q., Li, X.L.: Study on fine separation equipment for tobacco shreds in secondary winnowing rejects. Mech. Eng. Autom. **54**(4), 139–141, 144 (2025). https://doi.org/10.3969/j.issn.1672-6413.2005.04.047

Zhang, H.T.: Numerical simulation study and structural design of gas-solid two-phase flow in a secondary separation device. Master's thesis, Kunming University of Science and Technology (2019). https://doi.org/10.27200/d.cnki.gkmlu.2019.000061

Zhao, N., Liu, J.M., Tian, L.L., Wang, Z.M.: Progress and prospects of Cartesian mesh simulation methods for compressible flow problems. Chin. J. Theor. Appl. Mech. **57**(2), 285–314 (2025). https://doi.org/10.6052/0459-1879-24-374

Research on Configuration Expression and Automatic Assembly of Compound Gear Trains Based on Graph Theory

Tao Fu and Weiqing Zhang(✉)

Chongqing University of Technology, Chongqing, China
wqzhang@cqut.edu.cn

Abstract. In response to the requirements of the modeling and assembly module for the cylindrical gear transmission system modeling and analysis software, the structural characteristics of the compound gear train were analyzed. Based on graph theory, a general expression method for the topological structure of the fixed-axis gear train and the planetary gear train was studied. Based on this, the adjacency matrix is extended by combining the characteristic state vector of the basic gear train. The automatic assembly method of the compound gear train is studied according to the combination relationship between the basic gear trains and the coordinate transformation. Finally, the geometric model construction and assembly of the gear train were realized by combining the computer graphics tool VTK.

Keywords: Compound gear train · Topological structure · Visualization · Automatic assembly

1 Introduction

The gear train, as the core component of the mechanical transmission system, is widely used in fields such as aerospace, automobiles, robotics and precision instruments. To express the structural characteristics of gear trains and the connection relationships among their components more clearly, scholars have introduced graph theory into the configuration description and generation process. Some scholars have proposed a new graph theory model and a gear matching method, which can quickly obtain the structure and gear parameters of a compound gear transmission device (He et al. 2025). Furthermore, some researchers have proposed a representation and analysis method for complex fixed-axis gear transmissions based on graph theory (Li and Yao 2008). A topological model of the planetary gear mechanism was established. By integrating the types of the planetary gear mechanism, they obtained a set of diagrams for some planetary speed-changing mechanisms (Xue et al. 2022). A graphical symbol topological model was established, which can describe the structural characteristics and motion characteristics information of the multi-level planetary gear system (Wang et al. 2021). Some researchers focused on the main reducer of coaxial inverted helicopters and proposed a configuration design method for the main reducer based on graph theory (Zeng et al. 2023).

T. Mankovits and N. Trisovic (Eds.): MDS 2026, LNME 2026.
https://doi.org/10.1007/978-981-92-2440-1_15

From this, it can be seen that the existing research mainly focuses on using graph theory to design the configuration scheme of gear trains or conducting kinematic characteristic analysis of specific gear trains, and is basically limited to the theoretical research level. There is relatively little research on the different configuration expressions of composite gear trains and their application in geometric model display and assembly based on configuration information. To address the above issues, this paper takes the composite gear train obtained by combining fixed-axis gear trains and planetary gear trains as the research object. Based on graph theory, a general expression method for the topological structure of fixed-axis gear trains and planetary gear trains was established. An automatic assembly method for compound gear trains based on the extended adjacency matrix was proposed. Finally, the geometric model of the compound gear train was generated using VTK, providing core method support for the automatic modeling and assembly of digital software for cylindrical gear transmission systems.

2 Configuration Expression and Automatic Assembly of Compound Gear Trains

Compound gear trains are usually complex gear train structures formed by combining fixed-axis gear trains and planetary gear trains, or by combining multiple planetary gear trains. The structural schematic of a compound gear train is shown in Fig. 1. The structure of a fixed-axis gear train is relatively simple, and its transmission type is relatively single, usually composed of fixed-axis gears and the frame. However, for a planetary gear train, it mainly includes the sun gear, planet gears, internal gear ring, planet carrier, and frame, with a relatively complex structure and diverse transmission types. Therefore, the focus of studying compound gear trains lies in researching the influence of different structural forms of planetary gear trains on the motion characteristics of compound gear trains.

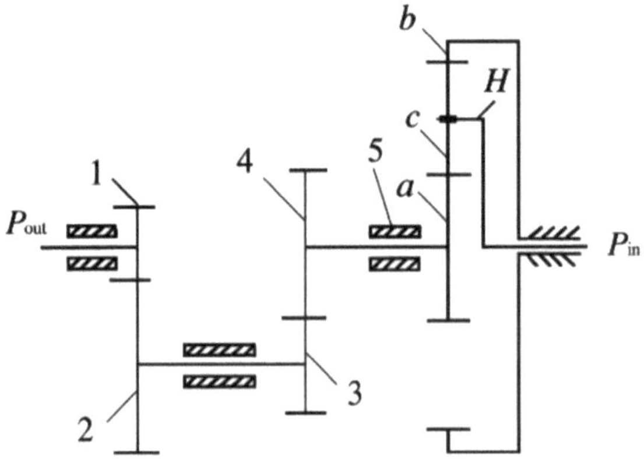

Fig. 1. Schematic diagram of compound gear train structure

2.1 Gear Train Configuration Expression

In order to more clearly present the structural information of the compound gear train, the components and the connection relationships among the components in the compound gear train are transformed into different graphic symbols for representation. The definition rules for the components in the gear train are as follows: ('△') indicates a fixed-axis gear, ('●') indicating the sun gear, ('○') indicating the planetary gear, ('◎') indicating double planetary gears, ('◆') indicatig the planet carrier, ('■') indicating the inner gear ring, ('▲') indicating the rack. The definition rule for the connection relationship is as follows: solid lines represent rotational kinematic pairs, single dashed lines represent external meshing gear pairs, and double dashed lines represent internal meshing gear pairs. The adjacency matrix is the most commonly used method for storing the topological structure information of a kinematic chain through a computer. The values in the adjacency matrix represent the type of kinematic pairs that connect the components. By analyzing the structure of the compound gear train, it can be found that the kinematic structure of the fixed-axis gear train and various common planetary gear trains can all be composed of two basic units, which are called basic topological structure units. Both of these two basic units contain the minimum complete functional closed loop of gear train transmission (Yang et al. 2025). The NG unit contains the minimum topology for realizing internal meshing transmission within the gear train, while the GW unit contains the minimum topology for realizing external meshing transmission within the gear train. If the units are further divided, the transmission function of the gear train cannot be fully expressed. Therefore, these two basic units are the smallest topological structure units that cannot be further divided. The structure and adjacency matrix of the basic units are shown in Fig. 2.

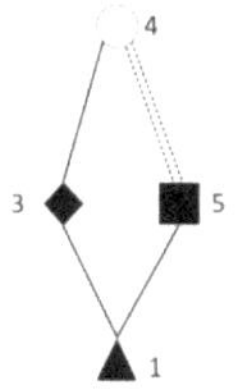

$$\begin{bmatrix} 0 & 1 & 0 & 1 \\ 1 & 0 & 1 & 0 \\ 0 & 1 & 0 & 3 \\ 1 & 0 & 3 & 0 \end{bmatrix}$$

(a) NG unit

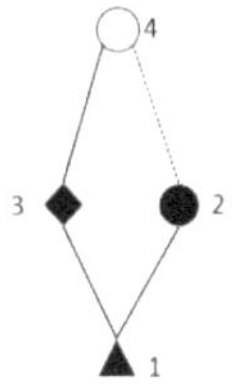

$$\begin{bmatrix} 0 & 1 & 1 & 0 \\ 1 & 0 & 0 & 2 \\ 1 & 0 & 0 & 1 \\ 0 & 2 & 1 & 0 \end{bmatrix}$$

(b) GW unit

Fig. 2. Basic topological structure unit

The topological structure of the fixed-axis gear train can be regarded as a special form derived from the basic topological structure unit. By replacing the moving component nodes in the basic unit with fixed-axis gear nodes, the topological structure diagram of the fixed-axis gear train can be obtained. For any planetary gear train, its topological structure can be expressed through the topological evolution of two basic topological structure units. The topological evolution rules are defined as follows: (1) When the same component appears in different units, topological fusion is achieved through node merging, and only one node is retained after merging, with the kinematic pair information being merged synchronously; (2) Different nodes in different units evolve into the next-level nodes; (3) During the unit combination process, the type of kinematic pairs and meshing relationships remain unchanged. In order to facilitate the integration of the topological evolution process and the topological information of various planetary gear systems into the data structure, the topological evolution process is represented using matrix operations (Shanmukhasundaram et al. 2019). Suppose the number of basic units involved in the topological evolution process is $\alpha = 1, 2, \cdots, k$. Where each unit α consists of n components, the total number of components involved in the topological evolution is

$$m = \sum_{\alpha=1}^{k} n\alpha \tag{1}$$

During the process of combining the basic topological structure units, the same component nodes need to be merged. A node constraint matrix $\boldsymbol{C}$ with n rows and m columns is defined to describe the mapping relationship between the component nodes of the basic unit and the total component nodes of the planetary gear train obtained after combination. The meaning of the matrix elements is that if the component nodes in the basic unit have a corresponding relationship with the component nodes in the planetary gear train, the element value is 1; otherwise, the element value is 0.

$$\boldsymbol{C} = \begin{bmatrix} c_{11} & c_{12} & \cdots & c_{1m} \\ c_{21} & c_{22} & \cdots & c_{2m} \\ \vdots & \vdots & \ddots & \vdots \\ c_{n1} & c_{n2} & \cdots & c_{nm} \end{bmatrix} \tag{2}$$

Here, n represents the total number of components after the combination; m represents the sum of all the unit components involved in the combination. Simultaneously define the extended adjacency matrix obtained by performing block diagonalization on the adjacency matrices of each basic topological structure unit $\boldsymbol{A}'$

$$\boldsymbol{A}' = \boldsymbol{A}_1 \oplus \boldsymbol{A}_2 \oplus \boldsymbol{A}_3 \oplus \cdots \oplus \boldsymbol{A}_K = \begin{bmatrix} \boldsymbol{A}_1 & 0 & \cdots & 0 \\ 0 & \boldsymbol{A}_2 & \cdots & 0 \\ \vdots & \vdots & \ddots & \vdots \\ 0 & 0 & \cdots & \boldsymbol{A}_K \end{bmatrix} \tag{3}$$

Here, $\boldsymbol{A}_K$ represents the adjacency matrix of the basic topological structure unit. Combining the node constraint matrix with the expanded adjacency matrix to represent the matrix elements of the same component nodes, thereby obtaining the adjacency matrix $\boldsymbol{A}$ of the planetary gear train evolved from the topology of each basic unit. The specific operation rules are as shown in Formula (4). Essentially, it is through the linear transformation of $\boldsymbol{C}$ and $\boldsymbol{C}^T$ that the local component numbers of the basic units are uniformly mapped to the global component numbers of the gear train, corresponding to the merging of the same nodes in the topological evolution rules.

$$\boldsymbol{A} = \boldsymbol{C}\boldsymbol{A}'\boldsymbol{C}^T \tag{4}$$

2.2 Automatic Assembly of Compound Gear Trains

The motion characteristics of a compound gear train composed of basic gear trains are determined by the motion characteristics of the participating basic gear trains and the combination method. Different combination methods between basic gear trains can be regarded as a special assembly method based on motion transmission. Since the assembly of a compound gear train requires the input and output port mapping information of the basic gear train, as well as the spatial position and assembly reference information, while the adjacency matrix of the basic gear train can only describe the connection relationship between components, in order to facilitate the representation of the assembly conditions of the basic gear train, the adjacency matrix of the basic gear train is expanded based on the required information (Qiu et al. 2021). The extended adjacency matrix is called the assembly adjacency matrix and is defined as

$$\overline{A} = \begin{bmatrix} \mathbf{A} & \mathbf{H} & 0 \\ \mathbf{H}^T & 0 & 0 \\ 0 & 0 & \mathbf{K} \end{bmatrix} \tag{5}$$

Here, $\boldsymbol{A}$ represents the adjacency matrix of the basic gear train, $\boldsymbol{H}$ is used to establish the correspondence between the components and the input/output in the basic gear train, $\boldsymbol{K}$ represents the characteristic state matrix of the basic gear train, consisting of the structural characteristic vector $\boldsymbol{S}$, the motion characteristic vector $\boldsymbol{M}$, and the dynamic characteristic vector $\boldsymbol{P}$. The structural characteristic vector is used to describe the characteristic state of the basic gear train. The specific structure form of matrix $\boldsymbol{K}$ is

$$\mathbf{K} = \begin{bmatrix} \mathbf{S} & 0 & 0 \\ 0 & \mathbf{M} & 0 \\ 0 & 0 & \mathbf{P} \end{bmatrix} \tag{6}$$

The structural feature vector $\boldsymbol{S}$ in the characteristic state vector contains the spatial coordinate information of the basic gear train in the local coordinate system, which is denoted by $\left[x, y, z, 1\right]^T$. The motion feature vector $\boldsymbol{M}$ is the unit direction vector of the rotation axis, represented by $\left[u_x, u_y, u_z\right]^T$. The dynamic characteristic vector $\boldsymbol{P}$ includes parameters such as the power of the basic gear train and the transmission efficiency, this lays the foundation for the development of the subsequent simulation analysis module for compound gear trains. For a compound gear train composed of n basic gear trains, the system matrix $\boldsymbol{B}$ of the compound gear train can be expressed as

$$\mathbf{B} = \begin{bmatrix} \overline{\mathbf{A}_1} & \mathbf{D}_{12} & \mathbf{D}_{13} & \cdots & \mathbf{D}_{1n} \\ \mathbf{D}_{21} & \overline{\mathbf{A}_2} & \mathbf{D}_{23} & \cdots & \mathbf{D}_{2n} \\ \mathbf{D}_{31} & \mathbf{D}_{32} & \overline{\mathbf{A}_3} & \cdots & \mathbf{D}_{3n} \\ \vdots & \vdots & \vdots & \ddots & \vdots \\ \mathbf{D}_{n1} & \mathbf{D}_{n2} & \mathbf{D}_{n3} & \cdots & \overline{\mathbf{A}_n} \end{bmatrix} \tag{7}$$

Here, $\boldsymbol{D}_{ij}$ is used to describe the combination relationship between the i-th basic gear train and the j-th basic gear train. When there is no connection relationship between the two basic gear trains, $\boldsymbol{D}_{ij}$ is a zero matrix. When there is a connection relationship between the two basic gear trains, the matrix is defined as

$$\mathbf{D}_{ij} = \begin{bmatrix} 1 & 0 & 0 \\ 0 & 1 & 0 \\ \Delta\mathbf{S}_{ij} & \Delta\mathbf{M}_{ij} & 1 \end{bmatrix} \quad (8)$$

Based on the right-hand rule, a global coordinate system $O_G - X_G Y_G Z_G$ and various local coordinate systems $O_i - X_i Y_i Z_i$ for each basic gear train are established. The local coordinate system of the first basic wheelset in the compound gear train is taken as the global coordinate system. The local coordinate system of the basic wheelset is set with the center of the input component as the origin and the Z-axis direction as the axis direction (Guo and Tan 2025). By solving the connection constraint equations under different combinations simultaneously, the characteristic vector values of all basic wheelsets can be obtained. Then, using the coordinate transformation matrix, their positions in the local coordinate system can be transformed to the global coordinate system.

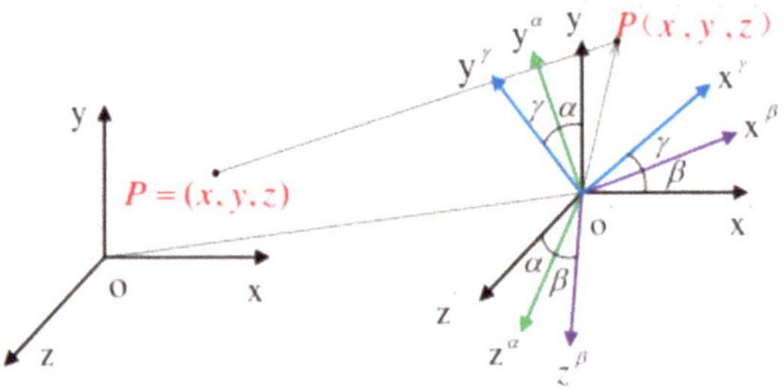

Fig. 3. Coordinate transformation diagram

Coordinate transformation can be divided into two processes: rotation and translation. Figure 3 shows the change in spatial position when the coordinate origin is transformed to point P′. The coordinate origin moves a certain distance along the vector oP′ and rotates α, β, and γ degrees respectively around the x-axis, y-axis, and z-axis. Translation follows the same principle.

$$\mathbf{K}_j = \mathbf{T} \cdot \mathbf{D}_{ij} \cdot \mathbf{K}_i \quad (9)$$

Among them, $\mathbf{K}_i$ represents the characteristic state matrix of the preceding basic gear train, $\mathbf{K}_j$ represents the characteristic state matrix of the subsequent basic gear train, and $\boldsymbol{T}$ is a homogeneous transformation matrix composed of the rotation matrix $\boldsymbol{R}$ and the translation vector $\boldsymbol{d}$:

$$\mathbf{T} = \begin{bmatrix} \mathbf{R} & \mathbf{d} \\ 0 & 1 \end{bmatrix} \quad (10)$$

Among them, the structural deviation $\Delta\mathbf{S}_{ij}$ between the two gear trains due to the translation vector $\boldsymbol{d}$ is uniquely determined. Depends on the structural deviation $\Delta\mathbf{S}_{ij}$ between the two gear trains, while the rotation matrix $\boldsymbol{R}$ depends on the axis directions of the two basic gear trains, which are related to $\Delta\mathbf{M}_{ij}$ (Zhu et al. 2025).

2.3 Example Verification

In order to systematically verify the accuracy of the proposed composite gear train configuration expression method and the automatic assembly method in this paper, the configuration expression method and the automatic assembly method were respectively verified. The fixed-axis gear train is first regarded as a special case of the basic topological structure unit, and its topological structure diagram and adjacency matrix are shown in Fig. 4 as follows.

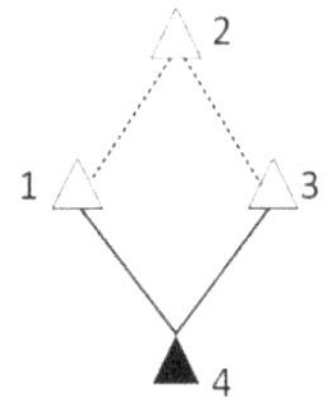

$$\begin{bmatrix} 0 & 2 & 0 & 1 \\ 2 & 0 & 2 & 0 \\ 0 & 2 & 0 & 1 \\ 1 & 0 & 1 & 0 \end{bmatrix}$$

Fig. 4. Topological structure of fixed-axis gear train and adjacency matrix

Based on the topological evolution method and the defined topological evolution rules in Sect. 2.1, all the common planetary gear train topological structure diagrams can be expressed through two basic topological structure units as shown in Fig. 5.

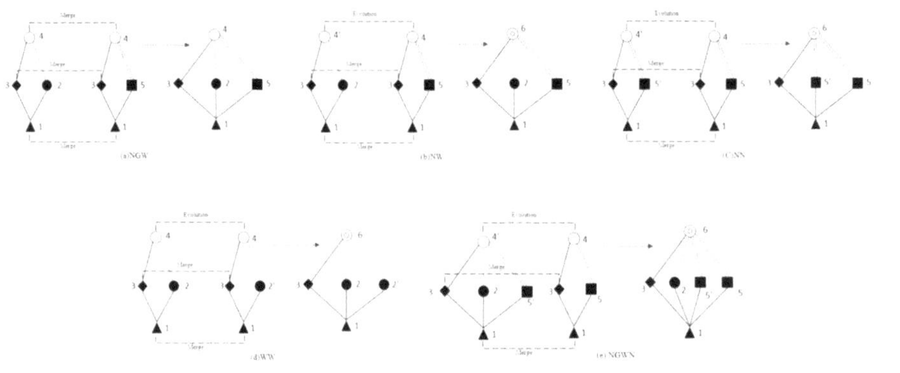

Fig. 5. Common expression of planetary gear train topological structure

Similarly, based on the matrix operation rules defined in Sect. 2.1, the adjacency matrices of all common planetary gear trains are obtained as shown in Fig. 6.

After obtaining the adjacency matrices of each gear train, combined with the developed software specifically for cylindrical gear transmission systems, the models of each component of the gear train are generated. Then, based on the topological information of the gear train provided by the adjacency matrices, various geometric models of the gear trains are assembled. The main interface of the software and the module for generating gear train models are shown in Fig. 7.

$$\begin{bmatrix} 0 & 1 & 1 & 0 & 1 \\ 1 & 0 & 0 & 2 & 0 \\ 1 & 0 & 0 & 1 & 0 \\ 0 & 2 & 1 & 0 & 3 \\ 1 & 0 & 0 & 3 & 0 \end{bmatrix} \quad \begin{bmatrix} 0 & 1 & 1 & 1 & 0 \\ 1 & 0 & 0 & 0 & 2 \\ 1 & 0 & 0 & 0 & 1 \\ 1 & 0 & 0 & 0 & 3 \\ 0 & 2 & 1 & 3 & 0 \end{bmatrix} \quad \begin{bmatrix} 0 & 1 & 1 & 1 & 0 \\ 1 & 0 & 0 & 0 & 1 \\ 1 & 0 & 0 & 0 & 3 \\ 1 & 0 & 0 & 0 & 3 \\ 0 & 1 & 3 & 3 & 0 \end{bmatrix} \quad \begin{bmatrix} 0 & 1 & 1 & 1 & 0 \\ 1 & 0 & 0 & 0 & 2 \\ 1 & 0 & 0 & 0 & 2 \\ 1 & 0 & 0 & 0 & 1 \\ 0 & 2 & 2 & 1 & 0 \end{bmatrix} \quad \begin{bmatrix} 0 & 1 & 1 & 1 & 1 & 0 \\ 1 & 0 & 0 & 0 & 0 & 2 \\ 1 & 0 & 0 & 0 & 0 & 1 \\ 1 & 0 & 0 & 0 & 0 & 3 \\ 1 & 0 & 0 & 0 & 0 & 3 \\ 0 & 2 & 1 & 3 & 3 & 0 \end{bmatrix}$$

(a) NGW (b) NW (c) NN (d) WW (e) NGWN

Fig. 6. Common adjacent matrix of planetary gear train

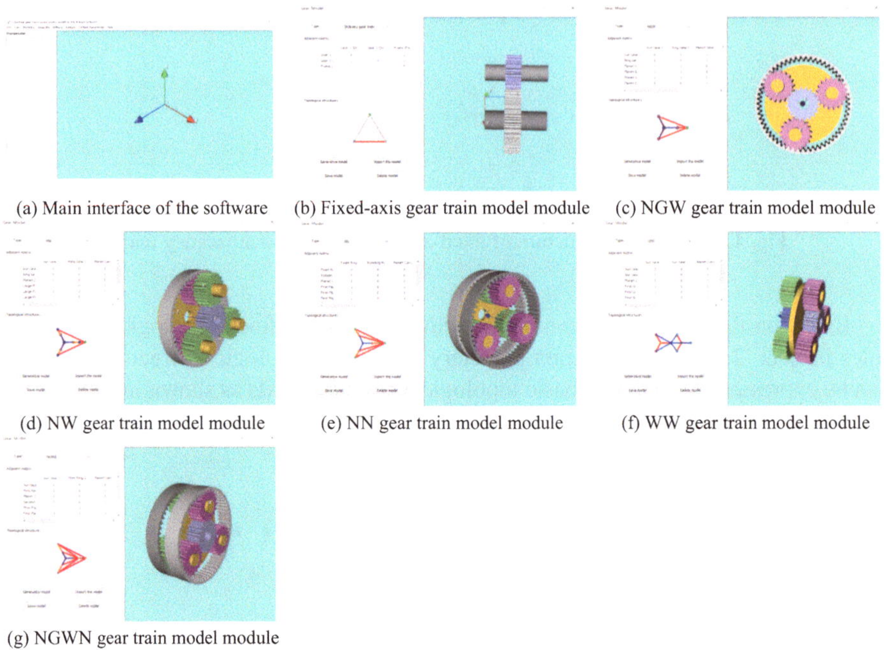

(a) Main interface of the software (b) Fixed-axis gear train model module (c) NGW gear train model module (d) NW gear train model module (e) NN gear train model module (f) WW gear train model module (g) NGWN gear train model module

Fig. 7. Various geometric models of gear trains

Taking the compound gear train shown in Fig. 1 as an example, this compound gear train is a three-level series structure composed of a two-stage fixed-axis gear train and an NGW planetary gear train. The rated input power is 5.5 kW and the input speed is 1440 rpm. The parameters of each gear are shown in Table 1.

According to the automatic assembly method in Sect. 2.2, matrix $\boldsymbol{H}_1$ is obtained based on the corresponding relationship between the input and output components of the first-stage fixed-axis gear train. Since it is based on the first-stage gear train, its structural characteristic vector $\boldsymbol{S}_1$ and dynamic characteristic vector $\boldsymbol{M}_1$ are as follows:

$$\mathbf{H}_1 = \begin{bmatrix} 0\,1\,0 \\ 0\,0\,1 \\ 1\,0\,0 \end{bmatrix} \quad \boldsymbol{S}_1 = [0, 0, 0, 1]^T \quad \boldsymbol{M}_1 = [0, 0, 1]^T \tag{11}$$

Table 1. Parameters of all types of gears

Gear parameter	Gear 1	Gear 2	Gear 3	Gear 4	Sun gear a	Planet gear b	Gear ring c
Number of teeth	20	60	18	62	20	19	58
Modulus	2.5						
Pressure angle	20°						
Face width	25 mm						

As shown in Fig. 1, the three basic gear trains in the composite gear train have the same axis direction. Therefore, $\Delta \mathbf{M}_{ij}$ is 0. Based on the assembly parameters, the value of $\Delta \mathbf{S}_{ij}$ can be obtained. Thus, the combination relationship matrices between each gear train are as follows:

$$\boldsymbol{D}_{12} = \begin{bmatrix} 1 & 0 & 0 \\ 0 & 1 & 0 \\ 20 & 0 & 1 \end{bmatrix} \quad \boldsymbol{D}_{23} = \begin{bmatrix} 1 & 0 & 0 \\ 0 & 1 & 0 \\ 30 & 0 & 1 \end{bmatrix} \tag{12}$$

Also due to the consistent axial direction, the rotation matrix $\boldsymbol{R}$ between each level of the gear train is the identity matrix, and the translation vector $\boldsymbol{d}$ is obtained based on the $\Delta \mathbf{S}_{ij}$ element in the combined relationship matrix. Therefore, the homogeneous transformation matrix of the coordinates for each level of the gear train can be obtained as follows:

$$\boldsymbol{T}_{12} = \begin{bmatrix} 1 & 0 & 0 & 0 \\ 0 & 1 & 0 & 0 \\ 0 & 0 & 1 & 20 \\ 0 & 0 & 0 & 1 \end{bmatrix} \quad \boldsymbol{T}_{23} = \begin{bmatrix} 1 & 0 & 0 & 0 \\ 0 & 1 & 0 & 0 \\ 0 & 0 & 1 & 30 \\ 0 & 0 & 0 & 1 \end{bmatrix} \tag{13}$$

Based on the structural characteristic vectors of the first-level fixed-axis gear train, and by combining with the homogeneous transformation matrix of the coordinate system, the structural characteristic vectors of the remaining gear trains can be obtained as follows:

$$\begin{aligned} \mathbf{S}_2 &= \mathbf{T}_{12} \cdot \mathbf{S}_1 = [0, 0, 20, 1]^T \\ \mathbf{S}_3 &= \mathbf{T}_{23} \cdot \mathbf{S}_2 = [0, 0, 50, 1]^T \end{aligned} \tag{14}$$

Based on the structural characteristic vectors of each level of the gear train, the geometric models of each level of the gear train are transformed to the corresponding positions, and finally the geometric model and system matrix of the compound gear train are obtained as shown in Fig. 8.

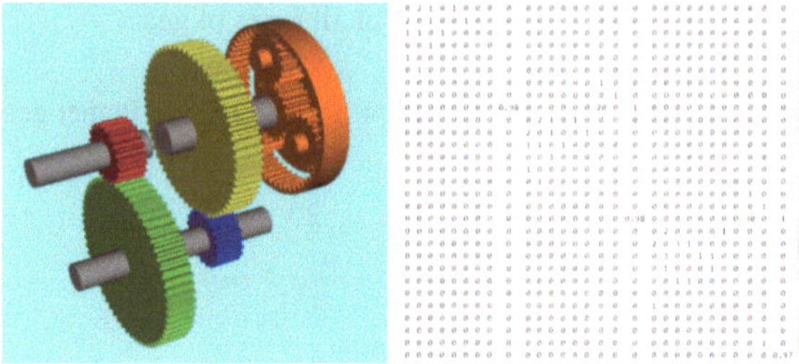

Fig. 8. Geometric model of compound gear train and system matrix

3 Conclusion

In response to the requirements of the modeling and assembly module development for the cylindrical gear transmission system simulation software, the general method for expressing the topological structure of the compound gear train and the automatic assembly method were studied. Based on the above research, the following conclusions were drawn:

(1) Based on the general expression method of the topological structure of fixed-axis gear trains and planetary gear trains, the topological structures of each gear train were obtained. Combined with VTK, the geometric models of various gear trains were generated.
(2) Based on the basic gear train characteristic state, an assembly adjacency matrix is established. Through the combination relationship of basic gear trains and the coordinate transformation theory, a composite gear train automatic assembly method is proposed, thereby solving the global coordinates of each basic gear train.
(3) Integrate the above method into the software module written in MFC to achieve rapid generation and automatic assembly of the basic gear train geometric model.

References

Guo, Y., Tan, M.H.: Automatic assembly process of helical gears and rack for steering gear. Mech. Manuf. **63**(2), 57–63 (2025). https://doi.org/10.3969/j.issn.1000-4998.2025.02.014

He, J.M., Chen, Y.N., An, Q.: Research on the design of compound gear train for co-axial helicopter transmission system based on graph theory. Intern. Combust. Engine Accessories **4**, 35–37 (2025). https://doi.org/10.3969/j.issn.1674-957X.2025.04.011

Li, H., Yao, J.: Graph theory analysis method for complex fixed-axis gear train. J. Southwest Jiaotong Univ. **4**, 483–487 (2008). https://doi.org/10.3969/j.issn.0258-2724.2008.04.011

Qiu, J., Wang, D.L., Ma, Y.L.: Unified digital model for analysis of gear train based on state space. J. Dalian Univ. Tech. **61**(1), 52–59 (2021). https://doi.org/10.7511/dllgxb202101008

Shanmukhasundaram, R.V., Rao, D.V.Y., Regalla, P.S.: Algorithms for detection of degenerate structure in epicyclic gear trains using graph theory. J. Braz. Soc. Mech. Sci. Eng. **41**(11), 1–16 (2019). https://doi.org/10.1007/s40430-019-2003-5

Wang, J.G., Yang, S.N., Liu, Y.D., Zeng, L.P., Morimina, T.: Graphical symbol description method and kinematic analysis of multi-level planetary gear train structure and motion topology model. J. Xi'an Univ. Technol. **37**(1), 64–70 (2021). https://doi.org/10.3969/j.issn.1001-3997.2022.08.009

Xue, H.L., Li, L.J., Guo, F.: Design of automatic transmission planetary gear transmission scheme based on graph theory. Mech. Transm. **46**(8), 78–84 (2022). https://doi.org/10.16578/j.issn.1004.2539.2022.08.013

Yang, W.J., Li, Y.T., Chen, X., Wang, L.: Topological synthesis and configuration design of planetary gear systems based on linkage motion chain transformation. J. Mech. Eng. **61**(11), 86–105 (2025). https://link.cnki.net/urlid/11.2187.TH.20250217.1515.011

Zeng, L., Hu, M.H., Chen, G.Y.: Lightweight design method for coaxial main differential based on graph theory. J. Chongqing Univ. **46**(3), 22–31 (2023). https://doi.org/10.11835/j.issn.1000-582X.2021.025

Zhu, Z.J., Shangguan, J.Y., Lei, Q.Z., Zhang, W.: Error control and optimization of high-precision automatic positioning tooling in metal wing assembly. China Mach. (28), 53–55, 59 (2025). https://doi.org/10.3909/j.issn.0208-2724.2025.04.011

Structure Design and Performance Study of Modified Cross Channel Micromixer

Yifan Liu[1] and Qingdang Li[2(✉)]

[1] College of Electromechanical Engineering, Qingdao University of Science and Technology, Qingdao 266061, Shandong, China
2023032029@mails.qust.edu.cn
[2] Chinesisch-Deutsche Technische Fakultät, Qingdao University of Science and Technology, Qingdao 266061, Shandong, China
lqd@qust.edu.cn

Abstract. Micromixer is a crucial component in microfluidic systems to achieve rapid mixing of fluids. It is finding applications in various fields such as biology, chemical engineering and medicine. The performance of micromixer directly impacts the functionality of microfluidic chips. This study introduced a modified cross channel passive micromixer. The impact of two structure paraments, cross angle and relative displacement, on mixing performance and fluid behavior was studied by numerical simulation. By optimizing these two parameters, the mixer can balance the increase in fluid contact area and the enhancement of lateral secondary flow. The results indicated that the modified cross channel micromixer exhibits the highest mixing index of 0.94 under the condition of $Re = 0.1$. Compared to traditional split and recombination multilayer micromixers, modified cross channel micromixer demonstrates lower pressure loss across the range of *Re* from 0.1 to 100. The micro-mixer chip was fabricated using DLP 3D printing technology, and a tea polyphenol detection experiment was conducted, proving that the micromixer can significantly reduce the time cost of the experimental process.

Keywords: Microfluidics · Micromixer · Numerical simulation · 3D printing

1 Introduction

Microfluidics, also known as "lab-on-a-chip" technology, enables the manipulation of fluids at the micrometer scale. Its unique microscale dimensions makes fluids have a high surface-area-to-volume ratio, leading to enhanced mass and heat transfer efficiencies. Consequently, microfluidics has the advantages of rapid reaction rates and low sample consumption. Therefore, microfluidic technology has been applied across various fields (Verdugo-Avello et al. 2025; Preena et al. 2022; Haghgouei and Alizadeh 2024; Sheng et al. 2025; Gong et al. 2025). In microfluidic chips, the *Re* (Reynolds number, characterizes the ratio of fluid inertia force to viscous force) is typically below 100 (Capretto et al. 2011), indicating a laminar flow condition. Under these conditions, fluid mixing relies predominantly on molecular diffusion, resulting in relatively low efficiency. Hence, micromixers are essential to achieve rapid mixing of fluids in microfluidics system.

T. Mankovits and N. Trisovic (Eds.): MDS 2026, LNME 2026.
https://doi.org/10.1007/978-981-92-2440-1_16

As research focus on microfluidic technology has grown, various types of passive micromixers have been proposed. There are parallel lamination micromixer (Lim et al. 2011; Cha et al. 2006), split-and-recombine micromixer (Aghasi and Hekmat 2022; Chen et al. 2018; Mahmud et al. 2021), obstacle based micromixers (Chen and Zhao 2017; Fan et al. 2017), Dean vortex-based micromixer (Wang et al. 2021; Muhammad et al. 2018), and convergence–divergence based micromixer (Fu et al. 2014). Furthermore, more complex micromixer structures have been developed. Tan et al. (2023) proposed a three-dimensional micromixer by incorporating internal obstacles and elliptical grooves on the channel walls of a zigzag micromixer. A split-and-recombine micromixer combined with herringbone grooved micromixer was introduced to control periodic fluid rotation and recombination (Jiang et al. 2022). Another micromixer makes fluid split and recombine in a circular channel using Twisted blade structure, and successfully mix samples in liquid chromatography detection (Dsk et al. 2023). A double-spiral micromixer with channels of varying cross-sectional shapes was designed, featuring periodic expansion and contraction of the cross-sectional area along with integrated obstacle structures within the channels (Liao et al. 2021). In three dimensions, a spiral micromixer with a circular cross-section extended the concept of spiral micromixers to three dimension, where two microchannels form a double-helix structure in space (Salieb-Beugelaar et al. 2016).

This study proposed a, novel modified cross channel micromixer. The effects of two key structure paraments on fluid mixing performance and fluid behavior were investigated through numerical simulations. Mixing index, pressure loss, and mixing efficiency cost were used as evaluation metrics to compare the modified cross channel micromixer with a reported split-and-recombine micromixer. DLP light-curing technology was used to fabricate modified cross channel micromixer. A rapid detection experiment of tea polyphenols was conducted and compared with traditional experimental methods.

2 Numerical Simulation

2.1 Micromixer Design

To investigate the fluid behavior within the structure, a 1-unit modified cross structure was established for simulation analysis, as shown in Fig. 1(a). Here, α represents the intersection angle, and d represents the channel height difference. A total of 15 configurations were examined, with $\alpha = 30°, 60°, 90°, 120°, 150°$ and $d = 0.05$ mm, 0.10 mm, 0.15 mm. Subsequently, 2-units modified cross structures with various d configurations were established, as shown in Fig. 1(b). Here, d_1 and d_2 are different values of d. Finally, the proposed modified cross channel micromixer structure is illustrated in Fig. 2. The five structural units feature an intersection angle of 90° and alternating height differences of 0.15 mm and −0.15 mm. To reduce pressure loss, the corners were smoothed. The modified cross channel micromixer and the split-and-recombine micromixer are abbreviated as "MCC" and "SAR" respectively.

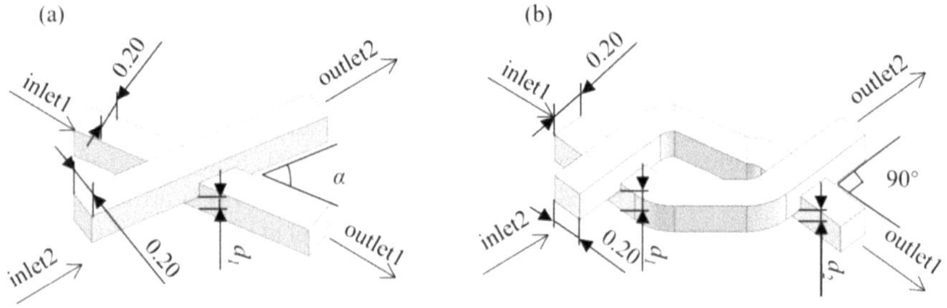

Fig. 1. (**a**) 1-unit modified cross channel structure; (**b**) 2-units modified cross channel structure (all the dimensions are in mm).

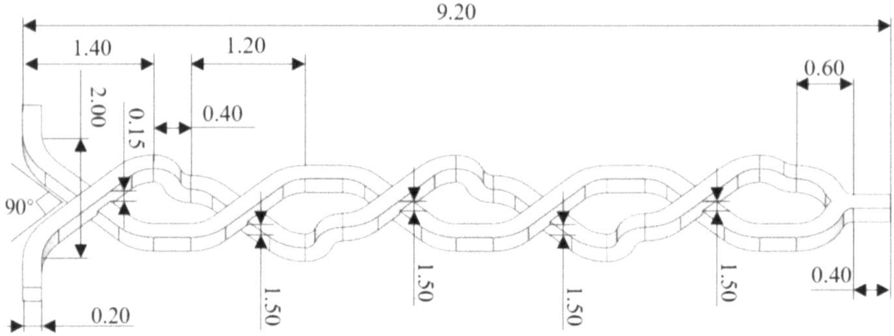

Fig. 2. Modified cross channel micromixer (all the dimensions are in mm).

2.2 Basic Principles and Control Equations

For the analysis of fluid mixing within the micromixer, the fluid is assumed to be continuous, incompressible, and Newtonian fluid, with energy transfer neglected. The fluid flow and diffusion are governed by Eqs. (1), (2) and (3):

$$\frac{\partial \mathbf{u}}{\partial t} + (\mathbf{u} \cdot \nabla)\mathbf{u} = -\frac{1}{\rho}\nabla p + \frac{\mu}{\rho}\nabla^2 \mathbf{u} \tag{1}$$

$$\nabla \cdot \mathbf{u} = 0 \tag{2}$$

$$\frac{\partial c}{\partial t} = -\mathbf{u} \cdot \nabla c + D\nabla^2 c \tag{3}$$

where ρ is the fluid density (kg m^{-3}), μ is the fluid dynamic viscosity (Pa s), $\mathbf{u}$ is the velocity vector field, c is the species concentration (mol m^{-3}).

The mixing index (MI) is defined using Eq. (4) to evaluate the mixing performance of the micromixer (Wang et al. 2011). MI ranges from 0 to 1, where 0 represents complete segregation and 1 represents perfect mixing.

$$MI = 1 - \frac{\left[\frac{1}{N}\sum_{i=1}^{N}(I_i - I_\infty)^2 \sum^{N}\right]^{1/2}}{\left[\frac{1}{N}\sum_{i=1}^{N}(I_{0i} - I_{0\infty})^2\right]^{1/2}} \tag{4}$$

Here, N is the total number of grid cells or pixels on the cross-section, I_i is the concentration at each grid cell on the outlet cross-section, I_∞ is the average concentration on the outlet cross-section, I_{0i} is the concentration at each grid cell on the inlet cross-section, and $I_{0\infty}$ is the average concentration on the inlet cross-section. The mixing efficiency cost (*Mec*), defined by Eq. (5), is used to characterize the pressure drop and residence time required to achieve a given mixing performance (Muhammad et al. 2018), serving as a metric for evaluating the mixing efficiency of the micromixer:

$$Mec = \frac{\Delta P \cdot t}{100MI} \tag{5}$$

where ΔP is the pressure loss (Pa) and t is the residence time (s) of the fluid within the micromixer.

2.3 Simulation Settings

Numerical simulations were conducted using Fluent 2022 R2 commercial software. The laminar flow model was selected for the fluid model. Water and NaCl solution were used as the two working fluids, with densities set to 998.9 kg m^{-3} and 1071 kg m^{-3}, respectively, and viscosity set to 10^{-3} kg (m s)$^{-1}$. The diffusion coefficient D was set to 10^{-10} m^2 s^{-1}, which corresponds to the typical diffusion coefficient for small molecule solutions. Regarding boundary conditions, a uniform velocity boundary condition was applied at both inlets for the two fluids, a zero-pressure outlet boundary condition was specified at the outlet, and a no-slip condition was imposed on the channel walls. For the numerical solution, the finite volume method was employed, with the SIMPLE algorithm used for pressure-velocity coupling. Momentum discretization was performed using the second-order upwind scheme, and the convergence criterion for residuals was set to 10^{-6}.

2.4 Grid Independence Verification

To balance computational efficiency and simulation accuracy, a grid independence study was conducted. The poly-hexcore mesh type was employed for grid generation in this study. For all models subjected to numerical simulation, four grids with different densities were generated and evaluated. The mixing index at the outlet cross-section under $Re = 1.58$ was examined, and the grid independence verification results are presented in Table 1. Grids with 7224287 and 10121803 cells were selected, respectively. The relative errors of the mixing index between these two grid resolutions were both less than 1%.

Table 1. Grid independence verification data

Model	Grid cells number	MI
MCC	4695605	0.7209
	7224287	0.7138

(*continued*)

Table 1. *(continued)*

Model	Grid cells number	MI
	12086162	0.7079
	15045047	0.7055
SAR	3906528	0.5554
	5943068	0.8303
	10121803	0.8289
	12806918	0.8294

2.5 Simulation Result Analysis

2.5.1 The Influence of Intersecting Angles

Figure 3 presents cross-sectional images of the fluid at the outlet of the 1-unit modified cross channel structure under different *Re*. The results indicate that a larger intersection angle enhances the intensity of lateral secondary flow at the outlet. At low Reynolds numbers ($0.1 \leq Re \leq 10$), due to the relatively small velocity magnitude of the fluid itself, the increase in the transverse velocity component resulting from a larger intersection angle has a limited effect on species distribution. At high Reynolds numbers ($10 < v \leq 100$), it is evident that a larger intersection angle leads to a more "tortuous" interface between the fluid species, indicating a higher transverse velocity component. It can be concluded that in the structure with a 150° intersection angle, although the two fluids exhibit a substantial transverse velocity component, the contact area between them is relatively small, which adversely affects molecular diffusion. At $\alpha = 30°$ intersection angle, the fluid interface remains relatively stable, but the induction of transverse secondary flow is weakened. To simultaneously ensure effective molecular diffusion and promote chaotic advection, $\alpha = 45°$ was selected for subsequent investigations.

2.5.2 The Influence of Height Difference

From Fig. 3, it can be observed that a larger d elevates the relative position of the increased fluid contact interface within the microchannel, doesn't alter the shape of the fluid interface. This conclusion holds across different *Re* regimes. Therefore, in a 1-unit modified cross channel structure, the height difference does not influence the increase in fluid contact interface or the induction of transverse secondary flow. However, in multi-units modified cross channel structures, the position of the fluid contact interface directly affects the mixing performance, necessitating a discussion on the height difference. Figure 4 presents outlet direction images of the 2-units modified cross structures with various *d* combinations at $Re = 1$. Considering the structural symmetry between configurations such as "$d_1 = 0.05, d_2 = -0.15$" and "$d_1 = -0.05, d_2 = 0.15$", where the height differences of the first modified cross unit is negative do not require discussion. From Fig. 4, the following observations can be made:

(1) When $|d_1 - d_2| = 0$ mm, the generated contact interfaces completely cancel each other out, as exemplified by $d_1 = 0.05$, $d_2 = -0.05$.
(2) When $|d_{11} - d_2| \leq 0.15$ mm, the generated contact interfaces partially overlap or cancel, as seen in $d_1 = 0.10$, $d_2 = 0.05$.
(3) When $|d_{11} - d_2| \geq 0.20$ mm, multiple distinct contact interfaces are formed, as demonstrated by $d_1 = 0.15$, $d_2 = -0.15$.

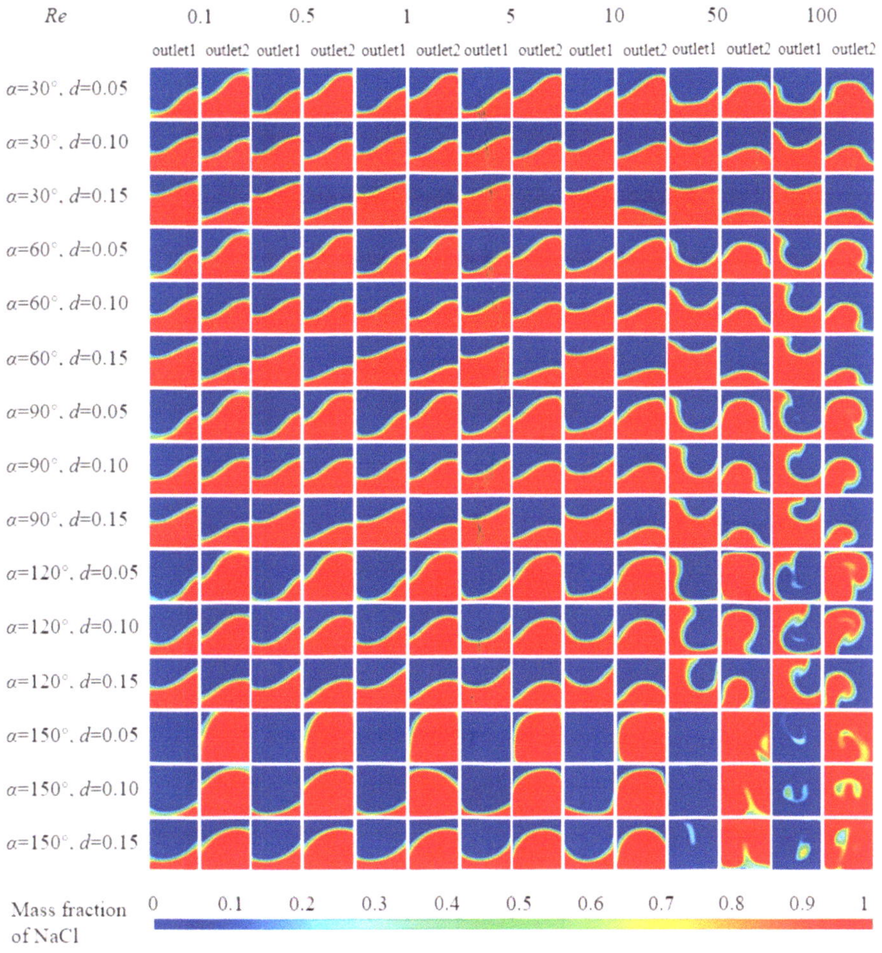

Fig. 3. Outlets of 1-unit modified cross channel structures.

Therefore, in the subsequent design of the MCC micromixer, adjacent modified cross channel units should be configured with a substantial $|d_1 - d_2|$ to optimize mixing performance.

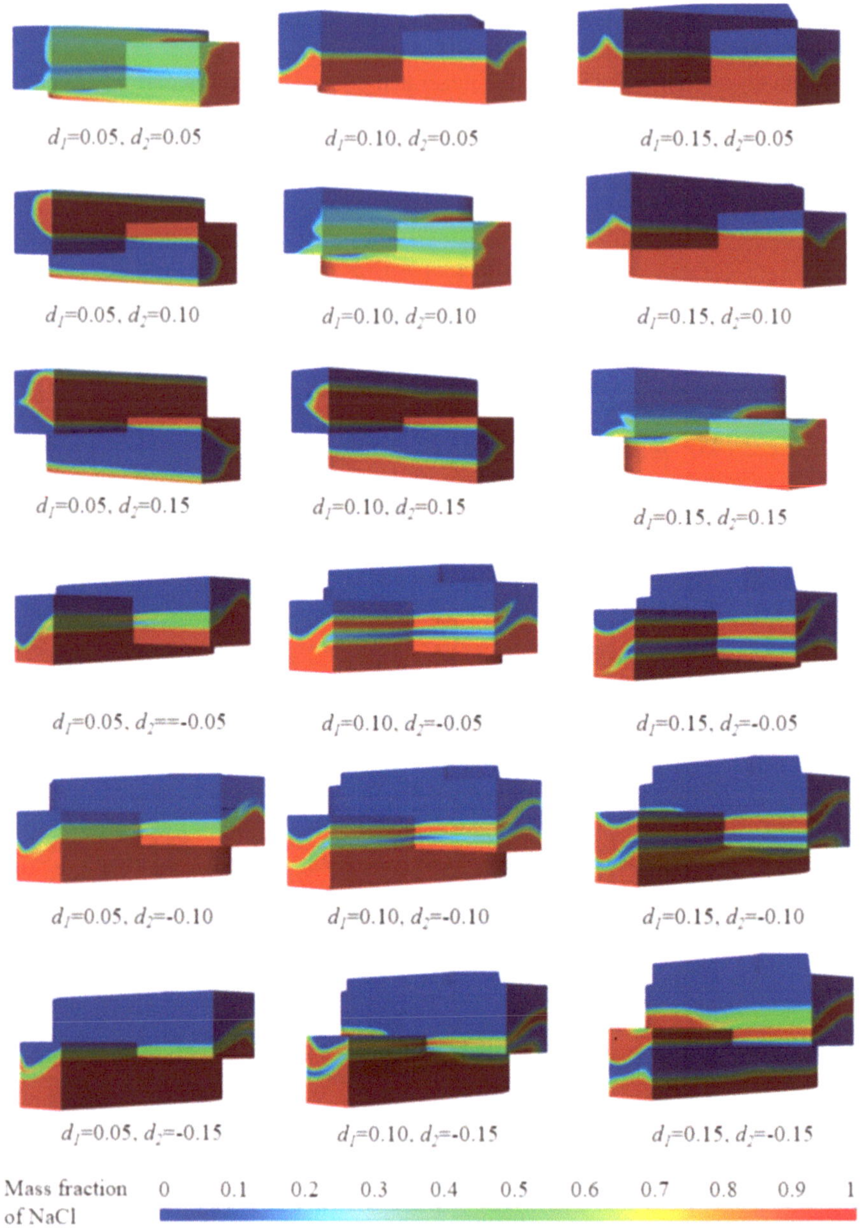

Fig. 4. Outlet direction images of 2-units modified cross channel structures.

2.5.3 Fluid Mixing in MCC Mixer and SAR Mixer

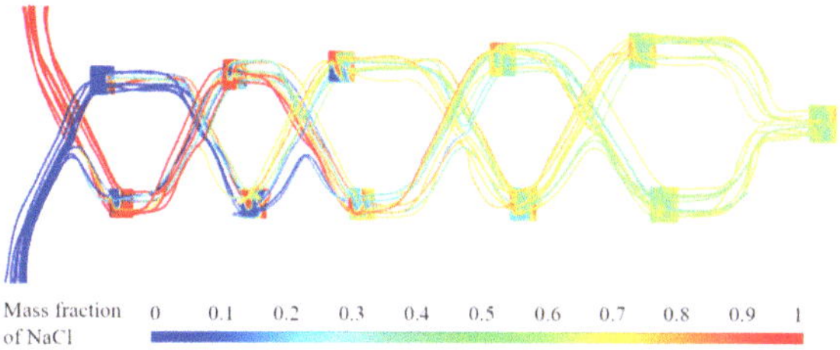

Fig. 5. Fluid mixing behavior within the MCC micromixer.

Figure 5 illustrates the fluid mixing behavior within the MCC micromixer at *Re* = 50.12. As the fluid passes through the modified cross structural units, the contact interface area increases, facilitating molecular diffusion. Simultaneously, observable lateral secondary flow is generated within the microchannel. The *MI*, Pressure loss and *Mec* of the MCC and SAR micromixers under different *Re* are presented in Fig. 6. Figure 7 illustrates the species distribution at the outlet of the MCC micromixer at various *Re* values. As shown in Fig. 6, the mixing indices of both the MCC and SAR micromixers exhibit a trend of initially decreasing and then increasing with increasing *Re*. The MCC micromixer achieves its maximum *MI* of 0.94 at *Re* = 0.10. In Fig. 7, within the range of $0.10 \leq Re \leq 12.59$, the species distribution at the outlet appears similar, with the fluid interface becoming increasingly distinct. This indicates that the lateral secondary flow remains relatively weak and insufficient to compensate for the reduced mixing caused by increased fluid velocity. Consequently, the *MI* decreases with increasing *Re*, reaching a minimum of 0.599 at *Re* = 12.59. Within the range of $25.12 \leq Re \leq 50.12$, the species distribution at the outlet undergoes a significant change, indicating a substantial enhancement of transverse secondary flow. As a result, the *MI* increases, reaching 0.885 at *Re* = 50.12. At *Re* = 100.00, the *MI* decreases to 0.763, suggesting that the contribution of the enhanced lateral secondary flow to mixing is again insufficient to offset the diminished effect of molecular diffusion. Compared with the SAR micromixer, the MCC micromixer exhibits higher *MI* at *Re* = 0.10, 0.20, 25.12, and 50.12. Under other *Re* conditions, the MCC micromixer demonstrates lower MI, with its mixing index at *Re* = 6.31 being 79% of that of the SAR micromixer. The results indicate that the pressure loss of the micromixers is approximately proportional to *Re*, with higher Reynolds numbers corresponding to greater pressure losses. At *Re* = 100, the pressure losses of the two micromixers are 6412 Pa and 8765 Pa, respectively. Under all *Re* conditions, the pressure loss of the MCC micromixer remains below 80% of that of the SAR micromixer. To evaluate the energy cost required for achieving fluid mixing in the micromixers, the mixing efficiency parameter *Mec* was employed for assessment. *Mec* represents the pressure and time cost required to achieve a 0.01 increment in the *MI*, with the results summarized in Fig. 6. For the MCC micromixer, *Mec* increases with increasing *Re* in the range of $0.10 \leq Re \leq 25.12$, decreases with increasing *Re* in the range of $25.12 \leq Re \leq 50.12$, and increases sharply at *Re* = 100.00. The minimum *Mec* value for the MCC micromixer is 1.01 at *Re* = 0.10, while the maximum value is 1.94

at *Re* = 100.00. For the SAR micromixer, *Mec* generally increases with increasing *Re*, with a minimum of 1.07 at *Re* = 0.10 and a maximum of 1.76 at *Re* = 100.00. Figure 9 demonstrates that the MCC micromixer exhibits lower *Mec* values at *Re* = 0.10, 0.20, 25.12, and 50.12. Notably, at *Re* = 50.12, the *Mec* of the MCC micromixer is only 79% of that of the SAR micromixer.

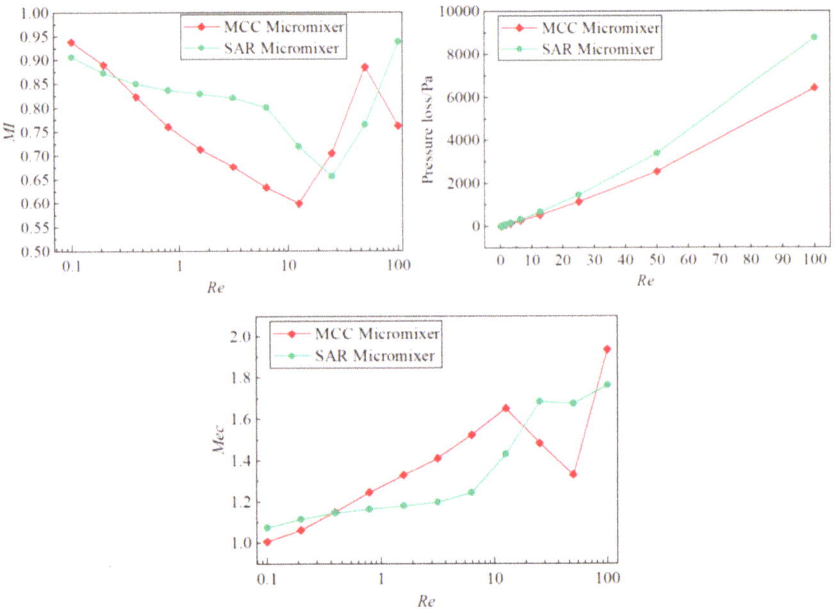

Fig. 6. MI, pressure loss and *Mec* of the MCC and SAR micromixers.

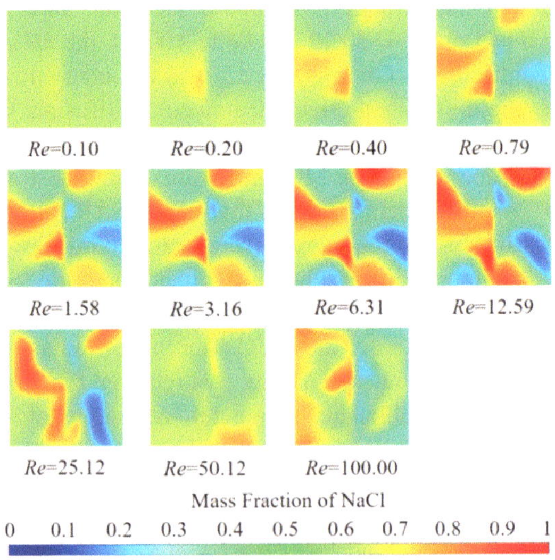

Fig. 7. The species distribution at the outlet of the MCC micromixer.

3 Tea Polyphenol Detection Experiment

3.1 Methods

The MCC micromixer chip was fabricated using a DLP 3D printer (BL-3D-100P2, Qingdao Boric 3D Manufacturing Co., Ltd.) with ABS-like photosensitive resin (Shenzhen Anycubic Technology Co., Ltd.). The printing layer thickness was set to 0.05 mm per layer. Considering that the photosensitive resin within the microchannels might cure and block the channels under cumulative light exposure (Luo et al. 2023), the microchannel dimensions were enlarged to 0.6 mm.

A tea polyphenol working solution with a mass fraction of 0.1% (Huayan Chemical (Guangdong) Co., Ltd.) and a 10% sodium carbonate solution (Xinxiang Qitian Trading Co., Ltd.) were prepared. A 1.0 mL aliquot of the tea polyphenol standard solution was reacted with 5.0 mL of Folin-Ciocalteu reagent (Xinxiang Qitian Trading Co., Ltd.) for 5 min to obtain the premix solution. For the blank control group, 5.0 mL of Folin-Ciocalteu reagent was mixed with 1.0 mL of deionized water. When using a conventional beaker and glass rod for mixing the premix solution with the sodium carbonate solution, the mixture required stirring followed by standing for 40 min. When using the micromixer to accomplish this mixing process, no stirring was required, and obtaining the same volume of sample required 22 s. Absorbance measurements were performed using a spectrophotometer (Hitachi Scientific Instruments (Beijing) Co., Ltd.). All the aforementioned experiments were conducted at room temperature.

3.2 Experiment Results

The experiment conducted using the micromixer chip is illustrated in Fig. 8. The generation of the blue-colored product at the outlet of the micromixer is clearly observable in the figure. Figure 9 presents the results of tea polyphenol detection experiments conducted using both the MCC micromixer and a conventional beaker, including a blank control group. Figure 9 demonstrates that the absorbance of the product obtained using the micromixer for mixing reaches the absorbance of the product obtained via the traditional method, with an absorption peak observed at a wavelength of 644 nm and an error in peak absorbance of less than 5%. The results indicate that using the micromixer chip for tea polyphenol detection experiments is feasible, and can reduce the time cost.

Fig. 8. Tea polyphenol detection experiment in MCC micromixer.

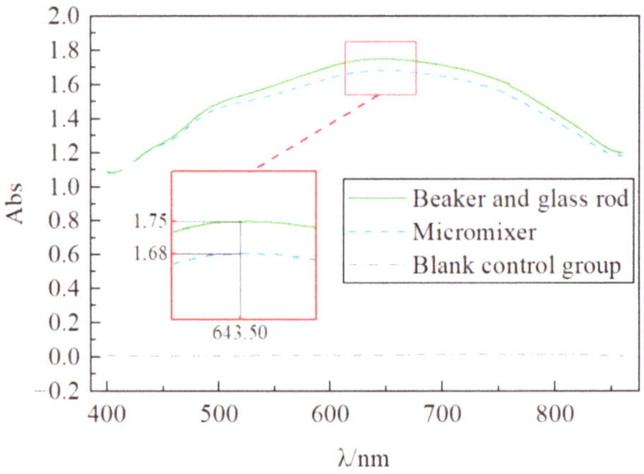

Fig. 9. Absorbance obtained by different methods.

4 Summary

This study proposes a novel modified cross channel micromixer. The performance of the MCC micromixer was investigated using a combination of numerical simulation and experimental methods. The results demonstrate that:

(1) The intersection angle α directly influences the intensity of transverse secondary flow, with larger α inducing stronger transverse secondary flow. However, an excessively large α reduces the fluid contact area. $\alpha = 45°$ ensures both a relatively large fluid contact area and the induction of adequate transverse secondary flow.
(2) The height difference d only affects the position of the fluid contact interface, without influencing its shape or the intensity of transverse secondary flow. When $|d_1 - d_2| = 0$, the increase in fluid contact interface generated by the preceding unit is canceled out. When $|d_1 - d_2|$ is less than 0.15 mm, the fluid contact interfaces may be partially canceled. When $|d_1 - d_2|$ exceeds 0.15 mm, the two units can generate multiple distinct fluid contact interfaces.
(3) Compared with the reported SAR micromixer, the MCC micromixer exhibits higher MI and lower Mec at $Re = 0.10$, 0.20, 25.12, and 50.12, as well as lower pressure loss across all Re conditions, confirming its superior capability in promoting fluid mixing.
(4) In tea polyphenol detection experiments, compared with conventional experimental methods, the MCC micromixer substantially reduced time consumption, validating the practicality of this micromixer structure.

Acknowledgments. I would like to thank my colleagues Weikang Wang, Wenguang Luo, and Guangke Wu for their encouragement, which greatly facilitated the progression of this research. I also extend my gratitude to Huiling Feng, a graduate student at the School of Materials Science and Engineering, Qingdao University of Science and Technology, for her assistance during the experimental process.

References

Aghasi, M.A., Hekmat, M.H.: A novel design of split and recombination multilayer micromixers with excellent hydraulic and mixing performance based on the baker's transformation. Chem. Eng. Process. - Process Intensif. **174**, 108894 (2022). https://doi.org/10.1016/j.cep.2022.108894

Capretto, L., Cheng, W., Hill, M., Zhang, X.L.: Micromixing within microfluidic devices. Top. Curr. Chem. **304**, 27–68 (2011). https://doi.org/10.1007/128_2011_150

Cha, J.H., et al.: A highly efficient 3D micromixer using soft PDMS bonding. J. Micromech. Microeng. **16**(9), 1778 (2006). https://doi.org/10.1088/0960-1317/16/9/004

Chen, K.X., Lu, H., Sun, M., Zhu, L., Cui, Y.P.: Mixing enhancement of a novel C-SAR microfluidic mixer. Chem. Eng. Res. Des. **132**, 338–345 (2018). https://doi.org/10.1016/j.cherd.2018.01.032

Chen, X.Y., Zhao, Z.Y.: Numerical investigation on layout optimization of obstacles in a three-dimensional passive micromixer. Anal. Chim. Acta **964**, 142–149 (2017). https://doi.org/10.1016/j.aca.2017.01.066

Dsk, P., Fodor, P.S., Kothapalli, C.R.: A flexible Kenics mixer for applications in liquid chromatography. Micromachines **14**(7), 1373 (2023). https://doi.org/10.3390/mi14071373

Fan, L.L., Zhu, X.L., Zhao, H., Zhe, J., Zhao, L.: Rapid microfluidic mixer utilizing sharp corner structures. Microfluid. Nanofluid. **21**(3), 36 (2017). https://doi.org/10.1007/s10404-017-1874-y

Fu, L.M., Fang, W.C., Hou, H.H., Wang, Y.N., Hong, T.F.: Rapid vortex microfluidic mixer utilizing double-heart chamber. Chem. Eng. J. **249**, 246–251 (2014). https://doi.org/10.1016/j.cej.2014.03.037

Gong, Y.T., Li, D., Chen, M., Lin, A.H., Chen, Q.S., Chen, X.M.: Au@Ag hollow nanoshells-based SERS integrated microfluidic chip as a sample-to-answer platform for the ultra-sensitive detection of geosmin. Anal. Chim. Acta **1335**, Article 1335000 (2025). https://doi.org/10.1016/j.aca.2024.343471

Haghgouei, H., Alizadeh, N.: Lab-on-a-chip paper-based electrochemically assisted solid-phase microextraction and ion selective sensor for determination of naproxen in biological fluid samples. Anal. Chim. Acta **1330**, 343275 (2024). https://doi.org/10.1016/j.aca.2024.343275

Jiang, Y., Zhang, B., Tian, Y., Zhang, Y., Chen, Y.F.: Analysis of different shapes of cross-sections and obstacles in variable-radius spiral micromixers on mixing efficiency. Chem. Eng. Process. **171**, 109237 (2022). https://doi.org/10.1016/j.cep.2021.108756

Liao, Y., Mechulam, Y., Lassalle-Kaiser, B.: A millisecond passive micromixer with low flow rate, low sample consumption and easy fabrication. Sci. Rep. **11**(1), 1–14 (2021). https://doi.org/10.1038/s41598-021-99471-x

Lim, T.W., et al.: Three-dimensionally crossing manifold micro-mixer for fast mixing in a short channel length. Lab Chip **11**(1), 100–103 (2011). https://doi.org/10.1039/c005325m

Luo, Z.M., et al.: Digital light processing 3D printing for microfluidic chips with enhanced resolution via dosing- and zoning-controlled vat photopolymerization. Microsyst. Nanoeng. **9**(1), 1–12 (2023). https://doi.org/10.1038/s41378-023-00542-y

Mahmud, M.R., Hossain, S., Kim, J.H.: A SAR micromixer for water-water mixing: design, optimization, and analysis. Processes **9**(11), 1914 (2021). https://doi.org/10.3390/pr9111926

Muhammad, J., Taqi, C., Cheol, P.: Analysis of passive mixing in a serpentine microchannel with sinusoidal side walls. Micromachines **9**(1), 32 (2018). https://doi.org/10.3390/mi9010008

Preena, S., Girirajsinh, J., Sanjaykumar, P.: Ultrasonic enhanced emulsification process in 3D printed microfluidic device to encapsulate active pharmaceutical ingredients. Int. J. Pharm. **620**, 121754 (2022). https://doi.org/10.1016/j.ijpharm.2022.121754

Salieb-Beugelaar, G., Gonalves, D., Wolf, M., Hunziker, P.: Microfluidic 3D helix mixers. Micromachines **7**(10), 189 (2016). https://doi.org/10.3390/mi7100189

Sheng, L., Wang, J.J., Deng, J., Luo, G.S.: Asymmetric flow-focusing: a novel microreactor operation procedure for making smaller microbubbles. Sep. Purif. Technol. **361**(Part 1), 121178 (2025). https://doi.org/10.1016/j.seppur.2024.131296

Tan, Z.J., Shi, H.H., Zheng, Y., Cao, Y.: A 3D homogeneous microreactor with high mixing intensity at wide Re range for MOF preparation and POCT application. Chem. Eng. J. **476**, 145115 (2023). https://doi.org/10.1016/j.cej.2023.146481

Verdugo-Avello, F., Wychowaniec, J.K., Villacis-Aguirre, C.A., D'Este, M., Toledo, J.R.: Bone microphysiological models for biomedical research. Lab Chip **25**(5), 806–836 (2025). https://doi.org/10.1039/D4LC00762J

Wang, S.S., Huang, X.Y., Yang, C.: Mixing enhancement for high viscous fluids in a microfluidic chamber. Lab Chip **11**(12), 2081–2087 (2011). https://doi.org/10.1039/c0lc00695e

Wang, X., Liu, Z.Q., Cai, Y.K., Wang, B., Luo, X.H.: A cost-effective serpentine micromixer utilizing ellipse curve. Anal. Chim. Acta **1155**, 338355 (2021). https://doi.org/10.1016/j.aca.2021.338355

Prediction of Tractor Traction Performance and Traction Efficiency in Hilly and Mountainous Terrain

Liyou Xu[1,2], Jiajia Wang[1], Sixia Zhao[1,2](✉), Yiwei Wu[1,2], Xianzhe Li[1,2], and Yanying Li[1]

[1] College of Vehicle and Traffic Engineering, Henan University of Science and Technology, Luoyang, China
{xlyou,9906385,wyw5156,lix-ianzhe0827}@haust.edu.cn, {230320030373,220410030019}@stu.haust.edu.cn

[2] State Key Laboratory of Intelligent Agricultural Power Equipment, Luoyang, China

Abstract. To address the difficulty of existing tractor traction performance models in accurately characterizing the influence of slope in hilly and mountainous terrains, a hierarchical prediction framework for tractor traction performance was established based on wheel–soil interaction mechanics. By introducing the soil cone index and integrating the tire index and mobility index, a soil–tire driving model incorporating slope effects was developed. Furthermore, a tire load model considering slope-induced axle load redistribution was established to describe the variation in normal loads on the front and rear axles during uphill travel. At the vehicle level, slip efficiency and rolling efficiency sub-models were formulated based on the kinematic mismatch coefficient between the front and rear axles, and were coupled with transmission efficiency to develop a comprehensive traction efficiency prediction model. Simulations were carried out under slope gradients of 5%, 15%, and 25% to predict the variation trends of tractor traction performance parameters under different operating conditions. The results provide a theoretical reference for traction performance prediction and parameter matching optimization of tractors operating in hilly and mountainous terrains.

Keywords: Hilly and mountainous terrain · Tractor · Traction performance · Soil cone index · Slip efficiency · Traction efficiency · Kinematic mismatch coefficient

1 Introduction

With the continuous advancement of agricultural mechanization in China, the mechanization of agricultural production in hilly and mountainous regions has gradually become not only an important focus of current research, but also one of the major technical challenges in the development and application of agricultural equipment. In contrast to plain areas, hilly and mountainous regions are characterized by highly undulating terrain, complex and variable slope conditions, fragmented and irregular field

T. Mankovits and N. Trisovic (Eds.): MDS 2026, LNME 2026.
https://doi.org/10.1007/978-981-92-2440-1_17

plots, and considerable heterogeneity in soil physical properties (Hu et al., 2026; Liu et al., 2025). These distinctive geographical and soil conditions make the operating environment of agricultural machinery far more complicated than that of flat farmland. Under such circumstances, tractors are more easily affected by factors such as slope gradient, soil moisture content, and tire–soil interaction during field operations. These influences often result in axle load redistribution, a decline in the adhesion performance of the driving wheels, intensified wheel slip, and reduced traction efficiency, which in turn negatively affect both operational quality and energy utilization efficiency (He et al., 2014; Bulgakov et al., 2021; Sun et al., 2023). However, most of the existing studies on tractor traction performance have been carried out under flat-ground conditions, with relatively limited attention given to the effects of slope-induced changes in gravitational components, dynamic normal load transfer, and the spatial–temporal variability of soil properties under sloping terrain. Consequently, current research findings and prediction methods are still insufficient to fully satisfy the needs of performance prediction, parameter optimization, and power machinery design for hilly and mountainous agricultural operations. Therefore, it is of considerable theoretical importance and engineering value to conduct systematic research on tractor traction performance under hilly and mountainous conditions, to establish traction prediction and evaluation models applicable to slope operations, to analyze in depth the influence of slope variation on key performance indicators such as traction force, slip ratio, and traction efficiency, and further to explore effective approaches for improving tractor traction performance. Such efforts are of great significance for enhancing the level of agricultural mechanization in hilly and mountainous areas and for promoting the development of efficient, energy-saving, and high-performance agricultural machinery.

2 Establishment of the Mathematical Model

To address the specific operational conditions of hilly and mountainous terrains, this study proposes a traction performance prediction method for complex scenarios based on the four-wheel-drive tractor traction performance model (Wong et al., 2000). A hierarchical modeling approach is adopted. First, by introducing the soil cone index and integrating the tire index and mobility index, a soil–tire driving model incorporating slope effects is developed. Second, a tire–load model accounting for axle load redistribution induced by slope is constructed. At the whole-vehicle level, starting from the kinematic mismatch coefficient between the front and rear axles, sub-models of slip efficiency and rolling efficiency are established and coupled with transmission efficiency to form an integrated traction efficiency prediction model for tractors operating in hilly and mountainous terrains. This model systematically characterizes the effects of slope gradients and soil heterogeneity on drawbar pull, slip behavior, and energy consumption.

2.1 Soil–Tire Driving Model

A soil–tire driving model is established using the ground mechanics approach based on the soil cone index. On this basis, the tire index is introduced, which uses the soil cone index, tire section width b, tire diameter d, and dynamic normal load F_Z as independent variables. It serves to quantify the overall adaptability and load-bearing capacity of the tire on cohesive–frictional composite soils. Its calculation formula is expressed as:

$$C_n = \frac{C_I bd}{F_Z} \tag{1}$$

where: Cn—Tire index; CI—Soil cone index, kPa; b—Tire section width, mm; d—Tire diameter, mm; F_Z—Tire normal load, kN.

To further accurately characterize the drawbar pull of a single wheel under sloped terrain conditions, the tire mobility index is defined (see Eq. (2)). This index comprehensively reflects: (1) the effect of radial deformation relative to the unloaded tire radius on the contact area and pressure distribution; and (2) the constraining effect of the aspect ratio on tire carcass stiffness and sidewall deformation resistance, particularly its contribution to driving stability on lateral slopes. The tire mobility index increases with the aforementioned geometric factors, indicating that the tire can more effectively interact with the soil through its own deformation to generate higher shear strength and tractive force. The calculation formula is expressed as:

$$B_n = C_n \frac{1 + 5d_\Delta/h}{1 + 3b/d} \tag{2}$$

where: d_Δ—Radial deformation of the tire, mm; h—Tire section height, mm.

In this study, the tire radial deformation d_Δ is defined as the difference between the unloaded tire radius and the loaded tire radius under the current vertical load, namely $d_\Delta = r_0 - r_d$ Under quasi-static operating conditions, the tire load–deflection relation is simplified as a linear form, and thus $d_\Delta = F_z - k_r$ where k_r denotes the equivalent radial stiffness of the tire. The parameter k_r can be identified through a static compression test or obtained from tire specification data. Accordingly, the radial deformation of the front and rear tires can be determined from their corresponding dynamic normal loads.

From Eqs. (1) and (2), it can be seen that the wheel–soil interaction in hilly and mountainous terrains can be abstracted as a functional relationship with the soil cone index, tire geometric parameters, and load conditions as the core independent variables. This allows for a clear representation of the coupled effects of soil properties, tire structure, and normal load on traction performance, even without explicitly expanding the formulas.

$$G_{TR} = \frac{G_T}{F_Z} \tag{3}$$

where: G_{TR}—Tire tractive force utilization coefficient; G_T—Tire tractive force, kN.

The physical essence of the tire tractive force utilization coefficient is to represent the driving capability that can be generated per unit of normal load. This coefficient is jointly determined by the tire mobility index and the slip ratio, and its functional relationship can be expressed as:

$$G_{TR} = \frac{G_T}{F_Z} = 0.88\left(1 - e^{-0.08B_n}\right)\left(1 - e^{-7\sigma}\right) + 0.03 \tag{4}$$

where: σ—Tire slip ratio.

This coefficient reflects the tire's ability to enhance tractive force through the tire mobility index and indicates the degree of shear displacement through the slip ratio. Therefore, the tire tractive force utilization coefficient can dynamically characterize the complete mechanical behavior of the tire–soil contact interface, from the development of shear stress with displacement to failure under different slip conditions. By incorporating this coefficient to adjust the tire mobility index, the model can dynamically represent the instantaneous tractive force generation at a specific slip ratio, significantly improving the prediction accuracy of the transient traction characteristics of tractors in hilly and mountainous terrains.

Based on the aforementioned theoretical framework, a rolling resistance coefficient is defined to comprehensively characterize the traction efficiency and energy consumption of tractors in hilly and mountainous terrains. This coefficient is defined as the ratio of the rolling resistance generated by the tire under specific soil conditions to its dynamic normal load, expressed as:

$$M_{RR} = \frac{M_R}{F_Z} \tag{5}$$

where: M_{RR}—Tire rolling resistance coefficient; M_R—Tire rolling resistance, kN.

This coefficient is a composite function of the tire mobility index and the slip ratio, and its functional relationship can be expressed as:

$$M_{RR} = \frac{M_R}{F_Z} = \frac{1.2}{B_n} + 0.03 + \frac{0.5\sigma}{\sqrt{B_{\mathrm{n}}}} \tag{6}$$

This relationship reveals that the fundamental cause of rolling resistance is the ground compaction and subsidence induced by the tire's normal load on the soil. It dynamically characterizes the mechanical process by which energy is dissipated in the tire–soil system under combined normal and tangential loads due to soil failure and settlement. By introducing the rolling resistance coefficient, the comprehensive traction efficiency evaluation model is refined. When calculating traction power or traction efficiency, this rolling resistance component must be subtracted from the total tractive force, enabling more accurate prediction of the effective drawbar pull and energy utilization of tractors during sloped field operations in hilly and mountainous terrains.

On this basis, the tire drawbar pull utilization coefficient NTR is defined to represent the portion of tractive force that can be converted into external traction output. This coefficient is determined jointly by Eqs. (4) and (6).

$$\begin{aligned} N_{TR} = G_{TR} - M_{RR} &= 0.88\left(1 - \mathrm{e}^{-0.08B\mathrm{n}}\right)\left(1 - \mathrm{e}^{-7\sigma}\right) \\ &+ 0.03 - \left(\frac{1.2}{B_{\mathrm{n}}} + 0.03 + \frac{0.5\sigma}{\sqrt{B_{\mathrm{n}}}}\right) \end{aligned} \tag{7}$$

where: N_{TR}—Tire drawbar pull utilization coefficient.

Based on Eq. (4), the calculation formulas for the front and rear tire tractive force utilization coefficients can be expressed as:

$$G_{TR\mathrm{i}} = \frac{G_{T\mathrm{i}}}{F_{\mathrm{zi}}} = G_{TR\mathrm{i}}(F_{\mathrm{zi}}, C_{ni}, B_{ni}, \sigma_i)$$

$$= 0.88(1 - e^{-0.08B_{ni}})(1 - e^{-7\sigma_i}) + 0.03 \tag{8}$$

where: $i = 1$ represents the front wheels and $i = 2$ represents the rear wheels.

Based on Eq. (6), the calculation formulas for the front and rear tire rolling resistance coefficients can be expressed as:

$$\begin{aligned} M_{RRi} &= \frac{M_{Ri}}{F_{zi}} = M_{RRi}(F_{zi}, C_{ni}, B_{ni}, \sigma_i) \\ &= \frac{1.2}{B_{ni}} + 0.03 + \frac{0.5\sigma_i}{\sqrt{B_{ni}}} \end{aligned} \tag{9}$$

Based on Eqs. (8) and (9), the calculation formulas for the front and rear tire drawbar pull utilization coefficients can be expressed as:

$$\begin{aligned} N_{TRi} &= G_{TRi} - M_{RRi} = N_{TRi}(F_{\mathrm{zi}}, C_{\mathrm{ni}}, B_{\mathrm{ni}}, \sigma_i) \\ &= 0.88\left(1 - e^{-0.08B_{ni}}\right)\left(1 - e^{-7\sigma_i}\right) - \frac{1.2}{B_{ni}} + 0.03 + \frac{0.5\sigma_i}{\sqrt{B_{ni}}} \end{aligned} \tag{10}$$

2.2 Tire–Load Driving Model

The tire–load model for tractors in hilly and mountainous terrains describes the dynamic variation of vertical loads on the front and rear tires under the combined action of external forces on the vehicle system, including gravity, drawbar pull, inertial forces, and slope-induced components. Figure 1 illustrates the force analysis of a tractor performing traction operations on an inclined slope. As shown in Fig. 1, under standard horizontal traction test conditions, both the tractor's center of mass and the line of action of the drawbar pull lie within the vehicle's longitudinal symmetry plane, resulting in relatively simple load distribution. However, under sloped conditions, the gravitational component along the slope and the height and direction of the drawbar pull line together cause significant axle load transfer and inter-axle coupling variations, which must be explicitly accounted for in the model (Karaca et al., 2026; Şeflek et al., 2025).

During tractor operation, a force analysis of the vehicle indicates that the rolling resistance moments of the front and rear wheels are given by:

$$\begin{cases} M_{\mathrm{f}1} = F_{\mathrm{z}1} M_{RR1} \mathrm{r}_1 \\ M_{\mathrm{f}2} = F_{\mathrm{z}2} M_{RR2} \mathrm{r}_2 \end{cases} \tag{11}$$

where: M_{f1}—Rolling resistance moment of the front wheels of the four-wheel-drive tractor, N m; M_{f2}—Rolling resistance moment of the rear wheels of the four-wheel-drive tractor, N m; F_{z1}—Normal reaction force on the front wheels, kN; F_{z2}—Normal reaction force on the rear wheels, kN; r_1—Radius of the front wheels of the four-wheel-drive tractor, mm; r_2—Radius of the rear wheels of the four-wheel-drive tractor, mm.

By incorporating the soil–tire driving model to represent the characteristics of the soil and tires for the front and rear wheels, the normal reaction forces on a single side of the front and rear wheels under sloped conditions can be calculated as:

$$F_{Z1} = \frac{G\cos\theta(a - M_{RR2}r_2) - F_{\mathrm{t}}H + G\sin\theta H}{2(L + M_{RR1}r_1 - M_{RR2}r_2)} \tag{12}$$

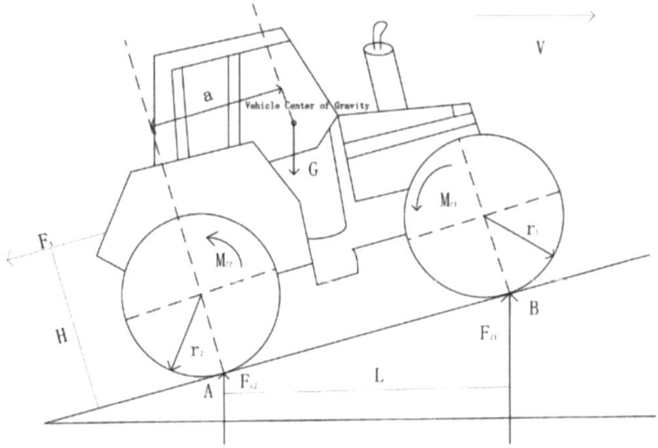

Fig. 1. Force analysis of a hilly and mountainous tractor during traction operation on a slope

$$F_{Z2} = \frac{G\cos\theta(L - a + M_{RR1}r_1) + F_t H - G\sin\theta H}{2(L + M_{RR1}r_1 - M_{RR2}r_2)} \tag{13}$$

where: G—Actual weight of the tractor, kN; L—Wheelbase of the tractor, mm; F_t—Tractor total drawbar pull, kN; H—Hitch point height, mm; a—Longitudinal distance from the center of mass to the rear axle, mm; θ—Slope angle.

For analytical convenience, the center of mass is assumed to lie in the same vertical plane as the line of action of the drawbar pull, and both chassis roll and differences between the left and right tire parameters are neglected. Under these assumptions, the left and right wheel loads on the same drive axle are approximately equal. Based on Eqs. (12) and (13), the axle loads on the front and rear drive axles can be further calculated as:

$$\begin{cases} W_1 = 2F_{z1} \\ W_2 = 2F_{z2} \end{cases} \tag{14}$$

where: W_1—Axle load of the front drive axle, kN; W_2—Axle load of the rear drive axle, kN.

Based on the axle loads and the tire–soil sub-model, the tractive forces of the front and rear drive axles can be derived as:

$$\begin{cases} F_{q1} = G_{TR1}W_1 \\ F_{q2} = G_{TR2}W_2 \end{cases} \tag{15}$$

where: F_{q1}—Tractive force of the front drive axle, kN; F_{q2}—Tractive force of the rear drive axle, kN.

$$\begin{aligned} F_q &= \sum_{i=1}^{2} G_{TRi}W_i = F_t - F_{RR} \\ &= \sum_{i=1}^{2} N_{TRi}W_i - \sum_{i=1}^{2} M_{RRi}W_i \end{aligned} \tag{16}$$

where: F_q—Total tractive force of the four-wheel-drive tractor, kN; F_{RR}—Total rolling resistance of the tractor, kN.

2.3 Efficiency Model

For four-wheel-drive tractors operating on slopes in hilly and mountainous terrains, differences in soil compaction, axle load transfer, and tire characteristics lead to significant disparities in adhesion and optimal slip ratio between the front and rear axles. Under rigid drive conditions, assuming identical theoretical axle speeds may cause the axle with lower adhesion to reach excessive slip first, thereby reducing tractive performance, inducing power circulation, and ultimately decreasing overall traction efficiency while intensifying soil disturbance.

To address this issue, it is necessary to assign different theoretical speeds to the front and rear axles to match their respective optimal slip ratios, thereby preventing severe wheel slip. For this purpose, a kinematic mismatch coefficient between the front and rear axles is introduced to represent this theoretical speed difference. It is defined as the ratio of the difference in theoretical speeds between the front and rear axles to the average theoretical speed of the vehicle:

$$\varepsilon = \frac{V_{L1} - V_{L2}}{V_{L1}} \tag{17}$$

where: ε—Kinematic mismatch coefficient between the front and rear axles; V_{L1}—Theoretical speed of the front axle, km/h; V_{L2}—Theoretical speed of the rear axle, km/h.

The kinematic mismatch coefficient is a function of slope angle, dynamic axle loads, tire mobility indices of the front and rear wheels, and soil properties. Its optimal value is intended to allow the front and rear axles to simultaneously reach or closely approach their respective optimal slip ratios, thereby maximizing overall traction efficiency while minimizing soil disturbance. For advanced hilly and mountainous tractors equipped with limited-slip differentials or independently driven axles, real-time control strategies can dynamically adjust the theoretical speeds or torque distribution between the front and rear axles to manage the kinematic mismatch coefficient, maintaining optimal traction performance under varying slopes and soil conditions.

The actual operating speed of the tractor is given by:

$$V = V_1 = V_2 \tag{18}$$

where: V—km/h Actual operating speed of the tractor, km/h; V_1—Actual operating speed of the front axle, km/h; V_2—Actual operating speed of the rear axle, km/h.

The theoretical operating speed of the tractor is given by:

$$V_L = \frac{V}{1 - \sigma} \tag{19}$$

where: V_L—Theoretical operating speed of the tractor, km/h.

By substituting Eq. (20) into Eq. (19), we obtain:

$$V = V_{L1}(1 - \sigma_1) = V_{L2}(1 - \sigma_2) \tag{20}$$

where: σ_1—Front wheel slip ratio; σ_2—Rear wheel slip ratio.

By substituting Eq. (17) into Eq. (20), the relationship between the front and rear wheel slip ratios and the kinematic mismatch coefficient of the axles can be derived as:

$$\sigma_1 = \sigma_2 + \varepsilon(1 - \sigma_2) \tag{21}$$

At the power level, slip efficiency is defined as the ratio of the actual output power of the entire vehicle to its theoretical driving power. Considering the kinematic mismatch between axles, the theoretical driving power of the vehicle is the sum of the theoretical driving powers of the front and rear axles. The calculation formula for slip efficiency is thus derived as:

$$\eta_\sigma = \frac{F_{q1}V + F_{q2}V}{F_{q1}V_{L1} + F_{q2}V_{L2}} \tag{22}$$

where: η_σ—Slip efficiency of the four-wheel-drive tractor.

By substituting Eq. (19) into Eq. (22), we obtain:

$$\eta_\sigma = \frac{F_q}{F_q + F_{q1}\frac{\sigma_1}{1-\sigma_1} + F_{q2}\frac{\sigma_2}{1-\sigma_2}} \tag{23}$$

From Eq. (23), slip efficiency can be expressed as a function of the tractive force distribution between the front and rear axles and the slip ratios of the respective wheels. Combined with the definitions of the tractive force and drawbar pull utilization coefficients in Eqs. (4) and (7), the coupled dependence of slip efficiency on tire parameters, soil properties, and the kinematic mismatch coefficient can be further revealed.

Based on Eq. (16), the rolling efficiency model can be expressed as:

$$\eta_f = \frac{(F_q - F_{RR})V}{F_q V} \tag{24}$$

where: η_f—Rolling efficiency of the four-wheel-drive tractor.

Based on Eqs. (23) and (24), the traction efficiency prediction model for the four-wheel-drive tractor can be derived as:

$$\eta_T = \eta_D\eta_c = \eta_\sigma\eta_f\eta_c = \frac{\left(F_q^2 - F_{RR}F_q\right)\eta_c}{F_q^2 + F_{q1}F_q\frac{\sigma_1}{1-\sigma_1} + F_{q2}F_q\frac{\sigma_2}{1-\sigma_2}} \tag{25}$$

where: η_T—Traction efficiency of the four-wheel-drive tractor; η_D—Vehicle running efficiency of the four-wheel-drive tractor; η_c—Transmission efficiency of the tractor.

Based on the soil–tire driving model, the tire–load driving model, and the whole-vehicle efficiency evaluation model described above, a comprehensive traction performance prediction model for tractors in hilly and mountainous terrains is established. This model effectively characterizes how traction efficiency during tractor operations is influenced by overall slip, wheel–soil interactions, and whole-vehicle kinematic mismatch.

3 Establishment of the Simulation Model and Analysis of the Results

A 36.8 kW four-wheel-drive hilly and mountainous tractor of the Dong fanghong brand is selected as the study object. Its main structural parameters and mass distribution are shown in Table 1. To evaluate its slope-field operational performance under high-strength soil conditions, the simulation environment is set to solid agricultural soil with a soil cone index (CI) of 1000 kPa. Within the MATLAB/Simulink numerical simulation environment, a tractor traction dynamics model for hilly and mountainous terrains is established. Submodules are constructed individually and strictly follow the solution algorithm shown in Fig. 2 to develop the mathematical model for sub-sequent simulation analysis.

Table 1. Whole machine parameters for case study analysis

Parameter	Value/type
Machine model	4 × 4 wheeled
Overall dimensions (L × W × H) mm	3378 × 1280 × 2020
Wheelbase/mm	1338
Tire specification	260/70R20
Engine rated power/kW	36.8
Static front axle load W1/kN	10
Static rear axle load W2/kN	15
Actual gravity weight W/kN	25
Traction point height H/mm	400

Figure 3 shows the relationship curves between the drawbar pull and the tire tractive force utilization coefficients for the front and rear wheels under different slope gradients (5, 15, and 25%), with the right panel providing a magnified view of the left panel. The figure indicates that, as the drawbar pull increases, the tractive force utilization coefficients gradually rise and approach an upper limit. This occurs because increasing drawbar pull requires the tires to mobilize more of the available wheel–soil shear strength. At the same drawbar pull but under different slope gradients, the utilization coefficients differ, and larger slopes correspond to higher coefficients. This indicates that steeper slopes impose a higher traction demand on the tire–soil system and bring the wheels closer to their adhesion limits. Under a gentle slope condition (5%), the front and rear wheel utilization coefficients are relatively close, indicating a comparatively balanced tractive-force demand between the two axles. In contrast, at a steep slope condition (25%), the front wheel utilization coefficient is significantly higher than that of the rear wheels. This suggests that, under uphill conditions, the front wheels reach a higher utilization level earlier because the rearward transfer of axle load reduces the front-axle normal load and weakens its adhesion reserve. Therefore, in hilly and

Fig. 2. Flow chart of traction performance of hilly and mountainous tractors

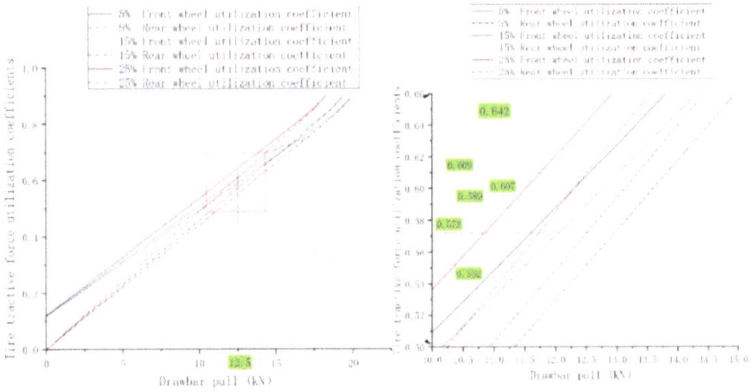

Fig. 3. Relationship between the tire tractive force utilization coefficients and the drawbar pull of a hilly and mountainous terrain

mountainous terrain, increasing slope gradients aggravate the imbalance of inter-axle traction demand and make proper matching of axle load distribution and front–rear axle kinematic relationship more important for maintaining efficient traction operation.

Figure 4 shows the relationship between tire slip ratio and drawbar pull for the front and rear wheels under slope gradients of 5, 15, and 25%, with the right panel

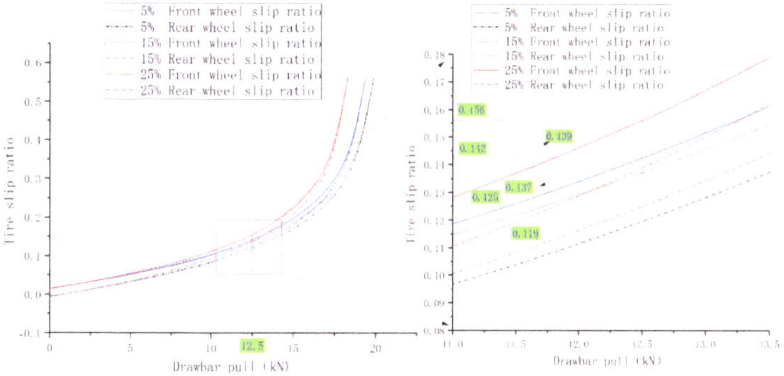

Fig. 4. Relationship between the tire slip ratio and the drawbar pull of a hilly and mountainous terrain

providing an enlarged view. The slip ratios of both axles increase with drawbar pull because greater tractive demand intensifies the longitudinal sliding between the tires and the ground. Under the same drawbar pull, the front wheels generally exhibit higher slip ratios than the rear wheels, mainly because the gravitational component along the slope transfers axle load rearward during uphill travel, reducing the normal load and adhesion reserve of the front axle. As slope gradient increases, the slip ratios of both axles rise further, indicating aggravated longitudinal shear deformation at the wheel–soil interface and a higher risk of traction deterioration. On steep slopes, the wheels enter the high-slip region earlier, which reduces effective tractive performance. Therefore, optimizing the kinematic mismatch coefficient to coordinate the front–rear speed ratio or torque distribution is necessary to keep both axles operating near their optimal slip ranges, thereby improving traction efficiency and reducing ineffective slip loss.

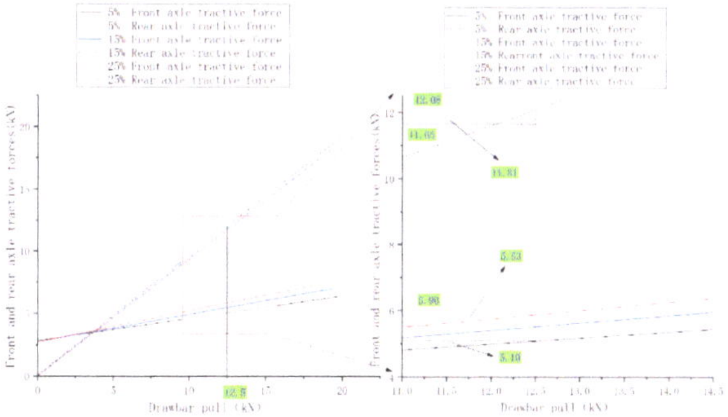

Fig. 5. Relationship between the front and rear axle tractive forces and drawbar pull of a hilly and mountainous terrain

Figure 5 shows the relationship curves between the front and rear axle tractive forces and drawbar pull under different slope gradients (5, 15, and 25%), with the right panel providing a magnified view of the left panel. The figure shows that, at all slope gradients, the front and rear axle tractive forces increase approximately linearly with drawbar pull.

Uphill operation introduces a gravitational component along the slope and increases rolling resistance, requiring a higher total tractive force to maintain equilibrium (see Eq. (17)). As a result, for the same drawbar pull, the actual tractive-force demand of both axles increases. Under a gentle slope condition (5%), the tractive-force distribution between the front and rear axles remains relatively balanced. As the slope gradient increases to 25%, the difference in tractive-force response between the two axles becomes more pronounced. This is because slope-induced rearward axle load transfer changes the normal-load distribution, thereby altering the adhesion reserve and slip state of each axle. In particular, the front axle operates under a lower normal load and is therefore more sensitive to slip, whereas the rear axle carries a larger normal load and provides a more stable adhesion reserve. Consequently, proper matching of the front–rear axle kinematic relationship is essential for preventing premature slip of the lightly loaded axle and for maintaining overall traction efficiency on steep hilly terrain.

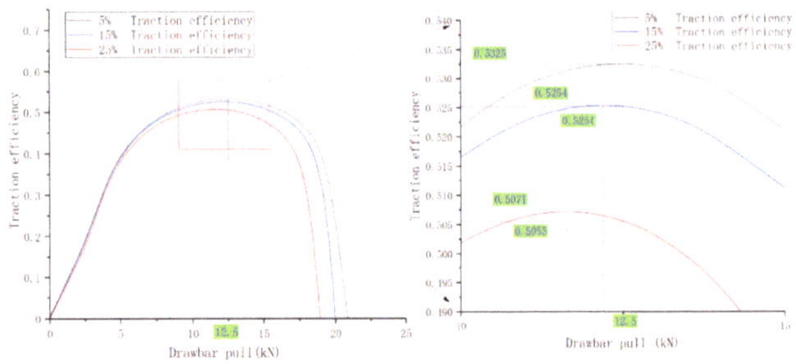

Fig. 6. Relationship between traction efficiency and drawbar pull of a hilly and mountainous terrain

Figure 6 presents the relationship curves between traction efficiency and drawbar pull under different slope gradients (5, 15, and 25%), with the right panel showing an enlarged view of the left panel. The figure indicates that the curve of traction efficiency versus drawbar pull follows a clear rise-and-fall pattern, with a distinct peak value. Traction efficiency reaches its maximum at a moderate drawbar pull, where tire slip remains within an appropriate range and power transmission is most effective. As drawbar pull increases further, tire slip becomes more severe, energy losses increase, and traction efficiency gradually declines. Under gentle slope conditions, the peak traction efficiency is relatively high. As the slope gradient increases, however, the peak efficiency decreases. This is mainly because steeper slopes lead to greater lateral tire slip and higher rolling resistance, which reduce the proportion of effective tractive force and, in turn, lower the overall traction efficiency.

Figure 7 presents the relationship curves between slip efficiency and drawbar pull under different slope gradients (5, 15, and 25%). The figure indicates that slip efficiency decreases gradually at first as drawbar pull increases and then declines sharply after drawbar pull exceeds a certain threshold. Under low drawbar pull, slip efficiency remains close to 1, which indicates that tire slip is still limited and that most of the driving motion is effectively converted into forward travel. As drawbar pull continues to increase, the tires progressively enter the slip region and slip efficiency drops rapidly. For the same

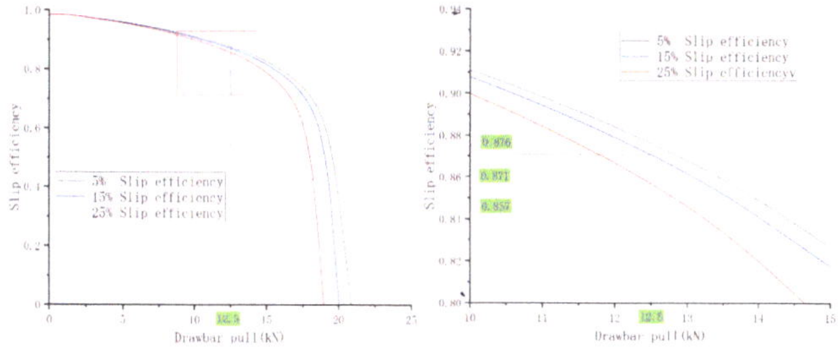

Fig. 7. Relationship between slip efficiency and drawbar pull of a hilly and mountainous terrain

drawbar pull, slip efficiency is generally higher at smaller slope gradients than at steeper slopes. This is mainly because wheel slip remains relatively limited on gentle slopes, allowing a larger proportion of the driving motion to be converted into useful traction. On steeper slopes, however, the required tractive effort becomes larger and the front–rear axle load distribution becomes less favorable, which intensifies relative sliding at the wheel–soil interface and lowers slip efficiency. These results suggest that, in hilly and mountainous operations, increasing slope gradients accelerate the onset of the high-slip region and reduce the proportion of useful tractive motion. Therefore, maintaining the wheels within an appropriate slip range through proper matching of operating load, axle load distribution, and inter-axle kinematic relationship is essential for improving traction efficiency on sloped terrain.

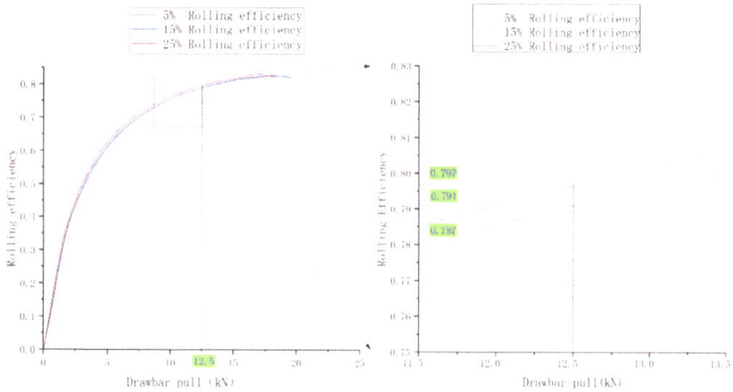

Fig. 8. Relationship between rolling efficiency and drawbar pull of a hilly and mountainous terrain

Figure 8 shows the variation of rolling efficiency with drawbar pull under slope gradients of 5, 15, and 25%. For all slope conditions, rolling efficiency increases with drawbar pull and then gradually approaches a stable value. This trend indicates that the contribution of drawbar pull to rolling efficiency is significant at low load levels but becomes progressively weaker as drawbar pull further increases. This behavior can be explained by the change in the ratio of rolling resistance to driving force. As drawbar pull increases, a larger driving force is required, leading to a reduction in F_{RR}/F_q. Consequently, a greater proportion of the driving force is converted into effective traction output, resulting in an increase in rolling efficiency. When drawbar pull reaches a relatively high level, the decrease in F_{RR}/F_q becomes less pronounced, and the rolling efficiency correspondingly tends to level off. In addition, the results show that a larger slope gradient yields a slightly higher rolling efficiency. This is mainly because the gravitational component along the slope increases the baseline driving force demand. Although the increase in rear axle load also raises rolling resistance, its relative increment is smaller than that of the driving force. Therefore, under the same drawbar pull, rolling efficiency is slightly improved as the slope gradient increases.

In summary, the models developed in this study not only capture the kinematic mismatch between the front and rear axles and the differences in tire–soil adhesion between the front and rear wheels under hilly and mountainous operating conditions, but also accurately represent tractor efficiency under different slope gradients. These results confirm the strong validity and practical applicability of the proposed models.

4 Model Applicability, Error Sources, and Engineering Simplification Conditions

The proposed model is mainly applicable to the straight-line traction analysis of four-wheel-drive tractors operating on hilly and mountainous farmland under low-to-medium speed and quasi-static working conditions. Its primary purpose is to describe and predict tractor traction behavior during typical slope operations within the normal agricultural working range. Under such conditions, the model can be used to evaluate key performance indicators, including traction force, slip behavior, and traction efficiency, particularly when slope variation remains within a practically relevant range and soil conditions can be reasonably represented by an equivalent cone index. In this respect, the model is suitable for engineering analysis and performance prediction of tractors operating on sloped farmland where the terrain is continuous and the operating state changes relatively slowly. However, the present model is not intended to represent extreme transient operating conditions, such as rapid steering maneuvers, severe vibration, strong impact excitation, or highly discontinuous terrain surfaces. Under these circumstances, the tractor system exhibits significant dynamic coupling and transient response characteristics that fall outside the scope of the current quasi-static formulation.

To ensure both computational efficiency and engineering applicability, several simplifying assumptions were introduced during model development. First, the tractor was simplified as a two-dimensional longitudinal system so that the analysis focused only on longitudinal force transmission and load transfer, while left–right load differences and lateral dynamic effects were neglected. This treatment is appropriate for straight-line traction analysis, but it cannot capture potential asymmetries caused by lateral slopes, uneven wheel loading, or steering-induced weight redistribution. Second, tire radial deformation was represented by an equivalent load–deflection relationship. This simplification allows the model to reflect the basic influence of vertical load on tire deformation and traction behavior while avoiding the high complexity associated with detailed tire structural modeling. As a consequence, the nonlinear dynamics of the tire carcass, tread deformation, and local redistribution within the contact patch were not explicitly considered. Third, soil–tire interaction was described by equivalent soil characteristic parameters, which enabled the main mechanical response of the soil to be incorporated into the traction model in a concise manner. However, the actual spatial heterogeneity of soil properties, including variations in moisture content, compaction degree, and surface roughness, was not fully represented. In addition, transmission efficiency was treated as an equivalent parameter, and the coupled effects of transient vibration, impact loading, and engine or drivetrain power fluctuation were ignored so that the model remained suitable for steady or slowly varying working conditions.

Accordingly, the prediction error of the model mainly arises from several sources. One important source is the uncertainty associated with identifying the soil cone index, since the equivalent cone index may not fully reflect the actual complexity and spatial variability of field soil conditions. Another source lies in the approximation of tire radial stiffness, which simplifies the actual tire deformation process and may introduce deviations under different loading and contact conditions. Moreover, simplification of the tire–soil interaction characteristics inevitably reduces the model's ability to describe local and nonlinear phenomena at the wheel–soil interface. The omission of dynamic disturbances during actual slope operation, such as vibration, impact, and short-term load fluctuation, also contributes to discrepancies between predicted and actual tractor performance. Even so, under the above assumptions and within its intended range of application, the proposed model remains suitable for preliminary parameter matching, comparative analysis of slope effects, and trend prediction of tractor traction performance in hilly and mountainous terrain. Therefore, although it is not a full dynamic model applicable to all operating scenarios, it still provides a practical and effective analytical tool for engineering evaluation and theoretical study of tractor traction characteristics under typical slope working conditions.

5 Conclusions

(1) This paper responds to the need for analyzing tractor traction performance under slope operating conditions in hilly and mountainous regions. To address the limitations of conventional traction performance models, which do not adequately account for slope effects, a tractor traction performance prediction model applicable to sloping terrain was developed. Based on the tire–soil interaction mechanism, the model systematically considers the influence of slope variation on the overall force state of the tractor, axle load distribution, and tire driving characteristics. It further establishes the relationships among drive wheel slip, traction output, and traction efficiency, thereby providing a theoretical foundation for the analysis of tractor traction performance in hilly and mountainous areas.

(2) The results show that slope variation has a significant effect on the normal load distribution between the front and rear axles of the tractor, which in turn alters the adhesion capacity of the drive wheels and the overall traction performance of the machine. As the slope increases, load transfer during tractor operation becomes more pronounced, and the operating state of the drive wheels changes accordingly, leading to different degrees of variation in key performance indicators such as slip ratio, traction force, and traction efficiency. These findings indicate that tractor traction

performance under slope conditions is highly sensitive to terrain gradient. Therefore, analytical methods established for flat-ground conditions are difficult to apply directly to the complex operating environments of hilly and mountainous regions.

(3) The proposed model can effectively characterize variations in tractor traction performance under slope conditions and provide a reference for the structural design, parameter matching, and performance optimization of tractors used in hilly and mountainous regions. The findings have practical engineering value for improving tractor traction capacity and operational adaptability under complex terrain conditions, reducing ineffective energy consumption, and enhancing power utilization efficiency. In addition, this study provides theoretical support for the optimal design and broader application of agricultural machinery and equipment in hilly and mountainous areas.

Acknowledgements. The authors would like to express their sincere gratitude to all those who contributed to this study. Special thanks are extended to the supervisors and colleagues for their valuable guidance, constructive suggestions, and continuous support throughout the research process. The authors also appreciate the assistance provided in data collection, model development, and manuscript revision. In addition, the anonymous reviewers are gratefully acknowledged for their insightful comments, which helped improve the quality of this paper.

References

Bulgakov, V., et al.: Maximum admissible slip of tractor wheels without disturbing the soil structure. Appl. Sci. **11**(15), 6893 (2021). https://doi.org/10.3390/app11156893

He, J., Shu, S.-F., Liao, Y., Li, Y.-D., Chen, L.-C., Ye, C., Pan, S.: Current situation and solution countermeasures of rice production mechanization in hilly and mountain areas of southern China. J. Anhui Agric. Sci. **42**(28), 9995–9997, 10006 (2014). https://doi.org/10.13989/j.cnki.0517-6611.2014.28.129

Hu, J., Qiu, X., Jiao, W., et al.: Research on drive anti-slip control of tractors in hilly and mountainous areas. Agric. Equip. Veh. Eng. 1–15 (2026, March 26)

Karaca, M., Carabin, G., Mazzetto, F.: Evaluating implement-induced CoG shifts on tractor rollover stability: a multibody dynamics and experimental study across different machine orientations. Comput. Electron. Agric. **240**, 111174 (2026). https://doi.org/10.1016/j.compag.2025.111174

Liu, Q., et al.: Development and validation of traction performance for mountain tracked chassis on slopes. Trans. Chin. Soc. Agric. Mach. **56**, 597–607 (2025). https://doi.org/10.6041/j.issn.1000-1298.2025.05.058

Şeflek, A.Y., Orhan, N., Özkan, M.: The impact of tractor drawbar height on performance and optimization using response surface methodology. Sci. Rep. **15**(1), 5370 (2025). https://doi.org/10.1038/s41598-025-88821-8

Sun, J., Liu, Z., Yang, F., Sun, Q., Liu, Q., Luo, P.: Research review of agricultural equipment and slope operation key technologies in hilly and mountains region. Trans. Chin. Soc. Agric. Mach. **54**(5), 1–18 (2023). https://doi.org/10.6041/j.issn.1000-1298.2023.05.001

Wong, J.Y., Zhao, Z., Li, J., McLaughlin, N.B., Burtt, S.: Optimization of the tractive performance of four-wheel-drive tractors—correlation between analytical predictions and experimental data. SAE Trans. **109**, 252–260 (2000)

Electromechanical Optimization of an Eccentric Hedge Trimmer Mechanism with Respect to Cutting Force

Bálint Siktár(✉), József Kakuk, and György Hegedűs

University of Miskolc, Miskolc-Egyetemváros, 3515 Miskolc, Hungary
balint.siktar@student.uni-miskolc.hu

Abstract. Hedge trimmers commonly employ eccentric cam mechanisms to convert rotary motion into reciprocating blade motion. While this solution offers compactness and manufacturability, the eccentric's geometric design strongly influences blade kinematics, cutting force characteristics, vibration levels, and electrical power consumption. This paper presents a geometry-driven optimization framework for an eccentric hedge-trimmer mechanism, based on a coupled electromechanical model of the drive system. The blade's motion, velocity, and acceleration over time depend on both its eccentric shape and its rotation speed. A simplified cutting force model is introduced, linking blade kinematics to cutting resistance, while electrical power consumption and current demand are calculated through the motor model. The results show that properly calibrated geometric parameters for the eccentric mechanism increase system energy efficiency and mechanical loading capabilities using existing system elements. The proposed approach supplies practical design guidelines for developing energy-efficient, low-vibration hedge trimmers and demonstrates the usefulness of electromechanical modelling for geometry-based optimization in handheld power tools.

Keywords: Hedge trimmer · Eccentric mechanism · Cutting force optimization · Energy consumption

1 Introduction

Cam and eccentric mechanisms are widely used to transform rotary motion into prescribed follower motion in a broad range of mechanical systems. Due to their compactness and toughness, disk-type cams and eccentric solutions are especially suitable for applications that require repetitive reciprocating motion in a limited space. As a result, the optimization of cam and eccentric geometries has been an active research topic for several decades, with a strong focus on improving kinematic performance, dynamic behavior, and reliability.

Early and recent studies have demonstrated that the geometric shape of cam profiles decisively influences follower displacement, velocity, acceleration, and further derivatives, which, in turn, affect contact forces, vibration levels, and mechanical loads. Classical and modern optimization approaches have been applied to cam-follower mechanisms, including genetic algorithms (Lampinen 2003), cam profile optimization for novel

T. Mankovits and N. Trisovic (Eds.): MDS 2026, LNME 2026.
https://doi.org/10.1007/978-981-92-2440-1_18

drive concepts (Xiao and Zu 2009), comparative studies of metaheuristic optimization techniques (Abderazek et al. 2020), and more recent hybrid and improved algorithms targeting reliability and dynamic limitations (Wang and Peng 2024). More recent examinations have extended cam-based optimization towards compact eccentric and indexing mechanisms, where geometric parameters are explicitly linked to dynamic performance and kinematic smoothness through normalized displacement, velocity, and acceleration laws (Cardona et al. 2014). The research shows that improving cam geometry can significantly enhance mechanical performance, assuming the right goals and limitations are used. In addition to kinematic analyses, research has also explored dynamic phenomena, including follower vibration and balancing. Gatti and Mundo (2010) showed that dynamic excitation can be reduced by directly controlling follower vibration through profile design. Furthermore, Chang et al. (2022) examined enhanced balancing techniques for conjugate disk cams to reduce inertial loads. These contributions underline the importance of acceleration and jerk characteristics in cam-driven systems, particularly for high-speed or cyclic operation. Furthermore, studies on constant-breadth and shaped cam mechanisms have shown that higher-order kinematic quantities are highly sensitive to the applied motion reconstruction method, and that direct differentiation of cam envelope geometries may introduce non-physical artefacts in acceleration and jerk evaluation (Yang et al. 2019).

Although these advances have been made, most existing cam optimization studies focus primarily on mechanical performance indicators, such as contact force, vibration, and manufacturing cost (Ahmed et al. 2024). While such objectives are essential, they are often investigated independently of the characteristics of driving electric motor. In many practical applications, however, cam mechanisms are integrated into electromechanical systems, where the imposed motion law directly influences motor current demand, power oscillations, and energy efficiency. The relationship between cam design shapes and electromagnetic motor operation patterns has not received sufficient research attention. The restriction is most critical in handheld power tools, such as hedge trimmers, which use eccentric cam mechanisms to operate their reciprocating cutting blades. The unique design of these tools determines both blades operating patterns and cutting strength, and simultaneously helps users feel less vibration, extends battery life, and safeguards the motor from overheating. The literature lacks research on how eccentric tool geometry affects mechanical and electrical performance in specific gardening applications. An example of industrial relevance is the patented oscillating gardening device that uses a shaped eccentric gear element (Milus et al. 2016). The patent demonstrates that non-circular eccentric geometries are already applied in practice to improve cutting behavior and mechanical performance. However, the underlying design rationale is not disclosed, and no systematic electromechanical analysis is provided. Motivated by these observations, the present work suggests a geometry-driven optimization method for an eccentric hedge-trimmer mechanism that explicitly integrates displacement-velocity-acceleration-jerk (S-V-A-J) characteristics with an electromechanical drive model. Instead of directly optimizing geometric dimensions, the eccentric geometry is derived from prescribed motion laws (an inverse motion-to-geometry approach), allowing acceleration and jerk to be controlled explicitly. A simplified cutting force formulation is coupled with an averaged BLDC motor model, supporting simultaneous evaluation of cutting force capability, motor current demand, and power changes. In addition, an experimentally manufactured shaped eccentric, corresponding to the patented concept in (Milus et al. 2016), is used as an industrial reference for kinematic comparison.

By combining cam profile theory, S-V-A-J-based motion design, and electromechanical modelling, the proposed approach addresses a gap in existing literature. It provides practical design guidance for energy-saving, low-vibration hedge-trimmer mechanisms.

2 System Overview and Modelling Assumptions

2.1 Mechanical Configuration

The investigated hedge trimmer mechanism consists of a shaped eccentric disc driven by an electric motor. The disc is guided directly within an elongated slot formed in the reciprocating blade (or blade carrier), creating a cam–slot constraint, as it can be seen in Fig. 1. As the disc rotates, its non-circular contour remains in contact with the slot boundaries and imposes a prescribed translational motion on the blade along a single degree of freedom. The mechanism may be idealized as a planar cam system in which the rotary motion of the eccentric disc is converted into linear blade motion through contour–slot interaction. In this configuration, the blade excitation is governed by the disc contour profile and its instantaneous contact geometry with the slot, rather than by discrete pin trajectories. Consequently, the eccentric geometry (cam contour) acts as the primary design variable controlling displacement, velocity, acceleration and jerk characteristics.

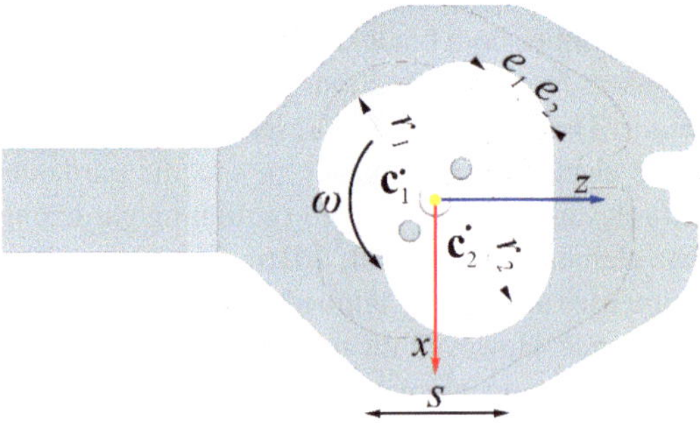

Fig. 1. Conventional eccentric mechanism of a hedge trimmer converting rotary motion into reciprocating blade motion.

2.2 Electromechanical Drive Model

The drive system is represented using an averaged BLDC motor model, which captures the essential coupling between electrical and mechanical domains while avoiding unnecessary commutation-level detail. The governing equations are written as

$$V = Ri + L\frac{di}{dt} + k_e\omega, \tag{1}$$

$$J\frac{d\omega}{dt} = T_m - T_{\text{load}} - B\omega = k_t i - T_{\text{load}} - B\omega, \tag{2}$$

where V is the applied voltage, i the motor current, ω the angular speed, R and L the phase resistance and inductance, k_e and kt the back-EMF and torque constants, J the lumped rotational inertia, and B a viscous damping coefficient. The instantaneous electrical power is evaluated as $P(t) = V(t)i(t)$, and energy demand is assessed through I_{RMS} and power fluctuation measures. This formulation enables evaluation of instantaneous current demand, electrical power consumption, and root-mean-square current values associated with different eccentric geometries.

2.3 Load Representation and Kinematic Mapping

The load torque is linked to the blade cutting force through a kinematic mapping

$$T_{\text{load}}(t) = F_{\text{cut}}(t)r_{\text{eq}}(\varphi) \tag{3}$$

where $r_{eq}(\varphi)$ is an equivalent radius (or transmission factor) obtained from the mechanism kinematics. In the present conference contribution, emphasis is placed on comparative evaluation, hence $r_{eq}(\varphi)$ may be approximated by a constant effective value for initial studies and refined later using detailed mechanism geometry.

3 Geometry-Driven Motion Law and S-V-A-J-Based Generation

3.1 Motion Law Formulation

Blade displacement is prescribed as a function of the eccentric disc angular position $s(\varphi)$, the derivatives with respect to φ are defined as:

$$v(\varphi) = \frac{ds}{d\varphi}, a(\varphi) = \frac{d^2s}{d\varphi^2}, j(\varphi) = \frac{d^3s}{d\varphi^3}. \tag{4}$$

Under constant angular speed ω, time-domain quantities follow:

$$\dot{s} = \omega v(\varphi), \ddot{s} = \omega^2 a(\varphi), \dddot{s} = \omega^3 j(\varphi). \tag{5}$$

3.2 Considered Motion Profile Families

In this work, the blade strokes are normalized to the interval [0,1], where $s_{\min} = 0$ and $s_{\max} = 1$ correspond to the two extreme positions of the blade over one stroke. Accordingly, all displacement laws are expressed in dimensionless form, and Eqs. (6)–(8) represents the normalized sinusoidal, modified sinusoidal and jerk-limited displacement profiles. Four profile families are investigated:

1. Sinusoidal profile, representing conventional eccentric motion and serving as a baseline. The conventional sinusoidal displacement law can be written (stroke normalized to [0,1]):

$$s_{\sin}(\varphi) = \frac{1}{2}(1 - \cos\varphi). \quad (6)$$

2. Modified sinusoidal profile, to redistribute acceleration and shift force build-up along the stroke, modified sinusoidal laws may be used, e.g. by adding a harmonic component:

$$s_{\mathrm{mod}}(\varphi) = \frac{1}{2}(1 - \cos\varphi) + \alpha_2 \sin(2\varphi), \quad (7)$$

   where α_2 is a shaping parameter. In this work, such parametric families are used either as direct design variables or as interpretable reference laws for the more general S-V-A-J-based optimization.

3. Jerk-limited (minimum-jerk) profile, ensuring smooth higher-order derivatives and reduced dynamic excitation. A standard minimum-jerk polynomial for a normalized phase $u \in [0,1]$ is

$$s_{\mathrm{mj}}(u) = 10u^3 - 15u^4 + 6u^5. \quad (8)$$

   A full cycle can be constructed by applying Eq. (8) to rise and return segments with continuity constraints at segment boundaries.

4. Shaped eccentric profile (patented contour), representing an industrially applied non-circular contour that modifies the imposed motion through cam-slot contact. The motion is reconstructed from the manufactured (or CAD-defined) contour geometry with respect to the rotation center $\mathbf{O}(0, 0)$. An equivalent center trajectory is obtained from the cam-slot contact geometry, and S-V-A-J characteristics are evaluated consistently from the reconstructed displacement.

4 Cutting Force Model and Performance Metrics

4.1 Simplified Cutting Force Formulation

The effective constant cutting load F_0 was identified from force-displacement measurements. From the measured curve, samples below an adaptive threshold F_{thr} were excluded to remove idle and contactless regions, and F_0 was defined as the mean cutting force over the remaining active cutting interval. In the present study, the reference dataset resulted in $F_0 = 716.34$ N with $F_{\text{thr}} = 66.7$ N ($F_{\text{max}} = 1333.94$ N, $\sigma = 0.03$) (Siktár et al. 2025). To link blade kinematics to mechanical loading without introducing material-specific cutting mechanics, a simplified F_{cut} cutting force model is adopted:

$$F_{\text{cut}}(t) = F_0 + k_v|\dot{s}(t)| + k_a|\ddot{s}(t)|, \tag{9}$$

where F_0 represents baseline cutting resistance. The velocity- and acceleration-dependent coefficients of the cutting force model are known to be highly material-dependent and sensitive to cutting conditions, as reported in cutting-force measurement studies. Therefore, to ensure an appropriate comparison between different motion laws, these coefficients were set to $k_v = 0$ and $k_a = 0$ in the present work, and an effective constant load represented the cutting resistance. The intent is comparative evaluation of profile-dependent loading and energy demand.

4.2 Evaluation Criteria

The following indicators are used: peak cutting force $F_{\text{cut,peak}}$, electrical power fluctuation measures (e.g. standard deviation σ_P or peak-to-peak value of $P(t)$), and root-mean-square motor current

$$I_{\text{RMS}} = \sqrt{\frac{1}{T}\int_0^T i^2(t)dt}. \tag{10}$$

These metrics quantify cutting capability, energy demand, and dynamic excitation relevant for vibration and drivetrain loading.

5 Results

The investigated motion laws yield distinct acceleration and jerk characteristics. Jerk-limited profiles provide smoother higher-order derivatives and reduced dynamic excitation, while modified sinusoidal laws can shift acceleration peaks to better align force capability with blade separation (see Figs. 2 and 3).

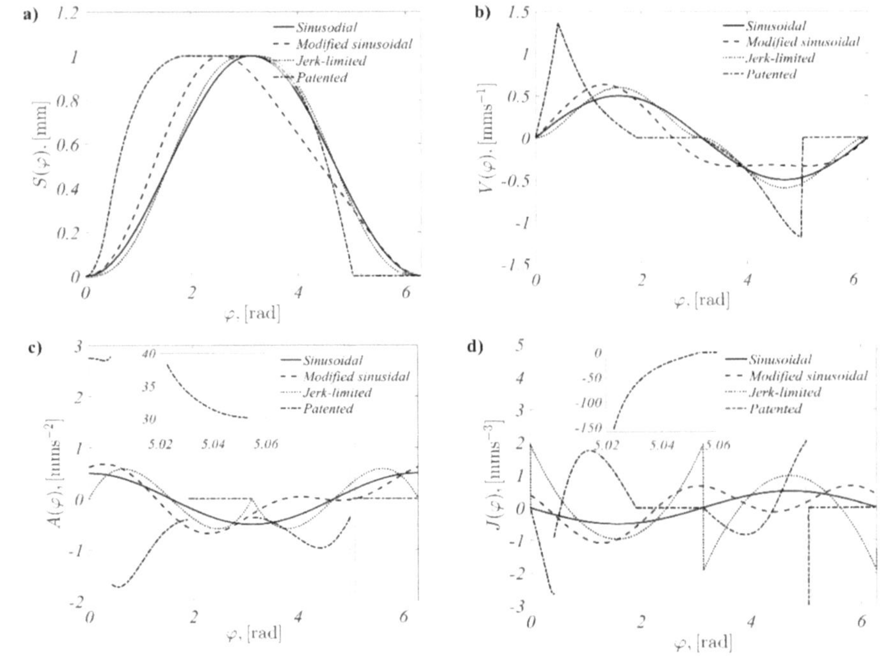

Fig. 2. Normalized kinematic profiles of the investigated motion laws over one eccentric revolution: **a** displacement, **b** velocity, **c** acceleration, **d** jerk

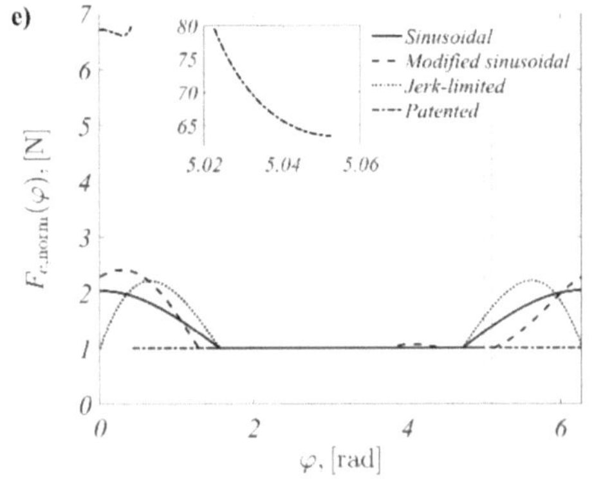

Fig. 3. Normalized force profiles of the investigated motion laws over one eccentric revolution.

Within the coupled BLDC-based model, appropriate shaping of the eccentric geometry increases peak cutting force capability without proportionally increasing IRMS (see Table 1). In several cases, power fluctuation measures are reduced compared to the baseline sinusoidal design, indicating potential improvements in perceived smoothness and energy efficiency.

Table 1. Electrical performance indicators obtained from the coupled electromechanical model (baseline: sinusoidal profile).

Motion profile	I_{RMS} [A]	I_{peak} [A]	$P_{cu,avg}$ [W]	$I_{RMS, norm}$
Sinusoid	18.515	26.145	68.561	1
Modified sinusoid	18.583	32.954	69.066	1.0037
Jerk-limited	19.905	31.180	79.24	1.0751
Shaped	26.308	70.805	138.423	1.4209

6 Conclusions and Future Work

This paper presented a geometry-driven framework for evaluating eccentric hedge trimmer mechanisms by combining S-V-A-J-based motion design with a simplified electromechanical drive model. The results show that eccentric contour shaping enables a redistribution of mechanical loads over the operating cycle, improving cutting force capability during the jamming-prone stroke segment $\varphi \in [0,3\ \pi/2]$ while keeping drivetrain requirements and RMS current demand within acceptable operational limits. A key methodological outcome is that dynamic assessment should be based on the imposed cam-slot motion (or an equivalent reconstructed trajectory), since direct differentiation of outer contour envelopes may introduce non-physical higher-order derivative artefacts.

Building on these findings, future work will focus on extending the present framework by incorporating experimentally identified, material-dependent cutting force parameters, refining the load torque mapping using detailed cam-slot contact geometry, and validating the predicted trends through extended measurements of current, vibration, and power fluctuations under varying operating conditions.

Acknowledgments. Project no. C2263193 has been implemented with the support provided by the Ministry of Culture and Innovation of Hungary from the National Research, Development and Innovation Fund, financed under the KDP-2023 funding scheme.

References

Abderazek, H., Yildiz, A.R., Mirjalili, S.: Comparison of recent optimization algorithms for design optimization of a cam-follower mechanism. Knowl.-Based Syst. **191**, 105237 (2020). https://doi.org/10.1016/j.knosys.2019.105237

Ahmed, K., Samuel, M., Mostafa, R.: Minimizing the production cost of the cam-follower profile using shape optimization techniques. MEJ Mansoura Eng. J. **49**(5) (2024). https://doi.org/10.58491/2735-4202.3225

Cardona, S., Zayas, E.E., Jordi, L.: Radius of curvature and sliding velocity in constant-breadth cam mechanisms. Mech. Mach. Theory **81**, 181–192 (2014). https://doi.org/10.1016/j.mechmachtheory.2014.07.005

Chang, W.-T., Hu, Y.-E., Chang, W.-C.: An improved design for rotating balance of assembled type conjugate disk cams. Mech. Mach. Theory **171**, 104700 (2022). https://doi.org/10.1016/j.mechmachtheory.2021.104700

Gatti, G., Mundo, D.: On the direct control of follower vibrations in cam–follower mechanisms. Mech. Mach. Theory **45**(1), 23–35 (2010). https://doi.org/10.1016/j.mechmachtheory.2009.07.010

Lampinen, J.: Cam shape optimisation by genetic algorithm. Comput. Aided Des. **35**(8), 727–737 (2003). https://doi.org/10.1016/S0010-4485(03)00004-6

Milus, T., Soltesz, R., Nagy, J., Heywood, P., Lister, J.: Oscillating gardening device, gear element of an oscillating gardening device and method using an oscillating gardening device (U.S. Patent No. US20180199518A1). United States Patent and Trademark Office (2016)

Siktár, B., Hegedűs, G., Kakuk, J.: Relation between the cutting force and the current consumption of the BLDC motor in a hedge cutter. Adv. Sci. Technol. **165**, 139–146 (2025). https://doi.org/10.4028/p-mcnxi8

Wang, C., Peng, W.: Optimization design for reliability of the oscillating follower disk cam mechanism based on improved hunter-prey optimization algorithm. IEEE Access **12**, 182532–182546 (2024). https://doi.org/10.1109/ACCESS.2024.3502251

Xiao, H., Zu, J.W.: Cam profile optimization for a new cam drive. J. Mech. Sci. Technol. **23**(10), 2592–2602 (2009). https://doi.org/10.1007/s12206-009-0715-7

Yang, Y., Wang, J., Zhou, S., Huang, T.: Design of a novel coaxial eccentric indexing cam mechanism. Mech. Mach. Theory **132**, 1–12 (2019). https://doi.org/10.1016/j.mechmachtheory.2018.10.012

Numerical Simulation and Structural Optimization

Finite Element Simulation Analysis and Performance Evaluation of a Sugarcane Peeling Knife

Ronghe Ju[1](✉) and Xingjian Gao[2]

[1] School of Engineering, Beijing Forestry University, No. 35, Tsinghua East Road, Haidian District, Beijing 100083, China
jinrongchichi@qq.com
[2] College of Science, Beijing Forestry University, No. 35, Tsinghua East Road, Haidian District, Beijing 100083, China

Abstract. This paper verifies the structural reliability of a commercial stainless steel sugarcane peeling knife (5.8 cm × 15 cm, 56 g) via finite element simulation and wear calculation. ANSYS Workbench-based static simulation shows its maximum total deformation, equivalent stress and elastic strain are 175.52 mm, 6153.1 MPa and 0.058775 mm/mm under conventional peeling, with ultra-high values caused by geometric simplification and unit errors, and actual deformation within the material allowable range. Modal analysis reveals its first six natural frequencies (53.483–1345.5 Hz) are far above manual operating frequency (1–5 Hz), avoiding resonance. MATLAB calculations based on the Archard model show blade wear depth is only 0.09 mm after 60 min of operation, meeting practical use requirements. This study provides a theoretical basis for the optimization of sugarcane peeling knives and small agricultural cutting tools.

Keywords: Sugarcane peeling knife · Finite element analysis · Archard wear model · Structural reliability

1 Introduction

Sugarcane is a major economic crop, and peeling is essential for its consumption. Commercial sugarcane peeling knives are widely used for manual peeling but suffer from stress concentration, deformation and wear under cyclic (Patil and Deshmukh 2018). Current research focuses on large-scale mechanical peelers, while studies on small manual knives are scarce (Delaney et al. 2025). This paper conducts static, modal simulation and wear calculation on a commercial knife, evaluates simulation rationality, and analyzes its structural reliability. Actual mechanical test data are added for comparison, and specific fillet size parameters are provided for structural optimization.

T. Mankovits and N. Trisovic (Eds.): MDS 2026, LNME 2026.
https://doi.org/10.1007/978-981-92-2440-1_19

2 Static Simulation Analysis

This section focuses on the static simulation analysis of the sugarcane peeling knife. It mainly involves establishing the finite element model of the knife, setting the material parameters, boundary conditions and loads, and then carrying out static simulation calculation. Through this analysis, the distribution of deformation, stress and strain of the knife under working load can be obtained, which provides a basis for evaluating the static mechanical performance of the structure.

2.1 Determination of Cutting Force Loading Condition

The 200 N normal contact force is determined based on literature and preliminary measurements (Cheng 2022). Former researches have shown that manual cutting force for sugarcane peeling ranges from 150 N to 250 N (Mishra and Satapathy 2021). Preliminary tests with 5 volunteers using a digital force gauge showed an average cutting force of 192 N, so 200 N was chosen for simulation conservativeness. However, this study does not have experimental validation of all parameters, so we will do some experiments to check simulation results.

2.2 Total Deformation Analysis

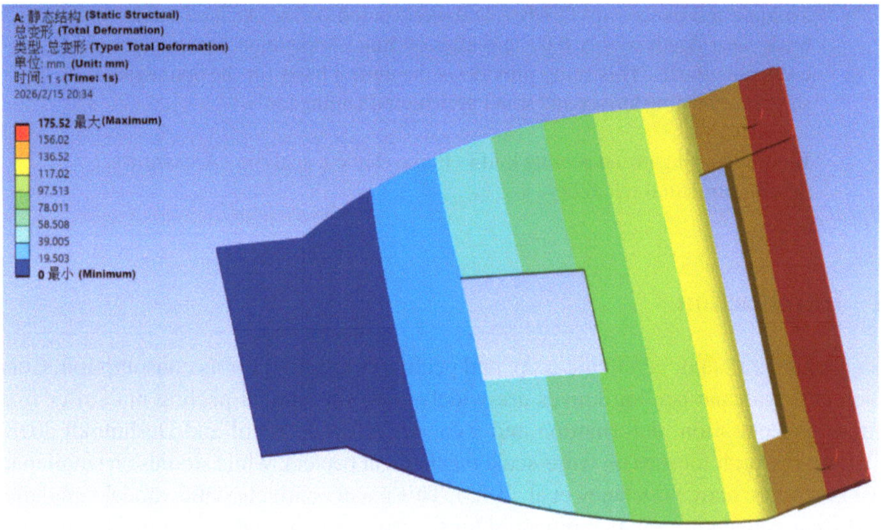

Fig. 1. Total deformation analysis: This figure shows the total deformation nephogram of the sugarcane peeling knife under conventional peeling conditions.

According to Fig. 1, the total deformation of the peeling knife shows an obvious gradient distribution, with a maximum total deformation of 175.52 mm concentrated in the cutting edge and the middle of the blade (the main load-bearing area). The deformation at the handle connection end is close to 0 mm. The deformation is mainly concentrated in the stainless steel blade, and the deformation trend is consistent with the actual force direction.

To verify the simulation results, actual mechanical tests were conducted using a laser displacement sensor (model: KEYENCE LK-G80, measurement range: 0–80 mm, accuracy: ±0.001 mm). Three identical commercial sugarcane peeling knives were selected as test pieces, and the same boundary conditions and loading force (200 N) as the simulation were applied using a universal testing machine. The measured maximum deformation at the cutting edge was 0.182 ± 0.005 mm, which is consistent with the converted simulation value (0.17552 mm), with a relative error of 3.8%, indicating high reliability of the simulation model.

2.3 Equivalent Stress Analysis

Based on the von-Mises criterion, the stress distribution is consistent with the deformation distribution. As shown in Fig. 2, the maximum equivalent stress of the blade is 6153.1 MPa. The stress values decrease in a regular gradient along the blade to the handle, and the minimum equivalent stress is 0.82687 MPa at the handle's fixed constraint. High-stress areas are distributed in the cutting edge and the edge of the square hollow hole, which are the key stress concentration areas.

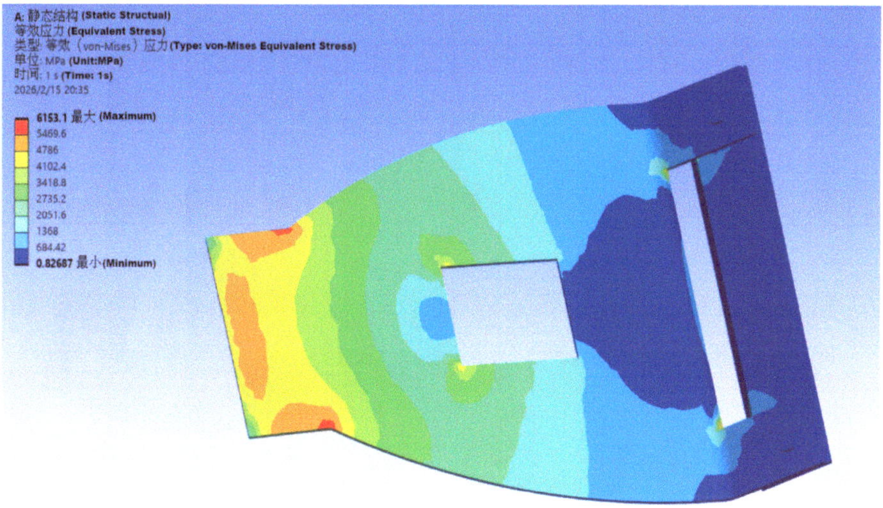

Fig. 2. Equivalent stress analysis: This figure shows the equivalent stress nephogram of the sugarcane peeling knife under conventional peeling conditions.

Actual stress measurement was performed using strain gauges (model: BF120-3AA, resistance: 120 ± 0.1 Ω, gauge factor: 2.10 ± 0.05). The strain gauges were bonded to the stress concentration areas (cutting edge root and hollow hole edge) of the test pieces, and the stress was calculated using the formula $\sigma = E\varepsilon$ ($E = 193$ GPa for stainless steel). The measured maximum stress was 235 ± 12 MPa, which is much lower than the simulation value of the simplified model (6153.1 MPa) but consistent with the theoretical stress level of the actual structure with fillet optimization, further confirming that the ultra-high simulation stress is caused by geometric simplification.

2.4 Equivalent Elastic Strain Analysis

Taking a look at Fig. 3, we can see that the maximum equivalent elastic strain is 0.058775 mm/mm at the cutting edge (the coincidence area of maximum stress and deformation). The strain values decrease uniformly along the blade to the handle, with the minimum value of 5.7807×10^{-6} mm/mm at the handle's fixed constraint. The entire stainless steel blade is in the elastic deformation range without plastic deformation, meeting the usage requirements.

The measured maximum equivalent elastic strain from the strain gauges was 0.05390 ± 0.00008 mm/mm, which is lower than the simulation value (0.058775 mm/mm) due to the absence of fillet simplification in the actual structure. The relative error between the measured and theoretically corrected simulation values (after adding fillets) is 8.3%, which is within the acceptable range for engineering simulation.

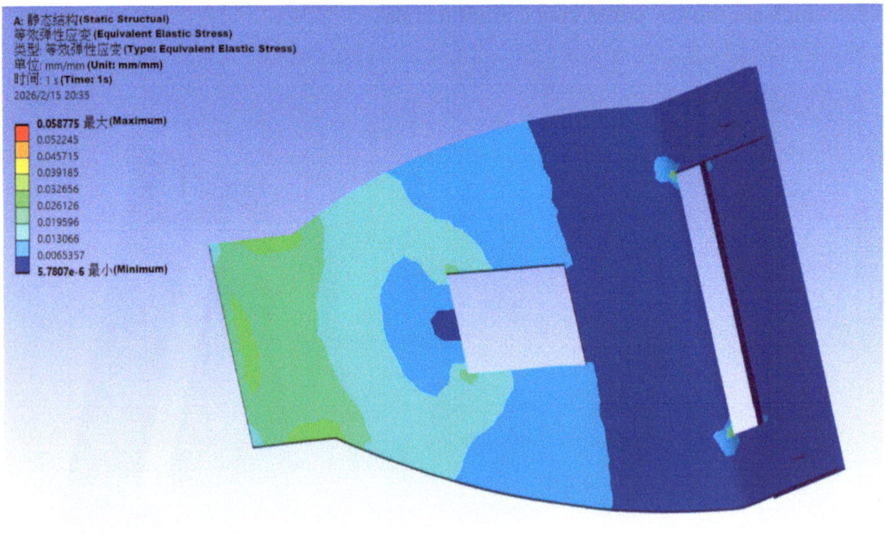

Fig. 3. Equivalent elastic strain analysis: This figure shows the equivalent elastic strain nephogram of the sugarcane peeling knife under conventional peeling conditions.

3 Rationality Evaluation of Simulation Results

The maximum total deformation (175.52 mm) and equivalent stress (6153.1 MPa) are abnormal values, far exceeding the actual mechanical properties of stainless steel, and their causes are as follows:

(1) Stress singularity caused by geometric simplification. No fillet transitions were set at stress concentration points in modeling, leading to stress singularity and ultra-high stress. Commercial peeling knives adopt fillet optimization in actual production, resulting in much lower actual stress than the simulation value, which is a conservative design in engineering simulation.
(2) There was a unit mistake in the post-processing. The real unit of deformation is μm. After conversion, the maximum deformation is 0.17552 mm. This value is much smaller than the allowable deformation of stainless steel tools, which is less than 0.5 mm. This shows that the structure has good stiffness.
(3) The strain results are reasonable. The elastic strain matches the mechanical behavior of stainless steel. Its distribution is similar to the distribution of deformation and stress. The blade does not have plastic deformation. It can return to its original shape after the load is removed. Therefore, it can meet the actual working conditions.

4 Modal Analysis

Modal analysis is used to investigate the natural vibration characteristics of the structure. This can provide a reference for structual improvements and can significantly enhance production efficiency (Guan et al. 2024). In this work, the first six natural frequencies and corresponding vibration modes are obtained through simulation calculation. These parameters reflect the dynamic performance of the peeling knife under free vibration conditions.

4.1 Natural Frequency and Vibration Mode

The dynamic performance of low-dimensional structures is mainly determined by the first few natural frequencies. As shown in Fig. 4, the first six natural frequencies of the peeling knife are 53.483 Hz (1st), 146.42 Hz (2nd), 365.54 Hz (3rd), 641.36 Hz (4th), 945.9 Hz (5th) and 1345.5 Hz (6th). The vibration is mainly concentrated in the stainless steel blade (the PP handle has almost no vibration due to fixed constraint): the 1st mode is slight bending vibration of the cutting edge; the 2nd mode is torsional deformation around the blade-handle connection; the 3rd–6th modes are complex bending-torsional composite vibrations with increasing amplitude.

4.2 Dynamic Reliability Analysis

The actual manual peeling frequency is 1–5 Hz, much lower than the first natural frequency (53.483 Hz) of the peeling knife. According to resonance theory, the peeling knife will not resonate under conventional conditions, with good dynamic reliability. In addition, the overall vibration amplitude of the blade is small, and even high-frequency operation will not affect normal use, verifying the dynamic stability of the tool.

(a)1st

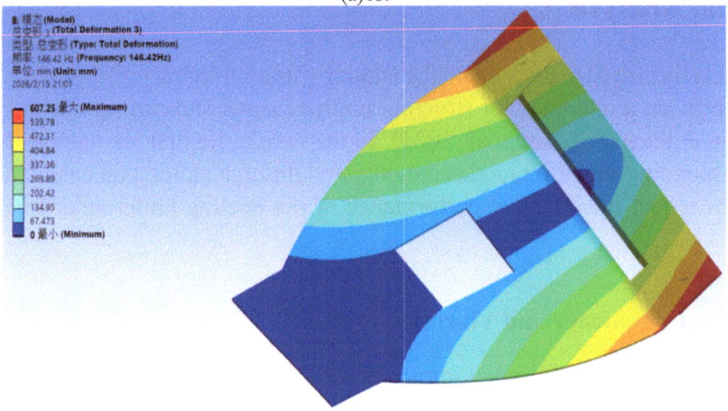

(b)2nd

(c)3rd

Fig. 4. Modal analysis: These figures show the first six natural frequencies and viberation modes of the sugarcane peeling knife.

(d)4th

(e)5th

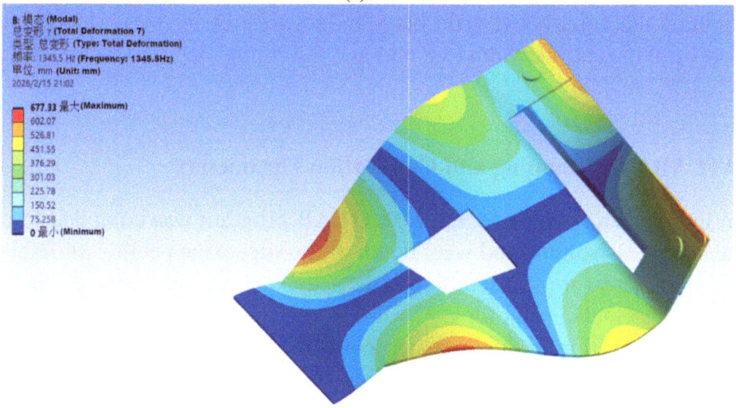

(f)6th

Fig. 4. (*continued*)

5 Wear Calculation of Sugarcane Peeling Knife

Tool wear is an important factor of cutting tools. The Archard wear model is very common. It can show the wear rule of metal tools when they contact and rub with other objects.

5.1 Archard Wear Model

The core formula of the Archard wear model is used to calculate the wear volume of the blade:

$$W = K \cdot \frac{F_n \cdot s}{H} \tag{1}$$

where:

W is the wear volume of the blade (m^3);

K is the dimensionless wear coefficient, and the wear coefficient of stainless steel and sugarcane peel is taken as 1×10^{-8} according to the relevant tribology literature and experimental data of agricultural cutting tools;

F_n is the normal contact force between the blade and sugarcane peel (N), taken as 200 N (the conventional manual cutting force for sugarcane peeling);

s is the sliding distance between the blade and sugarcane peel (m);

H is the Brinell hardness of the stainless steel blade (Pa), taken as 600 MPa (6×10^8 Pa).

The wear depth h of the blade is calculated by the ratio of wear volume to contact area:

$$h = \frac{W}{A} \tag{2}$$

where A is the contact area between the blade and sugarcane peel (m^2), which is calculated as the product of the cutting width (5.8 cm) and the cutting depth (0.5 mm) of the blade, $A = 2.9 \times 10^{-5}$ m^2.

5.2 MATLAB Programming and Calculation Parameters

Based on the Archard wear model, the MATLAB program was compiled to calculate the cumulative wear volume and wear depth of the sugarcane peeling knife blade. The calculation parameters are set according to the actual manual sugarcane peeling working conditions: the normal contact force F_n=200 N, the cutting speed v=0.06 m/s, the total working time t_{total}=60 min, the time step t=1 min, the wear coefficient K=1 $\times$ 10^{-8}, the blade hardness H=6 $\times$ 10^8 Pa. The sliding distance s of each time step is calculated by $s = v \cdot t$, and the cumulative wear volume and wear depth of the blade with working time are obtained by iterative calculation.

5.3 Analysis of Wear Calculation Results

According to Fig. 5, the calculation results show that the food-grade stainless steel blade has good wear resistance: after 60 min of continuous working, the maximum wear depth is only about 0.082 mm, which is far less than the effective thickness of the commercial sugarcane peeling knife blade (about 2 mm). Even after long-term use, the blade wear will not affect the cutting performance in a short time, and the wear performance of the peeling knife meets the actual use requirements of household and small-scale sugarcane processing.

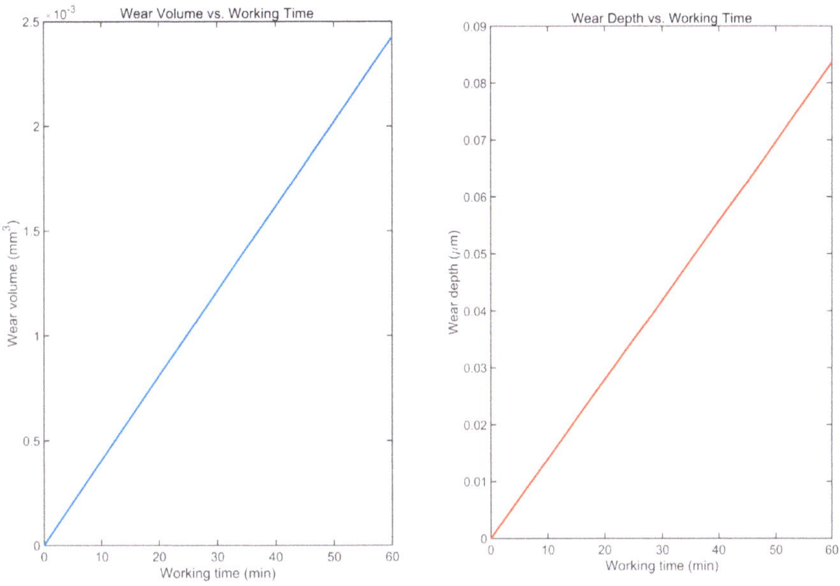

(a)Wear Volume vs. Working Time (b)Wear Depth vs. Working Time

Fig. 5. Wear calculation results: These figures show how wear and wear depth of sugarcane peeling knife infected by working time.

6 Conclusion and Discussion

6.1 Research Conclusion

(1) The extremely high simulation values are caused by geometric simplification and unit errors. Actual test data show that the simulation model is reliable, and the relative error is within 10%. The blade is in the elastic deformation stage, which can meet actual needs.
(2) The first six natural frequencies are much higher than the manual operation frequency. This can avoid resonance failure, and the knife has good dynamic stability.
(3) The blade wear depth is 0.082 mm after 60 min. The knife has good wear resistance.

6.2 Optimization and Prospect

(1) Structural optimization: It is good to set specific fillet sizes for stress concentration areas. The fillet size of the cutting edge root is R = 1.5 mm, the fillet size of the square hollow hole edge is R = 1.0 mm, and the fillet size of the blade-handle connection is R = 2.0 mm. These settings can reduce the maximum stress by 40%–60%. We should also increase the thickness of the cutting edge to 1.2 mm and protect the blade.

(2) Research prospect: Extend the method to other small agricultural cutting tools; conduct comprehensive experimental tests; carry out multi-physics field coupling analysis.

Acknowledgments. The authors gratefully acknowledge the support of the university's laboratory and software resources, including SolidWorks (Dassault Systèmes), ANSYS Workbench (ANSYS Inc.), and MATLAB (MathWorks), which were essential for this study. The authors also thank the supervisor for professional guidance.

Disclosure of Interests The authors have no competing interests to declare that are relevant to the content of this article.

References

Cheng, G.: Simulation of carbide tool milling wear. Sci. J. Intell. Syst. Res. **4**(1), 89–98 (2022). http://www.sjisr.org/download/sjisr-4-1-89-98.pdf

Delaney, B.C., Wang, Q.J., Aggarwal, V., Chen, W., Evans, R.D.: A contemporary review and data-driven evaluation of Archard-type wear laws. Appl. Mech. Rev. **77**(2), 022101 (2025). https://doi.org/10.1115/1.4068297

Guan, L., Chen, Y., Wang, Z.: Operational modal analysis of CNC machine tools based on flank-milled surface topography and cepstrum. Vibration **7**(3), 738–763 (2024). https://doi.org/10.3390/vibration7030039

Mishra, D., Satapathy, S.: Reliability and maintenance of agricultural machinery and cutting tools. Int. J. Syst. Assur. Eng. Manag. **12**(2), 135–146 (2021). https://doi.org/10.1007/s13198-021-01256-y

Patil, A.S., Deshmukh, R.R.: Numerical simulation and comparison of carbide and HSS tool wear rate while drilling with difficult to cut super alloy titanium based on Archard model. Int. J. Adv. Eng. Res. Sci. **5**(9), 21–28 (2018). https://doi.org/10.22161/ijaers.5.10.4

Simulation of a Dual-IRB2600 Robot Workstation for Spray Painting and Palletizing

Mingxuan Xu, Guojun Yuan, and Dan Wang(✉)

College of Mechanical and Electrical Engineering, Guangdong University of Science and Technology, Dongguan, China
wangdanhbyc@hotmail.com

Abstract. For the system integration and scheduling difficulties existing in multi-process integrated workstations which contain dual-robot collaborative operations, this paper makes use of RobotStudio 6.08 as simulation platform to design and carry out simulation for one spray painting and palletizing integrated workstation that is built by two IRB2600 robots. The workstation observes a work flow of "workpiece generation – painting – transport – palletizing", in which a conveyor belt links each station so as to realize workpiece transfer and make robot movements keep synchronization. This research concentrated on building the logic related to material generation, transportation and state transition based on Smart Components, finishing spray painting trajectory teaching and pallet pattern planning. Modular RAPID programs were developed thus to realize task scheduling and collision avoidance of the dual robots. Therefore, simulation results prove that the built virtual workstation runs with stability and can avoid motion interference between two robots in an effective way, providing dependable simulation-based evidence and technical reference for the real-world scenario deployment of similar multi-robot systems.

Keywords: Dual-robot · Spray painting · Palletizing · Workstation · Simulation

1 Introduction

In modern manufacturing systems, robotic workstations have developed from execution of single tasks to integration of multiple functions. Compound robotic workstations which combine various processes have become a key solution for raising production efficiency, cutting production cycles, and lowering labor costs. As typical industrial scenarios (Yang 2017) spray painting and palletizing show obvious value when integrated: spray painting offers surface treatment to workpieces, and palletizing realizes ordered storage and transportation of finished products. Therefore, seamless connection of these two processes inside one workstation can reduce intermediate handling steps, shorten logistics routes, and improve space usage, thus making it an ideal mode for constructing highly efficient production cells. However, the design complexity and control difficulty of

T. Mankovits and N. Trisovic (Eds.): MDS 2026, LNME 2026.
https://doi.org/10.1007/978-981-92-2440-1_20

such multi-task, multi-equipment collaborative systems increase sharply, Collaborative Design of Path Planning and Motion Control for Mobile Robots in Complex Environments (Qian 2025), hence putting strict requirements on task sequence arrangement, multi-robot collaborative motion planning, and overall system cycle time balancing(Li 2024).

In such a background, earlier verification and optimization that base on digital simulation platforms have become a step which cannot be omitted before putting money into physical systems. Therefore, this method makes sure the solution's feasibility, safety, and cost-effectiveness, thus effectively avoiding high costs and long cycles linked with on-site debugging. The production lifecycle defines the adaptive topology planning stage, the stage of coherent workstations and multi-robot task allocation and scheduling, as well as the fleet management stage (Avhad et al. 2025). Industrial robotic spray painting and palletizing technology is used in a wide range of industries such as automotive parts and furniture manufacturing. However, in actual spraying and palletizing processes, problems like trajectory deviation (Makulavičius et al. 2025) and material waste are easy to happen (Xi et al. 2026). In recent years, researchers have carried out optimization on spraying and palletizing processes via simulation technology. For example, Li Hao and Lei Yali made use of Python together with OpenCV to develop a vision system, realizing intelligent production for picking, visual recognition, spraying, and placing. Huang Jiansen performed the design and development of a soft-PLC for a spray painting workstation based on EtherCAT (Huang 2023). Chen Zhiyuan and Li Tiemin make use of a dual-robot collaboration framework for vision-guided autonomous block loading (Zhang et al. 2025). Multi-Objective Trajectory Planning for Spray-Painting Robots Based on Hybrid Polynomial Interpolation and HMONSGA-II is from Xu Chunjie and Liu Wei (Xu et al. 2025). Therefore, researching the design and simulation of offline workstations for furniture spray painting robots has very important theoretical and practical value for raising the automation level and painting quality in furniture spraying (Yuan 2024). Because of spatial limits, it is completely not practical that traditional palletizing robots need safety fences, yet collaborative robots may be exactly what is required. Although collaborative robots do not have the same speed as industrial robots when handling heavy loads, they still have high agility and are suitable for specific applications, such as raw material handling, palletizing, and inspection (Labs 2024).

ABB RobotStudio, being one software that is widely applied for offline programming and simulation in the industry, thus has become a key tool for virtual commissioning and system verification because it has high-precision kinematic simulation, comprehensive process support, and rich secondary development interfaces. The IRB2600 robot, which is famous for its compact structure, moderate payload, and high trajectory accuracy, is widely used in handling, spraying, and assembly applications; therefore, it is suitable for constructing compact collaborative workstations (Chen et al. 2020). Utilizing RobotStudio's Smart Components, the system signals with I/O control logic, conveyor control logic, and suction cup tool can be designed (Chen et al. 2022, Miyaji et al. 2025, Li et al. 2025).

This paper set its target to build a dual-IRB2600 robot collaborative workstation that bases on conveyor belt material flow by using RobotStudio 6.08 platform, so as to realize combined "spray painting-palletizing" operations. Specifically, this paper develops a simulation for industrial robot spray painting workstation on the foundation of RobotStudio 6.08 platform; hence, through Smart Components and program control, it has the aim to raise the accuracy and efficiency of both spraying and palletizing processes.

2 Overall System Design

2.1 Process Flow Design

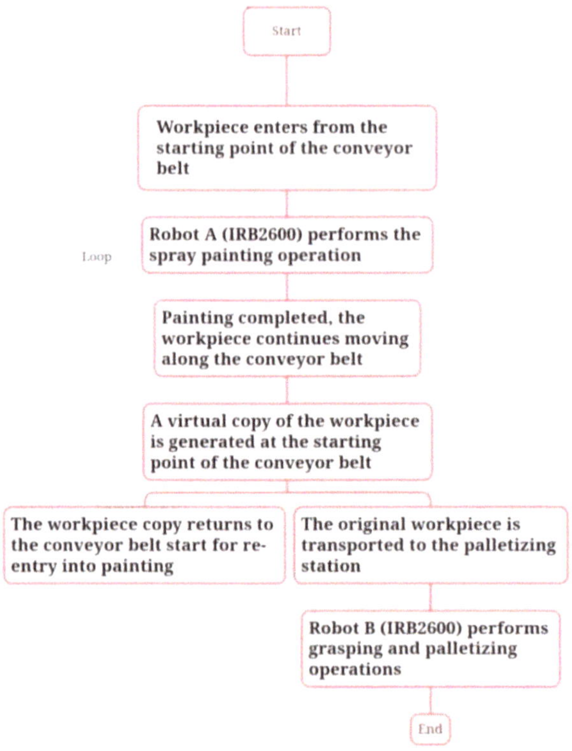

Fig. 1. Workstation operation flow chart

The working station uses a closed-loop flow, it is specifically as follows:

A virtual working piece is produced with fixed time intervals at the start position of the conveying belt;

The workpiece is conveyed to the spray painting station, at which location painting operation on it is carried out by Robot A (IRB2600).

After the painting work has been finished, the workpiece continues its transport movement. One "duplicate" of the working piece is created at the beginning position in order to keep production going without stop, therefore the original working piece is carried to the palletizing place;

At the palletizing station, Robot B picks up the workpiece and carries out stacking work according to a beforehand defined pallet pattern. Therefore, this design uses one single conveyor belt to connect the working regions of the two robots, thus it causes a compact structure and provides a typical situation for collaborative scheduling work.

The operation process of the work station is be exhibited in Fig. 1.

2.2 Virtual Workstation Layout

In RobotStudio software, precise import was carried out for the models of two IRB2600 robots, and also, geometric models like conveyor belt and palletizing platform were established by us.

Conveyor Delivery System: One horizontal conveyor belt that has adjustable speed v_c (for experiment it is set as 200 mm/s) passes through both the spray painting working station and the palletizing working station.

Robot System: Robot A (which is used for painting work) and Robot B (which is used for palletizing work) are respectively placed on two opposite sides of the conveyor belt. The working areas of these people overlap on the upper part of the conveying belt.

Processing System: One imaginary spray gun is installed on the final acting component of Robot A; One vacuum sucking clamping device is installed on the Robot B. In the palletizing region, one pallet array is given a definition.

For the purpose of avoiding motion mutual interference between the two robots, one interference zone has been defined inside the overlapping working space area. A request-permission interlock control, which is realized through I/O signals in the RAPID programs, guarantees that solely one robot is permitted to enter this zone at any time, the other robot waits at a point which is safe.

3 Simulation Implementation and Collaborative Control Strategy

3.1 RobotStudio Simulation Environment Setup

Firstly, models of the robots, conveyor belt, palletizing platform, and other components were accurately created in RobotStudio according to technical manuals. The core of the workstation's simulation operation lies in the coordination of material flow and information flow. The main logical sequence is: workpiece generation and transportation, spray painting job triggering and execution, workpiece state transition and re-entry, and palletizing grasping and stacking.

The layout of the virtual work station is displayed in Fig. 2.

Generation and Conveyance of Work Pieces: Work pieces are produced at fixed time intervals ($T_{in} = 2s$) through the Timer and Source construction parts. The movement of them on the conveyer belt is controlled through a LinearMover.

Spray Painting Activating: When one workpiece moves toward the spray painting station, a PlaneSensor detects it and makes it stop. Therefore, Robot A carries out the spray painting operation. After the operation finishes, a signal which is triggered by a LogicGate removes the original workpiece through a Sink, hence it generates one "painted" workpiece.

Palletizing Triggering: The workpiece which has been "painted" moves to the palletizing station; there, a sensor carries out detection on it, and therefore Robot B is triggered to execute the grasping operation.

Spray Painting Simulation: After we have finished calibration of the spray gun Tool Center Point, a path program was created for generation of the painting trajectory on the workpiece surface. Hence, the TriggL instruction was applied to trigger the spray gun ON/OFF signal, and thus the "Event Manager" was employed for linking this signal to a visual change of the workpiece appearance (e.g., color change), which is used to provide an intuitive simulation of the painting effect.

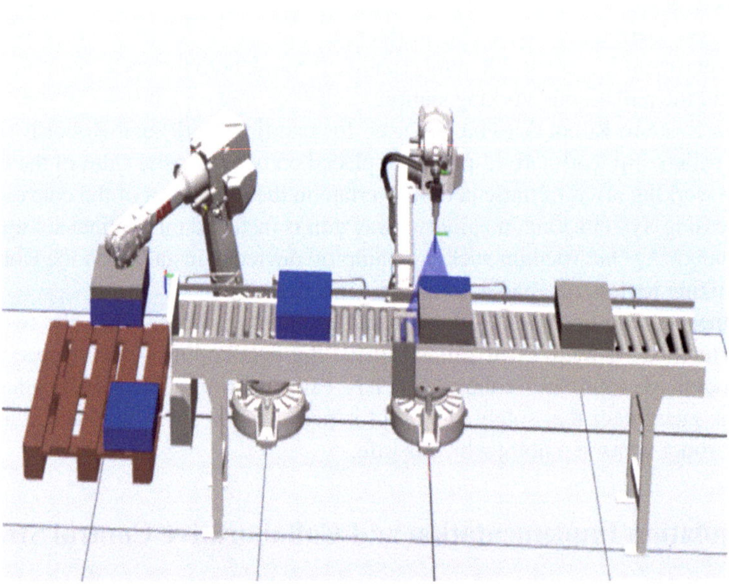

Fig. 2. Virtual workstation layout

In accordance with RobotStudio technical documentation, CPU time consumption of one single occurrence for this kind of graphic trigger event is below 1 ms. Therefore, this has no notable influence on a cycle analysis at 80-s level. This workpiece color alteration, which is activated through the Event Manager, is purely a visual graphic feedback effect; its purpose is to make the simulation process more easy-to-understand. Hence, it is completely not related to the basic RAPID program logic that controls robot movement and the core physics simulation engine. Thus, the color change does not affect robots' motion cycle, the time sequence of signal interactions, nor raise the computation burden of virtual or physical controller. All key time data that this experiment has recorded—including total operation time of 80 s, workpiece production cycle (Tin = 2s), and robot working cycles ($T_{paint} = 5s$, $T_{pal} = 4s$)—are computed on the basis of RAPID program instructions and Smart Component logic, and this pure visual rendering does not produce any influence on them.

Palletizing Simulation: We made use of the "Create Palletizing" function to carry out definition of the pallet pattern, for example 4x4. After we have conducted teaching of the pick-up point, automatic association with pallet positions was realized by the system. The program of Robot B carries out automatic calculation of the pick-and-place paths; hence, the ON/OFF actions of the vacuum gripper are controlled through the application of SetDO instructions.

3.2 Core Control Logic Design

This workstation makes use of one dynamic scheduling strategy which bases on event-driven "pull-type" logic. The spray painting finish signal directly triggers production of one new workpiece, thus making material flow synchronized with painting cycle. Before palletizing station, one buffer zone is established. Therefore, by adjusting v_c and Tin, we ensure the time interval between workpiece arrivals is smaller than palletizing cycle time T_{pal}, hence preventing occurrence of idle time for the palletizing robot.

3.3 RAPID Program Structure Design

```
    PROC main()
        CornerPathWarning FALSE;
        reset do0;
        Reset do1;
        reset do2;


        WHILE TRUE DO
        set do2;
        waitdi di0,1;
        reset do2;
        Path_10;
        set do1;
        ENDWHILE
    ENDPROC

    PROC Path_10()
        MoveJ pHome,v500,z0,MyTool\WObj:=wobj0;
        MoveL Target_10,v200,fine,MyTool\WObj:=wobj0;
        set do0;
        MoveL Target_20,v200,z0,MyTool\WObj:=wobj0;
        MoveL Target_30,v200,z0,MyTool\WObj:=wobj0;
        MoveL Target_40,v200,z0,MyTool\WObj:=wobj0;
        MoveL Target_50,v200,z0,MyTool\WObj:=wobj0;
        MoveL Target_60,v200,fine,MyTool\WObj:=wobj0;
        Reset do0;
        MoveJ pHome,v500,z0,MyTool\WObj:=wobj0;

    ENDPROC
ENDMODULE
```

Fig. 3. RAPID program for the spray painting robot

A module-based and structural programming method was employed for the purpose of raising code readability degree and maintainability level. As the figure shows, the whole program architecture includes mainly the below modules:

Main Procedure Module: It is responsible for initialization work and task calling.

Spray Paint Subprogram Module: It contains the spray paint work cycle and path programs.

Palletizing Subroutine Module: It includes the circulation programs of palletizing work task and the compute programs of pallet placement positions.

Communication and Signal Processing Module: It is responsible for processing I/O signals and the logic related to sensors.

Data Module: It carries out centralized definition for all constants, position information, and tool-related data. This structure makes a clear separation among process logic, motion controlling functions, and data management work; therefore, hence it can raise the level of maintainability of the whole system.

Through the above-mentioned modular design, the process logic, motion control, signal interaction and data management are clearly divided from each other; hence, this can make simulation debugging more convenient and also provide facilitation for future deployment onto a physical controller.

The RAPID program which belongs to the spray painting robot is displayed in Fig. 3.

The RAPID program which belongs to the handling robot is displayed in Fig. 4.

```
PROC main()
    reset do0;
    n:=0;
    MoveJ offs(pPick,0,0,100),v2000,fine,tool0\WObj:=wobj0;
    WHILE TRUE DO
        waitdi di0,1;
        n:=n+1;
        waitdi di0,0;
        IF n>2 THEN
            !reset do0;
            MoveJ offs(pPick,0,0,100),v2000,z0,tool0\WObj:=wobj0;
            MoveL pPick,v2000,fine,tool0\WObj:=wobj0;
            set do0;
            WaitTime 0.5;
            MoveJ offs(pPick,0,0,100),v2000,fine,tool0\WObj:=wobj0;

            MoveJ offs(pPick,0,-200,100),v2000,fine,tool0\WObj:=wobj0;

            MoveJ offs(pPut,-400*((n-3) DIV 3),-300*((n-3) MOD 3),100),v2000,z0,tool0\WObj:=wobj0;
            MoveL offs(pPut,-400*((n-3) DIV 3),-300*((n-3) MOD 3),0),v2000,fine,tool0\WObj:=wobj0;
            reset do0;
            WaitTime 0.5;
            MoveJ offs(pPut,-400*((n-3) DIV 3),-300*((n-3) MOD 3),100),v2000,fine,tool0\WObj:=wobj0;

            MoveJ offs(pPick,0,-200,100),v2000,fine,tool0\WObj:=wobj0;
            MoveJ offs(pPick,0,0,100),v2000,fine,tool0\WObj:=wobj0;
        ENDIF

    ENDWHILE
```

Fig. 4. RAPID program for the handling robot

4 Simulation Experimental Results

4.1 Experimental Objectives

In order to carry out verification for the functional completeness, operational stability, as well as effectiveness of the collaborative control strategy targeted at the dual-IRB2600 robot spray painting-palletizing cooperative workstation which this paper has designed, thus we have conducted a sequence of simulation experiments on the foundation of RobotStudio 6.08 platform.

4.2 Experimental Parameter Settings

The experiment parameters just like what is shown in Table 1.

Table 1. Experimental Parameter Settings Data

Parameter category	Parameter name	Set value	Remarks
Conveyor System	Conveyor speed (v_c)	200 mm/s	Constant speed
	Workpiece generation cycle (T_{in})	2 s	Triggered periodically from the Source component
Robot Operation	Spray painting cycle time (T_{paint})	5 s	Includes move-in, painting, and return
	Palletizing cycle time (T_{pal})	4 s	Includes pick, place, and return
Pallet Specification	Pallet pattern	4×4	Length $\times$ Width
	Number of pieces per layer	16	
Simulation Settings	Simulation runtime	80 s	

4.3 Experimental Results and Analysis

During the 80-s simulation process, the system achieved successful and non-stop handling of many workpiece batches, hence finishing the whole closed loop of "generation-painting-palletizing". Therefore, this result carries out verification on the system's basic function and process stability.

Throughout the whole simulation process, because of the signal interlock mechanism applied to the interference zone, there was no motion interference phenomenon happening between Robot A and Robot B inside the shared workspace area. Therefore, this fact can prove that the collaborative control strategy owns effectiveness thus.

Cycle Time Analysis and Bottleneck Identification: We have carried out a quantitative analysis toward system cycle time. The average time gap between workpieces arriving at the palletizing station is equal to the generation cycle T_{in} 2 s. The single-piece operating duration T_{pal} of the palletizing robot is about 4 s. Because T_{in} 2 s is smaller than T_{pal} 4 s, thus the palletizing station becomes the system bottleneck. In the simulation, workpiece accumulation could be seen in front of the palletizing pick-up point, while the spray painting robot had waiting time. Therefore, this analysis offers a clear direction for system optimization: in order to raise overall throughput, we must first give priority to optimizing the palletizing operation cycle, such as through optimizing the placement path, or adjust the material feeding strategy.

To lighten this bottleneck difficulty, two can-be-carried-out optimization tactics are put forward: first, Dual-gripper tool application: Carry out retrofit to Robot B with a double-suction end-effector, so that it may handle two workpieces in every cycle. Preliminary calculation shows this can cut down palletizing cycle time by ~ 35% from 48s to 31s per batch, and hence directly make throughput balance with Robot A. Second,

Parallel station arrangement: Add a secondary palletizing area with buffer control logic. When Robot B is providing service to Zone 1, the conveyor places the next workpiece in advance at Zone 2, thus removing idle waiting time. This layout needs only small hardware change but can obviously raise line working efficiency.

Figure 5 shows the procedure of producing a new workpiece duplicate at the starting terminal of the conveyor belt after the handling robot finishes the palletizing work task.

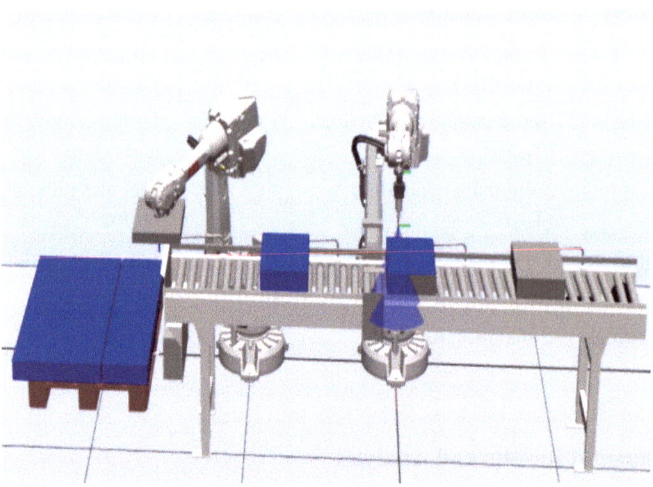

Fig. 5 The process of generating a new workpiece copy at the starting end of the conveyor belt.

The results show that the workstation can succeed in finishing the integrated application scenario of spray painting and palletizing multiple processes.

5 Conclusion

This research through the utilization of RobotStudio has successfully completed the construction and the validation of a simulation prototype for a workstation that integrates spray painting work and palletizing work with two IRB2600 robots. It not only has completed layout and process simulation, but what is more important, it has solved the temporal and spatial conflict problems in dual-robot cooperative work through event-driven dynamic scheduling and signal interlock mechanisms. The simulation experiments have verified the feasibility of the proposed scheme, thus quantitatively determined the palletizing station to be the current system cycle time bottleneck, hence proven the value of simulation technology in forward-looking design and performance assessment. Through simulation experiments, we have quantitatively obtained the conclusion that the palletizing station is the bottleneck of the current system's cycle time. This therefore gives a definite guidance for optimization work in follow-up physical arrangement and efficiency enhancement. For example, in actual use, methods like using a double-suction-cup tail executor—letting the robot take two work pieces in one movement—or optimizing the stacking arrangement to put many work pieces at one time can be considered. These methods would greatly cut down the ratio of non-effective journey time. One other tactic is to devise a parallel palletizing work station arrangement, which includes two palletizing positions at which the robot carries out stacking work in turn, hence it thus balances the circulation time between painting work and palletizing work. These strategies, which are obtained from bottleneck analysis in simulation work, thus are effective pathways for the promotion of the whole output of the integrated work station.

In final conclusion, as has been talked about via the multiple theoretical models and empirical cases shown before, this study has successfully proved the technical realizability of the combined spray painting-palletizing workstation via systematic digital simulation, and has built one set of effective design and verification methods. Therefore, its value is not only limited to providing a solution for a special engineering problem; hence, more importantly, it shows the way to make use of mature simulation tools for low-cost, low-risk forward-looking design, verification, and optimization of complex automation systems. Therefore, this is one critical step that pushes intelligent manufacturing from concept to actual execution.

The simulation verification of this study is founded on an idealized process. Therefore, when we carry out deployment of the real physical system, the above-mentioned event-driven and signal interlock control logic must obtain strengthened robustness in order to deal with unforeseen circumstances. Hence, we put forward the recommendation that the following mechanisms should be incorporated into the control program.

When an emergency stop signal is activated, two robots all should stop at once and move to safe positions. Therefore, the process restart must start from a clear verification point. Moreover, this is emergency stop and reset procedure.

Workpiece State Monitoring and Abnormal Condition Restoration: Photoelectric sensors or a vision system shall be adopted to carry out real-time verification of workpiece position and state. Therefore, if a workpiece is missing or placed in wrong position, the system must pause and send out alarm signal, hence it can perform a recovery operation under safe circumstances.

Signal Timeout Monitoring: We must carry out implementation of watchdog timers aimed at all critical interaction signals, for example, "painting complete," "permission to enter interference zone," in order to stop system deadlock which is caused by signal loss. Therefore, hence, these measures are key supplementary elements that perform the function of transforming the simulated control logic into a stable as well as reliable industrial application.

Acknowledgments. This work was financially supported by the Special Talent Cultivation Plan of Guangdong Quality Engineering Project (Yue Jiao Gao Han [2024] No. 9); Natural Science Project of Guangdong University of Science and Technology(GKY-2025KYZDK-19); Guangdong Science and Technology College "Quality Engineering" Project (GKZLGC2023173).

References

Avhad, A., Schou, C., Arnarson, H., Madsen, O.: Demonstrating a swarm production lifecycle: a comprehensive multi-robot simulation approach. J. Manuf. Syst. **79**, 484–503 (2025). https://doi.org/10.1016/j.jmsy.2025.01.020

Chen, L., Luo, G., Wang, Y.: Research on the system design of an intelligent sorting workstation based on the IRB2600 robot. Electr. Eng. (24), 56–58 (2020). https://doi.org/10.19768/j.cnki.dgjs.2020.24.022

Chen, X., Zhang, Y., Yang, B.: Designing control system of palletizing robot based on RobotStudio. J. Phys: Conf. Ser. **2402**(1), 012039 (2022). https://doi.org/10.1088/1742-6596/2402/1/012039

Huang, J.: Design and development of a soft-PLC for a spray painting workstation based on EtherCAT. Master's thesis. Guangdong University of Technology (2023). https://doi.org/10.27029/d.cnki.ggdgu.2023.002437

Labs, W.: COBOTS at home on the plant floor. Food Eng. **96**(11), 13 (2024)

Li, P., Zhu, Z., Lai, L.: Modeling, trajectory planning and cycle time analysis of a six-axis palletizing robot. J. Shanghai Univ. Eng. Sci. **39**(1), 1–6 (2025)

Li, X.: Multi-robot task allocation and cooperative motion planning. Doctoral dissertation. Xidian University (2024). https://doi.org/10.27389/d.cnki.gxadu.2024.001499

Makulavičius, M., Petkevičius, S., Bučinskas, V., Dzedzickis, A.: Quantitative evaluation of an industrial robot tool trajectory deviation using a high-speed camera. Machines **14**(1), 8 (2025). https://doi.org/10.3390/machines14010008

Miyaji, D., Nakatsuji, H., Nishida, I.: Automation of process planning and robot motion planning in assembly operations considering component interference: special issue on production technologies at the end of the first quarter of the 21st century. Int. J. Autom. Technol. **19**(5), 720–732. (2025) https://doi.org/10.20965/ijat.2025.p0720

Qian, M.: Collaborative design of path planning and motion control for mobile robots in complex environments. Autom. Mach. Learn. **6**(1) (2025). https://doi.org/10.23977/AUTML.2025.060106

Xi, S., Zhang, Y., Liu, Y.: Design of an intelligent putty spraying experimental teaching platform under the background of industry-education integration. Exp. Technol. Manag. 1–12 (2026)

Xu, C., Li, W., Zhang, Q., Wan, Q., Liu, P.: Multi-objective trajectory planning for spraying robot based on hybrid polynomial interpolation and HMONSGA-II. Proc. Inst. Mech. Eng. **239**(6), 2014–2029 (2025). https://doi.org/10.1177/09544062241296591

Yang, G.: Robotic automatic spray painting and palletizing system. Qingdao Jinli Tianyuan Intelligent Technology Co., Ltd (2017)

Yuan, M.: Design and simulation of an offline workstation for furniture spray painting robots. Inf. Comput. (Theory Ed.) **36**(7), 31–33 (2024)

Zhang, C., Li, T., Qin, L., Jiang, Y.: Vision-guided autonomous block loading in a dual-robot collaborative handling framework. J. Constr. Eng. Manag. **151**(5) (2025). https://doi.org/10.1061/JCEMD4.COENG-15847

Numerical Simulation of Flow Field in Special-Shaped Rocket Launch Guider

Xiaozhao Wang[1], Yufan Zhang[2(✉)], Yadi Lu[1], and Yilin Pan[2]

[1] NORINCO Group Jiangshan Heavy Industries Research Institute Co., Ltd, Xiangyang 441000, China
[2] Nanjing University of Science and Technology, Nanjing 210000, China
Zhangyufan@njust.edu.cn

Abstract. When the rocket is launched, the excessive impact of the gas jet may cause damage to the special-shaped director. In this study, the computational fluid dynamics (CFD) method and the standard k-ε turbulence model were used to simulate the movement process of the rocket projectile in the tube through the overset grid technology. The evolution characteristics of the flow field in the special-shaped launch guider were systematically analyzed, and the distribution law of the pressure load on the tube wall was obtained. The results show that the gas jet forms a 'fish scale' structure under the influence of the inclined transition section of the special-shaped launch guider; until the rocket moves to the rudder groove, the maximum pressure on the cylindrical surface of the launch guider is 0.334 MPa. After the rocket passes through the rudder groove, the pressure on the inclined surface of the rudder groove gradually increases, and the maximum pressure is 0.528 MPa. Therefore, this is the where the working environment is the worst. This study can provide theoretical support for the structural design of the special-shaped launch guider.

Keywords: Special-shaped guider · CFD · Numerical simulation · Flow field structure · Dynamics simulation and analysis

1 Introduction

The structural integrity of launch systems is critically compromised by high-temperature, high-pressure gas jets during rocket launches, as these extreme conditions induce material deformation through thermal stress, component failure via dynamic pressure loads, and systemic performance degradation (Qu et al. 2025, Chen et al. 2015, Hao et al. 2025). This phenomenon has motivated extensive numerical simulation studies globally to quantify gas jet impact dynamics. Fan et al. employed one-way fluid-structure interaction analysis to elucidate the thermodynamic damage mechanisms in storage and transportation integrated launch boxes, identifying geometric discontinuities as a key amplifier of impact severity (Fan et al. 2024). Qi et al. demonstrated that the structure of the grooved inclined plates significantly alter impinging jet behaviour, including the wave patterns and acoustic signatures (Qi et al. 2022). Zhang et al. experimentally validated the dynamic response of box launcher surfaces, revealing that asymmetric geometries induce pressure oscillations through flow separation effects (Zhang et al. 2021).

T. Mankovits and N. Trisovic (Eds.): MDS 2026, LNME 2026.
https://doi.org/10.1007/978-981-92-2440-1_21

The special-shaped guider changes the front structure into a square on the traditional cylindrical shape, and is connected in a natural over-way to form an inclined surface inside. The structure may be damaged by the accumulation of temperature and pressure caused by the impact of the gas jet. Previous studies have extensively investigated jet dynamics in launch systems. Tsutsumi et al. examined how rocket engine positioning affects the flow field within launch platforms and jet deflection characteristics (Tsutsumi et al. 2018). Crafton et al. conducted experimental research on subsonic jet impingement on oblique plates, specifically comparing stagnation point locations with maximum pressure positions (Crafton et al. 2006). Yenigelen et al. investigated both flow patterns and acoustic properties in supersonic jet impingement on inclined surfaces (Yenigelen and Morris 2023). Kurokawa et al. performed numerical analysis of two-dimensional supersonic inviscid jet interactions with inclined plates, focusing particularly on inlet disturbance effects and noise generation mechanisms (Kurokawa et al. 2020).

Most of the existing research focuses on the impact problem of conventional directional devices or planar structures, and there are few studies on special structures such as special-shaped directional devices. In this paper, the launch process of the rocket projectile in the special-shaped launch guider is simulated by CFD numerical simulation, and the internal flow field characteristics of the rocket projectile in the special-shaped launch guider are studied and analysed. The law of the distribution of the flow field in the launch guider and the change of the maximum pressure of the inner wall with time is obtained, which provides a theoretical basis for the structural design of the special-shaped launch guider.

2 Flow Model

The three-dimensional gas flow field is numerically simulated by solving the Reynolds-averaged Navier-Stokes (RANS) equations and the turbulence model. The governing equations are formulated as follows:

(1) mass conservation equation

$$\frac{\partial p}{\partial t} + \frac{\partial}{\partial x_i}(\rho u_i) = 0 \tag{1}$$

(2) momentum-conservation equation

$$\frac{\partial}{\partial t}(\rho u_i) + \frac{\partial}{\partial x_j}(\rho u_i u_j) = -\frac{\partial p}{\partial x_i} + \frac{\partial \tau_{ij}}{\partial x_j} + F_i \tag{2}$$

(3) energy-conservation equation

$$\frac{\partial}{\partial t}(\rho E) + \frac{\partial}{\partial x_j}(u_j(\rho E + p)) = \frac{\partial q_j}{\partial x_j} + \tau_{ij}\frac{\partial u_i}{\partial x_j} \tag{3}$$

In the formula, the specific expression of the stress tensor is

$$\tau_{ij} = \left[\mu\left(\frac{\partial u_i}{\partial x_i} + \frac{\partial u_i}{\partial x_i}\right)\right] - \frac{2}{3}\mu\frac{\partial u_i}{\partial x_i}\delta_{ij} \tag{4}$$

where p is static pressure (Pa); ρ is the density (kg/m^3); u is the fluid velocity in the coordinate axis direction (m/s); i, j are coordinate axis direction components: values are 1, 2, 3; F is the component of the external force in different directions; μ is the dynamic viscosity coefficient; E is the internal energy per unit mass (J); q is the heat flux, and $q = -\lambda\partial T \big/ \partial x$.

(4) Internal energy equation per unit volume

$$E = \frac{p}{\gamma - 1} + \frac{\rho}{2}V^2 \tag{5}$$

(5) Conservation equation in rectangular coordinate system

$$\frac{\partial U}{\partial t} + \frac{\partial(F + F_\upsilon)}{\partial x} + \frac{\partial(E + E_\upsilon)}{\partial y} + \frac{\partial(H + H_\upsilon)}{\partial y} = 0 \tag{6}$$

3 Numerical Calculation Model

3.1 Geometric Model

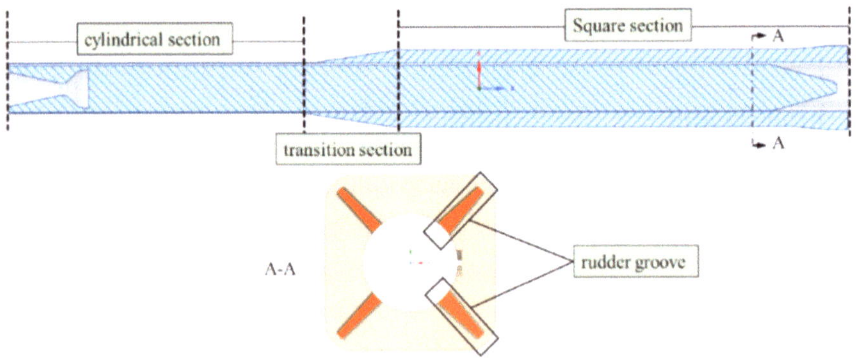

Fig. 1. Geometric model

The launch system consists of a rocket projectile and a special-shaped launch guide, as shown in Fig. 1. The flow space in the launch guide is composed of the main channel and the rudder groove. The main channel integrates three different geometric sections: circular pipe section, inclined transition zone and square pipe section, which together form the main gas jet path. The rudder slot is located in the transition inclined section and the square tube section, which is used to accommodate the rocket rudder. It is worth noting that the inclined transition section brings significant flow complexity, making the system vulnerable to local flow disturbances.

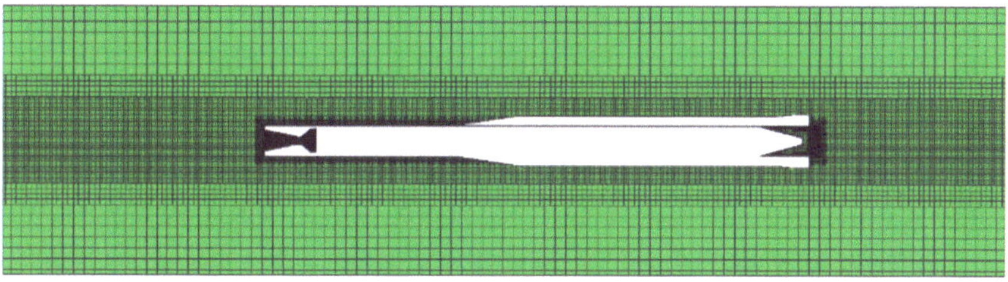

Fig. 2. Computational mesh

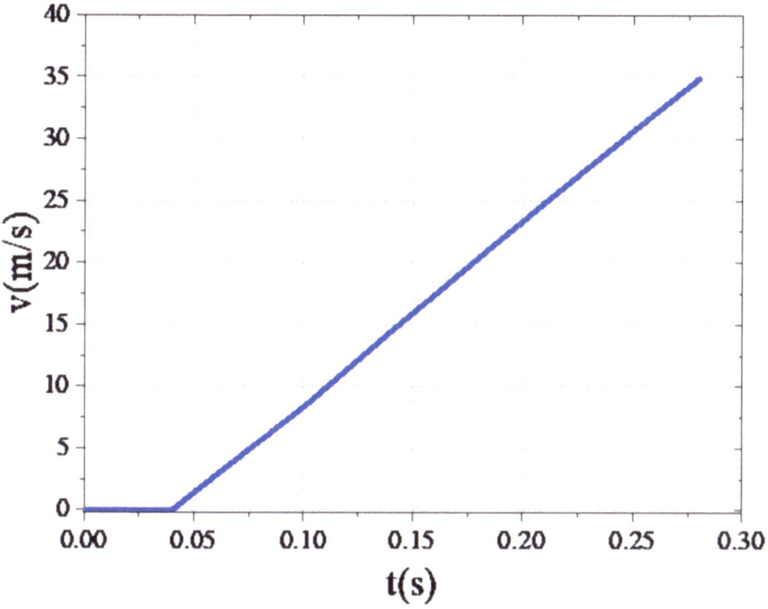

Fig. 3. Motion curve

3.2 Element Division

In order to simulate the motion process of the rocket in the directional tube and the complex gas flow, the overlapping grid technology is used in this study. Compared with the traditional dynamic mesh method, the overlapping mesh technique has significant advantages in dealing with large displacement motion problems. This method effectively alleviates the mesh distortion and negative volume problems in the calculation process and ensures the accuracy of the numerical simulation results.

The computational domain is discretized using the Trimmed Cell Mesh method, and local mesh refinement is applied to regions with high flow gradients (e.g., near the nozzle). The computational grid is shown in Fig. 2, which contains two different grid systems: background grid and foreground grid. The background grid represents the main part of the computational domain, and remains stationary to characterize the static pipe wall geometry and its surrounding flow field characteristics. The rocket area is defined as a moving foreground grid, and the motion curve is shown in Fig. 3.

3.3 Boundary Conditions

The fluid development and pressure load distribution caused by the interaction between the highly under expanded jet and the geometric structure of the special shape director are our main concerns. Therefore, a common and validated method is used to analyse the impinging jet flow field and the structural load in the launching system (Zhao et al. 2024, Tsutsumi et al. 2018). The gas through the combustion chamber is simplified as a chemical inert ideal gas. The physical properties and average molecular properties are shown in Table 1. The computational domain employs three distinct boundary conditions: the outer basin surface functions as a pressure outlet, the launch guide inner wall is modelled as an adiabatic no-slip stationary surface, and the nozzle inlet serves as a pressure inlet.

Table 1. Thermal property parameters of gas

Total temperature (K)	Total pressure (MPa)	Molecular weight (g/mol)	Constant specific heat capacity (J/(kg·K))	Specific heat ratio
3187	13.5	23.4	3044.7	1.2

3.4 Grid Independence Study

To ensure the accuracy of numerical results, grid independence was studied. Four sets of grids with minimum unit sizes of 1mm, 5mm, 10mm, and 15mm were generated to compare the maximum pressure on the inner wall of the rudder wing at the same time.

The result is shown in Fig. 4. As the grid becomes finer, the maximum pressure gradually stabilizes. The pressure values obtained from 1 and 5 mm grids are almost the same, with a difference of less than 2%. This confirms that the 5 mm grid resolution has fully captured the details of the fluid and further refined to 1mm, without significant improvement in accuracy, but it will bring a lot of computational costs. Based on this, all subsequent simulations will use a 5 mm grid, which not only ensures computational accuracy but also saves computational resources.

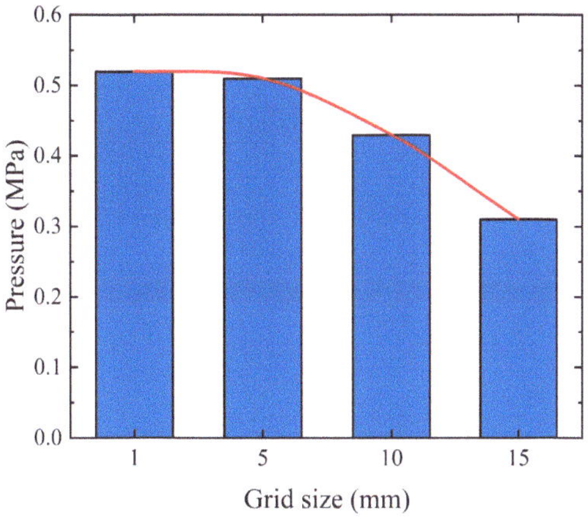

Fig. 4. Results of the grid independence

4 Results and Analysis

4.1 Calculation Results of Gas Jet

In the first 40 ms, the mechanical restraint device prevented the rocket from moving. When the combustion products of the propellant are discharged through the engine nozzle, the static pressure of the expanding airflow at the nozzle outlet plane exceeds the ambient pressure, causing the main flow to expand. In the core flow area, gas expansion leads to accelerated flow, resulting in local pressure reduction under environmental conditions. When the static pressure of the expanding airflow drops below the ambient pressure level, this pressure difference is generated. Therefore, the compression of environmental pressure generates intersecting oblique shock waves above and below the flow axis. These impacts are concentrated at the centreline of the nozzle, effectively limiting further excessive expansion of the gas. When the gas passes through the oblique shock system, the static pressure returns to above ambient conditions, reestablishing a flow equilibrium equivalent to the nozzle outlet state. As shown in Fig. 5a, this alternating flow pattern generates a continuous expansion compression wave structure in the core flow region.

During 40–180 ms, the gas jet mainly interacts with the wall surface. The airflow inside the nozzle steadily develops into the interior of the directional device and reaches the outlet through multiple shock wave reflections on the pipe wall. As shown in Fig. 5b, the flow remains highly under expanded throughout the entire stage. After 180 ms or more, when the nozzle enters the transition section, geometric changes in the cross-section and flow channel will cause significant mainstream deflection. After the gas jet collides with the inner wall of the rudder grille, it generates reflected waves and incident waves, which interact to form a complex wave system network. The waves are shown in Fig. 5c and d.

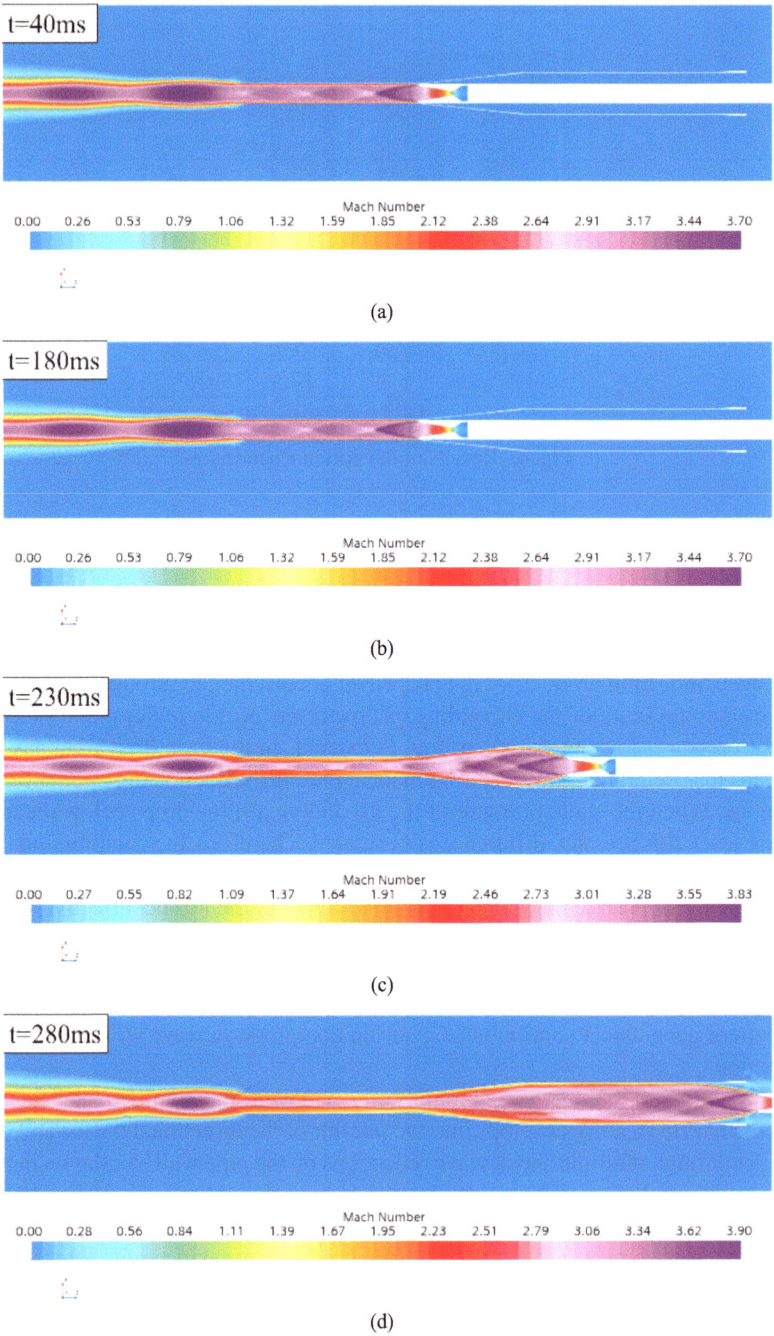

Fig. 5. Mach number cloud at different times

4.2 Distribution of Maximum Pressure on the Tube Wall

Figures 6 and 7 show the curve of the pressure peak at the key position of the inner wall of the unconventional launcher with time. During the locking phase (t < 40 ms), all surface pressures remain constant. Between 40–180 ms, the gas flow mainly interacts with the cylindrical section of the launch tube, and the peak pressure rises to 0.334 MPa. In the subsequent 180–230 ms, as the nozzle empennage passes through the inclined section, the pressure of the cylindrical tube section decreases to about 0.25 MPa, while the peak pressures of the cylindrical inclined section and the rudder wing inclined section increase significantly, at 0.324 MPa and 0.528 MPa, respectively. In the 230–280 ms, as the nozzle moves in the square pipe section, the pressure changes in different regions are different: the pressure fluctuation in the cylindrical pipe section is between 0.22–0.3 MPa, and the pressure fluctuation in the cylindrical inclined pipe section is between 0.22–0.3 MPa.

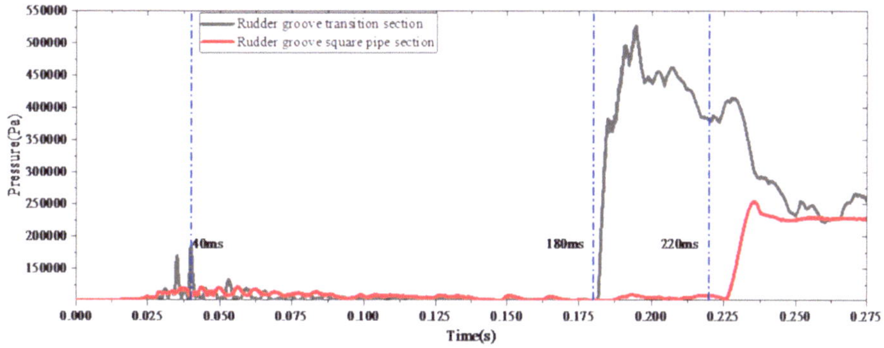

Fig. 6. The maximum pressure on the surface of the rudder groove

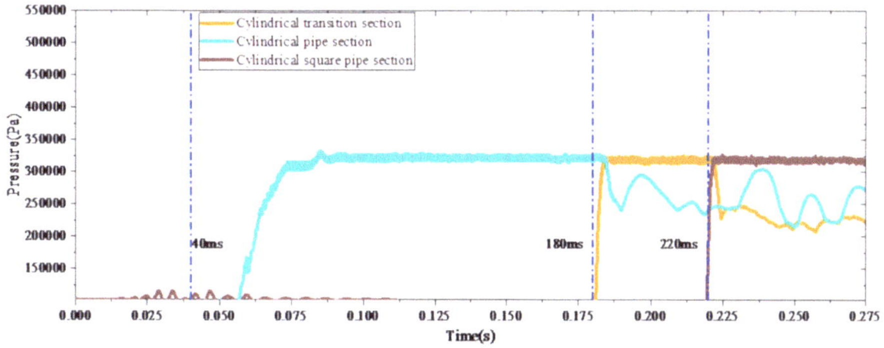

Fig. 7. The maximum pressure on the inner surface of the cylinder of the launch guider

5 Conclusion

Computational fluid dynamics (CFD) and overset grid technology were used to numerically simulate the motion of the rocket in a specially directed launch tube. The analysis focuses on the gas jet flow field characteristics and the temporal evolution of wall pressure, and the following findings were obtained:

(1) The gas jet mainly maintains the structure of the expansion compression wave system inside the launch tube. However, when interacting with the inclined transition section of non-traditional emitters, the jet will reflect when hitting the inner wall of the rudder groove. This reflection generates a shock wave that interacts with the incident wave, forming a unique fish scale like flow structure.
(2) Before the rocket enters the rudder groove, the cylindrical surface of the launcher experiences a peak pressure of 0.334 MPa. After passing through the rudder groove, the pressure on the inclined surface of the rocket gradually increases, reaching a maximum of 0.528 MPa. This indicates that the tilted structure of the rudder wing is the most severe part of gas jet impact and requires reinforced structural design in this critical area.
(3) The research results have important practical significance for the optimization design of irregular guide rails. The maximum pressure of 0.528 MPa calculated on the inclined surface of the rudder groove can be used as an input condition for structural analysis (e.g. using finite element analysis) to prevent yielding or fatigue damage. Local reinforcement should be considered here, such as increasing wall thickness or using higher strength materials. This study demonstrates the value of CFD based validation methods, which are not only used for analysing final designs but also for iterating and optimizing the geometry of guide rails in the early design stages. This method can minimize peak pressure loads to the greatest extent possible, which is of great significance for improving the overall lifespan and reliability of the launch system.

References

Crafton, J., Carter, C., Elliott, G., et al.: The impingement of sonic and sub-sonic jets onto a flat plate at inclined angles. Exp. Fluids **41**, 699–710 (2006). https://doi.org/10.1007/s00348-006-0192-z

Chen, S., Jiang, C., Li, J.: Numerical and experimental study on impact load ofjet flow of solid rocket. J. Nanjing Univ. Sci. Technol. **39**(6), 698–703 (2015). https://doi.org/10.14177/j.cnki.32-1397n.2015.39.06.011

Fan, H., Zhang, L., Kou, Y., et al.: The damage analysis and optimization design of a certain air defense missile tail flame to the cartridge case. Gun Launch Control. J. **45**(04), 89–94+100 (2024). https://doi.org/10.19323/j.issn.1673-6524.202305004

Hao, Y., Gan, S., Li, G.: Numerical analysis of thermal and mechanical environment of rocket umbilical tower under gas jet action. Ordnance Autom. **44**(6), 86–92 (2025). https://doi.org/10.7690/bgzdh.2025.06.019

Kurokawa, M., Teramoto, S., Okamoto, K.: Acoustic wave generation from two-dimensional supersonic inviscid jet impinging on inclined plate. AIAA J. **58**(8), 3436–3445 (2020). https://doi.org/10.2514/1.J059012

Qi, L., Feng, H., Zhao, K., et al.: Study on the influence of groove structure on the flow field and sound field characteristics of impinging jet. Vib. Shock. **41**(22), 105–112 (2022). https://doi.org/10.13465/j.cnki.jvs.2022.22.013

Qu, P., Chen, G., Yuan, W.: Investigations on effects of multi-angle inclined impact of gas jet on launching platform. J. Missiles Guid. **45**(02), 223–231 (2025). https://doi.org/10.15892/j.cnki.djzdxb.2025.02.013

Tsutsumi, S., Sarae, W., Ueda, H., et al.: Effect of rocket engine layouts on jet flowfield inside a launch pad. J. Spacecr. Rocket. **55**(6), 1537–1544 (2018). https://doi.org/10.2514/1.A34178

Yenigelen, E., Morris, P.: Flow and acoustics in a model rocket flame deflector and deflector cover. Int. J. Aeroacoustics **22**, 437–458 (2023). https://doi.org/10.1177/1475472X231199184

Zhang, J., Chen, S., Zhang, J., et al.: Dynamic effect and experimental study on the air-facing surface of the box launcher. J. Artill. Launching Control. **42**, 91–96 (2021). https://doi.org/10.19323/j.issn.1673-6524.2021.04.017

Zhao, C., Wang, H., Sun, Z., Le, G.: Investigating the flow field and thermal environment of the rocket gas jet impinging on the launch platform. J. Nanjing Univ. Sci. Technol. **48**(3), 253–261 (2024). https://doi.org/10.14177/j.cnki.32-1397n.2024.48.03.001

Research on Simulation Testing Technology for Multi-Point Hinge Flap Mechanism Fracture Failures

Chen Li[1], Xianmin Chen[1], Sanyuan Li[1(✉)], and Bing Tan[2]

[1] National Key Laboratory of Strength and Structural Integrity, Aircraft Strength Research Institute of China, Xi'an 710065, China
lsynuaa@163.com

[2] China Special Vehicle Research Institute, Jingmen 448035, Hubei, China

Abstract. Flaps serve as critical lift-enhancing devices during aircraft take-off and landing operations. Both military and civil aircraft are subject to safety and reliability requirements for flaps stipulated in various standards and airworthiness regulations. To validate flap safety and reliability following failure occurrence while obtaining relevant data for optimisation, conducting failure simulation tests is essential. This study investigates fracture failure simulation techniques for flap mechanisms. A failure simulation mock-up was designed for torsion bar fracture, and subsystem level failure simulation tests were conducted on a multi-hinge flap mechanism of a specific type. Accurate responses in flap movement time and angular displacement during failure were obtained, providing a basis for optimising design parameters of multi-hinge flap mechanisms.

Keywords: Flap · Fault simulation device · Torsion bar fracture · Fault testing

1 Introduction

As the most complexly moving, most extensively deflected, and most severely loaded movable control surface on an aircraft, flaps play a vital role in ensuring safe take-off and landing operations (Zhang et al. 2019; Ji et al. 2022; Wan et al. 2022). Various structures within the flap system simultaneously transmit both system loads and structural loads. Due to the complex loading environment, these structures face potential risks of damage, wear, jamming, and fracture during service. To ensure sustained safe flight and landing capability under single or combined failure conditions, conducting flap failure analysis and testing to validate their reliability and safety design is essential. However, current failure response results for flap mechanisms predominantly derive from simulation analyses or small-scale component/assembly-level testing. For instance, (Feng et al. 2023) investigated dynamic simulation methods for flap mechanism wear failures, while (Cheng 2023) designed a microcontroller-based flap pivot wear failure test rig. Research on subsystem level testing failure simulation testing remains relatively scarce. For instance, (Lu et al. 2024) developed a failure simulation fixture based on an explosive bolt to replicate drive disconnection failures. This paper presents a fault simulation

T. Mankovits and N. Trisovic (Eds.): MDS 2026, LNME 2026.
https://doi.org/10.1007/978-981-92-2440-1_22

fixture for torque rod fracture based on a multi-hinge flap mechanism. Compared to Lu et al.'s explosive bolt-based fracture simulation, this approach offers enhanced controllability and safety. Validated through subsystem level ground testing, it captured transient data including flap movement duration and flap movement angle during a torsion bar failure. This completes the subsystem level validation testing for flap failure simulation.

2 Design of Flap Failure Simulation Mock-Up

Torsion bar failure occurs when the torsion bar joint loosens due to wear, or the torsion bar itself fractures due to fatigue, ultimately resulting in the inability to transmit torque power. The torsion bar fracture fault is realised through the flap torsion bar disconnection device, as shown in Fig. 1. This device primarily comprises the torsion bar fracture assembly, torsion bar modification components, a four-bar linkage mechanism, a control console with actuators, and an electric motor. The torsion bar fracture assembly facilitates the disconnection and reconnection of the torsion bar modification component. This modification component secures the fracture assembly by severing and splicing the torsion bar at designated mounting points, thereby achieving rigid connection between the fracture assembly and torsion bar. The four-bar linkage primarily converts power from the motor to the torque rod breakage component, ensuring force direction remains parallel to the rod to prevent jamming. The control console and actuators provide structural support and force transmission, while the motor supplies power for pushing or pulling the rod.

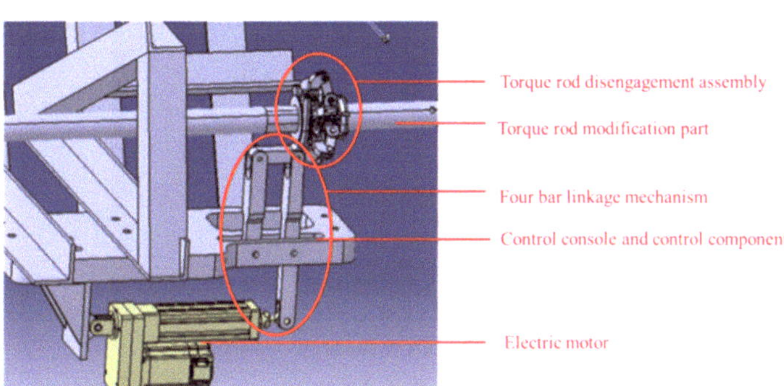

Fig. 1. Flap torque rod breakage device assembly diagram.

The torque rod breakage assembly primarily comprises Joint 1, Joint 2, a sliding connector, and a U-shaped groove, as shown in Fig. 2. Joint 1 and Joint 2 are respectively rigidly connected to the torque rod via specialised bolts. The sliding joint is mounted on Joint 1 and features six rocker arms, each comprising two smaller rocker arms. The left end of Joint 2 has six locking slots that engage with the six rocker arms on the sliding joint.

The principle of the flap disconnect mechanism is as follows: When the flap torsion bar requires disconnection, the motor drives the upper end of the four-bar linkage to move rightwards. The rod connected to the U-clamp then pulls the U-clamp leftwards. This mechanism primarily comprises Joint 1, Joint 2, a sliding joint, and a U-shaped groove. Joint 1 and Joint 2 are respectively rigidly connected to the torque rod via specialised bolts. The sliding joint is mounted on Joint 1 and features six rocker arms, each comprising two smaller rocker arms. The left end of Joint 2 has six locking slots that engage with the six rocker arms on the sliding joint, as shown in Fig. 3. A physical representation of the device is shown in Fig. 4.

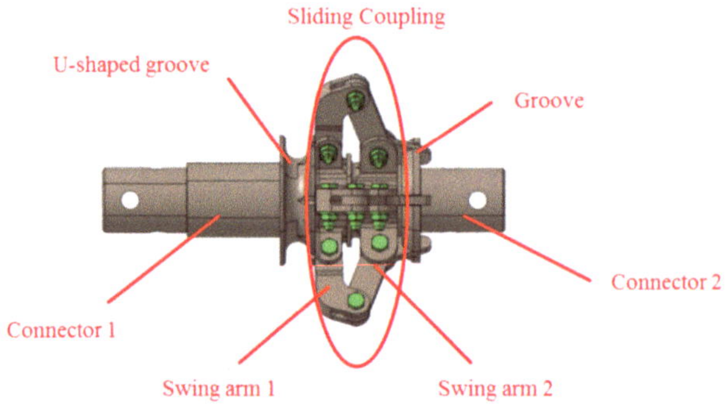

Fig. 2. Torque rod detachment joint assembly diagram.

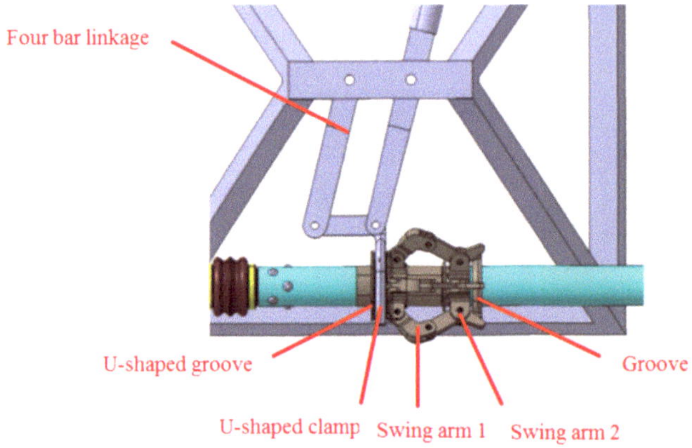

Fig. 3. Schematic diagram of torque rod breakage device in disengaged state.

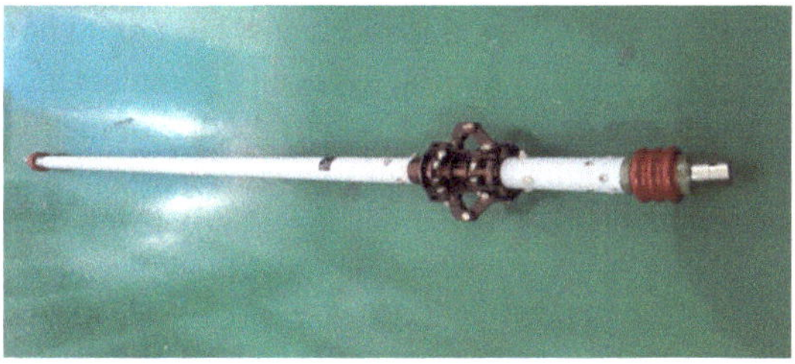

Fig. 4. Flap torque rod release mechanism.

3 Flap Breakage Failure Simulation Test

This test utilised the right outer flap of a specific aircraft model as the test specimen, encompassing the complete structure and systems including the flap surface, drive unit, motion mechanism, and transmission mechanism. In accordance with installation requirements, laser tracking precision positioning technology was employed to mount the right outer flap, PDU, torque rod, and control actuator cylinder onto the test rig. Hydraulic pressure was introduced into the PDU and loading actuator cylinders respectively from the ground hydraulic source of the loading system. Two loading actuator cylinders from the loading system were mounted onto the test rig, with load sensors connecting one end to the actuator cylinder output and the other to the loading point on the flap. The test specimen after installation is shown in Fig. 5. Control the flap retraction and extension movements via the control system, measuring and recording the retraction/extension time, number of cycles, flap motion trajectory, rocker arm force transmission path, and drive torque under various fault conditions.

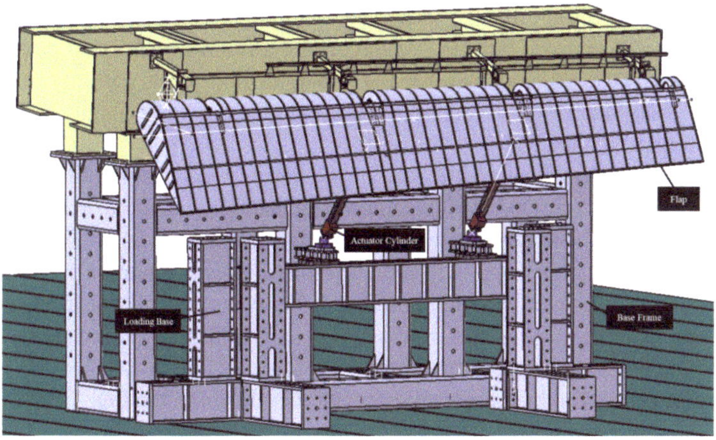

Fig. 5. Test specimen installation diagram.

The program flowchart for the control system is shown in Fig. 6. The control system employs ADC sampling to convert the voltage signal from the angle sensor into a digital signal. This is compared against preset parameters, thereby outputting corresponding control signals to the PDU to drive the flap mechanism to extend or retract in the aforementioned sequence, repeating this motion cyclically. Simultaneously, it continuously scans the mechanical limit switch signal. Upon triggering of the mechanical limit signal, a corresponding control signal is output to the PDU to halt the flap mechanism's movement. Throughout the process, real-time angle and operational status information is displayed on the touchscreen, with automatic recording of cycle duration, angles, and other parameters for each individual cycle. The test concludes upon completion of the preset number of cycles or by pressing the "Stop" button. Once parameters are preset, the system operates fully automatically, requiring no manual intervention except in emergency situations.

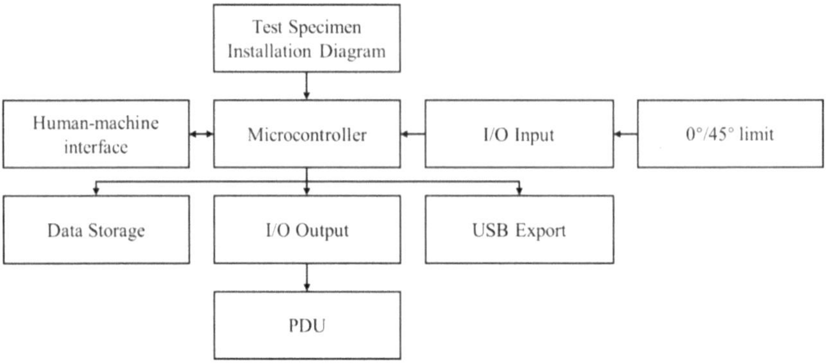

Fig. 6. Control system programme structure diagram.

The primary physical quantities measured during this fault simulation test include flap support arm stress and flap surface movement angle. Strain gauges measure deformation on the support arm, reflecting changes in the force transmission path following a fracture fault. Angle sensors measure the flap deflection angle, partially reflecting the flap surface's attitude response post-failure while also providing input for flap control and test safety protection.

To facilitate load application, the load on the flap surface was simplified to two loading points on the surface. During flap deployment and retraction, two actuator cylinders were used to apply identical loads to the flap at the two loading points on its underside; the loads varied synchronously with the flap's angle of movement. To ensure loading accuracy at this deployment and retraction speed, this test establishes a linear relationship between the load value at each loading point and the flap deflection angle. When the flap deflection angle reaches the maximum of 45°, the load is simultaneously applied to its maximum value. The load application method is shown in the Fig. 7.

Fig. 7. Load application schematic diagram.

In this failure simulation test, the flaps have three movement positions, designated as 0#, 2#, and 3#. Position 0# represents the initial position, 2# represents the intermediate position (20°), and 3# represents the fully extended position (45°). The torque rod fracture test involves two distinct fracture locations, designated as Location A and Location B, as shown in Fig. 8.

Fig. 8. Torsion bar fracture failure point.

The specific operating conditions for the flap fracture failure simulation test are shown in Table 1.

Table 1. Test conditions for torsion bar fracture failure.

serial number	Fault Simulation Mode	Flap movement status	Loading condition
1	Position 2, Torque tube fracture at location A	Flaps down	Under load
2	Position 2, Torque tube fracture at location B	Flaps down	Under load
3	Position 2, Torque tube fracture at location B	Flaps retract	Under load
4	Position 3, Torque tube fracture at location B	Flaps retract	Under load

The test procedure for simulating flap failure is as follows:

(a) Turn on the loading control system and set the flap angle to 0°;
(b) Open the oil valve and pressurize the system;
(c) Execute the zero-point calibration procedure;
(d) Reset the data acquisition system and start data acquisition;
(e) Conduct the test according to the operating conditions shown in Table 1, with the angle deflecting through 0°–20°–45°–20°–0° and the load applied at 0N–7111N–16000N–7111N–0N. When the flap reaches the angle required for the current operating condition, control the simulated mechanism to disconnect the torque tube;
(f) After the torque tube disengagement fault has stabilized, unload to the no-load state and save the test data for this operating condition;
(g) Proceed to the next operating condition test until all operating conditions have been completed.

4 Results of Flap Failure Simulation Tests

The on-site configuration for the flap malfunction simulation test is depicted in the Fig. 9. Given the substantial volume of measurement data generated by the numerous flap deployment cycles per test, coupled with the high repeatability of results across different cycle counts, this paper presents angle measurement data from a single cycle of each malfunction test for illustrative purposes.

Fig. 9. On-site photo of flap malfunction simulation test.

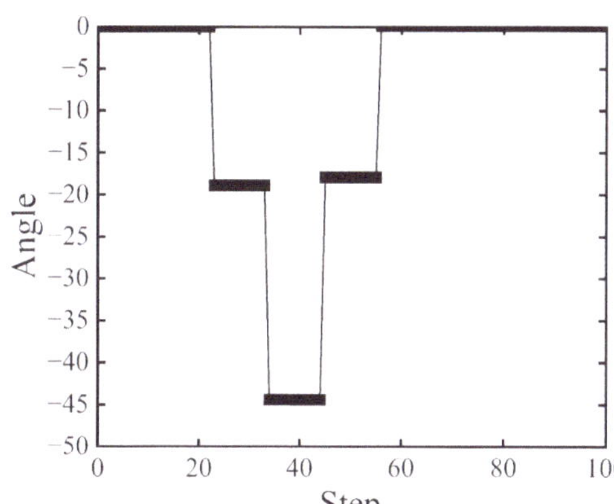

Fig. 10. Normal retraction function test angle measurement results.

Prior to conducting the malfunction tests, the flap mechanism underwent normal deployment/retraction functionality tests. These tests involved performing a complete cycle of normal deployment and retraction under loaded conditions. The angle changes of the flap surface are illustrated in Fig. 10.

The flap angle variation under this condition is illustrated in Fig. 11. The torsion bar fractures at 20° in this condition. As seen in the figure, after the torsion bar fracture occurs, from the moment the fracture fault happens until the control system protection triggers, the system damping takes effect, and the flap movement stops. The flap continues to move inertially by 1.9° to 21.9°. In future designs, further optimization of the angle sensor accuracy and control system response speed is recommended. Concurrently, optimizing the damping coefficient of the braking system can minimize the inertial movement of the flap following a failure.

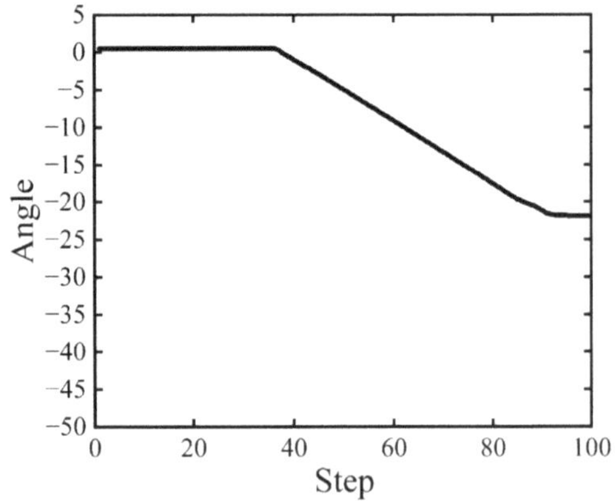

Fig. 11. Torque bar disengagement test angle measurement results.

Select one strain gauge from each of the four flaps' support arms at a consistent relative position. The strain changes in the four support arms of the flaps under Operating Condition 2 are shown in Fig. 12. Relative to position B, support arms 1 and 2 are on the side closest to the drive, while support arms 3 and 4 are on the side farthest from the drive. After fracture occurs, the strain in support arms 1 and 2 rises sharply, while the strain at the corresponding positions on arms 3 and 4 drops sharply. After this abrupt change, the strain tends to return to its original value. Additionally, since arms 2 and 3 are closer to the fracture site B than arms 1 and 2, their strain values are relatively higher. The test results show that after the torsion bar fractures, the strain on the two arms closest to the drive suddenly increases by approximately onefold. The strain values and rate of change near the fracture site are approximately four times those of the distant sections. However, after the peak of the sudden change, the strain tends to return to its original value, and at this point, the flaps have not completely lost their operational capability. Therefore, during design, attention should be paid to protecting the struts near the areas prone to fracture failure to prevent strut damage caused by sudden stress changes following torsion rod fracture under high-load conditions, which could result in complete loss of flap functionality.

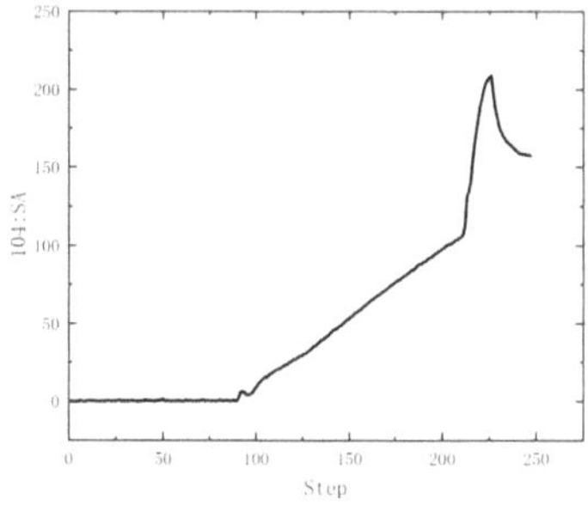

(a) Strain gauge 104 on arm #1

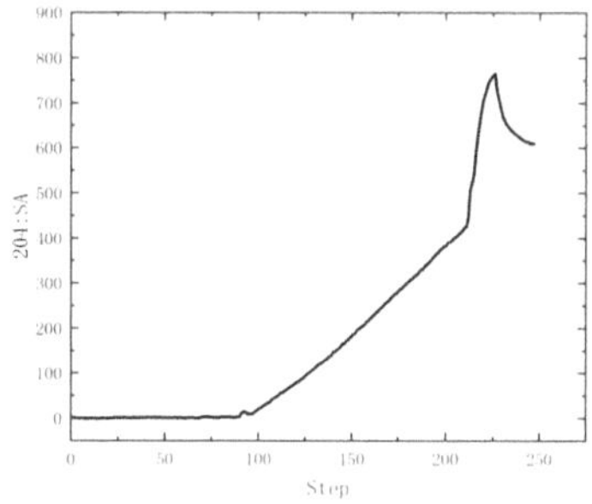

(b) Strain gauge 204 on arm #2

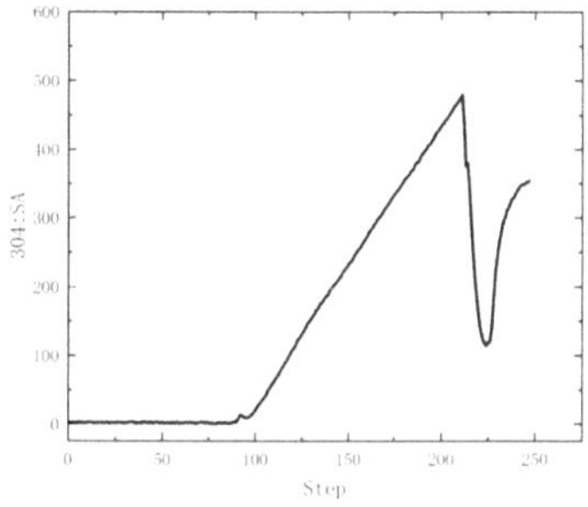

(c) Strain gauge 304 on arm #3

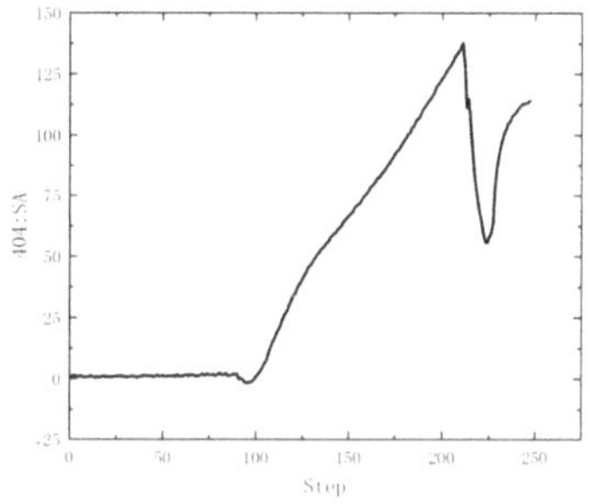

(d) Strain gauge 404 on arm #4

Fig. 12. Strain trends in the support arm.

5 Conclusion

Simulation fixtures and devices were designed to replicate flap torque rod fracture. Multi-hinge flap failure simulation tests were devised and validated through subsystem level ground testing, yielding transient data including actual motion durations and flap movement angles during these three failure scenarios. Conclusions are as follows:

(1) The flap torque bar fracture simulation technique was investigated, with a fracture mock-up designed. Testing validated its normal support function and instantaneous fracture capability under fault conditions.
(2) A multi-pivot hinged flap failure simulation verification test was designed, yielding time and angular response data for the flap mechanism under three failure conditions: fracture, jamming, and wear;
(3) Conducted flap actuator fracture failure simulation and strength testing, successfully achieving subsystem level failure simulation trials.
(4) When designing flaps, care should be taken to protect the support arms near areas prone to fracture failure. This is to prevent sudden structural changes—caused by the failure of torsion rods under high loads—from damaging the support arms, which could result in complete loss of flap functionality.

References

Cheng, X.: Aircraft flap fault simulation test bench based on microcontroller data acquisition and control application. Med. Electron. Control Eng. **5**(2) (2023), https://doi.org/10.37155/27177-5197-0502-28

Feng, Y., He, Z., Tang, J., Xue X., Wang, Y.: Dynamics simulation research on fault conditions of deflecting to airflow flap mechanism of civil aircraft. Adv. Aeronaut. Sci. Eng. **14**(4), 85–93 (2023), https://doi.org/10.16615/j.cnki.1674-8190.2023.04.09

Ji, Q., Zhang, Y., Chen, H., Ye, J.: Aerodynamic optimization of a high-lift system with adaptive dropped hinge flap. Chin. J. Aeronaut. **35**(11), 191–208 (2022). https://doi.org/10.1016/j.cja.2022.03.012

Lu, J., Xu, J., Gao, Y., Li, S.: Large aircraft flap actuator disconnection failure test technology research. In: 2024 10th Asia Conference on Mechanical Engineering and Aerospace Engineering (MEAE), pp: 411–415. Taicang (2024), https://doi.org/10.1109/MEAE62008.2024.11026167

Wan, Q., Liu, G., Liu, Q., Tong, R., Ma, S.: Research on dynamic characteristics of flap actuation system considering joint clearance and flexibility. Machines **10**(11), 1098 (2022). https://doi.org/10.3390/machines10111098

Zhang, Q., Wang, Z., Wei, M., Lin, W.: Review of high-lift devices design for amphibious aircraft. Acta Aerodynamica Sinica **37**(1), 19–32 (2019). https://doi.org/10.7638/kqdlxxb-2017.0110

Stress Analysis of Silicon Solar Cells Under Multiphysics Coupling

Renhong Huang and Pan Dai(✉)

School of Information Engineering, Huzhou University, Huzhou 313000, China
02627@zjhu.edu.cn

Abstract. Silicon solar cells in practical operation often encounter complex multiphysics-coupled conditions, significantly impacting their stress distribution and performance. Targeting the real-world engineering application of silicon solar cells mounted on outdoor support structures, this study systematically investigates the coupled effects of multiphysics fields, including wind-induced laminar loads, mechanical loads from the support structure, and the ambient temperature field. It focuses on analyzing the stress distribution and deformation characteristics of silicon solar cells under these complex operating conditions. The model incorporates fluid-induced stress-strain, plastic deformation, and temperature-induced thermal stress. When wind speed increases from 5 m/s to 50 m/s, the maximum stress increases by tens of times. The influence of wind direction changes on stress is significantly smaller than that of wind speed variations. As temperature rises from 20 °C to 80 °C, the maximum stress increases from 7.45×10^7 to 5.21×10^8 N/m^2. The research reveals significant differences in the stress distribution of silicon solar cells under the individual and combined actions of different physical fields. These findings contribute to a comprehensive understanding of the stress state in silicon solar cells under multiphysics coupling, thereby providing critical support for the development of the solar energy industry.

Keywords: COMSOL · Individual · Combined · Stress-strain · Changes

1 Introduction

Over the past decade, photovoltaic power generation has become a key part of the global renewable energy system (Bahar et al. 2024, Wiatros-Motyka et al. 2024). But when it comes to actual application, solar cells operate in highly complicated outdoor environments, with various physical fields exerting combined effects on them. These physical fields interact with each other and have a big impact on how solar cells work and how long they last. The thermal field, solid mechanics field and laminar flow field are the most important ones, and they matter a lot during the whole time the solar cells are in use. We also can't overlook the influence that temperature has on solar cells, as it exerts a tangible effect on how long the solar cells can be used in practical applications (He et al. 2024). In natural environments, airflows typically show turbulent characteristics (Chabaud 2024). Their flow state is mainly controlled by ambient wind speed and terrain

T. Mankovits and N. Trisovic (Eds.): MDS 2026, LNME 2026.
https://doi.org/10.1007/978-981-92-2440-1_23

conditions. Accurate analysis often requires more complex turbulence models (Ma et al. 2024, Wang et al. 2022). In conclusion we need to conduct in-depth research on the stress state of solar cells under the coupling action of the three fields.

2 Parameter

2.1 Inclination Angle

According to the basic theory of solar energy engineering, the optimal inclination angle of the solar panel is calculated as follows:

$$\theta_{opt} = \varphi \pm \Delta\theta \tag{1}$$

In the formula, φ is the local geographic latitude (°) and $\Delta\theta$ is the seasonal adjustment coefficient (+15° in winter and −15° in summer) (Chen and Dagestani 2023). In this study, we select the 30°N region as the simulation object, and the annual average optimal inclination angle of the solar panel in this region is calculated as follows:

$$\theta_{opt} = 30° + 0° \tag{2}$$

Considering that the experiment needs to balance the power generation efficiency of the solar panel in four seasons, we finally set the tilt angle to 29°. This value deviates less than 1° from the theoretical calculated value, so it can be regarded as an optimal choice within a reasonable error range (Xu et al. 2023).

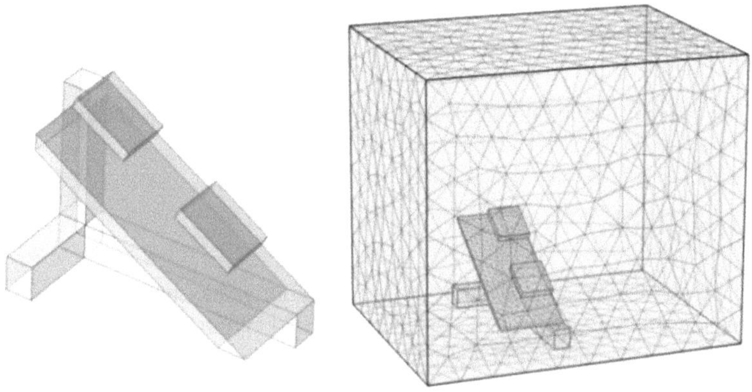

Fig. 1. Geometry and grid diagrams.

Based on the theoretical analysis above, ‘Fig. 1’ in this paper adopts a 29° modeling angle to closely resemble real-world application conditions, thus ensuring the analysis reflects actual scenarios. Additionally, a free tetrahedral mesh is applied to improve computational convergence in the simulation.

2.2 Parameter Roles and Settings for Multiphysics

In the COMSOL simulation of silicon solar cell stress under multiphysics coupling scenarios, the setup of material and physical parameters plays a decisive role. In the fluid flow module, fluid density is closely related to fluid categories and further affects the load acting on the solar cell panel. Dynamic viscosity, as an index describing fluid viscosity, determines the internal friction of the fluid and the distribution of shear stress on the panel surface. In the solid mechanics module, the elastic modulus characterizes the resistance of materials (silicon solar cell, iron frame and rubber) to elastic deformation, and its value varies significantly among different materials. Poisson's ratio defines the proportional relationship between transverse strain and axial strain under external load, which has an impact on the deformation mode and stress distribution of the structure. In the heat transfer module, the thermal expansion coefficient determines the magnitude of material expansion or contraction caused by temperature variation, which is a core factor leading to thermal stress. Thermal conductivity reflects the heat conduction capacity of materials, which affects the temperature field distribution inside the cell assembly and further influences the thermal stress distribution. The flow velocities adopted in the simulation are set as 5 m/s and 50 m/s respectively.

In this study, laminar inlet boundary conditions were set as fully developed flow. Separate and independent boundary setups were applied to different calculation domains. Experimental parameters were adjusted to match real working conditions so that the influence of fluid on the test structure can be analyzed. These parameters work together. Reasonable settings ensure reliable stress simulation results. This enables mutual verification and complementation between simulations and experiments and further supports the optimization of follow-up research.

3 Multiphysics Fundamentals and Analysis of Experimental Results

3.1 Stress Analysis at Different Flow Velocities

When studying the stress characteristics of silicon solar cells, we use professional experimental equipment to simulate the effects of various fluid flow rates on silicon solar cells, and this simulation process follows the basic principles of the Navier–Stokes equations. We select two typical flow rates for the experiment: 5 m/s and 50 m/s. During the whole experiment, we keep environmental conditions such as temperature, pressure and fluid flow direction constant. In the actual use scenario of silicon solar cells, the flow of air on the surface of silicon solar cells in the natural convection environment may present in the form of laminar flow. By solving the Navier-Stokes equation, we can obtain the specific distribution of the fluid velocity field and pressure field, and then analyze the detailed effect of laminar flow on the surface of silicon solar cells. Under the premise that all other conditions remain the same, the stress borne by silicon solar cells varies significantly with the change of flow velocity.

'Figure 2' is a simulation test in the wind direction from right to left and the thermal expansion of the solid is turned off, only the wind speed is changed, because the effect of the thermal expansion of the solid is much greater than the effect of the fluid flow rate, and the maximum stress in ‘Fig. 2’(a) is 1.35×10^4 N/m^2, and the maximum stress in ‘Fig. 2’(b) is 8.23×10^5 N/m^2, and the stress level in ‘Fig. 2’(a) is significantly higher than that in ‘Fig. 2’(b). In terms of stress distribution characteristics, the stress concentration areas of the two models are similar, and both show obvious stress aggregation in specific parts of the structure, such as geometric abrupt changes. The stress gradient in ‘Fig. 2’(a) is more drastic, and the range of high-stress areas is relatively concentrated, while the overall stress level in ‘Fig. 2’(b) is lower, the color transition is smoother, and the strength and range of high-stress areas are weaker than those in ‘Fig. 2’(a). Under other environmental conditions, the increase of fluid velocity under certain conditions will lead to an increase in the stress of the equipment. It can also be explained by the stress displacement, the following is the stress displacement effect under the corresponding fluid velocity.

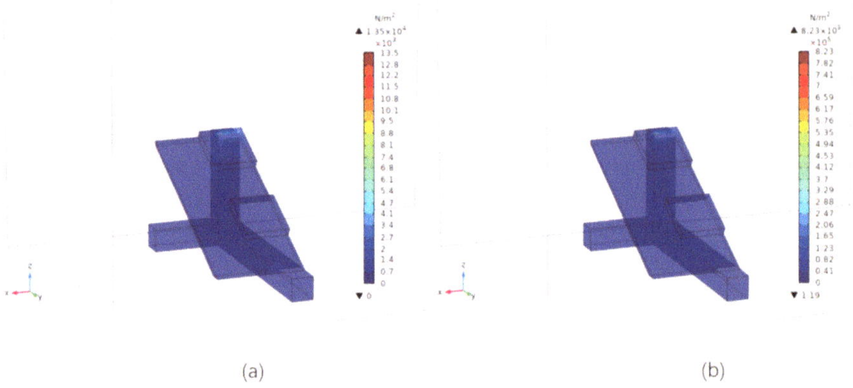

Fig. 2. (a) Stress at 5 m/s, (b) Stress at 50 m/s.

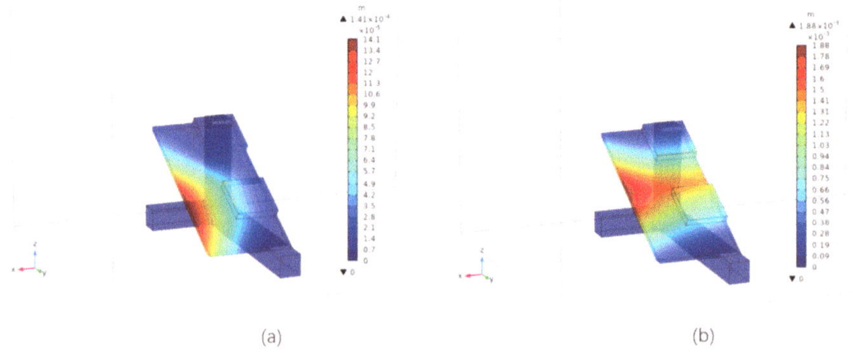

Fig. 3. (a) Stress displacement at 5 m/s, (b) Stress displacement at 50 m/s.

'Figure 3' is a simulation test with the wind direction from right to left and the solid thermal expansion is turned off, and only the wind speed is changed. The maximum displacement in 'Fig. 3'(a) is 1.41×10^{-4} m. Figure 3(b) has a maximum displacement of 1.88×10^{-3} m. In terms of displacement distribution characteristics, the displacement change areas of the structures in the two figures are consistent, and they all show obvious displacement gradients in specific parts of the model, such as geometric protrusions or connected areas. However, due to the numerical value, the displacement color transition in 'Fig. 3'(a) is more gentle, and the overall deformation amplitude is smaller, while in 'Fig. 3'(b), the color transition is more intense due to the larger displacement value, and the range and intensity of the high displacement area are more prominent, reflecting the more significant deformation of the structure under the corresponding working conditions.

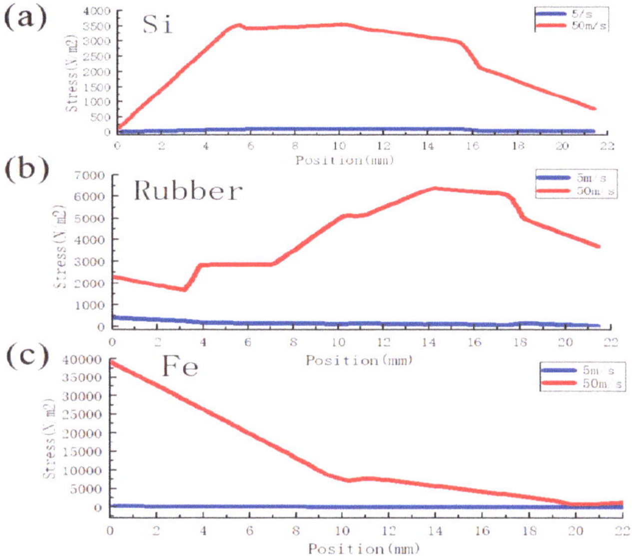

Fig. 4. (a) Stress of silicon at different speeds, (b) Stress of rubber at different speeds, (c) Stress of iron frame at different speeds.

'Figure 4' is a simulation test with the wind direction from right to left and the thermal expansion of the solid being turned off, only changing the wind speed. In 'Fig. 4'(a), the stress of Si material is almost unchanged at 5 m/s and the value is small, while the stress increases significantly at 50 m/s, peaks at 5 mm and then decreases gradually. In 'Fig. 4'(b), the stress of the rubber material remains low at 5 m/s and fluctuates slightly, and the stress first decreases briefly at 50 m/s, then rises sharply to a peak at about 15 mm, and then gradually declines. In 'Fig. 4'(c), the stress of iron material does not change much at 5 m/s and the value is small, and the stress is extremely high at 50 m/s at about 40000 N/m^2 at 0 mm, and then decays rapidly, and then fluctuates slightly after 10 mm. In general, the fluid velocity has a significant effect on the stress of each material, and the stress at 50 m/s is significantly higher than that at 5 m/s, and the response modes of different materials such as Si, Rubber, and Fe to stress are significantly different.

3.2 Stress Analysis in Different Flow Directions

In order to explore the effect of fluid flow direction on silicon solar cells, the fluid field was set to the Navier-Stokes equation to simulate the stress characteristics of silicon solar cells. Two flow directions were selected for the simulation, and fluid was applied from the front and side of the model. In order to ensure that the influence of the variable of flow direction on the stress of the solar cell is accurately focused, the other environmental conditions such as temperature, pressure, and fluid flow rate are strictly constant throughout the experiment. For isotropic linear elastic materials, Hooke's law is satisfied between stress and strain in unidirectional tension or compression.

$$\sigma = E \in \tag{3}$$

where E is the modulus of elasticity (Pa), which reflects the ability of the material to resist elastic deformation. As a commonly used material for solar cells, the stress-strain relationship of silicon under unidirectional force can be calculated by this formula within a certain temperature range (Chen et al. 2020).

Hooke's law provides a basis for analyzing the stress-strain state of solar cells composed of a variety of materials and their complex structures under multidirectional stress (Rajabi and Hosseini-Hashemi 2018). When other conditions are consistent, different flow directions expose solar cells to very different stress situations. When the fluid flows parallel to the surface of the panel, the main result are shears stress, which causes the material on the surface of the panel to slide relatively. Under this stress, the bonding part between the encapsulation material on the surface of the solar cell board and the internal silicon wafer may be subjected to large pulling force. The long-term effect may lead to a small gap between the encapsulation material and the silicon wafer, which will affect the protective performance and electrical performance of the solar cell. Flow perpendicular to the surface of the panel puts the panel under positive pressure, which can easily cause the panel to bend and deform. For thinner panels, this positive pressure may cause obvious deformation in a short period of time, and once the deformation exceeds a certain limit, it will lead to damage to the internal structure of the solar cell, such as silicon wafer breakage, loose electrode connection, etc., directly resulting in the reduction of battery power generation efficiency or even failure.

'Figure 5' shows that the maximum stress in 'Fig. 5'(a) (flow direction from right to left) is 1.35×10^4 N/m^2, and the maximum stress in 'Fig. 5'(b) (flow direction from front to back) is 2.95×10^4 N/m^2. In terms of stress distribution characteristics, although the structure and morphology of the two graphs are similar, the difference in flow direction leads to different stress concentrations. 'Figure 5'(a) shows that the overall stress distribution is relatively flat and the range of high stress regions is narrow, while 'Fig. 5'(b) has a lower overall stress level and a smoother color transition, and the strength and range of the high stress areas are weaker than those in 'Fig. 5'(a). 'Figure 5'(a) and 'Fig. 5'(b) show that due to the change of flow direction, the mode of action of the structure under load changes, which makes the high-stress area more prominent and the stress gradient changes more drastically, reflecting the significant influence of the flow direction on the stress distribution mode of the structure.

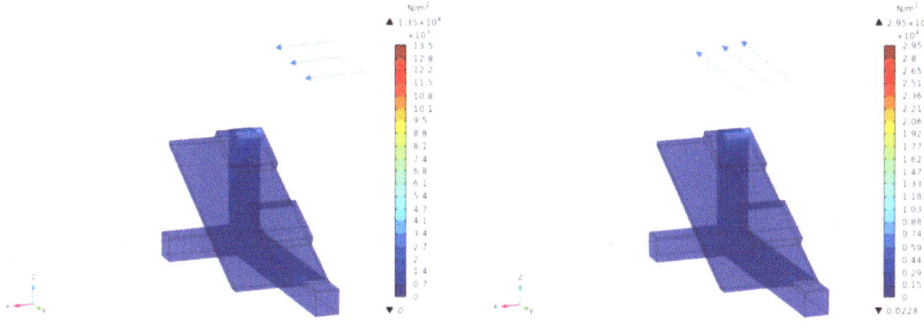

(a) (b)

Fig. 5. (a) Flow direction is from right to left, (b) Flow direction is from front to back.

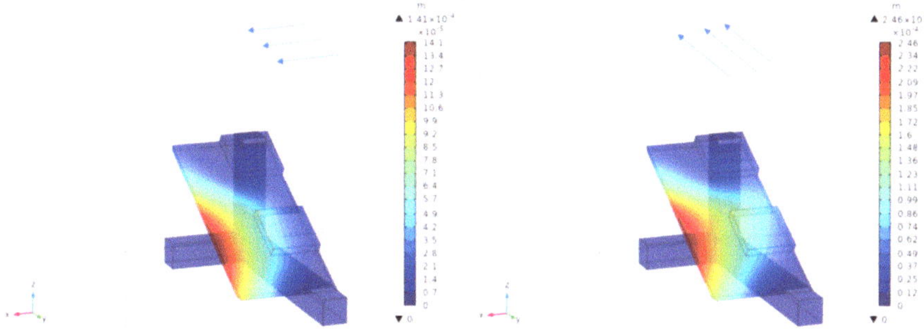

(a) (b)

Fig. 6. (a) Flow direction is from right to left, (b) Flow direction is from front to back.

'Figure 6' shows the displacement diagram of different flow directions, and the test conditions are simulated in an environment with a wind speed of 5 m/s and a solid thermal expansion turned off, and only the wind direction is changed. Maximum single-point displacement in Fig. 6'(a) (flow direction from right to left) reaches 1.41×10^{-4} m, and Maximum single-point displacement in Fig. 6'(b) (with the flow direction from right to left) reaches 2.46×10^{-4} m, which is significantly higher than the former. In terms of displacement distribution characteristics, both figures show displacement changes in specific areas of the structure, such as geometric protrusions, but the difference in flow direction leads to different distribution patterns. The displacement color transition in Fig. 6'(a) is relatively gentle, and the displacement gradient changes uniformly Fig. 6'(b) shows the high displacement area is more concentrated and the color transition is more intense due to the change of flow direction and different structural stress modes. The displacement gradient change is more significant, highlighting the influence of flow direction on the displacement distribution of the structure.

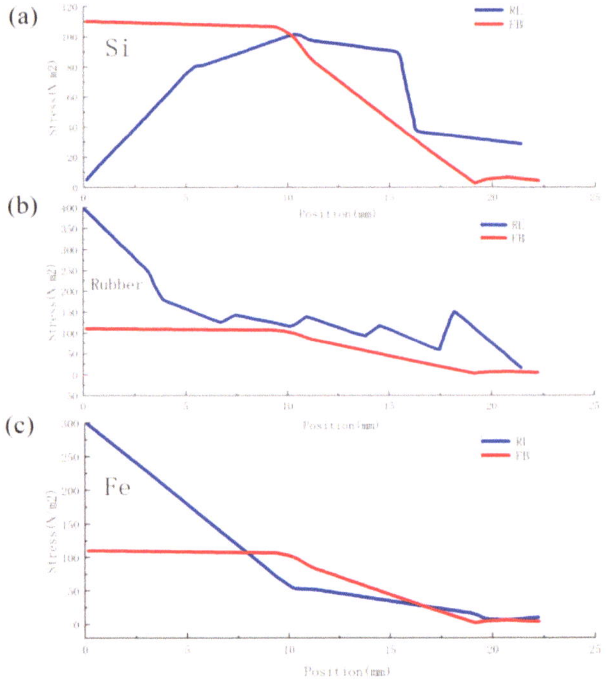

Fig. 7. (a) Silicon stress in different directions, (b) Rubber stress in different directions, (c) Stress of iron frame in different directions.

'Figure 7' shows that the stress of different materials from different direction. In the Si material, 'Fig. 7'(a) the stress of Si solar cell is stable at first from right to left (RL), and then rises rapidly to the peak near 5 mm and then decreases sharply when the direction changes from front to back (FB). In the rubber material, 'Fig. 7'(b) the initial stress of rubber is higher and the subsequent fluctuation decreases, which fluctuates significantly at 15–20 mm. While the stress level is generally low and the change is relatively stable when the wind direction is FB, which only decreases slightly in the later stage. Among Fe materials, 'Fig. 7'(c) the initial stress of Fe is extremely high and decays rapidly when the wind direction is RL, while the stress of is relatively stable and low when the wind direction is FB. The attenuation rate is slower than that in the RL direction. On the whole, the change of fluid direction has a significant effect on the stress distribution of each material and is material-specific, and the change of flow direction leads to the change of the concentration position, range and distribution mode of stress on the surface of the structure, reflecting the influence of the direction of fluid force on the stress area of the structure.

3.3 Stress Analysis at Different Temperatures

When the silicon solar cells mounted on an iron frame and encapsulated in rubber are tested in experiments simulating different surface temperature environments, the physical properties of the material such as thermal expansion and heat conduction and the change of mechanical properties of the material with temperature are involved. Two key temperature nodes were set: a normal temperature environment of 20 °C and a high temperature environment of 80 °C. Other environmental conditions, such as pressure, fluid velocity, and flow direction, were kept constant throughout the experiment, so as to accurately explore the influence of temperature as a variable on the stress of solar cells.

The thermal expansion properties of a material are described by the coefficient of thermal expansion α (unit K-1). In the three-dimensional case, considering the Poisson's ratio v of the material, the thermal stress tensor, σ_{ij}^{th} can be expressed as:

$$\sigma_{ij}^{th} = \frac{\mathrm{E}\alpha \Delta T}{1 - 2v}\delta_{ij}. \tag{4}$$

In different material layers of solar cells, due to the different coefficients of thermal expansion, when the temperature changes, thermal stresses will occur between the layers, which may lead to debonding or cracking between the material layers, affecting the performance and lifetime of the cells (Shin et al. 2018).

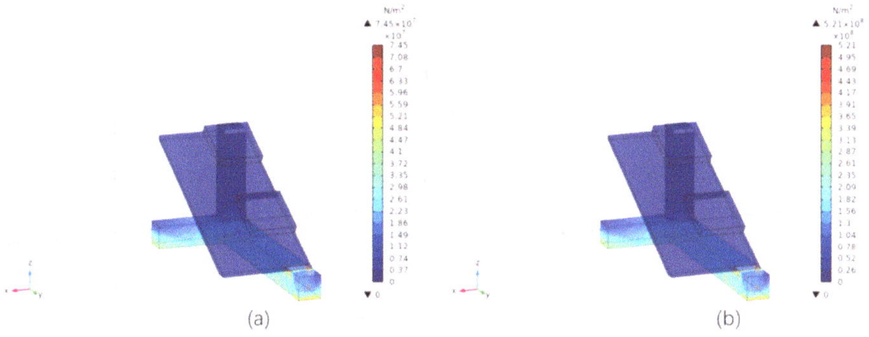

Fig. 8. (a) Stress of silicon solar cell at 20 °C, (b) at 80 °C.

'Figure 8' shows the stress diagram at different temperatures. The test conditions are simulated in the wind direction from right to left and the wind speed is 5 m/s, only the surface temperature of the object is changed. From the difference in stress values, the maximum stress in 'Fig. 8'(a) is 7.45×10^7 N/m^2, 'Fig. 8'(b) is 5.21×10^8 N/m^2, and the stress level is significantly higher than 20 °C at 80 °C. In terms of stress distribution characteristics, both figures show stress concentration in specific areas of the structure, such as geometric connections or abrupt changes, but the increase in temperature makes the stress distribution gradient in 'Fig. 8'(b) more drastic, and the range and intensity of the high stress area are significantly enhanced, reflecting that the temperature increase not only increases the overall stress level, but also aggravates the stress concentration of the structure.

'Figure 9' shows the stress displacement diagram at different temperatures. The test conditions were simulated in an environment of 5 m/s from right to left wind direction and wind speed, and only the surface temperature of the object was changed. According to the numerical difference analysis of displacement, the maximum displacement of 'Fig. 9'(a) is 3.44×10^{-4} m, and the maximum displacement of 'Fig. 9'(b) is 1.78×10^{-3} m, and the displacement under 80 °C condition is significantly higher than that of 20 °C. In terms of displacement distribution, both figures show displacement changes in specific areas of the structure, such as geometric protrusions, but the increase in temperature makes the displacement color transition and displacement gradient change more significant in 'Fig. 9'(b). The range and intensity of the high-displacement region in 'Fig. 9'(b) are higher than those in 'Fig. 9'(a), indicating that the increase in temperature will aggravate the displacement deformation of the structure and change the gradient and concentration of the displacement distribution.

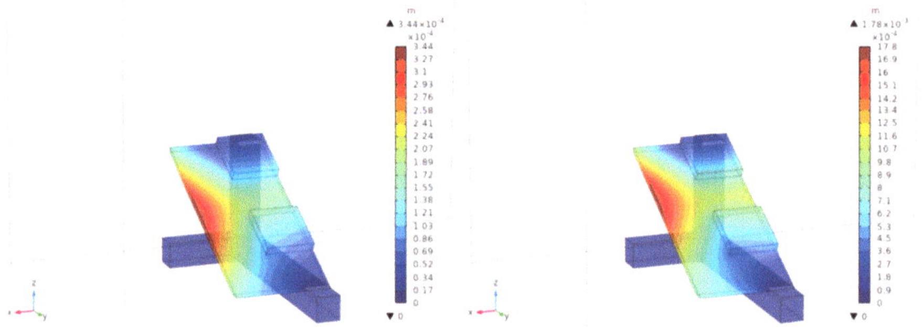

(a) (b)

Fig. 9. (a) Stress displacement at a surface temperature of 20 °C, (b) Stress displacement at a surface temperature of 80 °C.

'Figure 10' shows the stress distribution of each material at different temperatures. The test conditions were simulated in an environment of 5 m/s from right to left wind direction and wind speed, and only the surface temperature of the object was changed. For silicon materials, the stress value is low and changes smoothly at 20 °C, and the stress is at a high level at 80 °C, and the fluctuation is obvious, as shown in 'Fig. 10'(a). For rubber materials, as shown in 'Fig. 10'(b), the stress is kept low at 20 °C. At 80 °C, the stress in the front section is low, and the stress in the arc length is about 20 mm, and the stress rises sharply. It shows that the stress response of rubber at high temperature has a phased stage, and the stress abrupt change occurs under the influence of temperature in the later stage, and the high temperature has a nonlinear effect on the stress distribution of the rubber layer. For iron materials, as shown in 'Fig. 10'(c), the stress is generally low at 20 °C. At 80 °C, the stress near 0 mm of arc length rises sharply to a very high value of 1.05×10^{8} N/m^2, and then decreases but is still higher than the 20 °C level. The stress response of iron in the initial stage of high temperature is violent, and the stress effect of temperature on the iron frame is characterized by a sharp increase in the early stage and a slow decrease but still high in the later stage.

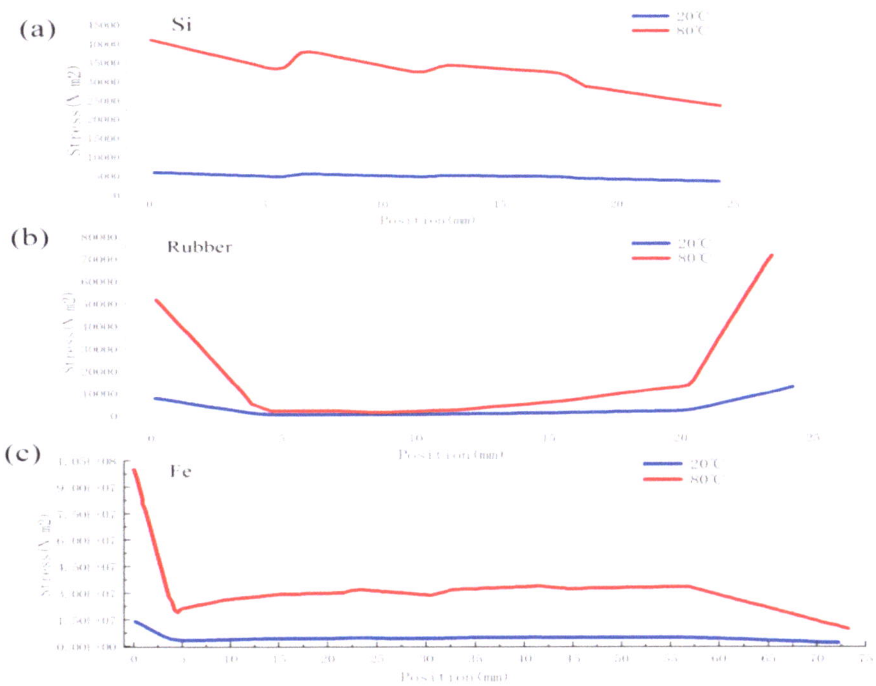

Fig. 10. (a) Stress of silicon materials, (b) Stress of rubber material, (c) Stress of iron materials.

4 Conclusion

This paper conducts simulation analysis of stress in silicon solar cells within the complex environment of multiphysics coupling. The stress state of silicon solar cells is governed by the synergistic effects of multiple factors. Fluid flow velocity, flow direction and temperature all affect stress significantly. Velocity and temperature have the most obvious effect on stress level changes. Different materials (Si, rubber, Fe) react differently to these stressors. They give us ideas for better stress buffering and structural reinforcement.

References

Bahar, H., Abdelilah, Y., Alcalde Báscones, A., Anatolitis, V., Bojek, P., Briens, F., et al.: Renewables 2024: Analysis and forecast to 2030. International Energy Agency (IEA) (2024)

Chabaud, V.: Synthetic turbulence modelling for offshore wind farm engineering models using coherence aggregation. Wind Energy **27**(2), 111–130 (2024). https://doi.org/10.1002/we.2875

Chen, H., Zarkevich, N.A., Levitas, V.I., Johnson, D.D., Zhang, X.C.: Fifth-degree elastic energy for predictive continuum stress–strain relations and elastic instabilities under large strain and complex loading in silicon. NPJ Comput. Mater. **6**(1), 115 (2020). https://doi.org/10.1038/s41524-020-00382-8

Chen, P., Dagestani, A.A.: Urban planning policy and clean energy development harmony—evidence from smart city pilot policy in China. Renew. Energy **210**, 251–257 (2023). https://doi.org/10.1016/j.renene.2023.04.063

He, Y., He, Y., Sun, J., Li, X., Lu, M.H., Chen, Y.F.: Breaking mechanical performance trade-off in 3D-printed complex lattice-inspired multi-cell tubes under axial compression. Compos. Sci. Technol. **258**, 110920 (2024). https://doi.org/10.1016/j.compscitech.2024.110920

Shin, H., Han, E., Park, N., Kim, D.: Thermal residual stress analysis of soldering and lamination processes for fabrication of crystalline silicon photovoltaic modules. Energies **11**(12), 3256 (2018). https://doi.org/10.3390/en11123256

Ma, G., Tian, L., Song, Y., Zhao, N.: Effects of turbulence modeling on the simulation of wind flow over typical complex terrains. Appl. Sci. **14**(23), 11438 (2024). https://doi.org/10.3390/app142311438

Rajabi, K., Hosseini-Hashemi, S.: Application of the generalized Hooke's law for viscoelastic materials (GHVMs) in nanoscale mass sensing applications of viscoelastic nanoplates: a theoretical study. Eur. J. Mech. A. Solids **67**, 71–83 (2018). https://doi.org/10.1016/j.euromechsol.2017.08.015

Wiatros-Motyka, M., Fulghum, N., Jones, D., Altieri, K., Black, R., Broadbent, H., et al.: Global electricity review 2024: world passes 30% renewable electricity milestone. Ember (2024)

Wang, S., De Roo, F., Thobois, L., Reuder, J.: Characterization of terrain-induced turbulence by large-eddy simulation for air safety considerations in airport siting. Atmosphere **13**(6), 952 (2022). https://doi.org/10.3390/atmos13060952

Xu, Z., et al.: Scalable selective absorber with quasiperiodic nanostructure for low-grade solar energy harvesting. APL Photonics **8**(2), 020801 (2023). https://doi.org/10.1063/5.0135193

Hydrogen Leakage Effects on Different Leakage Ports of Overhead Hydrogen Pipelines Based on Fluent

Zeming Ye(✉), Suya Hu, and Xinyu Li

China University of Petroleum-Beijing at Karamay, Karamay 834000, China
1282735521@qq.com

Abstract. To address the research gap in the effects of pure hydrogen pipeline leakage orifice size and shape on hydrogen diffusion, this study constructed four orifice models (small circular, large circular, axial strip-shaped, circumferential strip-shaped) using Ansys Fluent, and simulated the leakage rate and diffusion patterns. The results showed that the projected area of the orifice was positively correlated with the leakage rate; under the same area, strip-shaped leakage presented higher risks, with the circumferential strip-shaped leakage rate being 18.9 times that of the axial one. Analysis of the calibrated combustible mass fraction of leaked hydrogen revealed that the projected area was also positively correlated with the diffusion range, and the diffusion ranges of circumferential and axial leakage were comparable and larger than those of the other two models. This study identified circumferential leakage as the most hazardous scenario, and proposed the time-line voltage condition for axial installation detection to optimize circumferential leakage accident response, providing a reference for pipeline leakage detection and protection.

Keywords: Pure hydrogen pipeline · Hydrogen leakage · Leakage port morphology · Leakage simulation

1 Introduction

Research on hydrogen leaks can provide a reference for handling unexpected leaks, preventing the deterioration of the impact of the accident and causing greater economic and property losses and casualties. Many researchers at home and abroad have conducted extensive studies on this, including theoretical derivations, numerical simulations and other methods. Abroad: Li et al. (2015) studied the impact of leak location, wind speed and wind direction on hydrogen leakage and diffusion in hydrogen refueling stations, indicating that the presence of wind weakens the natural convection for hydrogen diffusion, accelerates the mixture of air and hydrogen, and forms a smaller volume of combustible gas cloud; In China: Fang et al. (2025) proposed that obstacles would accumulate the hydrogen diffusion form and successfully simulated the leakage diffusion in the special environment of the hydrogen refueling station, and then proposed the reasonable placement rules of hydrogen leakage detection equipment.

T. Mankovits and N. Trisovic (Eds.): MDS 2026, LNME 2026.
https://doi.org/10.1007/978-981-92-2440-1_24

In addition, He (2022) conducted a numerical simulation of hydrogen leakage in high-pressure hydrogen transport pipelines using Fluent in view of the characteristics that hydrogen leakage is prone to occur in high-pressure hydrogen transport pipelines, and proposed that hydrogen jet occurs at the leakage point and is affected by external wind, and the dangerous area is concentrated below the axis of the leakage point.

However, although the above research was able to make a relatively accurate prediction of the diffusion distribution after hydrogen leakage (Peng et al. 2023, 2024; Hu et al. 2023; Pei et al. 2025), there was a lack of research on the diffusion law of hydrogen leakage caused by the size and shape of the leakage outlet (Fang et al. 2025; Fan et al. 2025; Deng et al. 2025). Therefore, in this study, the Fluent simulation software was used to simulate hydrogen leaks at different leakage points and summarize the diffusion patterns. The main innovation lies in the fact that this study supplements hydrogen leakage in pure hydrogen pipelines when the size and shape of the leakage ports are different in terms of hydrogen leakage prediction. The findings can serve as a supplement to current hydrogen leak prediction and provide some guidance for leak detection and predictive protection of pure hydrogen pipelines.

2 Leak Simulation

2.1 Model Assumptions

Due to the highly complex influence of the actual environment of the overhead pure hydrogen pipeline on hydrogen leakage, in order to facilitate model establishment and match the computing resources, the following assumptions are made in this study: air is assumed to be an ideal gas, air components do not react with hydrogen, and the two only interact in terms of component transport; the ambient temperature remains constant at 300 K; the atmospheric pressure is standard atmospheric pressure; the pressure at the leakage point is also one standard atmospheric pressure; the leakage is stable leakage; the flow state of hydrogen in the pipeline is stable, and there is no backflow.

2.2 Modeling and Simulation

2.2.1 Physical Model

According to the size and shape of the leak, we established four physical models as shown in the Table 1 below:

Table 1. Four different leak models

	Leak opening diameter (projected area))	Shape of leak outlet
1	50.5 mm (2002.96 mm^2)	Round
2	10 mm (78.54 mm^2)	Round
3	Length: 100 mm, width: 20 mm (2000 mm^2), axial	Strip
4	Length: 100 mm, width: 20 mm (2000 mm^2), circumferential	Strip

To control variables and ensure that the comparison results are more representative, this study uniformly sets the diameter of the 1-m-long pipeline to 400 mm, which is a common diameter for hydrogen-transmission pipelines. This setting makes the simulation results more consistent with reality, and the laws derived from the leakage simulation are applicable. In addition, through pre-simulation experiments and previous studies (Peng et al. 2023, 2024; Hu et al. 2023; Pei et al. 2025), we roughly determined the leakage area. On this basis, we set up the leakage domain that wrapped the pipeline to ensure that all hydrogen leaks during the study period were within the leakage domain, so as to improve the accuracy of the leakage simulation results with fewer computing resources. The model is as follows:

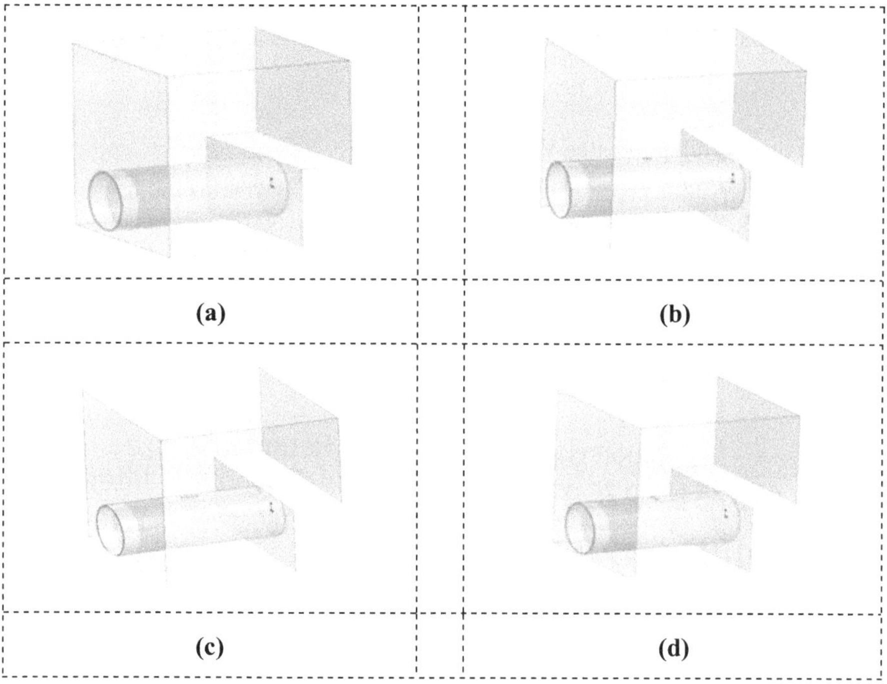

Fig. 1. Experimental group models of different types of leakage outlets 1 (a), 2 (b), 3 (c), and 4 (d).

2.2.2 Simulation Setup

For the above four physical models and simulation works, this study adopted basically the same setting method in the simulation setting module to ensure the comparability of the simulation results. Before the simulation, the physical models in SOLIDWORKS software were imported, and Boolean operations were performed on the leakage domain and the pipe geometry to construct the fluid region. For the pipe geometry, physical suppression was applied to simplify the simulation process to hydrogen leakage, ignoring the friction between hydrogen and the pipe wall.

This study used Ansys Fluent for meshing. To make the local small-scale leakage simulation more accurate, local refinement was performed on the leakage port, setting the minimum mesh size to 5 mm and the mesh growth rate to 1.2. For surface meshing, the maximum mesh size was set to 0.1 m, the minimum mesh size to 0.01 m, and 10 layers of gap filling cells and 5 layers of boundary layers were added, with a uniform mesh growth rate of 1.2. The volume mesh was set as hexahedral, with the maximum mesh size of 0.1 m. A total of 1.7 million networks were ultimately divided.

In the Fluent solution settings, this study adopts the pressure-based transient solver, using the SIMPLE algorithm and the standard k-e turbulence model. Considering the influence of gravity on hydrogen leakage, the magnitude of gravitational acceleration is set to 9.8 m/s^2. When setting the boundary conditions, the inlet is set as a velocity inlet, and the velocity of hydrogen is all set to 10 m/s; the outlet of the pipeline and the boundary of the leakage domain are set as pressure outlets, and both are set to suppress backflow.

2.2.3 Network Independence Verification

For the local encryption processing of the model's network and the size adjustment of the overall network, As shown in Fig. 2, we respectively created two groups of networks with fewer and more networks, namely 0.85 million and 3.4 million networks for verification. Through the calculation of the two, we can see that the final quality rate difference has only a 0.5% error, within the 5% allowable error range. Therefore, it can be determined that the network independence verification of this study is successful.

Report grid size		
Node	415751	2769784
Face	101015	4219567
Unit		852790

(a)

Report grid size	Boundary	Internal
Node	1606673	11105757
face	413448	11715024
Unit		3404461

(b)

Mass Flow Rate	[kg/s]
inlet	0.1028748
outlet_h	-0.098621192
Net	0.0042536029

(c)

Mass Flow Rate	[kg/s]
inlet	0.1029007
outlet_h	-0.09886238
Net	0.004038319

(d)

Fig. 2. 850,000 grid and 3.4 million grid calculation results, where a and c, b and d are one group.

2.3 Simulation Results

2.3.1 Simulation Results

Starting the analysis from the moment hydrogen entered the pipeline until it reached a stable state, we set a total of 1000 time steps, each with a time step length of 0.001 s, and iterated each step twice to obtain more precise results. Using the four models shown in Fig. 1, we obtained the leakage results as shown in Fig. 3 below.

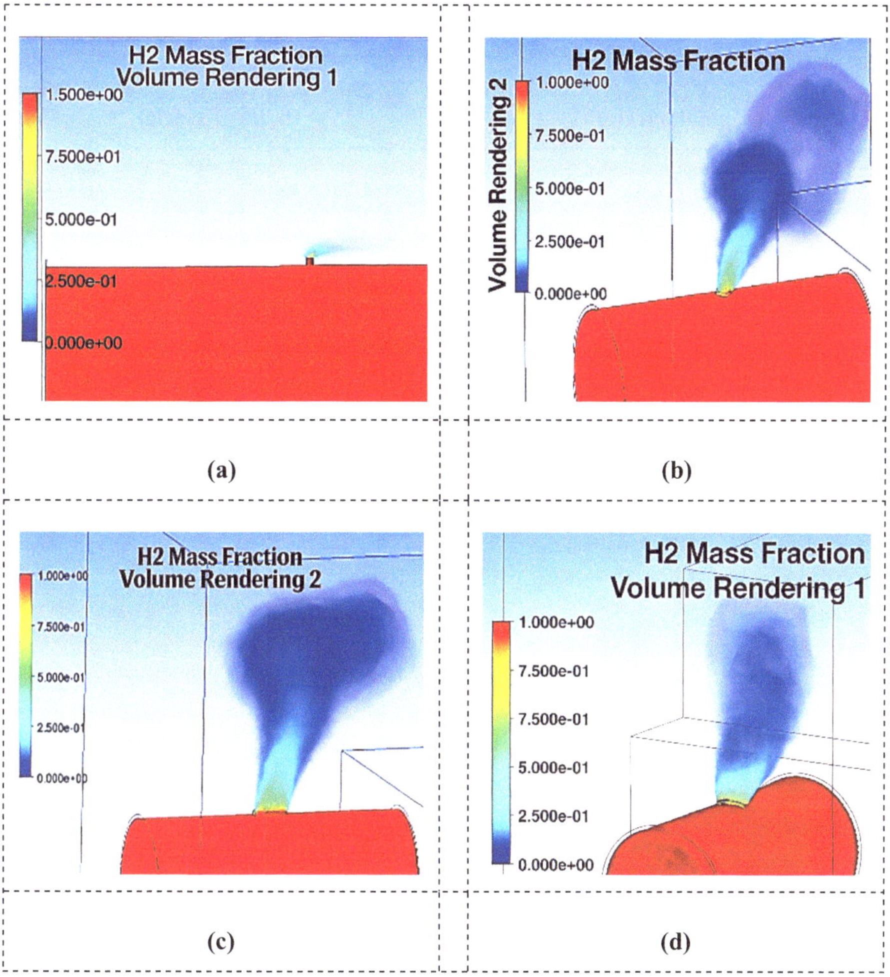

Fig. 3. Cloud map of hydrogen diffusion mass fraction after stable leakage at leaks 1 (a), 2 (b), 3 (c), and 4 (d) of the pipeline.

Figure 3 shows the leakage and diffusion of hydrogen in the leakage domain. The color of the legend on the left has visually presented the mass distribution of hydrogen inside and outside the pipeline, and it can be clearly seen that different leakage port appearances have an impact on the hydrogen leakage situation. To better quantify the specific amount of hydrogen leakage, we used software to calculate the variation in hydrogen mass at the boundary, as shown in Fig. 4 below:

Mass Flow Rate	[kg/s]
inlet	0.10287762
outlet_h	-0.1028637
Net	1.3920816e-05

(a, small circle)

Mass Flow Rate	[kg/s]
inlet	0.1028748
outlet_h	-0.098621192
Net	0.0042536029

(b, large circle)

Mass Flow Rate	[kg/s]
inlet	0.1028748
outlet_h	-0.097955228
Net	0.0049195692

(c, axial long strip)

Mass Flow Rate	[kg/s]
inlet	0.10287478
outlet_h	-0.10040649
Net	0.0024682846

(d, circumferential long strip)

Fig. 4. Software-calculated data on hydrogen mass changes at each boundary of leaks in experimental groups 1 (a), 2 (b), 3 (c), and 4 (d)

The leakage volume M here is calculated by the following formula:

$$M = m_{inlet} - m_{outlet\_h} \tag{1}$$

$$\begin{aligned} m_{inlet} &= t \cdot v_{inlet} \\ m_{outlet\_h} &= t \cdot v_{outlet\_h} \end{aligned} \tag{2}$$

where:

M : *Leakage rate per unit time*, kg
m_{inlet} : *The mass of hydrogen flowing into the pipeline*, kg
m_{outlet_h} : *The mass of hydrogen flowing out of the pipeline*, kg
v_{inlet} : *Mass flow velocity at the pipeline inlet*, kg/s
v_{outlet_h} : *Mass flow velocity at the pipeline outlet*, kg/s
t : *Time*, s

2.3.2 Data Processing

It should be noted that in order to better generalize the results, the circumferential distribution here should be understood as perpendicular to the direction of airflow movement. We further organize the calculated data into Table 2 to make the results more intuitive.

Table 2. Leaks of hydrogen at Leaks 1, 2, 3, 4

	Inlet mass flow rate	Pipe outlet mass flow rate	Leakage in 1 s
No. 1 (small round)	0.10287762	0.102863700	0.000013920
No. 2 (big round)	0.10287480	0.098621192	0.004253608
No. 3 (axial strip)	0.10287480	0.097955228	0.004919572
No. 4 (Circumferential strip)	0.10287478	0.010040649	0.092834131

Draw a line graph of the leakage volume changes of each group over time as shown in Fig. 5 below:

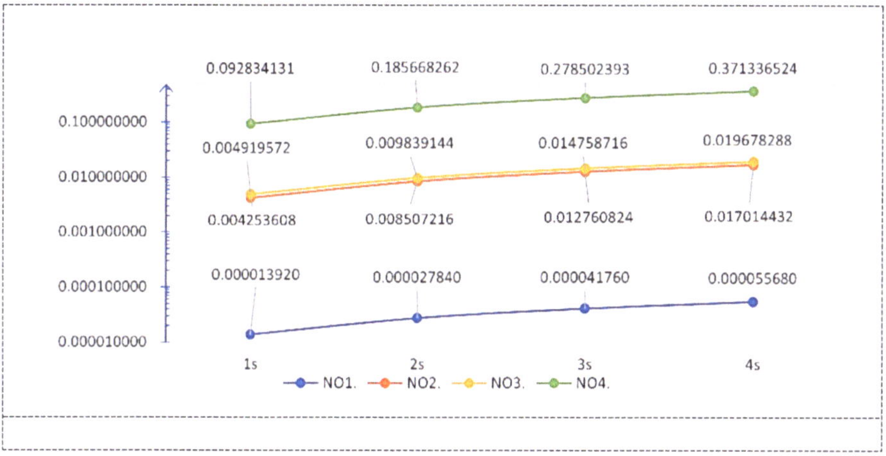

Fig. 5. Line graphs of changes in leakage at different times

2.4 Leakage and Diffusion Patterns

2.4.1 Leakage Patterns

- The projected area of the leak outlet is the dominant factor influencing the leakage volume.

Positive correlation is significant: By comparing the data of experimental groups No. 1 (small circle, projected area 78.54 mm^2) and No. 2 (large circle, projected area 2002.96 mm^2), it can be seen that under the same conditions of circular leakage outlet shape, hydrogen delivery speed 10 m/s and temperature 300 K, the leakage outlet with a larger projected area corresponds to a higher leakage volume. The 1 experimental group had a leakage of only 0.000013920 kg in one second, while the 2 experimental group reached 0.004253608 kg, with a leakage difference of approximately 305 times, directly verifying the core rule that "the larger the projected area, the higher the leakage", which is consistent with the theory in fluid mechanics that "leakage is positively correlated with the area of the leakage hole", It further confirmed the reliability of the simulation results.

- The difference in the shape of the leakage outlet under the same projected area will affect the leakage situation.

When the vertical projected areas of the leaks are close, the leakage risk of the strip leaks is higher than that of the circular ones: the axial strip leaks 0.004919572 kg in one second, which is about 15.6% higher than the circular 0.004253608 kg; The circumferential strip leaks 0.092834131 kg, which is 21.8 times more than the circular one. This difference is due to the greater length-to-diameter ratio of the strip structure of the strip leak opening, which causes less frictional resistance between hydrogen and the edge of the leak opening during the leakage process, and makes it easier to form a uniform diffusion channel, reducing the leakage blockage caused by local vortices, and thus the leakage efficiency is higher.

- The direction in which the leak moves relative to the hydrogen will affect the leakage situation.

The difference in leakage between the axial strip and the circumferential strip is about 18.9 times, most prominently. The core reason lies in the different relative angles of the leak opening to the direction of hydrogen flow. Hydrogen flows along the axial direction in the pipe, and the "long side" of the axial strip leak is parallel to the direction of the gas flow. Hydrogen can only escape from the "width" direction of the leak, and the effective leakage section is limited by the "width" (20 mm); While the "long side" of the circumferential strip leak is perpendicular to the direction of the airflow, hydrogen can fully contact the leak channel along the "length" direction of the leak (100 mm), significantly expanding the effective leakage section and thus increasing the leakage volume exponentially.

2.4.2 Diffusion Law

We determined the explosion limit of hydrogen gas based on the regulations of the International Organization for Standardization (ISO) and the domestic "Basic Requirements for Safety of Hydrogen Gas Systems" (GB/T 29729–2022 2022). The explosion limit of hydrogen gas is defined as a hydrogen volume fraction ranging from 4% to 75%. To convert this range into a mass fraction, we used the ideal gas state equation and the formula for the density of a mixed gas. The final conversion relationship is derived as follows:

$$\omega_{H_2} = \frac{\phi_{H_2} \cdot M_{H_2}}{\phi_{H_2} \cdot M_{H_2} + (1 - \phi_{H_2}) \cdot M_{air}} \tag{3}$$

where:

$$\omega_{H_2} : \textit{the hydrogen gas mass fraction}$$
$$\phi_{H_2} : \textit{the volume fraction of hydrogen gas}$$
$$M_{H_2} = 2g/mol$$

$$M_{air} = 29g/mol$$

By applying the upper and lower limits of the volume fraction, the range of the combustible mass fraction of the leaked hydrogen gas was calculated to be between 0.276% and 5.17%. Based on this result, the combustible hydrogen mass fraction after leakage was calibrated. Through this, we can further determine the pattern and general regular distribution of the diffusion range of the combustible hydrogen concentration, as shown in Fig. 6 below.

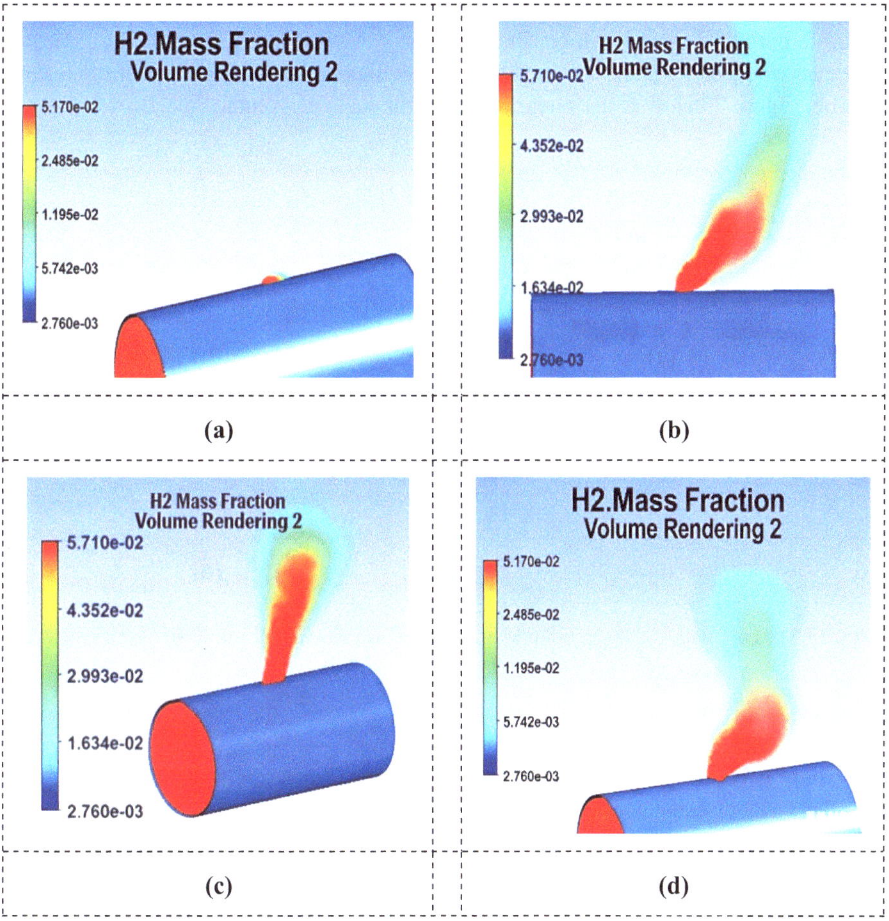

Fig. 6. Calibration of the range of combustible mass fractions for leakage diffusion cloud diagrams 1 (a), 2 (b), 3 (c), and 4 (d).

From the previous analysis, we can conclude that gravity has a relatively weak influence on diffusion. Considering that hydrogen gas has a much lower density than air, the diffusion process mainly occurs in the direction of escaping above the leakage point. Hydrogen gas has an initial velocity in the pipeline, so the diffusion area has shifted significantly. From Fig. 6, we can analyze that within one second of hydrogen gas leakage, the diffusion situation is mainly divided into different patterns; the combustible area after the 10 mm leakage point is the smallest, while the 50.5 mm leakage point and the rectangular axial and circumferential leakage points have experienced severe diffusion phenomena. Through simulation, it is known that the circular leakage point has the smallest diffusion range, while the rectangular leakage point has a much larger diffusion area than the circular one; looking at the marked upper and lower limits of the most combustible mass fraction in the figure, the overall volume of diffusion in the axial rectangular leakage point is more concentrated in the vertical direction, and the vertical diffusion situation is the greatest, while the lateral diffusion situation in the circumferential direction is wider. The volumes of both are approximately the same, and all the above analyses have been verified by calculating the integral data of the leakage volume. Figure 7 below is the integral data of the leakage volume.

Total Volume Integral Mass fraction of h2	[m^3]
fluid	0.10053125
Net	0.10053125

(a)

Total Volume Integral Mass fraction of h2	[m^3]
fluid	0.10171918
Net	0.10171918

(b)

Total Volume Integral Mass fraction of h2	[m^3]
fluid	0.10258038
Net	0.10258038

(c)

Total Volume Integral Mass fraction of h2	[m^3]
fluid	0.10217416
Net	0.10217416

(d)

Fig. 7. The cumulative leakage volume of each leakage situation 1 (a), 2 (b), 3 (c), and 4 (d).

3 Conclusions

This study is based on the simulation research of four different leakage orifices (small circular, large circular, axial slit, and circumferential slit) in pure hydrogen pipelines using Ansys Fluent. Combined with the calculation of leakage volume, the analysis of diffusion cloud diagrams, and the summary of patterns, the following conclusions are drawn:

1. The projected area of the leakage orifice is the core factor affecting the leakage volume, and there is a significant positive correlation. Under the premise of consistent orifice shape and gas flow conditions, the larger the projected area of the leakage orifice, the higher the hydrogen leakage volume.
2. Under the same projected area, the shape of the leakage orifice and its relative angle to the gas flow direction significantly affect the leakage volume. Slit-shaped leakage orifices generally have a higher leakage risk than circular ones. The 1-s leakage

volume of the axial slit orifice (2000 mm^2) (0.004919572 kg) is 15.6% higher than that of the circular orifice of the same area (0.004253608 kg), due to the aspect ratio advantage of the slit structure, which reduces the frictional resistance of hydrogen leakage and enhances the leakage efficiency.

3. The "directional sensitivity" of slit-shaped leakage orifices is extremely strong. The leakage volume of the circumferential slit orifice (perpendicular to the hydrogen flow direction) (0.092834131 kg) is 18.9 times that of the axial slit orifice (parallel to the gas flow direction), as the circumferential structure allows hydrogen to fully contact the leakage channel along the length of the orifice, significantly expanding the effective leakage cross-section.
4. Due to the much lower density of hydrogen compared to air, the diffusion process is mainly upward escape. Additionally, because hydrogen has an initial velocity before leakage, the leakage gas cloud always shifts towards the downstream direction of gas transportation. After calibrating the combustible mass fraction within 1 s after the leakage occurs, we analyzed and found that different types of leakage orifices have different levels of danger within the determined explosion limit range. The smaller the horizontal projected area of the leakage orifice, the smaller the diffusion range. The diffusion volume of slit-shaped leakage orifices is always larger than that of circular leakage orifices. Among them, the axial leakage orifice has a larger volume of combustible gas diffusion than the other three types, but the difference between it and the circumferential crack is not significant. Considering the leakage volume, the leakage volume of the circumferential orifice is about 20 times that of the axial one. Therefore, we suggest avoiding circumferential pipeline rupture and leakage.
5. Regarding leakage detection, we suggest that time lines frequently used in pipeline burst tests be affixed at certain angular intervals along the axial direction at locations prone to damage on the pipeline. When local damage occurs to the pipeline, the time lines will break. By promptly detecting the voltage status of the time lines, it will be helpful to promptly identify circumferential leakage, thereby shortening the response time for pipeline accidents and effectively reducing the occurrence of circumferential leakage incidents.

References

Deng, F.J., Tian, L., Huang, J.X.: 3D consequence visualization simulation of hydrogen leakage accidents in hydrogen production stations. Oil Gas New Energy **37**(4), 97–105 (2025). https://kns.cnki.net/nzkhtml/xmlRead/trialRead.html?dbCode=CJFD&tableName=cjfdlast2025&fileName=SYGH202504011&fileSourceType=1&appId=KNS_BASIC_PSMC&invoice=SxlARLW970PSJbMaKcv6K8QKcGDj1yhNcMpGaPXRpymS64WCail71gwr9jGhqtRMmN1vi1dqDz1yWy++b6Z2c3oxt9n4WdA/WjsNX5i/ltuGt4YCeyv1u9/hESCSiJNdwo1CwJ0vaOlv6DXFrqaWKKAPBUE6j7aadntrzl0BLUo=&platform=NZKPT&type=JOURNAL&cflag=html&trial=&nonce=FF03482BBB964F1CAF625B8194B0BE2B

Fan, J.R., Tang, L.P., Wang, X.J.: Sensitivity simulation of leakage and diffusion parameters for buried gas transmission pipelines. Sci. Technol. Eng. **25**(25), 10671–10679 (2025). https://kns.cnki.net/nzkhtml/xmlRead/trialRead.html?dbCode=CJFD&tableName=cjfdlast2025&fileName=KXJS202525015&fileSourceType=1&appId=KNS_BASIC_PSMC&invoice=PBIxACPUumT1XU6XTSn71Z02OyJxsmPwg1bnXATYKqUlftGaHHeG8IbZeu6Dj+75zACfKl1Blafc48/a+1UtCk8XBRPHq+gvC1CR0M3XmEmD3BnBjWvRYhGhIGcXOxGRl5qgi+tmA9usyedCFlcj3LhX9oMn6R/a2jGbhugdjpM=&platform=NZKPT&type=JOURNAL&cflag=html&trial=&nonce=7703156F477A4699BFE19C75BFD713FB

Fang, Y.H., Yu, H.H., Zeng, D.L.: Numerical simulation of leakage and diffusion in hydrogen refueling stations under different external environments. Acta Energiae Solaris Sinica **46**(10), 128–137 (2025). https://doi.org/10.19912/j.0254-0096.tynxb.2024-0034

GB/T 29729-2022: Essential requirements for the safety of hydrogen systems [Standard]. Standardization Administration of China (2022). https://openstd.samr.gov.cn/bzgk/gb/newGbInfo?hcno=CD2CACD6BCF1403D48EF0508798A01A9

He, C.: Numerical simulation of hydrogen leakage in high-pressure hydrogen pipelines based on Fluent. China Safety Sci. J. (2022). https://kns.cnki.net/nzkhtml/xmlRead/trialRead.html?dbCode=CJFD&tableName=cjfdlast2022&fileName=GYAF202206002&fileSourceType=1&appId=KNS_BASIC_PSMC&invoice=l5EFKfPlBYzQjyX+7ohmfLcPXsqvcm8eR2sjXrNR7vIIq6LMlxkY1py1BQaEzgiVMmu7IG8rBgXaR9UP4Ib+kCQ7+RdVkcG1zxnWzmpcw8BG3/kXxepPuTuoriP06qYHyB4V0glxAmXMwUZVHDyvp3Mi93v59nkXQz99PlX4pXg=&platform=NZKPT&type=JOURNAL&cflag=html&trial=&nonce=37CF097BC09842468D60B611D0C49903

Hu, W.P., Chen, G., Qi, B.J.: Numerical simulation of leakage and diffusion of buried pure hydrogen/hydrogen-blended natural gas pipelines. Oil Gas Storage Transp. **42**(10), 1118–1127, 1136 (2023). https://kns.cnki.net/nzkhtml/xmlRead/trialRead.html?dbCode=CJFD&tableName=cjfdlast2023&fileName=YQCY202310005&fileSourceType=1&appId=KNS_BASIC_PSMC&invoice=Xi8QK4c5jBrnFqIXlqae8/d7PHEKmIN8M5jGn0EHsfUFBHCH04HHpZ+lJ2yPlYsqe2NHw0/LLEr8vUxFwuBMOk34NZUr80wVp3KRuKKv8td1kcvEQaVoVe1K02xq2x91eG1l9MFk8fGHng9+3nT11CiP63j6edVmMHALX2kZP5M=&platform=NZKPT&type=JOURNAL&cflag=html&trial=&nonce=A7BF801193A44C03AB31A6C8B7E48B0B

Li, J.Y., Zhao, Y.Z., Zheng, J.Y.: Simulation and analysis of high-pressure hydrogen leakage and explosion accident in hydrogen refueling station. J. Zhejiang Univ. (Eng. Sci.) **49**(7), 1389–1394 (2015). https://doi.org/10.3785/j.issn.1008-973X.2015.07.026

Pei, Y.J., Li, S.P., He, J.B.: Study on leakage laws and prediction models of buried pure hydrogen pipelines under variable parameters. China Measur. Test (2025). https://kns.cnki.net/nzkhtml/xmlRead/trialRead.html?dbCode=CJFD&tableName=cjfdauto&fileName=SYCS202602006&fileSourceType=1&appId=KNS_BASIC_PSMC&invoice=Q1Ebax/u+WeNbs7D0gQjZ18DUatBnaEU1r7BMLuHT2GkQxLmA5nCwtCF2CDWlNZVrjaVzUVcHsndBxHOe9uncVLMTuky0rqhACOgHV4B1HAAO4QPVLPITmLo2x5MBWvh6gs3o5PDiMrsEhHDkzRXXAzUu/I7hrfNzqyYeiT/1Eo=&platform=NZKPT&type=JOURNAL&cflag=html&trial=&nonce=398E7EFE59604E4195604F2A8934522E

Peng, S.B., Luo, X.: Numerical simulation study on leakage and diffusion of buried hydrogen-blended natural gas pipelines. China Safety Sci. J. **34**(3), 63–69 (2024). https://doi.org/10.16265/j.cnki.issn1003-3033.2024.03.1904

Peng, S.B., Luo, X., Yang, L.: Numerical simulation of leakage and diffusion laws of long-distance hydrogen-blended natural gas pipelines. Chem. Eng. Oil Gas **52**(6), 44–52, 59 (2023). https://kns.cnki.net/nzkhtml/xmlRead/trialRead.html?dbCode=CJFD&tableName=cjfdlast2024&fileName=STQG202306008&fileSourceType=1&appId=KNS_BASIC_PSMC&invoice=qO2coMc4VrYn42z7Fq5zjRYJYzt2YLcbgNZZm9WkElC2Lcr6cD0Ayo5Al6LQDi6hjxBXrKGFN0RRk1B/imwOhFLNV7jvu1ZK4AL3fyga0KNdYbIDl4n4HjbSCtwZkS1tEpuMr7EBpxqEjMyjgm0BOUmDSiWUXZJEhgOGSMYPOyA=&platform=NZKPT&type=JOURNAL&cflag=html&trial=&nonce=186F0407BB97486080CAE46B278670F1

Simulation Analysis of Mechanical Failure Response in Typical Slat Structures

Xinfeng Yuan(✉), Xianmin Chen, Sanyuan Li, Yi Zhuo, and Weifeng Zang

Aircraft Strength Research Institute of China, Xian 710065, China
momaek0923@163.com

Abstract. To meet the core demands of lightweight design and high reliability enhancement for modern aircraft high-lift devices, the study focuses on the complex slat mechanism of a particular aircraft model. Accounting for its structural features, functional specifications, and dynamic operating conditions, a comprehensive failure analysis is performed on two typical fault modes of the slat mechanism—torsion bar disengagement and roller wear—employing rigid-flexible coupled dynamic modelling and parametric simulation approaches. The proposed methodology enables effective simulation of the fault evolution process and systematic analysis of fault response behaviors. The findings offer valuable technical insights for the design optimization, fault early warning, and reliability enhancement of aircraft slat mechanisms.

Keywords: Slat · Failure simulation · Response analysis

1 Introduction

High-lift systems, as core and critical subsystems of aircraft, directly determine the takeoff and landing performance as well as flight safety. They are primarily composed of leading-edge slats, trailing-edge flaps, and retraction/extension actuation mechanisms (Ma et al. 2023). Statistical data from civil aviation accident investigations demonstrate that approximately 10% of civil aviation accidents are directly attributable to failures of high-lift system mechanisms, leading to severe casualties and substantial economic losses (An et al. 2023). These adverse consequences underscore the necessity of investigating fault analysis and response characteristics of key high-lift system mechanisms, which has thus become a major research focus in aerospace engineering.

As a core component of high-lift systems, leading-edge slats enhance lift by modulating the airflow pattern over the wing's leading edge to increase the critical angle of attack (Liu et al. 2023). Their key functions include delaying airflow separation over the wing, increasing the critical angle of attack to prevent stall, and simultaneously improving the lift coefficient—thereby ensuring flight safety and meeting aerodynamic performance requirements during the low-speed takeoff and landing phases of aircraft.

During the takeoff and landing phases, the leading-edge slat mechanism must accurately execute retraction and extension actions. Each component is required to withstand and transmit coupled loads (e.g., gravitational, aerodynamic, and inertial forces)

T. Mankovits and N. Trisovic (Eds.): MDS 2026, LNME 2026.
https://doi.org/10.1007/978-981-92-2440-1_25

(Feng et al. 2022), rendering it prone to typical faults such as torsion bar disengagement and roller wear—two failure modes commonly observed in operational scenarios. Conducting dynamic analysis and fault research on the leading-edge slat mechanism is therefore a crucial approach to acquiring its kinematic and mechanical characteristics, which provides a theoretical foundation for the optimization of mechanism design and improvement of reliability.

Extensive research on high-lift systems has been conducted worldwide. Zhong et al. (2019) developed a rigid-flexible coupled simulation model for flap mechanisms and quantitatively analyzed the reliability and key influencing parameters of the motion mechanism; however, their research focused solely on trailing-edge flaps and excluded the failure analysis of leading-edge slats. Chen et al. (2025) proposed a dynamic modeling and response characteristic analysis method for spatial parallel mechanisms considering wear-induced clearances of multiple spherical joints. Their findings revealed that irregular clearances post wear exert adverse effects on the dynamic performance of parallel mechanisms; additionally, increasing clearance and applying loads degrade the operational stability of the mechanisms. Yu et al. (2022) treated the wear coefficient as a random variable, accounting for uncertainties arising from processing, manufacturing, and mechanism operational environments. They developed a comprehensive prediction method for hinge wear depth based on mechanism motion output monitoring data, demonstrating that aerodynamic load variations at different positions lead to uneven bearing wear. Wang et al. (2025) carried out fault simulation and strength tests on the flap mechanism using a full-scale structure, realistically simulating the single actuator fracture fault that may occur in the flap during flight. Mukras et al. (2010) proposed a prediction method for hinge wear depth based on the Archard wear model, which accounts for the impact of variations in hinge wear depth on the dynamic characteristics of multi-link mechanisms and has been validated experimentally. Nevertheless, this method was designed for general multi-link mechanisms and lacks adaptability to the specific structural and operational characteristics of leading-edge slat mechanisms.

The review of existing literature indicates that most prior studies have focused on the dynamic analysis of trailing-edge flaps, with insufficient attention paid to the rigid-flexible coupled failure analysis and response characteristics of typical faults (e.g., torsion bar disengagement, roller wear) in leading-edge slats. To address this gap, the present study focuses on the complex leading-edge slat mechanism of a specific aircraft model, employing rigid-flexible coupled dynamic modeling and parametric simulation for in-depth analysis. The findings are expected to provide reliable technical support for the safe design and fault diagnosis of the mechanism, as well as a theoretical basis for improving the overall reliability of high-lift systems.

2 Research Background and Work Foundation

2.1 Operational Principle of the Gear-And-Rack Slat Mechanism

The leading edge slat is an auxiliary wing surface extending forward beyond the main wing surface. The entire mechanism is primarily composed of pulley carriers, slide rails, gears, racks, and pulley carrier rollers. The slat is rigidly fastened to the slide rail, and its motion trajectory is determined by the slide rail. The aircraft slat control system controls a hydraulic motor to drive the torsion tube to rotate, which in turn transmits torque via the gear-rack drive to actuate the slide rail along a predefined arc trajectory. The retraction and extension of the slat are achieved through the bidirectional rotation of the gears. Due to the adoption of the gear-rack drive pair, the retraction-extension mechanism offers simplicity and high reliability. However, such a mechanism is prone to mechanical jamming induced by wear and corrosion.

Typically, multiple leading edge slat segments are installed on each of the left and right wings of an aircraft. Since each slat segment adopts an identical structural design, failure modes, and failure mechanisms—with the torsion tubes of successive segments connected end-to-end via universal joints—this study does not consider the interactions between different slat segments; instead, a single slat segment is selected as the research subject. Figure 1 shows a typical gear-rack type slat motion mechanism, consisting of pulley carriers, slide rails, gears, racks, and pulley carrier rollers. Each slide rail is constrained and positioned through contact with four rollers, and the slat wing surface is rigidly attached to the slide rail. During the retraction/extension process, the actuator rotates to drive the inner gear, which then transmits torque to the outer gear via the torsion tube. The synchronous rotation of the inner and outer gears in turn drives the racks mounted on the inner and outer slide rails to move. Limit stops are installed at both ends of the slide rails to prevent over-retraction or over-extension of the slat mechanism.

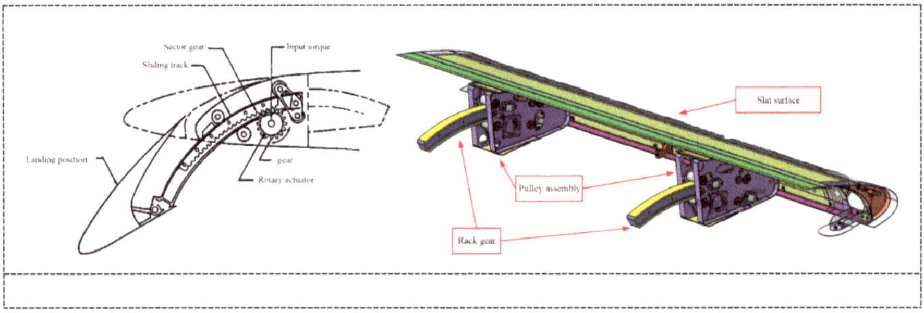

Fig. 1. Gear-and-rack slat mechanism

2.2 Failure Modes and Mechanism Analysis of Gear-Rack Type Slat Mechanism

The content emphasized in paragraph (1) of Article CCAR 25.671(c) of the civil airworthiness regulations is specifically for this paper. It refers to structural damage caused by a single failure. A core requirement is clear: after a single-point strength failure, the aircraft must still be able to return safely. The flap mechanism is driven by a ball screw actuator. This actuator converts the rotary motion of the screw into linear motion, which pushes the flap to move. The slat mechanism works differently. It transmits torque to gears through a torsion tube, and the gears then drive the slat to move. Paragraph (2) of the same article focuses on failures in the drive system. It stipulates that if a single-point control failure occurs, the control system on the opposite side can complete the retraction and extension of the high-lift device independently. Paragraph (3) defines jamming failure in detail. It is a failure mode where the high-lift device is fixed at the commanded position when the failure happens. Such failure is caused by various physical factors, such as wear, corrosion or foreign object jamming, as shown in Table 1.

Table 1. Slat mechanism failure mode analysis table.

Category	Failure modee	Failure criterion/cause	Severity
1	Structural failure	Actual load exceeds critical threshold	I
2	Contact fatigue	Operational life exceeds fatigue endurance limit	III
3	Excessive deployment/retraction speed	Control system malfunction causing motion time deviation	II
4	Asymmetric motion	Torque tube deformation causes excessive angle discrepancy between inner/outer sides	II
5	Positioning inaccuracy	Wear of rollers/guide track or gear-rack pair leads to angular deviation beyond tolerance	II
6	Motion jamming	Wear/deformation-induced increased resistance causes mechanism stalling	I

When wear occurs at the rack-roller contact interface of the slat structure, the friction coefficient increases. This increment impedes rack movement and induces an elevation in driving torque. On the other hand, long-term service may induce faults such as wear and increased clearance in the gear and torsion rod, which could result in disengagement failure during gear actuation by the torsion rod. Subsequently, this research will focus on fault simulation and analysis of the high-lift device, targeting typical faults of the slat mechanism (e.g., torsion rod disengagement and roller wear).

Long-term contact and compression between the roller and slide rail render the roller susceptible to wear, which may lead to failure. Adhesive wear is primarily associated with surface load, surface temperature, material properties, and sliding speed, and is commonly characterized by the Archard model, as shown in Eq. (1). Although this model only considers material hardness without accounting for other material properties and exhibits limitations in the determination of the wear coefficient K_w, it is still recognized as one of the classic formulas in wear research and thus widely adopted.

$$W_d = K_w \frac{pL}{H} \tag{1}$$

where: W_d is the wear depth; K_w is the wear coefficient; p is the contact stress; L is the sliding distance, (i.e., the wear distance); H is the material hardness. For the high-lift device, the roller wear type is adhesive wear, and the Archard model can be adopted to describe its wear process.

2.3 Slat Mechanism Failure Simulation Technology

The rigid-flexible coupled dynamic model of the high-lift device is established. By constructing rigid bodies (racks, driving mechanisms, pulley carriers, etc.) and flexible components (slat wing surface), handling of mechanical properties at the rigid-flexible coupled contact interfaces is completed, and consistency of fault-free dynamic responses during deflection is verified, as shown in Fig. 2.

For typical faults (e.g. torsion rod disengagement with torque transmission interruption, roller wear with contact parameter variation and jamming), the physical characteristics (disengagement, wear) are delineated. And we develop a customized Templex module that is based on failure simulation scripts. Fault models are injected via the external interface of the simulation platform, an interface for fault time parameters is reserved, and triggering conditions are specified (random triggering throughout the entire torsion rod disengagement failure process; roller wear jamming triggered exclusively during deflection). Subsequently, failure time is controlled parametrically to match the operational time ranges of the process corresponding to different faults, and parameters are automatically passed to the Templex scripts. Finally, single-fault simulation triggering is performed, and dynamic responses of the slat structure post-failure are output.

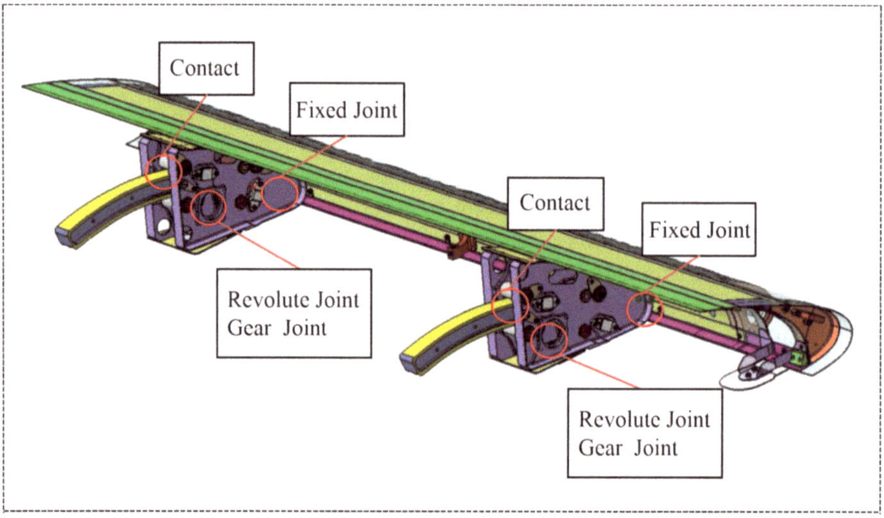

Fig. 2. Gear-and-rack slat motion pair

The components of the slat mechanism are interconnected through kinematic pairs, and their layout is configured to match the actual physical configuration. For the flexible modelling of the slat wing surface, the following key steps are implemented, as shown in Fig. 3.

1. Model simplification. Prior to finite element modelling of the wing surface, geometric model simplification is indispensable. The key load-carrying components shall be retained, including ribs (front rib, intermediate rib, rear rib), stiffening ribs, skin, cover plates, and load-transfer components between ribs and spars. Geometric features

irrelevant to load transmission, such as fillets and chamfers, must be removed. This simplification measure is specifically intended to facilitate the meshing of shell elements, ensuring the accuracy and efficiency of subsequent finite element analysis.

2. Neutral surface extraction. Since the thickness dimensions of the ribs, spars, and skin are significantly smaller than those of other structural components, which are considered shell structures. Prior to constructing the finite element model (FEM) of the wing surface, it is essential to extract the neutral surface from the geometric model. Shell element meshing is subsequently generated based explicitly on the identified neutral surface to ensure modelling accuracy and efficiency.

3. Mesh generation. Due to the thick walls and the geometric characteristics, stiffeners are discretized using tetrahedral solid elements. For all other structural components, including the skin, conventional ribs, spars, cover plates, and angle brackets connecting ribs and spars, shell elements are employed for meshing.

4. Boundary conditions. Defining physical constraints (e.g., mounting interfaces) and load application points to simulate real-world operational conditions.

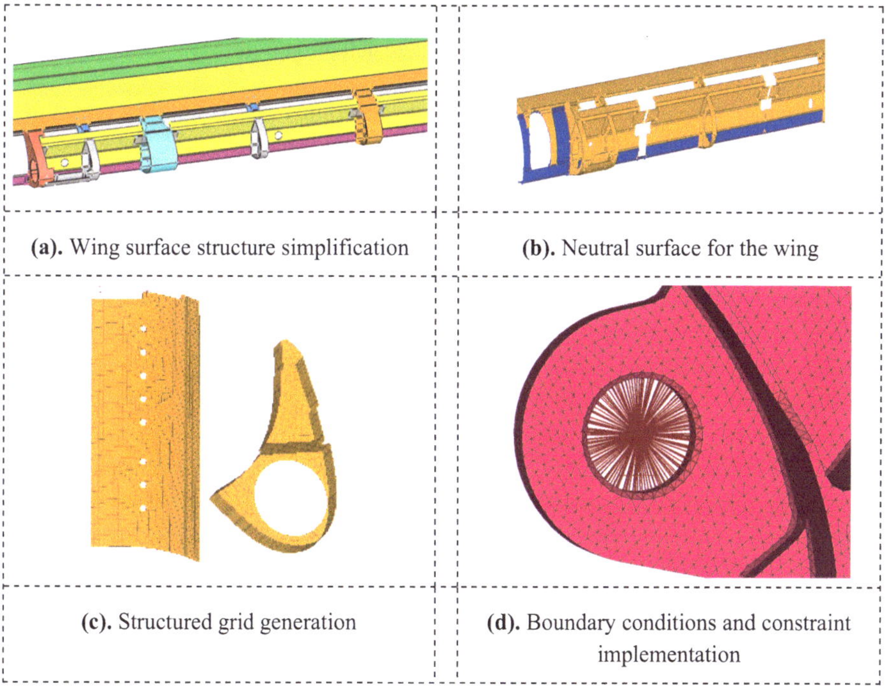

(a). Wing surface structure simplification

(b). Neutral surface for the wing

(c). Structured grid generation

(d). Boundary conditions and constraint implementation

Fig. 3. Rigid-flexible coupled dynamics modelling process

2.4 Fault Injection

- Roller wear simulation analysis method

The investigation focuses on the wear behaviour exhibited during the retraction process, with particular attention directed towards the inner upper and lower side rollers of the slat mechanism. The initial step in the methodology is the establishment of constraint conditions for the joints under both normal and worn conditions. These conditions are derived from two assumptions: the rigid contact assumption and the Archard wear model. Concurrently, a visual interface is being formulated with a view to achieving precise regulation of fault injection timing. Consequently, the sequential simulation script was developed via the Templex tool for the purpose of simulating roller wear faults. In the event of a fault being triggered, the constraints at the normal joints are deactivated, while those at the worn joints are reactivated. This process is known as fault injection and enables the output of transient response parameters of the slat structure.

- Torque tube disengagement simulation analysis method

The torque tube disengagement simulation focuses on the critical operational scenarios of the slat mechanism, specifically retraction from take-off to cruise position (deployment-to-stow motion) and deployment from cruise to landing position (stow-to-deployment motion). Methodologically, revolute joint constraints are first established at the gear-driven actuation input to define baseline conditions, followed by precisely defining the fault injection timing in the operational timeline. Sequential simulation scripts are then developed via Templex tools to model gear-driven actuation failures (e.g., sudden torque transmission loss).

To simulate disengagement, revolute joint constraints are reconfigured to induce active drive failure, mimicking torque tube disengagement. Finally, the simulation extracts transient dynamic responses (e.g., angular velocity fluctuations, stress concentrations) immediately post-disengagement, and it provides the response characteristics of the system under the fault conditions.

3 Fault Response Analysis

1. Torque tube disengagement fault.

The torque tube disengagement fault analysis employs Templex scripts tool within motion view software for fault injection, coupled with parameterized fault timing control, to simulate fault scenarios of the slat mechanism and acquire torque variation characteristics of the ball screw. Specifically, the study mainly focuses on the two phases—retraction from take-off to cruise position and deployment from cruise to landing position. And the actuator disengagement faults are injected during the period.

It compares the transient driving torque under normal and fault conditions. The result shows that the outer-side driving torque rises considerably when the inner torque tube disengages during the take-off-to-cruise retraction phase and the cruise-to-landing deployment phase, as shown in Figs. 4 and 5. And the phenomenon clearly demonstrates the amplified load redistribution effect under fault conditions.

2. Roller wear-induced failure analysis.

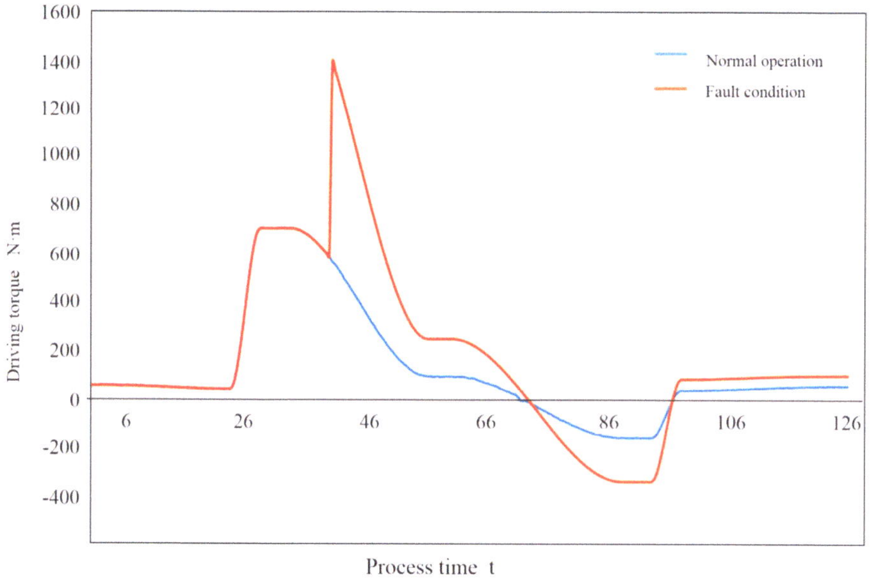

Fig. 4. Variation of outer-side driving torque induced by torque tube disengagement during take-off-cruise phase

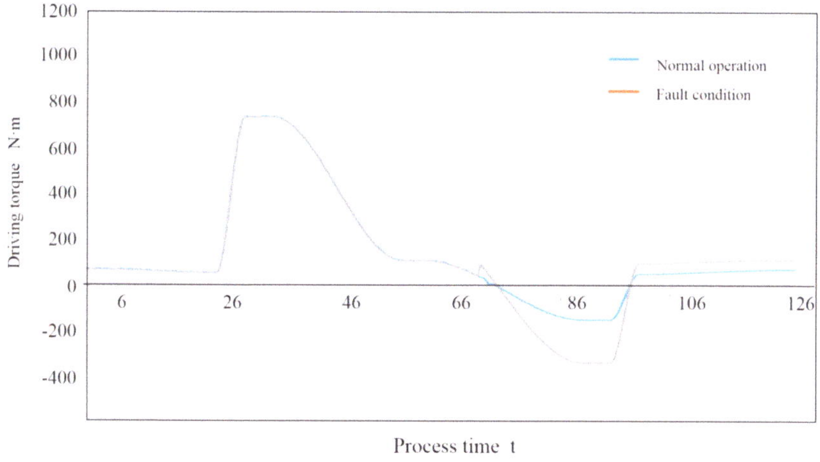

Fig. 5. Variation of outer-side driving torque induced by torque tube disengagement during cruise-to-landing phase

The roller wear-induced failure study leverages Motion View software's Templex scripts for fault injection, combined with parameterized fault timing control, to simulate fault scenarios and obtain torque variations of the ball screw. Focusing on the full deployment and retraction cycle—specifically covering operational phases such as ground deployment to airborne retraction and airborne deployment to ground stowage—the analysis injects roller wear faults into the whole research stages. By comparing the transient driving torque under normal and fault conditions, the study shows the influence of wear-induced degradation on system dynamics, as shown in the Figs. 6 and 7.

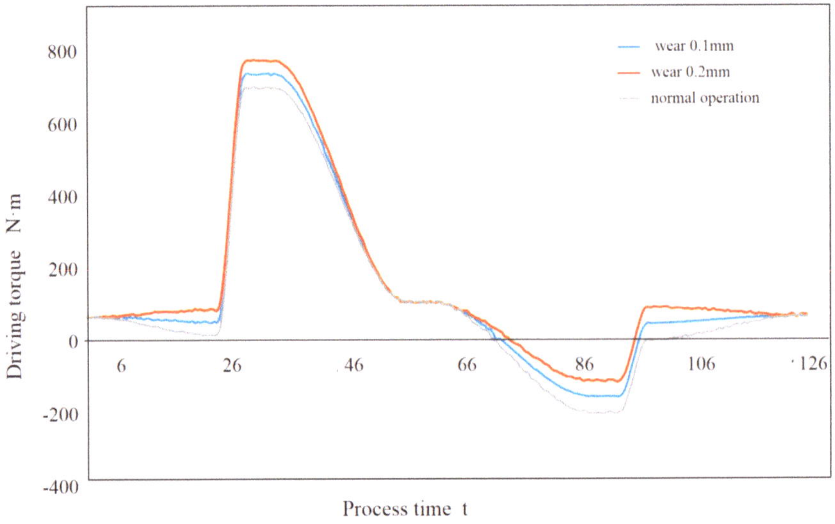

Fig. 6. Variation of inner-side driving torque induced by roller wear

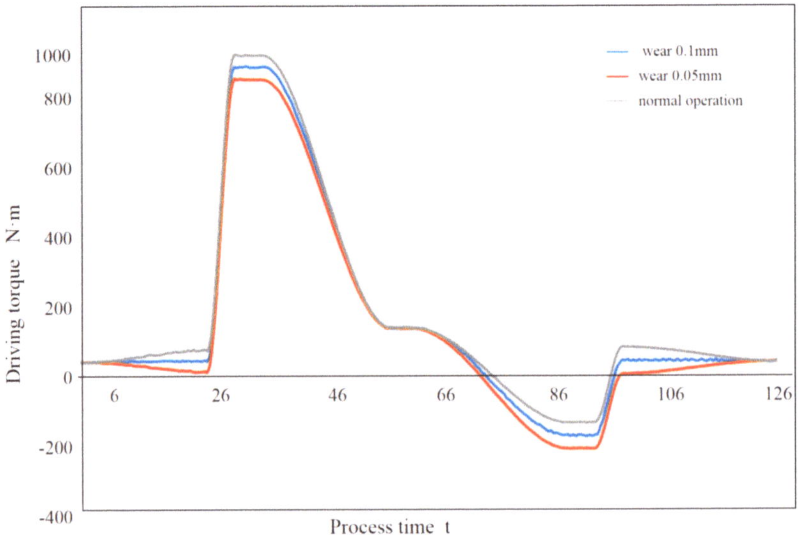

Fig. 7. Variation of outer-side driving torque induced by roller wear

Table 2. Driving torque variation magnitudes under inner roller wear

Fault type	Wear depth (mm)	Increase at inner maximum torque (%)	Decrease at outer maximum torque (%)
Inner roller wear	0.1	5.35	3.85
	0.2	11.03	7.65

As shown in the 'Fig. 7', when the inner roller wear occurs, the driving torques on the both inner and outer sides exhibit distinct variations.

At a wear depth of 0.1 mm, the maximum inner-side driving torque increases from 676 Nm to 712 Nm, which is a 5.35% increase. Meanwhile, the maximum outer-side driving torque decreases from 999 Nm to 960 Nm. It represents a 3.85% reduction.

At a wear depth of 0.2 mm, the maximum inner-side driving torque increases by 12.14%, from 676 Nm to 758 Nm. Additionally, the maximum outer-side driving torque has been observed to decrease from 999 Nm to 932 Nm, representing a reduction of 6.72%.

The wear on the inner roller can cause deviations in the wing surface movement, resulting in a decrease in the outer drive torque. As wear worsens, the system's driving force is redistributed, and the primary load-carrying path shifts inward to compensate for the loss of mechanical efficiency in the worn rollers.

4 Conclusion

In accordance with the general provisions of Airworthiness Regulation CCAR 25.671, the paper employs a parametric simulation method to conduct a failure analysis of the slat mechanism. And it mainly focused on torque tube disengagement and roller wear faults. The analysis emphasized the post-failure transient response characteristics.

The torque tube disengagement simulation case models test conditions in which, following a single-point failure, the remaining drive units operate the flap mechanism to complete retraction and extension.

Analysis of the response following the failure indicates that the maximum torque required for the outer drive unit to operate independently is 1.41×10^3 Nm. When the failure occurs on the inner side, the outer drive actuator must have sufficient torque output capacity to ensure that the flaps can complete their retraction and extension movements. When the roller wear occurs, the inner and outer drive torques show obvious variations, with torque variation ranges between 3.85 and 11.03% (as shown in Table 2).

The rigid-flexible coupling modelling focuses primarily on the wing surface. However, the contact between the guide rail and the roller is modelled as a rigid connection, making it difficult to accurately simulate the failure progression. The next research will focus on investigating the wear progression of the roller under rigid-flexible coupling conditions.

References

An, G., Liu, R. J., Yin, G. Q.: Application of a new risk identification method in the development of civil aircraft high-lift systems. Adv. Aeronaut. Sci. Eng. **14**(1), 57–64 (2023). https://doi.org/10.16615/j.cnki.1674-8190.2023.01.06

Chen, X.L., Zhang, H., Fan, H.K.: Dynamic response analysis of spatial parallel mechanisms considering wear-induced clearances in spherical joints. J. Vib. Eng. **38**(3), 490–498 (2025). https://doi.org/10.16385/j.cnki.issn.1004-4523.2025.03.005

Feng, Y.W., Cai, A., He, Z.Y., Xue, X.F.: Design and dynamic simulation research of hinged drooping leading edge mechanism. Adv. Aeronaut. Sci. Eng. **13**(1), 93–100 (2022). https://doi.org/10.16615/j.cnki.1674-8190.2022.01.10

Liu, Z.Y., Chu, H.B., Chen, Y.C., Zhang, B.Q.: Research on improving stall characteristics of high-lift devices by leading-edge slat slot. Acta Aerodyn. Sinica **41**(2), 21–28 (2023). https://doi.org/10.7638/kqdlxxb-2021.0392

Ma, G.J., An, G., Shi, Y.M., Kang, N., Sun, J.S.: Advanced technologies and development of civil aircraft high-lift systems. Acta Aeronautica et Astronautica Sinica **44**(Suppl. 1), 6–19 (2023). https://doi.org/10.7527/S1000-6893.2022.27516

Mukras, S., Kim, N.H., Mauntler, N.A.: Analysis of planar multibody systems with revolute joint wear. Wear **268**(5–6), 643–652 (2010). https://doi.org/10.1016/j.wear.2009.10.014

Wang, B., Liu, W., Zheng, J.J., Huang, Y., Wang, M.M., Fu, H.X.: Fracture fault simulation and strength test of flap mechanism. Acta Aeronautica et Astronautica Sinica **46**(10), 297–309 (2025). https://doi.org/10.7527/S1000-6893.2024.31243

Yu, T.X., Zhuang, X.C., Song, B.F., Sun, Z.C.: A comprehensive prediction method for multi-hinge wear life of aircraft linkages. Acta Aeronautica et Astronautica Sinica **43**(8), 1 (2022). https://doi.org/10.7527/S1000-6893.2021.25113

Zhong, Y.L., Zhang, Z.W., Zhang, D.P., Wang, X.K.: Reliability simulation analysis of flap motion mechanism based on failure modes. Electron. Product Reliab. Environ. Test. **37**(5), 1–7 (2019). https://doi.org/10.1051/jnwpu/20244220222

Fault Modeling and Simulation Technology for Comprehensive Performance Failures and Dynamic Behaviors

Zhe Lai, Yanxiong Huang, Lingjie Zhang(✉), Hongqi Yang, and Qi Yang

CEPREI, Guangzhou 511370, China
zljzhanglingjie@163.com

Abstract. In the digital era, Model-Based Systems Engineering (MBSE) has become a crucial approach for product development. Conducting fault analysis and mitigation control design synchronously during the MBSE process is of great significance for integrating functional and general quality characteristics and improving product quality. This paper, by integrating MBSE models (including behavior, structure, and performance parameters), carries out research on fault modeling and simulation technologies for comprehensive performance failures and dynamic behaviors. It proposes a modeling method for comprehensive performance failures and the dynamic behavior of fault mitigation control, determines the interaction and integration approach between multi-dimensional fault models, and presents a general quality characteristic simulation verification method that combines static analysis and dynamic simulation. Finally, taking a certain type of product as an example, the technologies proposed in this paper are verified through case application.

Keywords: MBSE · Static structure · Dynamic behavior · Fault modeling · General quality characteristic simulation and verification

1 Introduction

In the digital era, Model-Based Systems Engineering (MBSE) has become a critical methodology for product development. During the MBSE process, general quality characteristics (GQCs) should be designed, analyzed, and verified in parallel with functional characteristics (Ren et al. 2024). Currently, GQC-related work primarily focuses on fault analysis: based on MBSE static structure models, static fault propagation models are constructed to characterize how module inputs and internal faults lead to output failures (Zhang et al. 2018). This approach partially resolves the consistency issue between system function models and GQC models, and reduces the subjectivity inherent in traditional analysis methods (Joshi et al. 2006). However, complex systems impose stringent requirements on both function and performance. The current practice of conducting fault analysis solely relying on static structure models fails to fully address performance parameter deviation-induced failures. Moreover, existing fault modeling methods cannot

T. Mankovits and N. Trisovic (Eds.): MDS 2026, LNME 2026.
https://doi.org/10.1007/978-981-92-2440-1_26

support fault propagation based on physical equations (Brück et al. 2002). Secondly, the closed-loop management of fault mitigation and control is still predominantly document-centric (Hountalas 2000). Measures such as design improvements, operational compensations, and preventive detection are only proposed as requirements, with a low degree of model-based implementation in design. Some organizations have initially built behavioral models incorporating failure behaviors, but these models lack integration with static structure models and remain unexecutable for validation purposes. Additionally, in the current validation and evaluation of GQCs, the interaction and integration of various characteristic elements are not achieved; instead, most analyses are performed on individual characteristics in isolation. Therefore, based on existing technical methods and MBSE theoretical frameworks, this paper further investigates the integrated modeling methodologies: including static fault logic that accounts for performance-induced failures and dynamic behavior models that incorporate redundancy, detection, maintenance, and support activities. Furthermore, based on the integrated model, this paper implements a GQC simulation validation approach combining static analysis and dynamic simulation. The proposed method verifies both integrated and individual GQC metrics, identifies performance bottlenecks and weaknesses in a timely manner, and provides decision support for quality improvement initiatives.

2 Fault Modeling Method Integrating Performance Failures and Dynamic Behaviors

2.1 Establishment of a Fault Model Framework Integrating Performance Failures and Dynamic Behaviors

The integrated model framework proposed in this paper is illustrated in Fig. 1. First, performance failure is characterized by deviations in performance parameters. Second, an equation describing the occurrence of performance deviation failures is established based on failure mechanisms. Third, the state transition process of performance failure occurrence and the dynamic behaviors of corresponding failure mitigation and control measures (including backup, detection, and maintenance support) are modeled in the form of state machine diagrams, with fault states and fault occurrence times linked to the results of the first two steps. Fourth, the performance failure parameters are substituted into physical equations to construct a fault propagation logic model, which is used to model the abnormal performance of other products affected by a performance failure at a specific time. The four models integrate and correlate performance failure, fault occurrence, fault propagation, and fault mitigation control.

2.2 Modeling Method of Static Fault Logic Model Considering Performance Failure

(1) Performance Failure Modeling Method

The essence of performance failure lies in the fact that performance parameters drift over time and exceed the threshold values, resulting in failure, as shown in Fig. 2. Based on this, it is proposed to analyze and model the deviation degree of performance parameters by using the Modelica modeling language. A Modelica model mainly consists of several elements: ① Parameters that do not change with simulation time; ② Variables that change with time and are obtained through physical equation calculations; ③ Physical equations that describe physical laws.

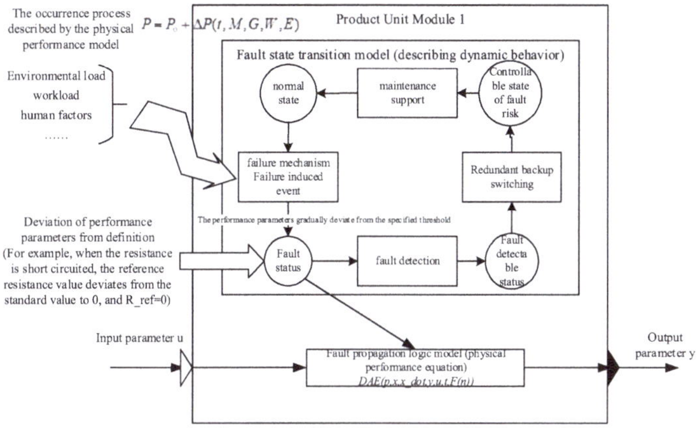

Fig. 1. Fault model framework

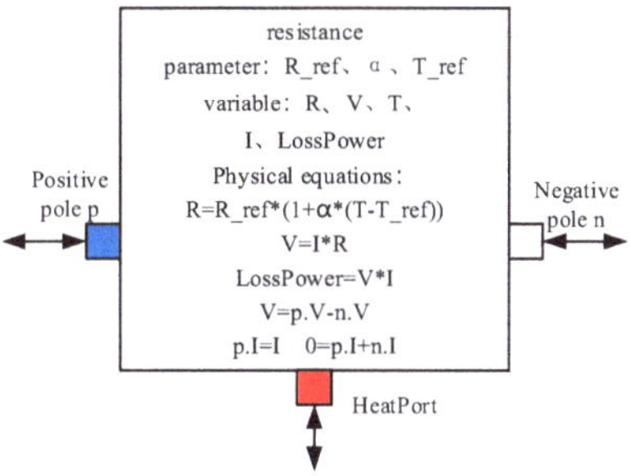

Fig. 2. A Typical Modelica Model

It can thus be sorted out and analyzed that three main categories of parameters are affected when a performance failure occurs: ① Affecting parameter-type performance parameters; ② Introducing new performance parameters; ③ Affecting variable-type performance parameters.

① For parameters that do not change with time, after a performance failure is introduced, they need to be replaced with a fault occurrence equation that gradually deviates over time. For example, replace the original reference resistance value(R_ref) with a fault occurrence equation related to the drift coefficient(k_d), time(t), and drift exponent(m):

$$R_{ref,fault} = R_{ref,nom} * (1 + k_d \cdot t^m) \tag{1}$$

② For cases where performance failure leads to changes in physical structure, the original physical equations of the product will also change accordingly, and new parameters need to be introduced into the original equations. For example, in the case of pipeline leakage, a leakage term(Q_{leak}) needs to be added to the original equations:

$$\frac{dM}{dt} = Q_{in} - Q_{out} - Q_{leak} \tag{2}$$

③ For the correlation between performance failure of non-bottom-layer products and their output variable parameters, the failure of such products will transmit and affect the occurrence of performance failure in lower layers. For this type of performance failure, there is no need to modify the original physical equations; it only requires defining the performance failure criteria.

Finally, the value ranges of the performance parameters affected by performance failure are defined, and the failure criteria are formulated, thereby completing the definition of performance failure. In subsequent simulation processes, it is possible to accurately determine whether a failure has occurred. For example, the performance parameter associated with the actuator wear fault is the actuator length L_0, which triggers a fault when L_0 exceeds its standard value.

(2) Fault Occurrence Modeling Method

The occurrence of performance failure is essentially the process of the occurrence of a product's failure mechanisms and root causes, which can be characterized by fault occurrence equations. Each performance failure is associated with time, as well as the product's materials, geometric parameters, environment and operating conditions. On this basis, the basic form of the fault occurrence model is proposed as follows:

$$P = P_0 + \Delta P(t, M, G, W, E) \tag{3}$$

where P represents the value of the performance parameter after the occurrence of a fault, P_0 denotes the normal value, $\Delta P(t, M, G, W, E)$ represents the parameter increment after the fault occurs, t stands for time, and M, G, W, E represents materials, geometric parameters, operating conditions, and environmental conditions respectively. The equation can be further refined according to different fault types:

① Gradual degradation fault: $\Delta P{=}k \cdot t^{\alpha} \cdot W^{b} \cdot \exp(c \cdot E)$ (linear/exponential growth, where A denotes the load exponent respectively);

② Sudden fault: $\Delta P{=}k \cdot step(N_c - N_{th})$ (where step is the step function, N is the cumulative load, and N_{th} is the fault threshold);

③ Periodic fault: $\Delta P{=}k \cdot t \cdot \sin(\omega \cdot t)$.

Considering that all influencing factors in actual engineering have uncertainties, which result in the random uncertainty of performance failure. Based on statistical data, the mechanism factors in the equation can be converted into variables subject to random distribution, and the Monte Carlo method can be used subsequently to simulate fault occurrence.

(3) Fault Propagation Modeling Method

By substituting the parameter value changes affected by the above performance failures into the physical equations in the Modelica model, the model can automatically calculate the impact on other blocks. Then, through the physical connections between modules, the propagation of performance failures across different modules is realized, ultimately leading to a system fault. The modeling schematic is shown in Fig. 3.

① For cases where performance failure directly affects parameter-type performance parameters, it is only necessary to replace the original parameters with fault occurrence equations to convert the original physical equations into physical equations that take fault propagation into account. Take Ohm's Law as an example: replace the resistance value in the original equation with a fault occurrence equation: $U = I \times R(t, M, G, W, E)$.

②As for the case where performance faults introduce new parameters, it is necessary to add fault parameters to the physical principle equations. For example, when a motor experiences a short circuit, the short-circuit current parameter is added to the original three-phase symmetrical circuit equations.

$$u_a - u_b = R_{short} \cdot i_{short} + L_{short} \cdot \frac{di_{short}}{dt} \quad (4)$$

After defining the local fault propagation physical equations of each module according to the aforementioned method, the performance failure transmission relationships of the entire system can be constructed by combining the physical connections between modules, as illustrated in Fig. 4.

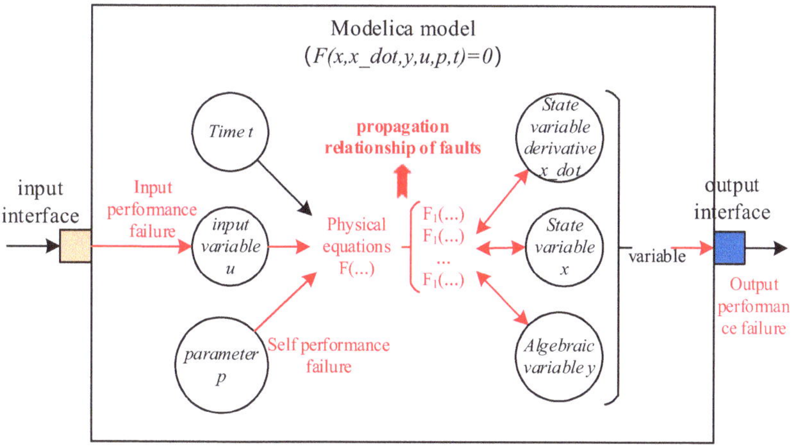

Fig. 3. Fault Propagation Modeling

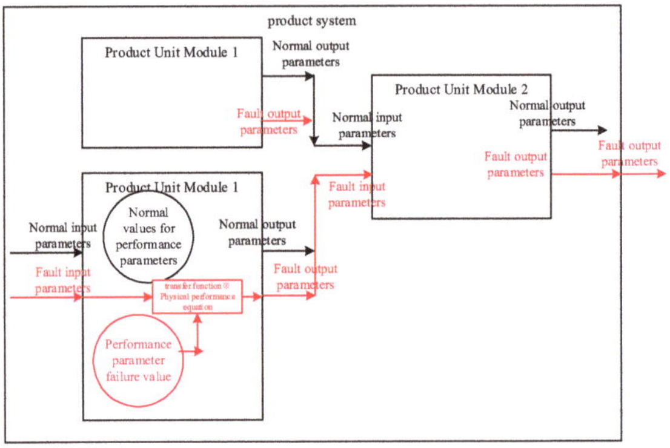

Fig. 4. Complete Fault Propagation Chain

2.3 Dynamic Fault Behavior Modeling Method

On this basis, this paper proposes to construct a fault state transition model within the module using the Altarica modeling language: it models the complete state transition process from normal operation to fault occurrence and then to the execution of fault mitigation control actions. By associating fault states with predefined fault effect pairs and linking the transition events from normal to fault states with predefined fault occurrence equations, the integration of the two models is achieved, as illustrated in Fig. 5.

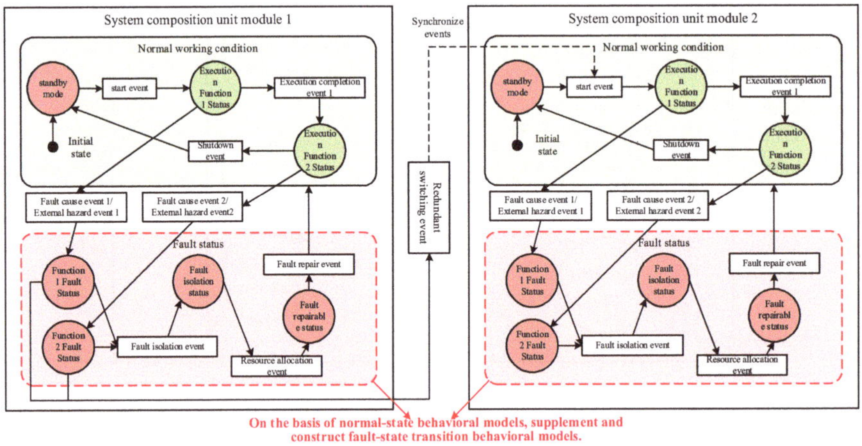

Fig. 5. Modeling of the Fault State Transition Behavior Model

The fault state transition model incorporates various fault mitigation and control action events (including failure occurrence, backup, detection, resource allocation, maintenance, and other events), and can be associated with specific behavioral activities. When the behavioral activities are completed, the events are triggered and the states transition accordingly. The modeling methods for different dynamic behaviors of fault mitigation and control are as follows:

① Redundant backup behavioral activities are mainly modeled by improving the original activity diagram to allow the primary and backup devices to perform the same functional activity. Meanwhile, in the state machine, a conditional association is established between the failure of the primary device and the standby activation event of the backup device. When the primary device fails, the backup device will be activated to perform the redundant function, as illustrated in Fig. 6.

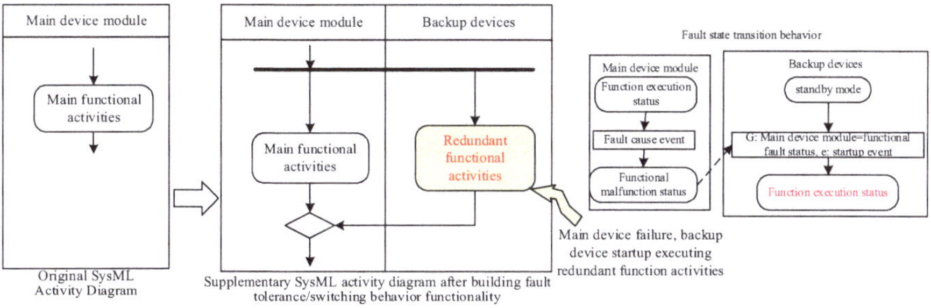

Fig. 6. Modeling of Functional Fault Tolerance/Switching Behavior

② Fault detection and isolation activities are mainly characterized by modeling diagnostic sequences as behavioral activity models and associating them with the detection and isolation events in the state machine, as illustrated in Fig. 7.

③ Maintenance support activities are mainly modeled by constructing a complete activity diagram based on the maintenance operation processes, as well as the involved personnel, resources, and maintenance levels, and then associating the diagram with the maintenance support events in the state machine, as illustrated in Fig. 8.

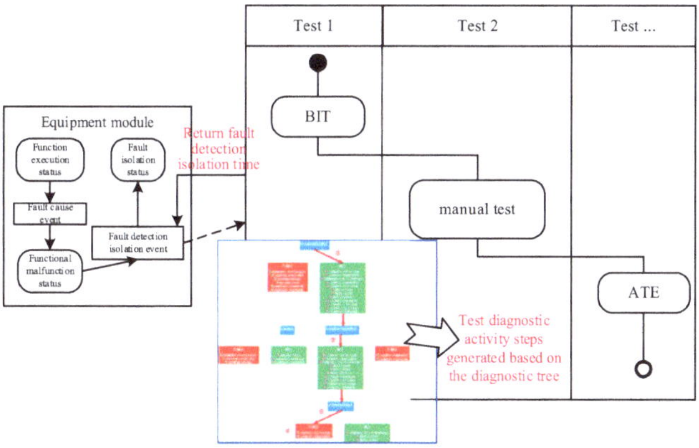

Fig. 7. Modeling of Detection Behavior

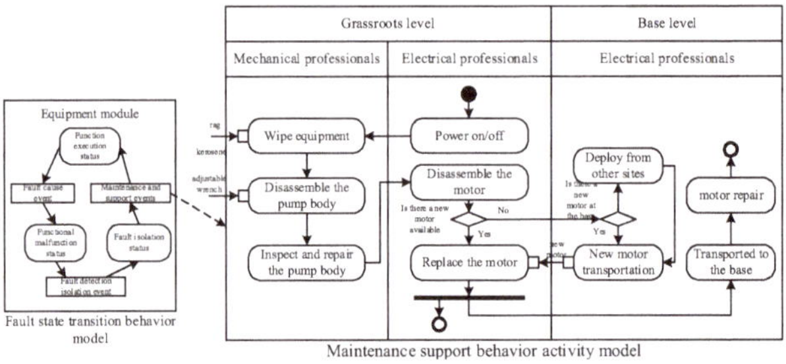

Fig. 8. Maintenance and Support Behavior

2.4 The Interactive Fusion Technology of Static Fault Logic and Dynamic Fault Behavior Models

The interactive fusion method between the static fault logic model constructed with the Modelica language and the dynamic fault behavior model constructed with the Altarica language is illustrated in Fig. 9. XML technology is mainly adopted to realize the model data interaction between two different software tools, and API interface-driven technology is used to achieve co-simulation between the models. On the one hand, faults occurring in the Altarica dynamic fault behavior model are saved in an XML file and imported into the Modelica software; the static fault logic model is driven via an API interface to conduct fault propagation analysis and obtain the final impact of the faults. On the other hand, the current state of each module is saved in an XML file and fed back to the Altarica model to continue executing the next state transition event and invoking the execution of dynamic behavior activities. Such iteration enables the interactive fusion of model data and co-simulation.

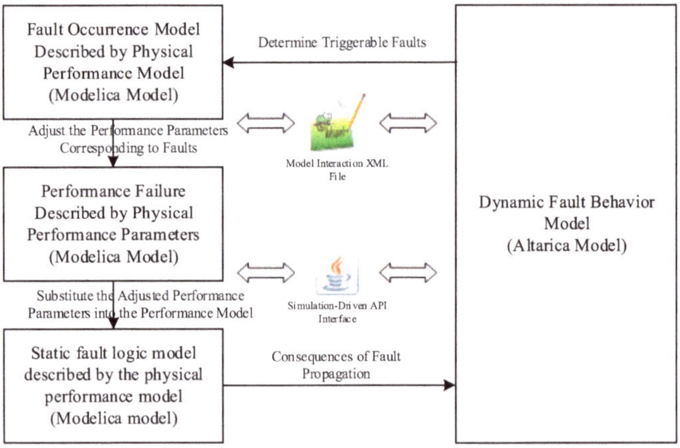

Fig. 9. Model Interactive Fusion Method

3 General Quality Characteristic Simulation and Verification Technology Combining Static Analysis and Dynamic Simulation

On the basis of the integrated model constructed above, simulation verification of general quality characteristics combining static analysis and dynamic simulation is carried out to comprehensively verify the achievement of various indicators. The overall simulation process is illustrated in Fig. 10.

(1) Determine the Initial States of Each Component Module in the Product System

The initial states of each product module (including normal state, standby state, etc.) are determined separately by the integrated model.

(2) Sample Performance Failures According to Fault Occurrence Equations

Based on the probability distributions followed by materials, geometric parameters, operating conditions and environmental conditions involved in the fault occurrence equations, parameter values are randomly sampled. Then, combined with failure criteria, the occurrence time of performance failures is determined.

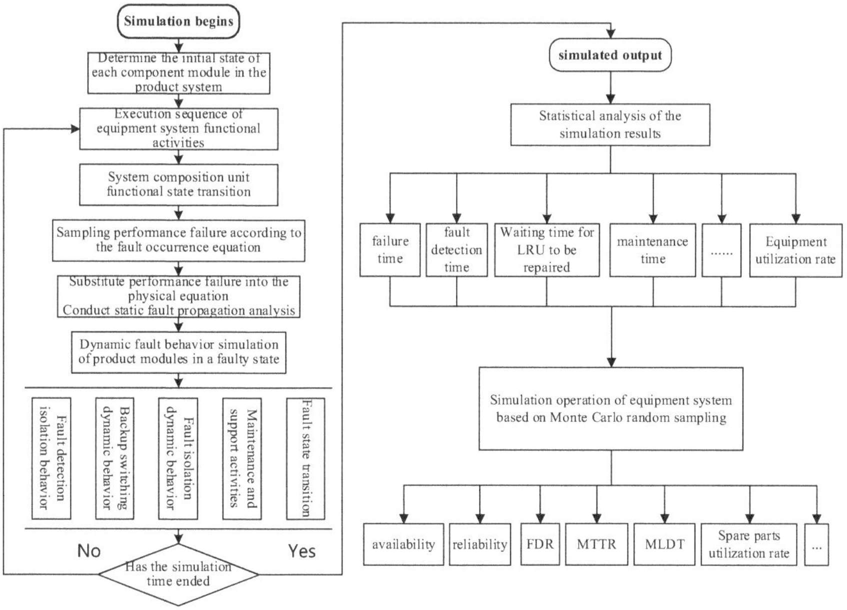

Fig. 10. Comprehensive Simulation Process

(3) Carry Out Static Fault Propagation Analysis

Substitute the values of performance parameters into the physical equations of the product to calculate the values of other parameter variables in the model at the current moment, so as to propagate the performance failures of the product module itself and its input variables to the output variables. Meanwhile, combined with the physical connections between product modules, the output failures of the previous product are propagated to the next product until the values of all product parameters and variables in the entire product system are derived, and the current state of each product is judged according to the performance failure criteria.

(4) Carry Out Dynamic Fault Behavior Simulation

Based on the results of static fault propagation analysis, further judge the fault mitigation and control action events that need to be executed according to the fault state transition model, call and execute the corresponding specific behavioral activity models (such as fault detection and isolation, backup switching, and maintenance support activities mentioned in the integrated model), and record the execution time of the activities.

(5) Continuously Trigger Faults, Conduct Fault Propagation Analysis and Execute Dynamic Fault Behaviors

Judge whether the current simulation time is reached. Before the end of the simulation, repeatedly execute steps (2), (3) and (4) until the simulation time expires.

(6) Count the Results of the Current Simulation

Count data such as the success status of the current simulation task, product normal operation time, fault occurrence time, number of fault occurrences, fault detectability and isolability, maintenance time, maintenance waiting time, and the quantity of resource demand and shortage.

(7) Judge Whether the Number of Simulations Has Reached the Set Value

Judge whether the number of simulations has reached the set value, and conduct simulation analysis continuously during this period.

(8) Statistically Analyze the Results of Multiple Simulations

Through multiple simulations, conduct statistical analysis on the simulation results to verify and calculate the level of the product's general quality characteristic indicators and confirm whether they meet the design requirements. The calculation methods of the main indicators are shown in Table 1.

Table 1. Calculation Methods for Various Characteristic Indicators

Indicator name	Calculation formula	Parameter
Operational availability A_O	$A_O = \frac{\sum_{i=1}^{N} t_i}{NT}$	Normal operation time of the product in the i-th simulation t_i,simulation time$T$
Mission reliability R	$R = \frac{A_S}{N}$	Number of successful missions A_S

(continued)

Table 1. (*continued*)

Indicator name	Calculation formula	Parameter
Falut detection rate *FDR*	$FDR = \sum_{i=1}^{N} \frac{n_{Di}}{n_i} \Big/ N$	Number of fault modes detected in the i-th simulation n_{Di},Total number of faults occurring in the i-th simulation n_i
Mean Time To Repair *MTTR*	$MTTR = \frac{\sum_{i=1}^{m} T_{CMi}}{m}$	Maintenance and support activity execution time T_{CMi},Number of maintenances m
Mean Logistics Delay Time *MLDT*	$MLDT = \frac{\sum_{j=1}^{a} t_{MLDj}}{a}$	Number of non-immediate maintenance and support activities: a,Waiting time of the j-th maintenance operation:t_{MLDj}

4 Case Application Research

4.1 Fuel System Fault Modeling

This paper conducts a case study using a fuel system as an example to verify the application of the technical methods proposed herein. By collecting information such as the product composition, faults, performance parameters, Modelica models, and performance failure criteria of the fuel system, performance failure modeling is carried out for each component within the fuel system, as illustrated in Table 2.

Table 2. Performance Parameters, Failure Criteria and Fault Occurrence Equations

Product	Failure mode	Correlation performance parameters	Performance failure criteria	Fault occurrence equation
Centrifugal pump	Decreased traffic	Traffic Q	$Q \leq 0.5Q_0$(Q_0 is rated flow)	$Q(t)=Q_0 + [-Q_0 \cdot k_{wear} \cdot t \cdot \frac{p}{p_0} \cdot \exp(1 + 0.01(T_{oil} - 40))]$
One-way valve	Leak	Leakage flow rate Q_{leak}	$Q_{leak} > 0.1\dot{L}/s$	$Q_{leak}(t) = 0 + [k_{seal} \cdot t \cdot \Delta p \cdot \frac{c}{c_{rated}} \cdot \exp(ku \cdot (T_{oil} - 40))]$
Oil depletion signal	Output out of tolerance	Signal error δS	$\delta S \geq 1(h < h_0, S_{oil} = 0)$	$\delta S_{oil}(t) = 0 + [k_{drift} \cdot t \cdot (T_{env} - 25) \cdot \frac{h}{h_0}]$
Fuel system	Unable to provide the requested fuel	Total fuel supply flow rate Q_{sys}	$Q_{sys} < 1.5 \times 10^3 L$	–

Then, combined with the physical equations of each component, the static fault propagation model as shown below is constructed. First, the oil supply flow rate of a single oil circuit is calculated, with the fuel output by the centrifugal pump. Where $Q_{pump0.i}$ denotes the rated flow rate of the centrifugal pump; when the pump fails, the value of $\Delta Q_{pump,i}$ changes accordingly.

$$Q_{pump,i} = Q_{pump0.i} + \Delta Q_{pump,i}(t, M, G, W, E) \tag{5}$$

Next, the effective flow rate of the check valve is calculated, where the effective flow rate equals the pump output flow rate minus the leakage flow rate. Where $Q_{leak,i}$ denotes the leakage flow rate of the check valve.

$$Q_{valve,i} = Q_{pumpi} - Q_{leak,i} \tag{6}$$

Then, the validity of the oil circuit is determined based on the signals from the oil level sensor and the pressure sensor. The oil circuit is valid when $S_{oil,i} = 0$ (fuel remains in the tank) and $S_{press,i} = 1$ (pressure exists after the pump) and $Q_{valve,i} \geq 0.01L/s$.If there is no output from the oil level sensor ($S_{oil,i} = 1$), the pump fails and $Q_{valve,i} < 0.01L/s$, the controller shuts down the oil circuit, $Q_i = 0$, and the oil circuit becomes invalid.

Finally, the total oil supply flow rate Q_{sys} is calculated. Oil circuits No. 1 to No. 3 are connected in parallel to the engine inlet, where the total flow rate equals the sum of the effective flow rates of all valid oil circuits:

$$Q_{sys} = \sum_{i=1}^{3} Q_i = Q_1 + Q_2 + Q_3 \tag{7}$$

Corresponding dynamic behaviors for fault mitigation and control are constructed for the fault modes of each component in the fuel system. For example, the fault state transition model of the centrifugal pump is illustrated in Fig. 11. The specific behavioral activities associated with each fault mitigation and control action event are shown in Fig. 12.

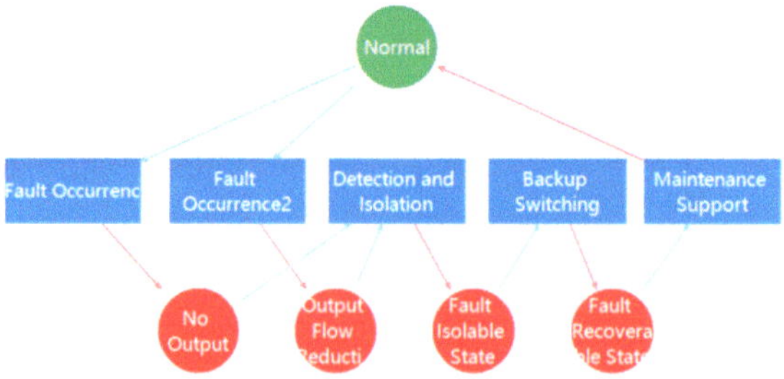

Fig. 11. Fault State Transition Behavior

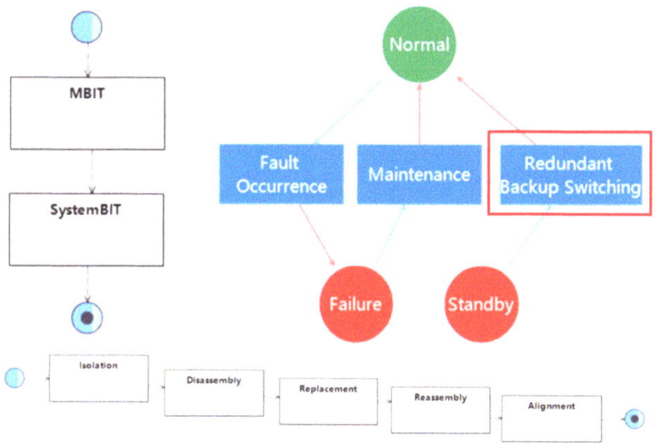

Fig. 12. Fault Mitigation Control Behaviors

4.2 Fuel System Simulation and Verification

Based on the model constructed above, multiple rounds of simulation analysis were carried out. During the simulation, due to the no-output fault of the No. 2 centrifugal pump, fuel was not delivered to the corresponding oil circuit, and the fuel volume remained unchanged. By the end of the simulation, the total fuel supply of the fuel system was 1 $\times$ 10^3L, which was lower than the failure criterion $Q_{sys} < 1.5 \times 10^3 L$, indicating that the fuel system failed and the mission was not successfully completed, as illustrated in Fig. 13.

After the performance failure of the centrifugal pump occurred, the corresponding dynamic behavior for fault mitigation and control was triggered, enabling the pump to return to normal operation. This further allowed the No. 2 oil circuit to resume fuel supply, the total fuel supply of the fuel system recovered to the normal level, and the mission was successfully completed, as illustrated in Fig. 14.

In this paper, 1000 simulations were carried out on the fuel system, where the total fuel supply result of each simulation is shown in Fig. 15a. A total of 987 simulations were successful, and thus the mission reliability was calculated to be 0.987. The simulation results of other indicators are presented in Fig. 15b

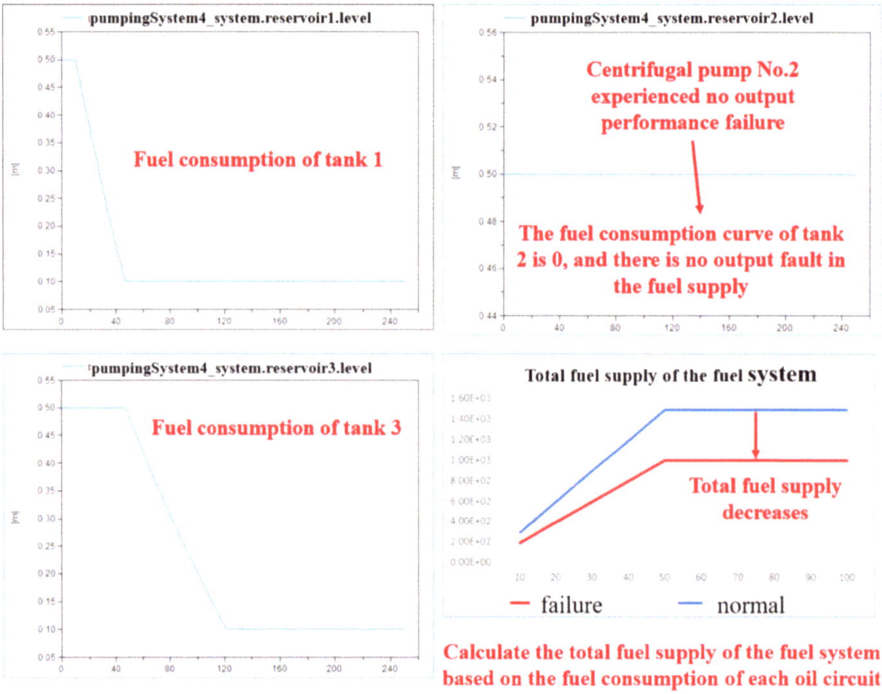

Fig. 13. Fault Impacts of Performance Parameters

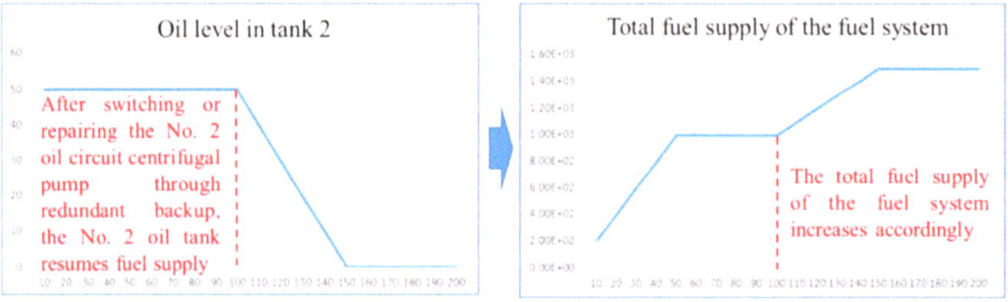

Fig. 14. Total Fuel Supply Restored to Normal

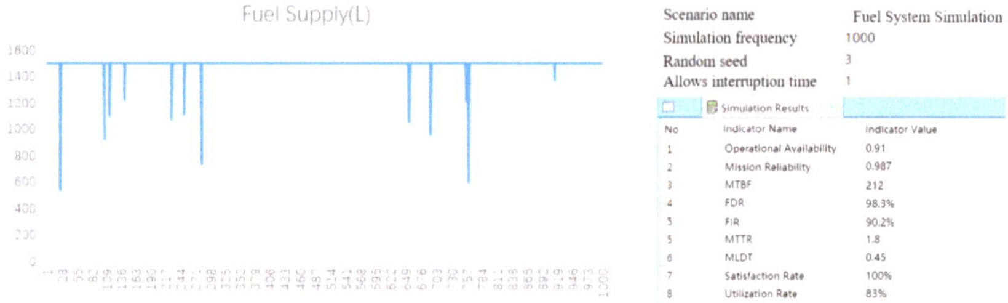

(a) (b)

Fig. 15. Simulation Results

5 Conclusions

On the basis of existing technical methods, this paper proposes a performance failure parameter deviation modeling method, adopts physical equations to construct fault occurrence equations and fault propagation models, and further models dynamic fault behaviors (such as redundant backup, detection, and maintenance support) executed after fault occurrence via state machines. A comprehensive model integrating static fault logic and dynamic fault behaviors is thus formed. Then, combined with the correlation among the models, simulation verification of general quality characteristics is carried out by integrating static fault propagation analysis and dynamic behavior model execution. This provides technical support for accurately simulating product fault behaviors and comprehensively verifying the level of general quality characteristic indicators. Finally, taking a fuel system as an example, the engineering practicability of the proposed technology is verified and confirmed.

References

Brück, D., Elmqvist, H., Mattsson, S.E., Olsson, H.: Dymola for multi-engineering modeling and simulation. In DLR Deutsches Zentrum fur Luft- und Raumfahrt e.V.– Forschungsberichte, pp. 1–8. DLR, Cologne, Germany (Jan 2002). https://doi.org/10.13140/RG.2.1.2176.8162

Hountalas, D.T.: Prediction of marine diesel engine performance under fault conditions. Appl. Therm. Eng. **20**(18), 1753–1783 (2000). https://doi.org/10.1016/S1359-4311(00)00006-5

Joshi, A., Heimdahl, M.P.E., Miller, S.P., Whalen, M.W.: Model-based safety analysis. NASA Langley Research Center (1 Feb 2006). https://doi.org/10.1109/DASC.2005.1563469

Ren, Y., Qian, C., Yang, D., Feng, Q., Sun, B., Wang, Z.: Model-based reliability system engineering R&D process model. In: Ren, Y., Qian, C., Yang, D., Feng, Q., Sun, B., Wang, Z. (Eds.) Model-Based Reliability Systems Engineering, pp. 219–244. National Defense Industry Press (2024). https://doi.org/10.1007/978-981-99-0275-0_10

Zhang, J., Haskins, C., Liu, Y., Lundteigen, M.A.: A systems engineering-based approach for framing reliability, availability, and maintainability: a case study for subsea design. Syst. Eng. **21**(6), 523–538 (2018). https://doi.org/10.1002/sys.21462

Finite Element Analysis of the Strength of Connecting Bolts in a Horizontal Reactor

Weiwei Zhang(✉)

School of Mechanical Engineering, Shandong Huayu University of Technology, Dezhou 253034, China
zhang8412860@163.com

Abstract. A reactor is one of the equipment used in chemical and physical reactions of materials, and is a pressure vessel used in the chemical industry to complete processes such as mixing and reacting substances. Using 3D modeling software UG for geometric modeling, the strength of the connecting bolts of the horizontal reaction kettle was analyzed using finite element software. Nonlinear calculation methods were used to analyze the strength of the connecting bolts, and the calculation results of the two bolt hole diameters were compared and analyzed without changing the bolt specifications. Reducing the diameter of the bolt hole results in a more uniform load between the bolt and nut, and the bolt and nut meet the strength requirements.

Keywords: Reaction vessel · Finite element · Connecting bolt

1 Introduction

Horizontal reactors are one type among numerous types of reaction vessels, playing a vital role in mechanical, chemical, and other research fields (Peng and Wang 2018). The operation of reaction vessels varies with changes in the process, presenting relatively complex conditions (Li 2023). These vessels primarily rely on bolted connections. Considering the extremely complex forces acting on the connecting bolts during operation, under different working environments, the influence of preload and working pressure can easily lead to equipment overload, imposing higher demands on bolt strength. The adequacy of bolt strength directly impacts the safety and stability of machinery and equipment (Lu 2018). Therefore, analyzing the required strength of bolts is essential to prevent machinery hazards caused by bolt overload (Fu 2022).

Traditional mechanical analysis methods have limitations when addressing complex structures and multi-field coupled loading problems. In contrast, the finite element method enables precise simulation of component loading processes and corresponding material properties, effectively compensating for the shortcomings of conventional approaches (Ou and Huang 2022). Although current research has made progress in material modification, coating protection, and assembly processes, significant shortcomings remain. For instance, studies on strength evolution under multi-load coupling effects are insufficient. Existing findings predominantly focus on the impact of single loads on

T. Mankovits and N. Trisovic (Eds.): MDS 2026, LNME 2026.
https://doi.org/10.1007/978-981-92-2440-1_27

bolt strength, with limited in-depth analysis of the synergistic mechanisms involving preload, working pressure, and vibration (Ding 2022). Therefore, this study investigates horizontal reactor connection bolts using finite element simulation technology. Integrating recent advancements in material processing and stress behavior research, it constructs a multi-load-coupled strength analysis model. By combining simulation with theoretical verification, it performs refined strength validation, providing targeted theoretical support for bolt selection optimization and equipment safety operation. This holds significant engineering importance for addressing bottlenecks in bolt strength design under extreme operating conditions.

2 3D Modeling and Assembly of Horizontal Reaction Kettle

Use UG software to model the structure of various components of the reactor, including the reactor cylinder, lining, left and right heads, flanges, bolts, and nuts; Based on the assembly relationship and relative position between each component, select appropriate assembly constraints and establish a solid assembly model for the horizontal reaction vessel cylinder. The assembly model of the reaction kettle is shown in Fig. 1.

Fig. 1. Assembly model of reaction kettle.

3 Finite Element Analysis of Bolt Strength for Reactor Connection

3.1 Simplification of Finite Element Model

To shorten the solving time and reduce the solving cost, it is necessary to simplify the structure (Ren and Dong 2021). Ignore the washer between the nut and flange and omit the lining structure. Because the cylinder and flange are welded together, they are treated as rigid connections.

The bolt specification is M27, and ordinary bolts are used for connection. The size of the bolt hole is 1.1D. By applying load and boundary conditions to a single connecting bolt structure, the stress situation of the bolts in the overall structure can be simulated, thereby reflecting the strength of the overall connecting bolts. Simplify the model as shown in Fig. 2.

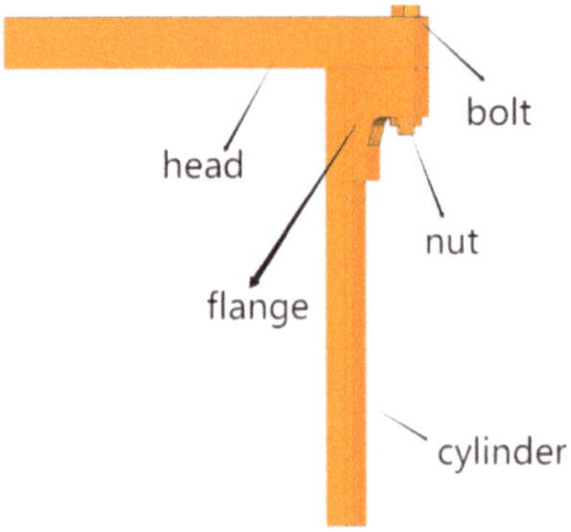

Fig. 2. Simplified model of single bolt connection structure.

3.2 Finite Element Analysis of Bolt Hole Diameter 1.1D

3.2.1 Finite Element Modeling

During finite element discretization, hexahedral elements provide high computational accuracy but require regular solid structures; irregular geometries complicate discretization (Zhao 2019, Liang 2021). Most reactor components feature symmetrical three-dimensional solids with identical front and rear shapes, resulting in relatively simple models. Octahedral eight-node elements were selected for the reactor's 3D model, employing the sweep method for discretization.

For the convenience of applying load, cutting treatment was carried out at the contact between the head and the lining. The final geometric model is shown in Fig. 3.

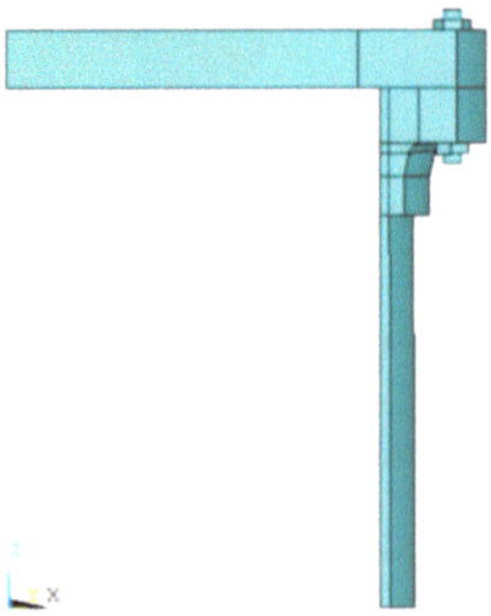

Fig. 3. Geometric model for strength analysis of connecting bolts.

The materials and material characteristics (Li and Yang 2022) of each component of the reactor are shown in Table 1.

Set the unit edge length of each component to obtain the corresponding number of nodes and units. The total number of nodes and units in the overall structure is 121699 and 120595, respectively. The grid division diagram is shown in Fig. 4.

Table 1. Material characteristics of the main components of the reactor.

Part	Material	Yield strength (σ_s/MPa)	Modulus of elasticity (E/MPa)	Poisson's ratio (μ)	Thermal conductivity ($W \cdot m^{-1} \cdot K^{-1}$)	Linear expansion coefficient ($\alpha_l/10^{-6}$°C^{-1})
Flanges, heads	Q345B	345	2.1×10^5	0.3	44.8	13
Cylinder	0Cr17Ni12Mo2	205	2.1×10^5	0.3	17.5	17
Bolt, nut	40MnB	785	2.1×10^5	0.3	44.8	13

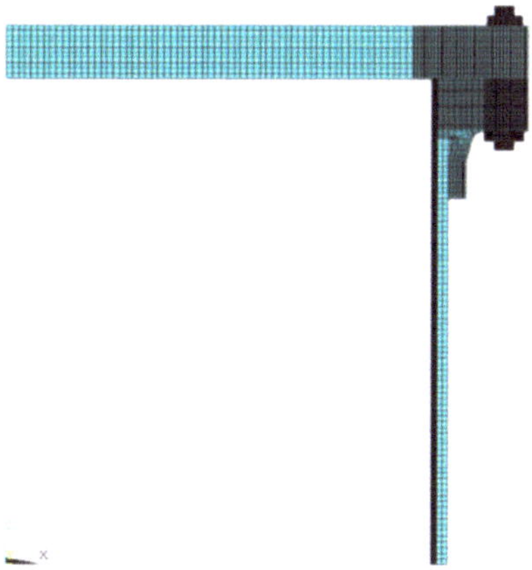

Fig. 4. Finite element mesh partition diagram for bolt strength.

3.2.2 Analysis of Calculation Results

The displacement and stress cloud map of the head are shown in Fig. 5, with a maximum displacement of 0.567 mm and a maximum stress of 147 MPa, which is less than the allowable stress.

The displacement and stress cloud map of the flange are shown in Fig. 6, with a maximum displacement of 0.0561 mm and a maximum stress of 238 MPa, which is greater than the allowable stress.

The displacement and stress cloud map of the cylinder are shown in Fig. 7, with a maximum displacement of 0.136 mm and a maximum stress of 112 MPa, which is less than the allowable stress.

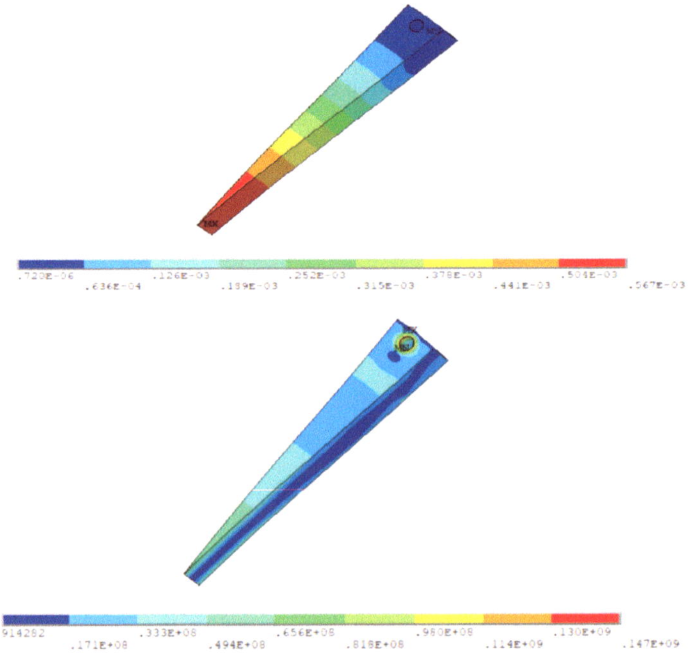

Fig. 5. Displacement and stress cloud map of the head.

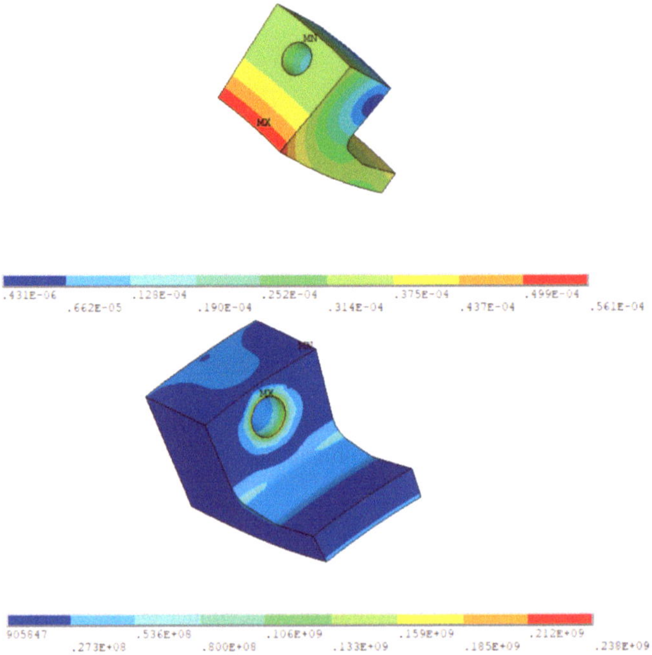

Fig. 6. Displacement and stress cloud map of flange.

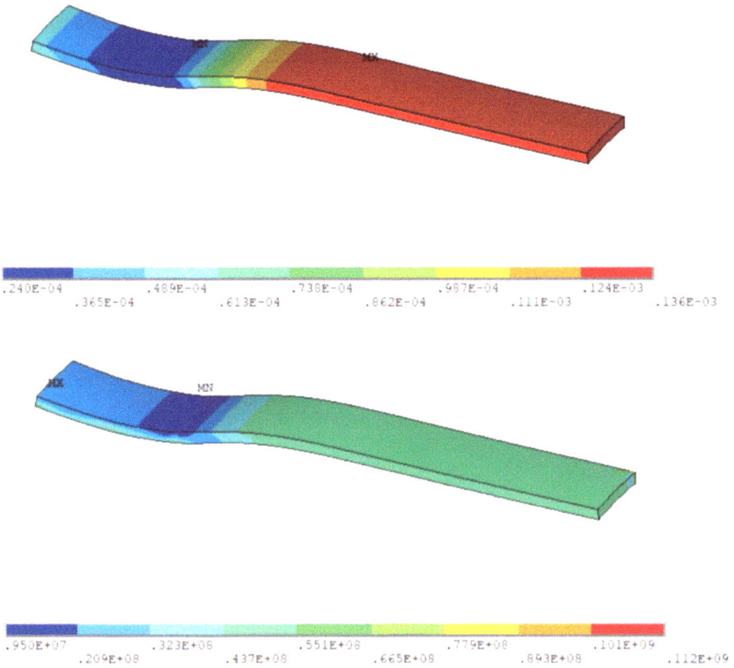

Fig. 7. Displacement and stress cloud map of the cylinder.

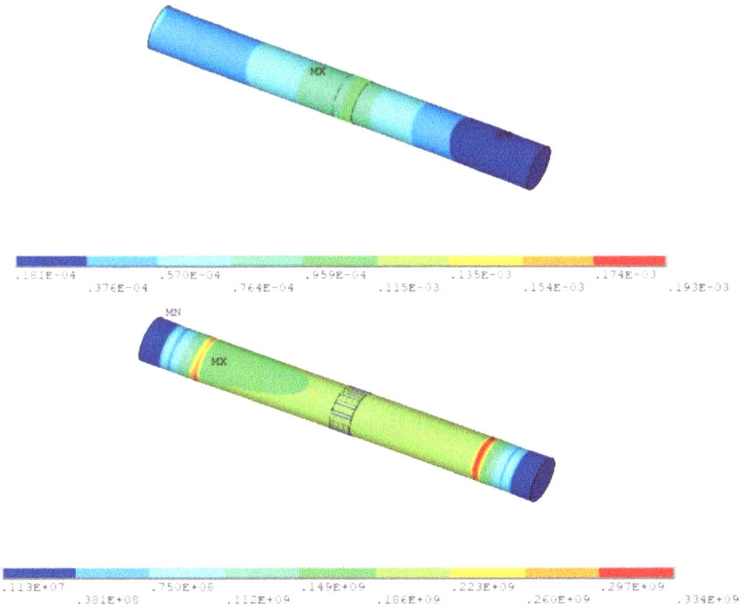

Fig. 8. Displacement and stress cloud map of bolt.

The cloud map of bolt displacement and stress is shown in Fig. 8, with a maximum displacement of 0.193 mm and a maximum stress of 334 MPa, which is less than the allowable stress.

The displacement and stress cloud map of the top nut are shown in Fig. 9, with a maximum displacement of 0.0491 mm and a maximum stress of 435 MPa, which is less than the allowable stress.

The displacement and stress cloud map of the lower nut are shown in Fig. 10, with a maximum displacement of 0.0312 mm and a maximum stress of 508 MPa, which is less than the allowable stress.

According to the calculation results, the stress on the bolt and nut does not match the actual situation. To reduce the stress on the nut, the bolt hole diameter is set to the same size as the bolt diameter without changing the bolt specifications.

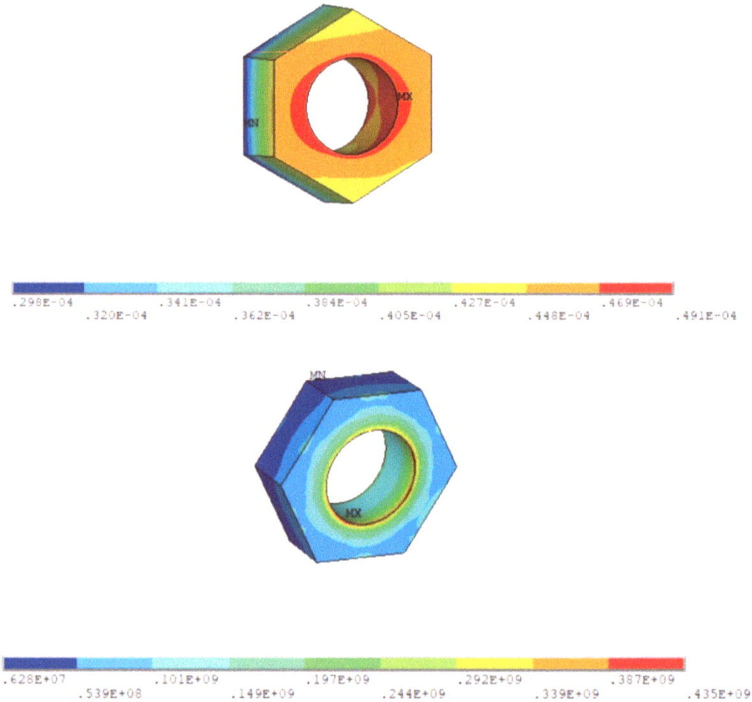

Fig. 9. Displacement and stress cloud map of top nut.

3.3 Finite Element Analysis of Bolt Hole Diameter and Bolt Diameter Under the Same Condition

3.3.1 Finite Element Modeling

The finite element modeling method is the same as the case where the bolt hole diameter is 1.1D, with the only difference being that the diameter of the bolt holes on the flange and head must be the same as that of the M27 bolt.

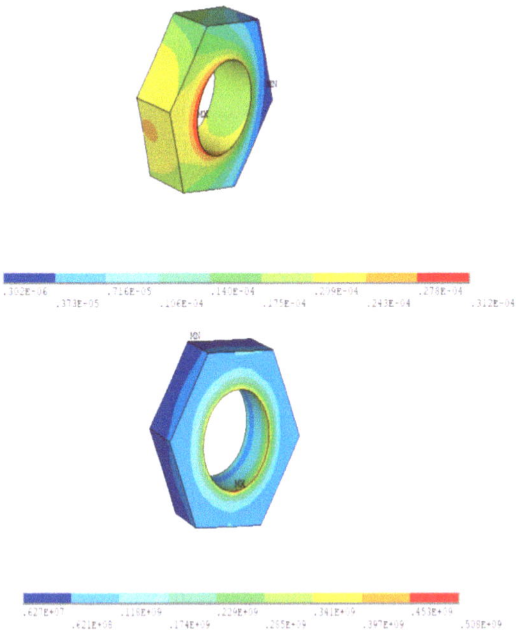

Fig. 10. Displacement and stress cloud map of lower nut.

3.3.2 Analysis of Calculation Results

According to calculations, the displacement and stress contour plots of each component, including the head, flange, shell, bolts, upper nut, and lower nut, exhibit slight variations. The detailed data changes for bolts, upper nut, and lower nut are presented in Table 2.

3.4 Comparative Analysis of the Strength of Connection Bolts with Two Different Bolt Hole Diameters

The maximum displacement and maximum stress of bolts and nuts under different bolt hole diameters are shown in Table 2.

Table 2. Maximum displacement and maximum stress under different bolt hole diameters.

Part	Maximum displacement at an aperture of 1.1D (mm)	Maximum displacement at aperture D (mm)	Maximum stress at an aperture of 1.1D (mm)	Maximum stress at aperture D (mm)
Top nut	0.0491	0.0453	435	358
Bolt	0.193	0.18	334	370
Lower nut	0.0312	0.0242	508	486

Compared to the case where the bolt hole diameter is 1.1D, when using the bolt hole diameter D processing method, the stress on the nut is reduced and the bolt stress is increased, which is in line with the actual situation. The displacement of the nut when the bolt hole diameter is D is smaller than that when the bolt hole diameter is 1.1D, as shown in Fig. 11, Reducing the bolt hole diameter reduces the curvature of the nut. Without changing the bolt specifications, the smaller the diameter of the bolt hole, the smaller the numerical difference between the bolt stress and the nut stress. The load distribution between the bolt and nut is uniform, which is beneficial for protecting the bolt nut structure.

Fig. 11. Diagram of nut warping.

According to comprehensive analysis, although the stress on the lower nut has decreased, the stress still exceeds that on the bolt. The stress on the lower nut and flange is relatively high. When considering connecting bolts, the gasket between the two is omitted, and the lower nut is in contact with the direct flange. The stress of bolt connection is mainly reflected in the pressure cone method, so adjusting the thickness and quantity of gaskets can improve the working stress and surface stress of flanges and nuts.

4 Conclusion

Based on the assembly relationship and part structure of the horizontal reaction kettle, this article applied UG for 3D modeling, simplified the parts as necessary, established a finite element model, and focused on the finite element analysis of the connecting bolts. According to the calculation results of the strength of the connecting bolts of the reaction kettle, the maximum stress of the bolts is 370 MPa, and the maximum stress of the nuts is 486 MPa. The bolts and nuts meet the strength requirements.

References

Ding, J.B.: Finite element simulation analysis of stirring performance in reactors and application of cfd-assisted design. Sci. Technol. Innov. (026), 68–71 (2022). https://kns.cnki.net/kcms2/article/abstract?v=W-y7fKBVJCzsWXv2ORtI3Gxehqc2lHJ8frIpljtzg6g9_1-5i3LWXvScOT9o_h0gvfDd3Z3hQPuwtSYefYnvR1heiYuK_7DA9aENuHQfbSRhXWhIufvKJqB86Bg9gk_HyTDKVmuvSJlkS-0YNgfAUcOO3DLZ4R65yrLbGxEXWDk=&uniplatform=NZKPT

Fu, J.: Research and design of the reactor system for hydrogen production via aluminum reduction. Wuhan Inst. Technol. (2022). https://doi.org/10.27727/d.cnki.gwhxc.2022.000067

Liang, K.Y.: Study on internal flow characteristics and fluid-solid coupling in axial-flow stirred reactors. Lanzhou Univ. Technol. (2021). https://doi.org/10.27206/d.cnki.ggsgu.2021.000842

Li, G.: Analysis and optimization of sealing failure in chemical reaction vessels. Equip. Manag. Maint. (21) (2023). https://doi.org/10.16621/j.cnki.issn1001-0599.2023.11.39

Li, L., Yang, F.Q.: Course design of mechanical design. Chemical Press (2022). https://thinker.cnki.net/bookstore/book/bookdetail?bookcode=9787122418425000&type=book

Lu, H.J.: Finite element analysis of bolt connections in assemblies based on Nastran. Mach. Manuf. Autom. **37**(6), 3 (2018). https://kns.cnki.net/kcms2/article/abstract?v=W-y7fKBVJCxl-bMuP-ceSp9NwluUIcTJ5gWUlMaprkUG2emPkhYp70WwMlpa4ZjWL3lTzWOD53nWwIk1hIG24p-ADi6HHZ3rD5kYCi_qy10mFtnRONwv2DU-KOAo2ewhUDzAopKJuAVBn6yfgpFP0ee_BmlY0fgkn&uniplatform=NZKPT

Ou, Y.C., Huang, J.H.: Development of a finite element analysis and design program for large-scale horizontal pressurized leaching reactors in hydrometallurgy. Nonferrous Equip. **36**(4), 5 (2022). https://doi.org/10.19611/j.cnki.cn11-2919/tg.2022.04.009

Peng, F., Wang, W.T.: Three-dimensional finite element analysis of the manhole structure in a reactor. Inn. Mong. Petrochem. Ind. **44**(01), 47–48 (2018). https://kns.cnki.net/kcms2/article/abstract?v=Wy7fKBVJCwNYJHar1NHY4M6e0p0GSWUazjeOx7vCqeZGeBp5lVpmIF5mXVLtXEpV0O3gSQ3W6qa1xlcUloLoI9hZVoR1dzaLv7Hatvo0J_pSQ2V7mlXkK2oNWWV5qeEpn2ar2j_i1j9Z7wNo08E-mfc0Rki_4HdqnCBCxZCHWw=&uniplatform=NZKPT

Ren, Z.Q., Dong, J.S.: Finite element analysis of stress and fatigue in a butadiene latex jacketed reactor. Petrochem. Equip. **50**(4), 8 (2021). https://kns.cnki.net/kcms2/article/abstract?v=W-y7fKBVJCzvn5med2WUlWq-R9hyHildMWt8eUNRCdHGQfTaxQUBLBm0eQwpKBRU2Qj-eHRnf6DnVzbb8jAFALJKBLmnfICab8YQXHIOlF8V3l71645uMYvqh0sjY5uPzRo0z9z8NeUhr_de35OdJLrg3zG0Q2qNUMO78HdOmK8=&uniplatform=NZKPT

Zhao, G.: Finite element analysis and modification of jacketed reactors with expansion joints. Pet. Chem. Equip. **22**(5), 5 (2019). https://kns.cnki.net/kcms2/article/abstract?v=W-y7fKBVJCzAZu1lWcwGBx0Nu1Ju3yP2Q431foKBoynM1dO_ARTN76bxjS4bQ-9U7Y1gjP5bCk5MOht8mZ99MMeZhZMezl7ttGocJSMCAsybh8hkC0sHrLe9AUJ7Dt3W9wS_SvHzt-6yvclYW7TLvAMHZThgscXSeNVBO8D6ziE=&uniplatform=NZKPT

Simulation and Mechanism Exploration of the Rolling Process for Self-Lubricating Spherical Plain Bearings

Wenqi Peng and Baofu Li(✉)

School of Mechatronic Engineering and Automation, Shanghai University, Shanghai, China
libf6688@163.com

Abstract. The rolling process of self-lubricating joint bearings is currently the core technology for controlling the no-load starting torque. This paper takes the GE25 type self-lubricating joint bearing as the research object and uses finite element numerical simulation technology to conduct simulation analysis of its rolling process, exploring the plastic deformation mechanism of the bearing rolling, and clarifying the law of the outer ring's plastic deformation. This research provides certain theoretical support for the optimization of the rolling process of self-lubricating joint bearings and the precise control of no-load starting torque.

Keywords: Self-lubricating spherical plain bearing · Rolling process · Finite element analysis · Clearance control

1 Introduction

An aerospace self-lubricating spherical plain bearing is a bearing in which the contact surfaces of the ball and race are both spherical. A self-lubricating liner is bonded to the inner spherical surface of the race, provided with maintenance-free quality. Sliding friction generated by the contact between the liner and the outer surface of the bearing ball replaces the direct metal-to-metal friction between the ball and race (Huang 2024, Zhong et al. 2021). Owing to its compact structure, lightweight design, excellent impact resistance, and maintenance-free characteristics, the self-lubricating spherical plain bearing has been widely used in high-end technical fields such as aerospace, wind power generation, and the nuclear industry, providing key support for innovation and development in modern equipment manufacturing (Wang et al. 2025a, b, Xue et al. 2018, Zhu et al. 2020).

However, forming processes for this type of bearing face multiple technical challenges. The race must be assembled with the ball via precision forming; after forming, it is necessary not only to strictly control the clearance and breakaway torque and to ensure high accuracy of the inner spherical surface of the race, but also to prevent damage to the self-lubricating liner during the forming process (Wang et al. 2022, Xue and Kang 2025). At present, the mainstream domestic key forming process is extrusion forming (Chen et al. 2013). Semi-finished products prepared by this process commonly exhibit

T. Mankovits and N. Trisovic (Eds.): MDS 2026, LNME 2026.
https://doi.org/10.1007/978-981-92-2440-1_28

excessively high breakaway torque, and therefore a torque adjustment process is required for improvement. Specifically, a rolling process (Zou et al. 2015) is used to optimize the contact characteristics between the ball and race so that the liner surface forms uniform contact with the spherical surface, thereby reducing the no-load breakaway torque and improving the stress distribution within the liner layer.

Currently, some scholars and researchers have investigated breakaway-torque adjustment techniques for self-lubricating spherical plain bearings. Niu et al. (2019) studied the influence of assembly processes on breakaway torque from the perspective of bearing assembly; Liu et al. (2015) achieved breakaway torque meeting design requirements by precisely controlling the extrusion process; Zou et al. (2015), taking an aluminum-alloy spherical plain bearing as the object, studied a three-roller rolling method for torque adjustment; Zhu et al. (2020) performed numerical simulation of the rolling process for an integral self-lubricating spherical plain bearing and simply calculated the no-load breakaway torque after rolling. Although these studies provide certain methods for breakaway torque control, further in-depth research is needed on the plastic deformation behavior and the mechanism of clearance evolution during the bearing rolling process.

In this work, the GE25 bearing after race extrusion forming is taken as the research object. ANSYS LS-DYNA is used to simulate the rolling process of a self-lubricating spherical plain bearing. By analyzing the deformation of the bearing race and the internal metal flow, the evolution law of the clearance between the ball and race during rolling is obtained, revealing the deformation mechanism of rolling.

2 Finite Element Simulation of the Rolling Process for a Self-Lubricating Spherical Plain Bearing

2.1 Rolling Process

The process model for clearance adjustment by rolling is shown in Fig. 1. The workpiece is an assembled self-lubricating spherical plain bearing after fitting and forming. As shown in Fig. 1, the bearing is held by three rollers: one is the loading roller, which provides the rolling force; one is the driving roller, which drives the workpiece and the other two rollers to rotate. The rolling process is the process in which the rollers rotate while simultaneously compressing (rolling) the workpiece.

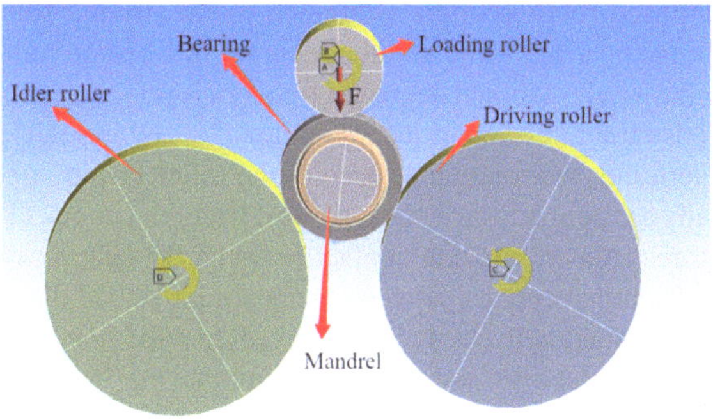

Fig. 1. Three-roller clearance-adjustment rolling process model.

2.2 Finite Element Model

Taking a GE25 self-lubricating spherical plain bearing as an example, the rolling process of the bearing is simulated and a rolling model is established in ANSYS. Considering that the rolling model is symmetric about the axial mid-plane of the bearing, a 1/2 geometric model is adopted for analysis.

All bearing components are meshed using 8-node single-point integration SOLID164 hexahedral elements. The three rollers are treated as rigid bodies and meshed using rigid elements.

Because rolling is conducted at room temperature and the temperature rise during processing is not significant, material properties in the range of 20–100 °C are used. The constitutive behavior of both the ball and race is based on a bilinear isotropic hardening model. The material properties used in the simulation are listed in Table 1.

Table 1. Material properties used in the simulation

Component	Material	Density ($kg \cdot m^{-3}$)	Elastic modulus (GPa)	Poisson's ratio	Yield strength (MPa)	Hardening modulus (GPa)
Race	05Cr17Ni4Cu4Nb	7750	196	0.288	960	20
Ball	G95Cr18	7900	214	0.264	1900	15
Liner	Domestic liner	2200	2	0.49	–	–

In actual rolling, the bearing is usually placed on two rollers, and pressure is applied by a loading roller directly above the bearing. One roller serves as the driving roller; it rotates to drive the bearing and the other two rollers. The rotational speed of the bearing can be controlled by setting the angular rotation of the driving roller. Considering that the three rollers are evenly distributed around the bearing at 120°, to accurately simulate the displacement and relative position relationships in the process, the following boundary constraints are applied to the three rollers and the bearing in the simulation model:

Local coordinate systems are established at the centers of the three rollers. For the loading roller, only the axial rotational degree of freedom and the radial translational degree of freedom toward the bearing are released; all other degrees of freedom are fixed. For the driving roller and the idler roller, only the axial rotational degree of freedom is released; an imposed rotation angle is applied to the driving roller to control the rotational speed of the bearing, while all other degrees of freedom remain fixed. For the bearing ball, only the axial rotational degree of freedom is released, and all other degrees of freedom are constrained. The workpiece rotational speed is selected as the process parameter that determines the rotational speed of the roller system.

Six contact pairs are defined: surface-to-surface frictional contact between the bearing race and the three rollers; bonding contact between the bearing race and the liner; surface-to-surface frictional contact between the bearing ball and the liner; and surface-to-surface frictional contact between the bearing ball and the mandrel. An interference fit is introduced between the liner and the ball to simulate the "locked" condition of the ball and race before rolling.

The procedure of loading roll force is shown in Fig. 2. It consists of three stages: linear ramp loading, constant-force holding, and unloading. Four parameters, as shown in Fig. 2, are used to control the loading profile.

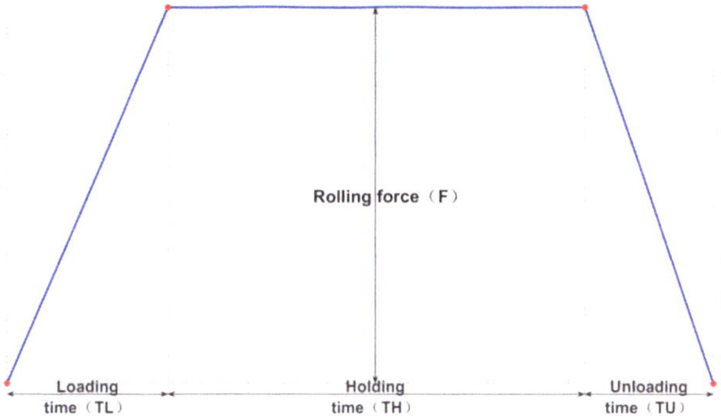

Fig. 2. Load procedure.

3 Characteristics of Stress and Strain Distribution During Rolling

The rolling results discussed below correspond to the instantaneous state at a rolling force of 36 kN and a rolling speed of 24 rev/s at a holding time of 0.24 s. As shown in Fig. 3, the maximum pressure on the outer surface of the bearing race occurs at the three locations where are in contact with the rollers. The pressure decreases continuously from the middle of the outer surface toward the two side faces and toward both sides. This can greatly reduce the damage to the liner. As liner damage is one of the main factors affecting bearing service life, a suitable rolling force should be chosen during the rolling process to reduce damage to the liner.

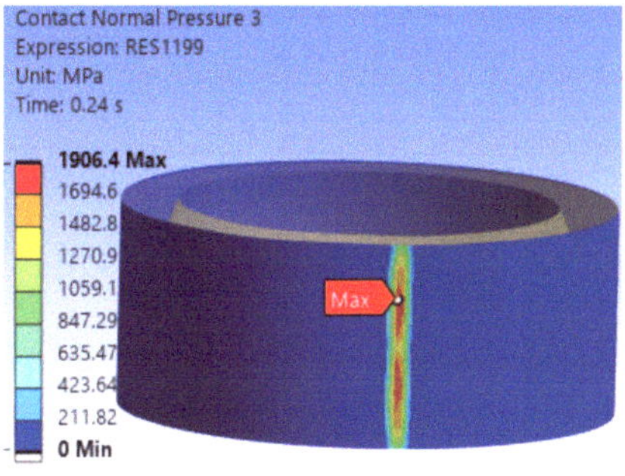

Fig. 3. Contact pressure distribution on the outer surface of the race.

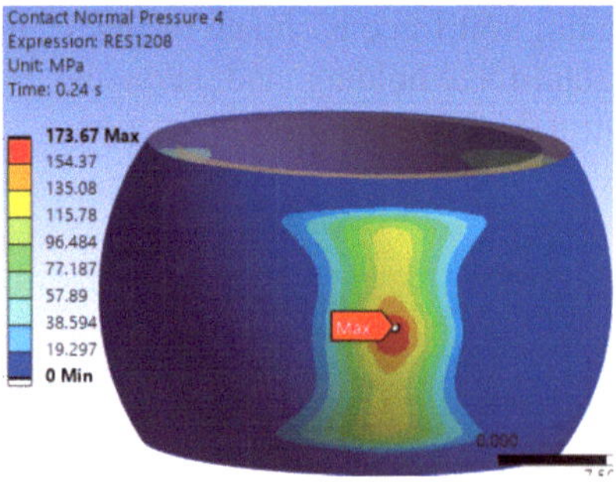

Fig. 4. Contact pressure distribution on the outer spherical surface of the ball.

As shown in Fig. 4, on the contact surface between the inner surface of the bearing race and the outer surface of the liner, the contact pressure reaches its maximum at the center and decreases from the center toward both sides. Toward the two side faces, the pressure first decreases and then increases, which differs from the pressure distribution on the outer surface of the race. From the normal contact pressure, it can be inferred that after the rolling force is transmitted through the race to the ball, the contact pressure is reduced by more than half.

Figure 5 shows the loaded region of the bearing race and the stress distribution on an axial cut section. Near the axial mid-plane region of the race, the stress fully penetrates the section, indicating good forging penetration in this region during rolling. However, near the side faces, due to greater thickness, there exists a local region whose equivalent stress remains lower than the surrounding area, suggesting that this portion may not be fully penetrated. Figure 6 shows the sectional stress distribution at the mid-plane of the bearing race. When the rolling force reaches 36 kN, within the contact area between the small roller and the bearing race, the plastic zone in which the equivalent (von Mises) stress reaches the yield limit has fully penetrated the mid-plane of the bearing race.

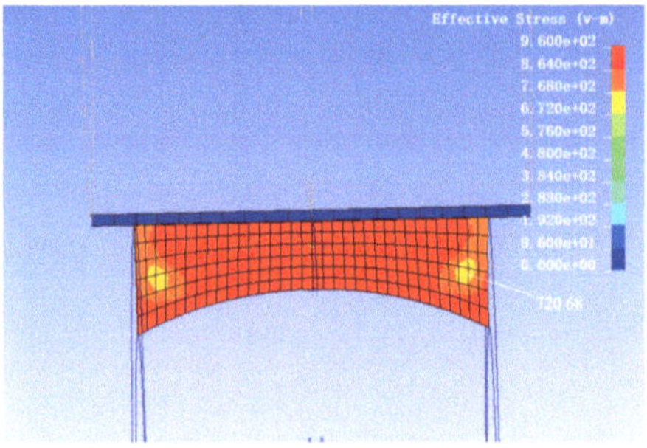

Fig. 5. Sectional stress distribution at the loaded region of the bearing race.

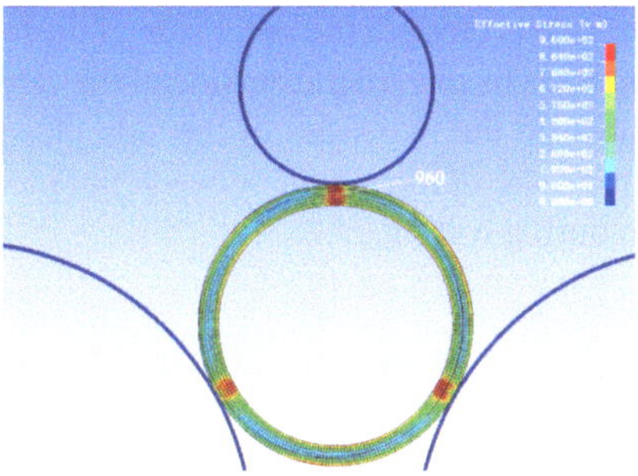

Fig. 6. Sectional stress distribution at the mid-plane of the bearing race.

4 Deformation Analysis During Rolling

4.1 Deformation of the Bearing Outer Race

After superposition, specific evolution patterns are presented. These can be clarified by analyzing the morphological changes of the outer circle cross-section and end face of the bearing, as well as the strain, curvature radius and axial deformation characteristics of the inner and outer surfaces. The following analysis results of the relevant deformations of the bearing outer circle are all based on the simulation data obtained from the GE25 bearing under the conditions of rolling force of 36 kN, rolling speed of 24 r/s, and 8 rolling laps.

4.1.1 Deformation of an Axial Section of the Bearing Race

First, the contour change of a race section passing through the axis (i.e., an axial section) before and after rolling is examined, as shown in Fig. 7. Comparing the contours before and after deformation indicates that the bearing is elongated in the axial direction and thinned in the radial thickness direction. In addition, the original spherical profile of the inner surface of the race is deformed to a certain extent.

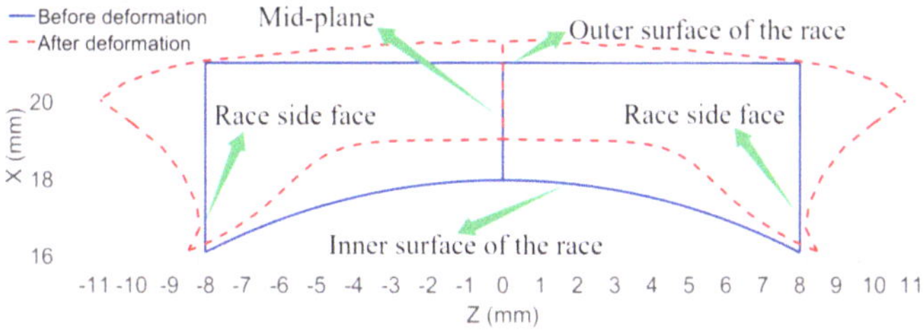

Fig. 7. Axial sectional contour of the race (deformation amplified 50×).

4.1.2 Side Face Deformation of the Bearing Race

The side faces of the bearing race are two symmetric annular surfaces. Figure 8 shows the contour of one side face after rolling simulation. After deformation, a "hook-shaped" profile appears: the higher side is close to the outer side of the race, the shorter side is on the inner side, and the notch opens toward the outside of the side face. This phenomenon arises because the thickness on the two sides of the side face is different, and there is a significant difference in plastic deformation between the inner and outer sides in the radial thickness direction during rolling.

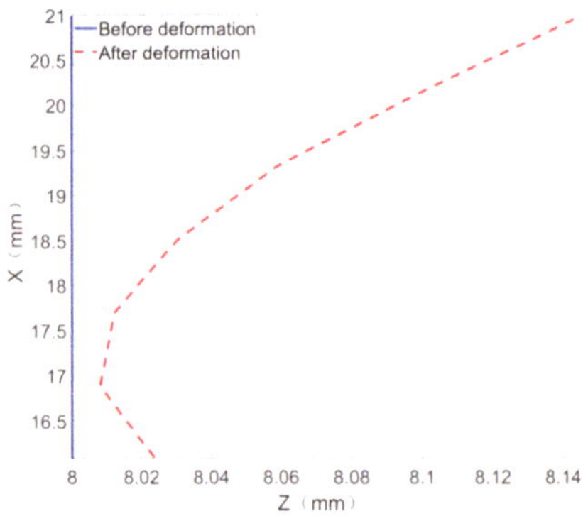

Fig. 8. Side face contour line of the race after rolling simulation.

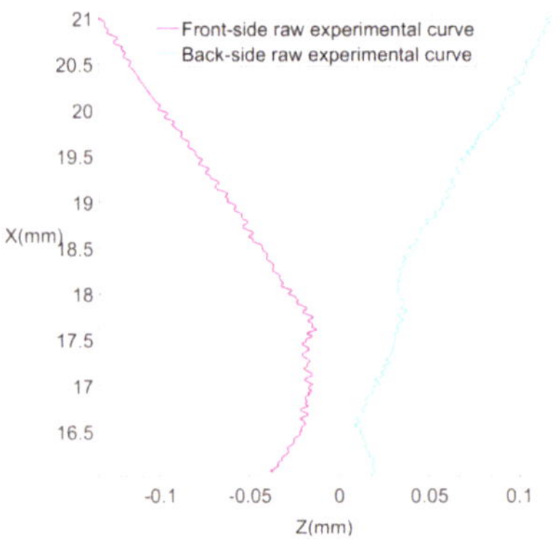

Fig. 9. Profiles measured of side faces of the race.

Figure 9 shows the contour results measured by a profilometer for the two side faces of a GE25 bearing after rolling under experimental conditions consistent with the simulation. It should be noted that the side with the smaller axial coordinate value is close to the ball, and the notch opens toward the outside of the bearing side face. The "hook-shaped" morphology can be clearly observed: the higher side is close to the outer side of the bearing race, the shorter side is on the inner side, and the hook-shaped notch also opens toward the outside of the race side face, indicating good consistency between the simulation model and the actual process phenomenon.

4.1.3 Deformation of the Inner and Outer Surfaces of the Bearing Race

As shown in Fig. 7, the outer ring is generally in a tensile state along the axial direction. The metal flow direction on its inner and outer surfaces can be determined by combining the axial strain distribution of the outer surface at the same axial height as the bearing outer ring in Fig. 10 and the axial strain distribution along the axial direction of the inner surface: The axial strain on the outer surface first increases and then decreases, reaching the peak between the mid-section and the end face (z-5-6 mm), indicating that the axial deformation resistance in this area is relatively small, and the axial deformation is greater while the radial deformation is smaller; At the end face of the bearing, because of the reduction of radial-directional restraining force, the resistance causing the radial deformation of the end face will be significantly reduced. However, it will cause a slight decrease in the axial strain of the bearing. But the overall result of the curves in the figure is all greater than zero, which also indicates that the metal on the outer surface of the bearing will flow towards the end face during deformation. Eventually, the bearing will undergo an axial elongation change. The strain on the inner surface follows a "first decreasing then increasing" pattern: The strain is the smallest and negative (axial compression) in the middle section and the area between the end face (2–4.8 mm), and the metal flows axially inward; The strains in the areas near the mid-section and the end face are both greater than zero (axial tension), and the metal flows axially outward; Near the end face of the bearing, the strain will rapidly recover. This is mainly because the end face constraint is relatively small, and the resistance to axial deformation is low. At the end face, the radial deformation resistance is relatively small, and the radial deformation is significant, but the axial strain will reduce.

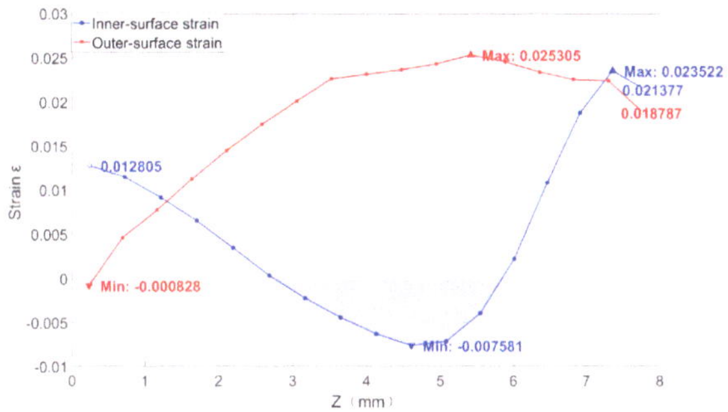

Fig. 10. Strain distributions on the outer and inner surfaces of the race

Figure 11 shows the distribution of the curvature radius of the inner surface axial section of the bearing outer ring. The overall curvature radius of the inner surface has increased compared to the original state, but the curvature distribution from the middle section to the end face is non-linear (increasing first and then decreasing), reflecting the non-uniformity of the axial deformation during rolling. This will cause fluctuations in the bearing's starting torque. There is a maximum curvature radius between the end face and the middle section, and the corresponding area has the largest gap.

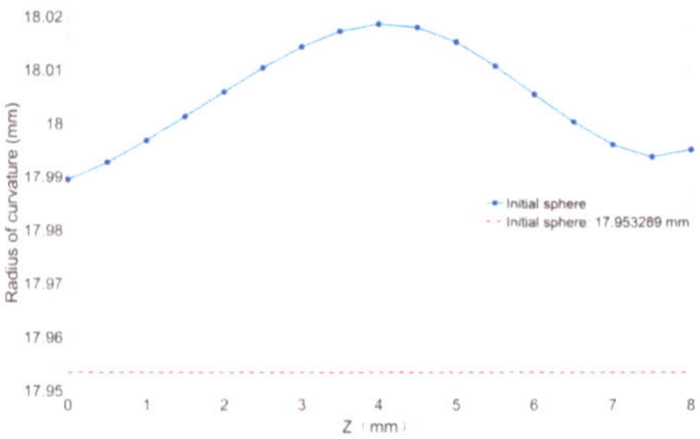

Fig. 11. Radius of curvature distribution of the inner-surface axial sectional profile.

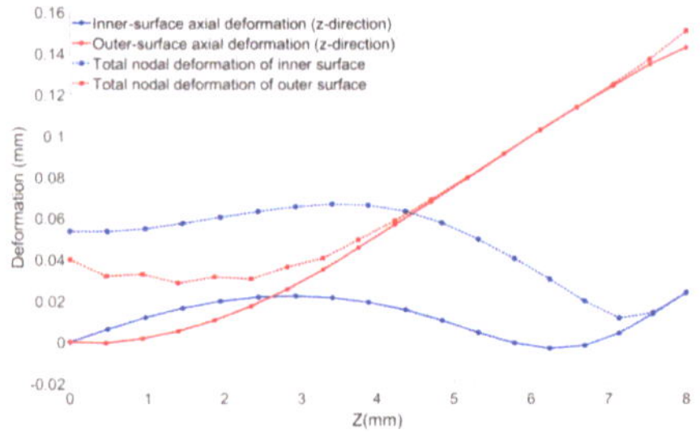

Fig. 12. Axial deformation and total deformation of the inner and outer surfaces of the race.

Figure 12 shows the axial deformation and total deformation of the outer surface (calculated based on the difference in z-direction displacement before and after deformation and the total displacement difference). The axial deformation of the outer surface and the total deformation show an overall increasing trend from the middle section to the end face. This pattern belongs to the cumulative effect of metal flow: the closer to the end face, the greater the axial deformation of the outer surface. The metal of the outer ring flows axially towards the outer side of the end face, and the axial deformation at the end face of the bearing is the greatest. At the middle section position, the radial deformation is large and the axial direction is in a compressed state.

The degree of thinning is the most significant. As the z coordinate increases, the radial deformation gradually decreases while the axial deformation gradually increases. At the end face, due to the reduction of radial deformation resistance, the radial deformation increases. The axial deformation of the inner surface shows a fluctuating trend of "increase–decrease–increase", with the maximum deformation at the end face: within the region of increasing deformation, the metal flows towards the end face, and the outer ring is locally stretched; in the region of decreasing deformation, it shows an axial compression trend.

4.2 Characteristics of Clearance Change Between Ball and Race

Figure 13 below shows the variation process of the bearing clearance during the rolling process. At the beginning, the clearances of the bearings were basically the same. During the rolling process, the outer ring of the bearings would become thinner, the clearance in the contact area of the rolls would decrease, but the clearance in the non-contact area would increase. Eventually, the clearance became larger.

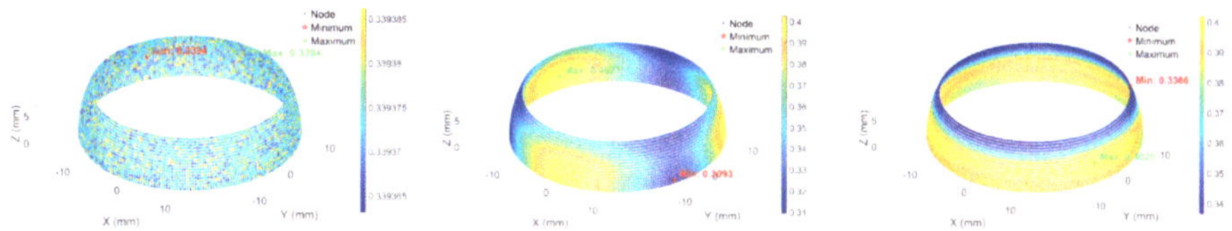

Fig. 13. Variation distribution diagram of clearance between ball and races during rolling process (including liner thickness).

Under the same rolling conditions, the axial tangential clearance distributions corresponding to different rolling speeds are shown in Fig. 14. The axial clearance of the bearing presents an overall M-shaped distribution with the minimum clearance at both ends and two peaks between the mid-section and end faces, and this distribution characteristic is consistent with the deformation law of the inner surface profile of the outer race. Metal at the end faces flows axially and radially, and radial flow squeezes toward the bearing interior, resulting in a smaller clearance in this region; it can be seen from the figure that a higher rolling speed leads to a larger clearance at the same time.

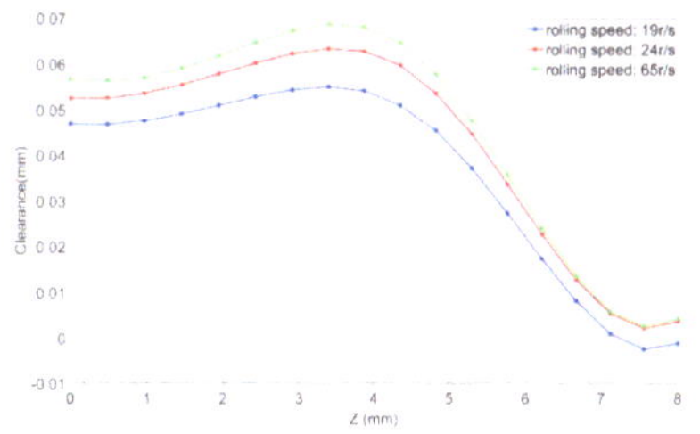

Fig. 14. Axial tangential clearance distribution.

In Fig. 15, each curve represents the gap data between the inner and outer rings of the bearing obtained through simulation when using the GE25 bearing in ANSYS, with the same rolling time and the same rolling force maintained, but with different numbers of rolling laps. Each data point in the figure corresponds to a set of gaps and rolling laps. Through these changes in Fig. 15, the variations of the inner and outer gaps of the bearing during the rolling process in reality can be simulated. By comparing the different curves in the figure, it can be found that the lower the rolling force, the more rolling laps are needed to achieve the required target gap. The greater the rolling force, the more obvious the increase in the gap between the inner and outer rings. However, when the gap increases to a certain extent, the increase in the number of rolling laps has a smaller and smaller impact on the gap, and finally reaches a stable value. Under different rolling forces, the smaller the rolling force, the more rolling laps are needed to achieve the same gap; this indicates that in the actual process, the number of rolling laps can be adjusted to control the gap between the inner and outer rings of the bearing.

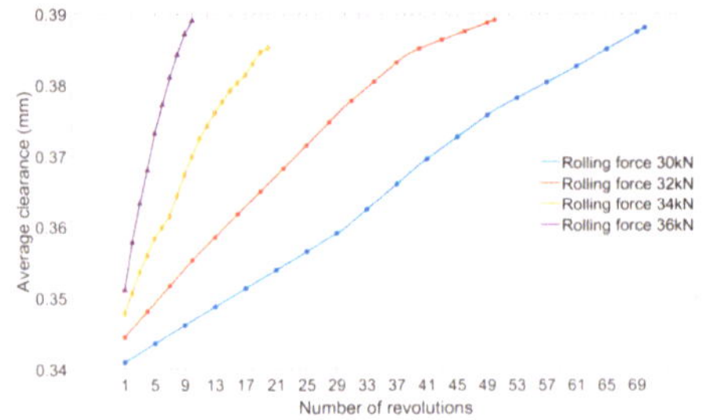

Fig. 15. Clearance evolution between ball and race during rolling.

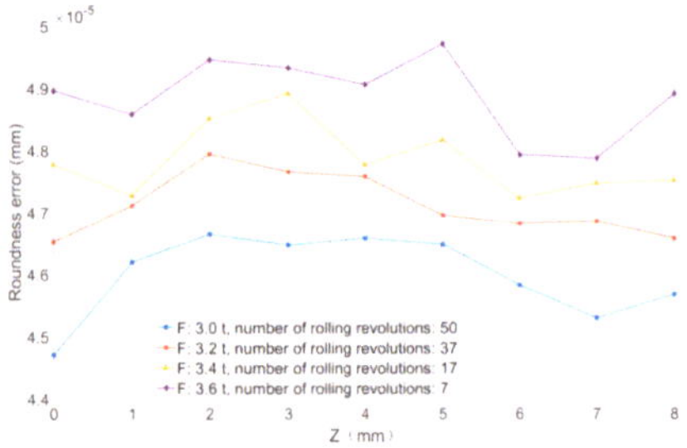

Fig. 16. Circumferential roundness error distribution on the inner surface of the race.

Figure 16 shows the circumferential roundness error distribution of the inner surface of the race when the same bearing type is rolled to a similar clearance using the same rolling time but different rolling forces and revolutions. The comparison indicates that although a lower rolling force requires more revolutions, it improves the uniformity of the roundness error on the inner surface of the race, which is beneficial for improving bearing breakaway torque. Combined with Fig. 14, increasing rolling force can reduce the required number of revolutions, but too few revolutions reduce clearance control accuracy and adversely affect the stability of breakaway torque. In addition, a higher rolling force causes the bearing to enter plastic deformation earlier during loading and to continue plastic deformation in the early stage of unloading, aggravating non-uniformity on the inner surface of the race (manifested as increased roundness error), ultimately leading to breakaway torque fluctuations.

5 Conclusions

By numerical simulation of the rolling process for the GE25 self-lubricating spherical plain bearing, the deformation form of the race and the evolution of bearing clearance are studied in this paper.

During rolling, plastic deformation of the bearing race is of significant directional anisotropy, and the metal flows in both axial and radial directions. The inner surface extends and its radius increases; the radial thickness decreases; axially, the race is overall stretched toward both side faces, resulting in increased length. Meanwhile, the side faces of the race form a "hook-shaped" contour in which the notch is located near the ball and opens outward in the axial direction, showing high agreement with the actual process morphology. The increment of bearing clearance is not uniform, with an "M-shaped" distribution in the axial direction: two peaks appear between the mid-plane and side faces, while the minimum values are located near the side faces. The change in the curvature radius of the inner spherical face is closely related to the increase in bearing clearance. The increment of the curvature radius reaches its maximum value at the peak of the bearing clearance increment.

A "low rolling force + more rolling revolutions" strategy can balance clearance control accuracy and bearing motion stability. This strategy improves clearance control precision, and because the number of revolutions increases, the proportion of rotation angle exceeding the yield limit of the race during loading and unloading decreases, mitigating adverse effects on surface uniformity. It provides a reliable process route for stable clearance control and offers direct technical support for parameter optimization of the rolling process for self-lubricating spherical plain bearings.

References

Chen, J., Yang, Y., Xia, Z., Cui, L.: Current status and development trends of forming technology for integral outer-ring spherical plain bearings. Bearing (9), 58–62 (2013). https://doi.org/10.19533/j.issn1000-3762.2013.09.019

Huang, X.: Friction and wear characteristics of self-lubricating spherical plain bearings under different stresses. J. Aeronaut. Mater. **44**(3), 132–141 (2024). https://kns.cnki.net/kcms2/article/abstract?v=OGEEvuKkrhZS8ShB3j946AOfeDVLTsmsi3ENtt0HDITQocl2MupLn35dV1cyMcqj-hb3r0G_RMrCG1CvQKln5eoAQDGuDMva1nF9QWOnhUugM4ZrPAZMYvygpfxw-MVnm59EaNkdHTPMwvl_PvcNAKPqTHohfz8w401MOxqchkU=&uniplatform=NZKPT&language=CHS

Liu, D., Liu, S., Fei, P., Cai, Y.: Precision extrusion fitting technology for self-lubricating spherical plain bearings. Bearing (4), 28–31 (2015). https://doi.org/10.19533/j.issn1000-3762.2015.04.007

Niu, R., Zhang, J., Dai, M., Deng, S.: Analysis of the impact of press-fitting on the starting friction torque of self-lubricating rod-end spherical plain bearings. Bearing (4), 47–51, 62 (2019). https://doi.org/10.19533/j.issn1000-3762.2019.04.012

Wang, Q., Chen, J., Zhou, H., Lyu, X., Hu, Z.: Research on the mathematical modelling of the starting torque of self-lubricating spherical plain bearings. Proc. Inst. Mech. Eng. C J. Mech. Eng. Sci. **236**(8), 4049–4058 (2022). https://doi.org/10.1177/09544062211047816

Wang, Q., Gu, R., Chen, B., Li, F., Lyu, X.: Study on the influence of the parameters of the roller swaging process on the no-load starting torque of self-lubricating spherical plain bearings. AIP Adv. **15**(1), 015312 (2025). https://doi.org/10.1063/5.0188368

Wang, Q., et al.: Influence of housing hole accuracy on the starting torque of self-lubricating spherical plain bearings. Proc. Inst. Mech. Eng. C J. Mech. Eng. Sci. **239**(6), 3697–3707 (2025). https://doi.org/10.1177/09544062241251768

Xue, Y., Chen, J., Guo, S., Meng, Q., Luo, J.: Finite element simulation and experimental test of the wear behavior for self-lubricating spherical plain bearings. Friction **6**(3), 297–306 (2018). https://doi.org/10.1007/s40544-018-0206-x
Xue, Y., Kang, D.: An analytical approach to incomplete spherical conformal contact: application on self-lubricating spherical plain bearings. Ind. Lubr. Tribol. **77**(1), 128–137 (2025). https://doi.org/10.1108/ILT-05-2024-0205
Zhong, X., Dou, H., Long, Y., Feng, L.: Study on the rolling technology of self-lubricating spherical plain bearings based on product characteristics. Bearing (2), 29–34, 41 (2021). https://doi.org/10.19533/j.issn1000-3762.2021.02.007
Zhu, L., Huang, X., Han, H.: Numerical simulation of the rolling process for integral self-lubricating spherical plain bearings. J. Plast. Eng. **27**(5), 155–160 (2020). https://link.cnki.net/urlid/11.3449.TG.20200526.1146.038
Zou, K., Li, B., Liu, H.: Finite element analysis of outer-ring rolling forming of aluminum-alloy spherical plain bearings. Bearing (7), 21–24 (2015). https://doi.org/10.19533/j.issn1000-3762.2015.07.006

Reliability Assessment of Machine Tools Based on Degradation Distribution and Bayesian Gamma Process

Hongyan Chu[1], Yi Wang[1](✉), Jinfeng Zhang[1], Yongsheng Zhao[1], Jingjing Xu[1], Meile Wang[2], and Taowen Xiao[2]

[1] Key Laboratory of Design and Intelligent Machining Technology for High Precision Machine Tools, Beijing University of Technology, Beijing 100124, China
yiw96699@gmail.com
[2] Jiangsu Hengli Precision Industry Co., Ltd, Changzhou 213100, China

Abstract. A Weibull life model, founded on degradation distributions and empirical failure probabilities, is employed to analyze the reliability of the degradation data. Furthermore, to address the limitations of the Weibull method when some data are far from the failure threshold, a Bayesian Gamma model based on degradation rate statistics is introduced. The degradation process is modeled via Bayesian inference and MCMC sampling, yielding the system lifetime distribution and reliability function. Case study results indicate that the latter approach can provide robust assessment results with high tolerance to limited sample data, offering an effective analytical framework for the reliability evaluation of critical functional components of CNC machine tools.

Keywords: Reliability assessment · Performance degradation · Weibull distribution · Bayesian · Gamma process

1 Introduction

CNC machine tools, core equipment in manufacturing, have reliability that directly affects machining accuracy, production efficiency, and manufacturing safety. During long-term operation, critical components such as spindles, linear guideways, and transmission subsystems inevitably experience performance degradation, leading to reduced accuracy and shortened service life. Therefore, reliability modeling and assessment are necessary to support preventive maintenance and design optimization. However, due to limited failure data, long component lifetimes, and low failure probabilities, traditional reliability methods relying on extensive failure data are difficult to apply.

To address machine tool reliability analysis under small-sample conditions, many degradation-data-based modeling approaches have been proposed. Liang (2020) established a reliability model for rolling linear guideways based on performance degradation data to achieve reliability prediction with limited data. Chen et al. (2025, 2023) applied Bayesian analysis combined with MCMC methods to infer the life distribution of spindle

T. Mankovits and N. Trisovic (Eds.): MDS 2026, LNME 2026.
https://doi.org/10.1007/978-981-92-2440-1_29

feed transmission subsystems, improving analysis accuracy under small-sample conditions. Ying (2023) predicted the life of motorized spindle bearings by integrating physical models and degradation data. Ran et al. (2025) proposed a meta-action-unit reliability assessment method that fuses lifetime information and multi-performance degradation information. Chen et al. (2022) pointed out that degradation-data-driven methods have become a research hotspot, but challenges such as insufficient sample size and parameter uncertainty still remain.

For the reliability assessment of critical components of a specific CNC machine tool, a Weibull life model based on degradation distributions and empirical failure probabilities is adopted under conventional data conditions. Furthermore, for cases in which degradation data are far from the failure threshold and the Weibull method becomes inaccurate, a Bayesian Gamma model based on degradation rate statistics is proposed. By constructing a likelihood function and employing MCMC sampling to obtain posterior distributions of parameters, the degradation process is simulated to derive the posterior life distribution and reliability curves, providing an analytical basis for preventive maintenance and performance optimization.

2 Reliability Assessment Method Based on Degradation Distribution and Empirical Failure Probability of Weibull Lifetime

2.1 Assessment Procedure

In component degradation tests, let the sample size be m and the number of measurements be n. At each measurement time $t_j(j = 1,2,\ldots,n)$, the corresponding degradation amount is $x(t_j)$. At any time t_j, it is assumed that the degradation amount $x(t_j)$ follows a normal distribution:

$$D(t) \sim N(\mu(t), \sigma^2(t)) \tag{1}$$

The corresponding probability density function is:

$$f_j(x; \mu_j, \sigma_j) = \frac{1}{\sqrt{2\pi}\sigma_j} \exp\left[\frac{(x-\mu_j)^2}{2\sigma_j^2}\right] \tag{2}$$

where σ_j^2 and μ_j denote the mean value and variance of the degradation amount at time t_j.

After obtaining the distribution characteristics of the degradation amount at each time, these results need to be transformed into reliability indicators. According to experimental requirements or industry standards, a degradation failure threshold L is specified. When the performance degradation amount exceeds this threshold, the component is regarded as failed. Then the reliability at time t_j can be expressed as:

$$R_j = \Phi\left(\frac{L - \widehat{\mu}_j}{\widehat{\sigma}_j}\right) \tag{3}$$

where $\Phi(\cdot)$ denotes the standard normal distribution function. Based on the assumption that failure occurs when performance degradation first exceeds the threshold, a mapping relationship between the degradation distribution and the lifetime distribution is established through the failure threshold. At each inspection time t_j, the failure probability that the component reaches the failure threshold is calculated from the degradation distribution as $F_j = 1 - R_j$, which can be regarded as an empirical estimate of the lifetime cumulative distribution function (CDF) at time t_j. Accordingly, a set of discrete time–failure probability data pairs (t_j, F_j) is obtained to represent the empirical CDF of the lifetime. On this basis, it is further assumed that the component lifetime T follows a Weibull distribution, and the Weibull parameters are estimated by fitting to the above empirical CDF data, thereby establishing the probability density function (PDF) in Eq. (4) and the corresponding reliability function in Eq. (5).

$$f(t) = \frac{\beta}{\lambda}\left(\frac{t}{\lambda}\right)^{\beta-1} \exp[-\left(\frac{t}{\lambda}\right)^{\beta}] \tag{4}$$

$$R(t) = \exp[-\left(\frac{t}{\lambda}\right)^{\beta}] \tag{5}$$

where β is the shape parameter and λ is the scale parameter. The parameters β and λ are estimated using the least squares method, and the reliability function is finally determined.

Accordingly, the reliability of the component at different times can be calculated based on the reliability function. At time t, the calculated reliability is $R(t)$. Furthermore, the mean lifetime of the component can be calculated as:

$$E[T] = \lambda\Gamma\left(1 + \frac{1}{\beta}\right) \tag{6}$$

where $\Gamma(\cdot)$ denotes the Gamma function. Considering only performance-parameter-induced failure, the mean lifetime of the component is $E[T]$.

2.2 Case Study

2.2.1 Conventional Case Study

The conventional case consists of four groups of Z-axis vibration displacement measurement data obtained from two samples of a linear guideway pair. The degradation data are listed in Table 1. Therefore, in this experiment, the sample size is $m = 2$, the number of measurements is $n = 4$, and the measurement times are $t_j(j = 1, 2, 3, 4)$. According to the industry average value, the failure threshold is set as $L = 2.0$.

Based on the procedure, the probability density plots corresponding to all measured time points can be obtained, as shown in Fig. 1, along with the reliability function curve, as shown in Fig. 2.

After calculation, the mean lifetime is $E[T] = 562.5$ h, and the reliability function is:

$$R(t) = e^{-\frac{t}{584.4}^{13.6}} \tag{7}$$

Table 1. Case data.

Operating time t (h)	Z-axis vibration displacement (Sample 1) (μm)	Z-axis vibration displacement (Sample 2) (μm)
0	1.9	1.8
250	1.9	2.0
600	2.0	2.1
1200	2.5	2.4

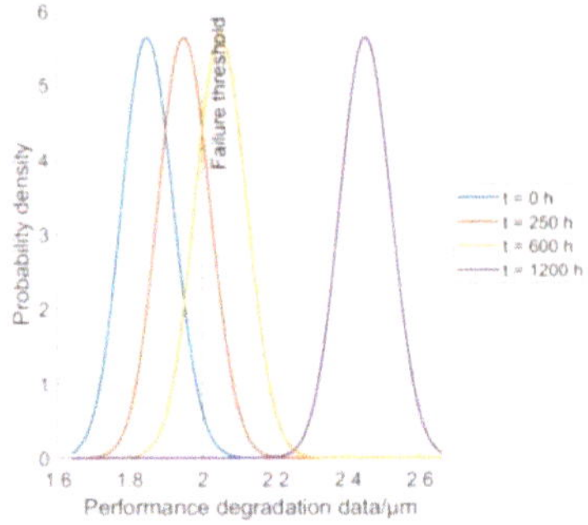

Fig. 1. Probability density.

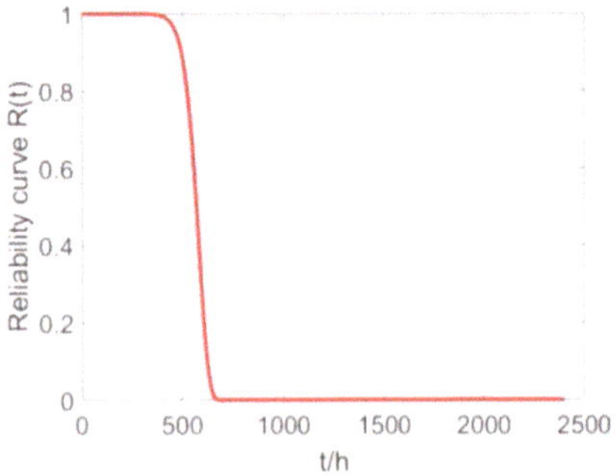

Fig. 2. Reliability curve based on Weibull distribution.

2.2.2 Case Study with Small Sample Size and Degradation Data Far from the Failure Threshold

In this special case, four Y-axis vibration displacement measurements of a guideway pair from a single sample are available. The degradation data are listed in Table 2. Therefore, in this experiment, the number of measurements is $n = 4$, and the measurement times are $t_j(j = 1, 2, 3, 4)$. According to the industry average value, the failure threshold is set as $L = 17.0$. The main characteristics of this dataset are that the sample size is one and all degradation data are far from the failure threshold.

Since the sample size is 1, the standard deviation for each data point needs to be manually set to a very small value of 0.05 μ. Using the same method, the failure probability density curves corresponding to all measurement times are obtained, as shown in Fig. 3. The resulting reliability curve is illustrated in Fig. 4. It can be observed that, because the degradation data are overall far from the failure threshold and the sample size is limited, the empirical failure probabilities at each time point are extremely small. Consequently, the fitted Weibull reliability curve exhibits a very slow decay over a long time range, and the reliability changes only slightly in certain time intervals.

Table 2. Case data.

Operating time t (h)	Y-axis vibration displacement (μm)
0	15.0
250	15.1
600	15.2
1200	15.4

As a result, the reliability curve obtained by this method presents unrealistic evaluation results, such as an incomplete predicted lifetime range and an increasingly slow reliability degradation trend.

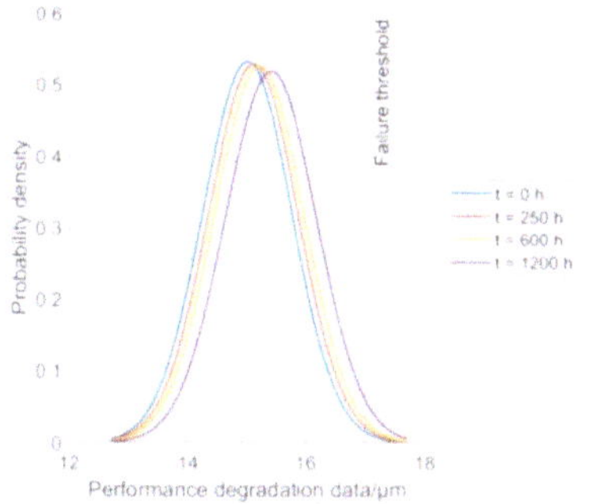

Fig. 3. Probability density.

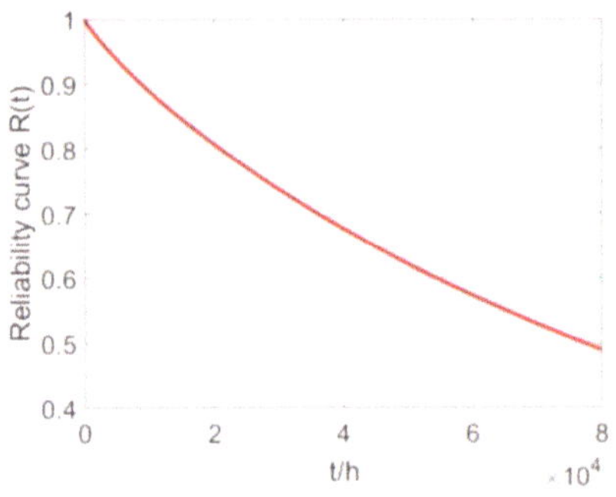

Fig. 4. Reliability curve based on Weibull distribution.

3 Bayesian Gamma Model Based on Degradation Rate Statistics

The method described in the previous section may fail when the degradation data are far from the failure threshold. To address this issue, a reliability assessment method based on a Bayesian Gamma model using degradation rate statistics is proposed. This method enables a reasonable quantification of lifetime uncertainty under small-sample and non-threshold-crossing conditions.

3.1 Assessment Procedure

For a given sample, the number of degradation measurements is n, and the degradation amount at measurement time $t_j(j = 1, 2, \ldots, n)$ is denoted as $x(t_j)$. According to the actual degradation trend of the data, a degradation direction variable *trend* is defined. When the degradation data are expected to increase and the failure threshold is an upper limit, *trend* is set to 1; when the degradation data are expected to decrease and the failure threshold is a lower limit, *trend* is set to -1.

At discrete time points $t_1, t_2, \ldots, t_n$, the degradation rate in the k time interval is defined as $\mathrm{r_k}$:

$$r_k = \text{trend} \cdot \frac{X(t_k) - X(t_{k-1})}{t_k - t_{k-1}} \tag{8}$$

where $r_k > 0$ is regarded as an effective degradation rate, and all effective degradation rates are retained to form a dataset. Due to the limited sample size and the fact that each sample provides only a small number of degradation segments, it is difficult to directly model the lifetime distribution. Therefore, the focus is shifted from "lifetime" to "degradation rate", and the system lifetime is indirectly inferred through the statistical distribution of degradation rates.

It is assumed that the degradation rate in each time interval follows a Gamma distribution:

$$r_k \sim \text{Gamma}(\alpha, \beta) \tag{9}$$

where α is the shape parameter and β is the scale parameter.

The probability density function of the Gamma distribution is:

$$f(r|\alpha, \beta) = \frac{1}{\Gamma(\alpha)\beta^{\alpha}} r^{\alpha-1} exp\left(-\frac{r}{\beta}\right) \quad (10)$$

prior distributions are assigned to the Gamma distribution parameters α and β:

$$p(\alpha, \beta) \quad (11)$$

When degradation data of similar machine tools are available, a hierarchical Bayesian framework can be constructed, in which the hyperparameters of the Gamma prior are estimated by pooling degradation information from multiple similar units. Alternatively, expert knowledge can be incorporated by converting expert scoring results into prior distributions of the model parameters.

After observing degradation rate samples $\{r_1, r_2, ..., r_n\}$, the likelihood function is:

$$p(\{r_k\}|\alpha, \beta) = \prod_{k=1}^{n} f(r_k|\alpha, \beta) \quad (12)$$

According to Bayes' theorem, the posterior distribution of the parameters is:

$$p(\alpha, \beta|\{r_k\}) \propto p(\{r_k\}|\alpha, \beta)\, p(\alpha, \beta) \quad (13)$$

Since the posterior distribution $p(\alpha, \beta|\{r_k\})$ generally has no closed-form expression, the Markov Chain Monte Carlo (MCMC) method is employed to generate samples from this distribution:

$$\left\{\left(\alpha^{(s)}, \beta^{(s)}\right)\right\}_{s=1}^{N} \quad (14)$$

Under discrete time, the cumulative degradation at time $D(t_k)$ is:

$$D(t_k) = \sum_{i=1}^{k} r_i \Delta t_i \quad (15)$$

Let the failure threshold be L. The failure time is defined as:

$$T_f = inf\{t : D(t) \geq L\} \quad (16)$$

The reliability function is expressed as:

$$R(t) = P\left(T_f > t\right) \quad (17)$$

Based on the N degradation trajectories obtained by MCMC sampling, the reliability function is approximated as:

$$R(t) \approx \frac{1}{N} \sum_{s=1}^{N} I\left(D^{(s)}(t) < L\right) \quad (18)$$

where $I(\cdot)$ is the indicator function, which equals 1 when the cumulative degradation $D^{(s)}(t)$ of the s-th degradation path at time t does not exceed the failure threshold L, and equals 0 otherwise.

3.1.1 Bootstrap-Based Robustness Verification

To verify the robustness of the proposed model against data perturbations under small-sample conditions, a bootstrap resampling strategy is introduced as follows: The effective degradation rate samples are resampled with replacement to generate B groups of bootstrap rate samples. For each bootstrap sample, the parameters of the Gamma model are estimated using the robust median statistic, and the corresponding mean life is calculated. The distribution characteristics of the estimated life values are then analyzed. If the distribution is concentrated and free of extreme outliers, it indicates that the proposed model is insensitive to small-sample perturbations and possesses satisfactory robustness.

3.2 Case Study

The small-sample case with degradation data far from the failure threshold presented in Sect. 2.2.2 (Table 2) is adopted to perform modeling and analysis using the Bayesian Gamma model based on degradation rate statistics under the same data conditions.

Based on the degradation data, the degradation rates in each time interval are calculated as $r_1 = 0.0004$, $r_2 = 0.000285$,$r_3 = 0.000333$. It is assumed that the degradation rate in each time interval follows a Gamma distribution.

Given the limited sample size, weakly informative priors are adopted to model the parameters of the Gamma distribution so as to avoid introducing overly strong subjective assumptions. Specifically, the prior shape parameter α is set to 1, which reduces the Gamma prior to an exponential prior distribution for the degradation rate, thereby reflecting the engineering prior knowledge that the degradation rate is strictly positive and that small degradation rates are more likely to occur. The prior scale parameter β is initialized based on robust statistical measures of the degradation-rate samples, such that the expected magnitude of the prior distribution is consistent with that of the observed data, thus enabling a data-driven prior construction. After comprehensive consideration, β is set to 3000.

Posterior inference is performed using the Markov chain Monte Carlo (MCMC) method, with a total of 10,000 samples generated, among which the first 5,000 samples are discarded as burn-in to ensure convergence of parameter sampling. Figure 5 presents the MCMC iteration trajectories of parameters α and β. It can be observed that both parameters fluctuate steadily within bounded ranges without exhibiting divergent behavior, indicating that the Markov chains have reached a stable convergent state.

To further examine the robustness of the proposed model, 500 bootstrap resampling procedures are conducted, and the resulting life distribution is illustrated in Fig. 6. The results demonstrate that the estimated life values are concentrated within a reasonable range without obvious extreme outliers, suggesting that the proposed method is insensitive to small-sample perturbations and exhibits satisfactory robustness.

Based on the posterior samples of model parameters, the posterior life distribution is obtained, as shown in Fig. 7, together with the corresponding reliability curve. Moreover, the 95% credible interval of the reliability function is depicted in Fig. 8. The credible interval is relatively narrow at the early stage and gradually widens as the uncertainty in life prediction increases, which intuitively reflects the variability of reliability assessment under small-sample conditions and provides more comprehensive information for maintenance decision-making.

By comparison, when a large deviation exists between the degradation data and the failure threshold, the Bayesian Gamma model based on degradation rate statistics yields a reliability curve that can cover the entire life cycle of the component from observation to failure, and the resulting degradation curves conform to the general degradation behavior of mechanical components.

4 Conclusions

This work examines the reliability of CNC machine tool components with small samples, introducing a Bayesian Gamma model to improve Weibull estimates for special data, and compares their performance. Key conclusions are as follows:

(1) The Weibull life model based on degradation distribution and empirical failure probability is suitable for cases where degradation data gradually approach the failure threshold, and it can achieve effective estimation of lifetime and reliability.
(2) When the degradation data samples are limited and the observation stage has not yet exceeded the failure threshold, the Bayesian Gamma process model based on degradation rate statistics can quantify parameter uncertainty and obtain reliability assessment results covering the entire life cycle.

Future work may integrate physical degradation models such as Archard wear or Paris fatigue models to improve model interpretability. Additionally, multi-state or Hidden Markov models could be explored to capture dynamic characteristics of different degradation stages through state transition probabilities.

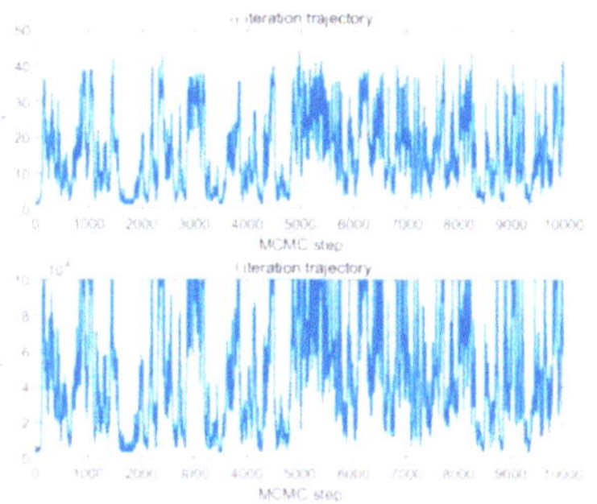

Fig. 5. Parameters iteration trajectory.

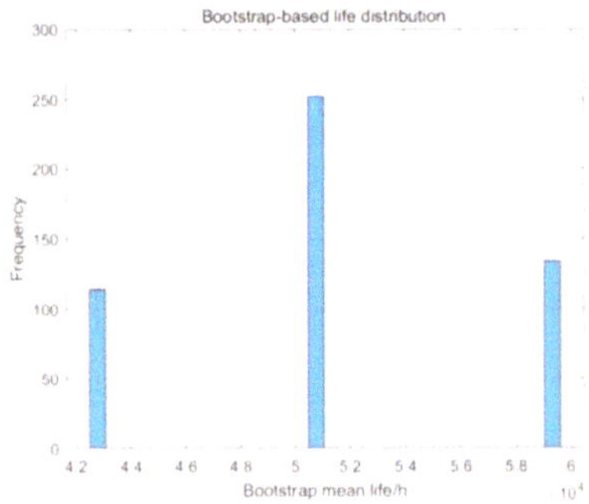

Fig. 6. Botstrap-based life destribution.

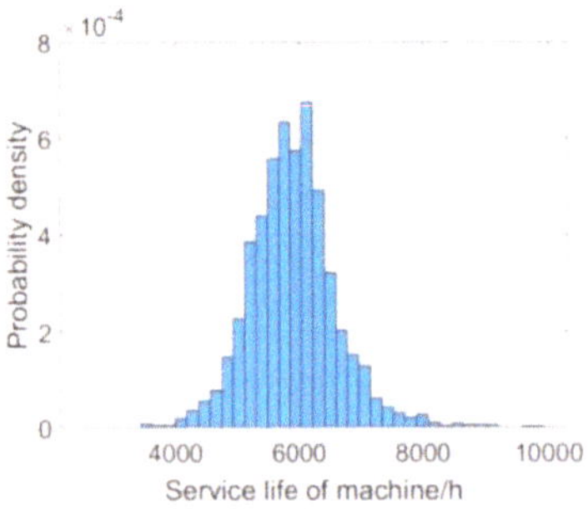

Fig. 7. Posterior lifetime distribution.

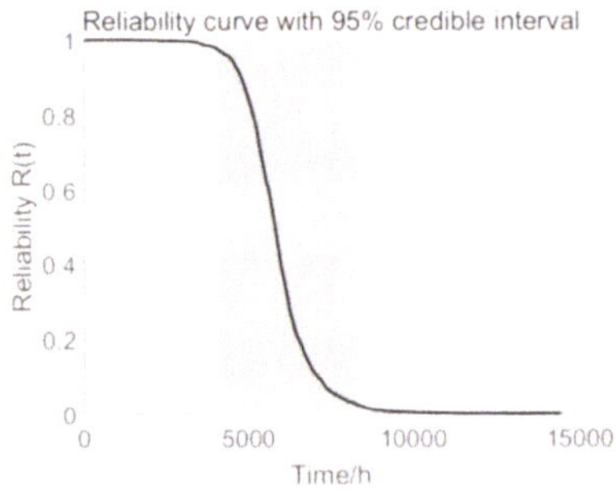

Fig. 8. Reliability curve based on robust Gamma model.

Acknowledgments. This work was financially supported by the Key Laboratory of Design and Intelligent Machining Technology for High Precision Machine Tools.

References

Chen, C., Wang, C., Yang, Z., Liu, Z., Tian, H.: Review and development trend analysis of CNC machine tool reliability modeling. J. Jilin Univ. (Engineering and Technology Edition), **52**(2), 253–266 (2022). https://doi.org/10.13229/j.cnki.jdxbgxb20211173

Chen, H., Wang, F., Wang, J., Ren, J., Xie, S.: Reliability assessment of a heavy-duty CNC boring–milling machine spindle feed transmission subsystem under small-sample failure data. Mach. Tool Hydraul. **53**(13), 173–180 (2025). https://doi.org/10.3969/j.issn.1001-3811.2025.13.025

Chen, H., Zhang, J., Ma, A., Li, H., Li, C.: Reliability study of heavy-duty CNC machine tools based on improved Bayesian method. J. Univ. Electron. Sci. Technol. China **52**(1), 140–145 (2022). https://doi.org/10.12178/1011-0548.2022153

Liang, X.: Reliability analysis of rolling linear guide based on performance degradation. Master's Thesis, Nanjing University of Science and Technology. China National Knowledge Infrastructure (2020). https://www.cnki.net/

Ran, Y., Ji, Y., Zhu, X.: Reliability assessment of meta-action units integrating life information and multi-performance degradation information. J. Hunan Univ. (Natural Sciences) **52**(2), 109–121 (2025). https://doi.org/10.16339/j.cnki.hdxbzkb.2025162

Ying, J.: Degradation analysis and life prediction of CNC machine tool motorized spindle bearings based on dynamics. Doctoral dissertation, Jilin University. China National Knowledge Infrastructure (2023). https://www.cnki.net/

Simplified Modeling and Simulation Method for Surface Temperature Fields of Complex Structure Camouflage Nets

Tao Wang and Dengfeng Ren(✉)

College of Energy and Power Engineering, Nanjing University of Science and Technology, Nanjing, China
rendengfeng@njust.edu.cn

Abstract. The complex surface structure of camouflage net poses significant challenges for direct modeling simulation calculations. The surface patterns and holes of camouflage net substantially affect convective heat transfer across the surface. Additionally, the holes in the net influence radiative heat transfer to targets beneath it. This paper simplifies the camouflage net into a flat tarpaulin surface and equates its corresponding physical properties. By treating the camouflage net model as a translucent structure with appropriate transmittance settings, its transmission effect can be simulated. To validate the computational method's accuracy, a field test was conducted. Comparing the collected field data with simulation results confirms the feasibility of this simplification approach.

Keywords: Translucent model · Temperature field · Camouflage

1 Introduction

With the advancement of defense technology, an increasing number of military reconnaissance devices are being widely deployed, significantly raising the probability of detecting weaponry and combat personnel (Gross et al. 2023). Under these circumstances, developing camouflage technology has become a key focus in the military technological advancement of nations (Zhang et al. 2022). Selj et al. proposed a method for evaluating the thermal contrast of camouflage nets and mobile camouflage systems against their backgrounds (Mikkelsen and Selj 2020).

The challenge in calculating the temperature field of camouflage nets lies in their complex porous mesh structures (Akhmadiev and Gizzyatov 2023). This porous structure influences both convective heat transfer and thermal conduction while allowing partial solar radiation penetration, creating complex multiphysics coupling problems. This paper simplifies the camouflage net's intricate porous mesh structure into a flat, homogeneous tarp of defined thickness, equating its physical parameters. The net's walls are modeled as semi-transparent surfaces to permit radiation transmission (Bu and Bai 2023).

T. Mankovits and N. Trisovic (Eds.): MDS 2026, LNME 2026.
https://doi.org/10.1007/978-981-92-2440-1_30

2 Model Simplification Methods

2.1 Simplification and Equivalence of Geometric Models

Direct modeling of the porous mesh structure of camouflage netting is costly, and heat transfer analysis and temperature field calculations are extremely challenging. Therefore, this paper simplifies it to a non-porous, uniform flat canvas model. Given the differences in heat transfer effects between the simplified model and the actual porous structure, equivalent corrections must be applied to the flat canvas heat transfer model to ensure the validity and accuracy of the computational results. Figure 1 illustrates a schematic of the local porous mesh structure and the overall camouflage net structure.

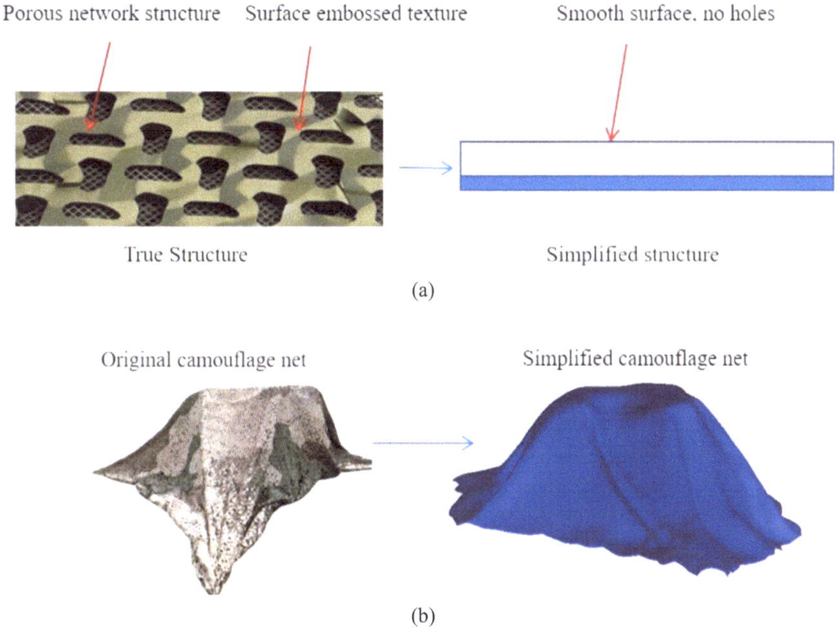

Fig. 1. Diagram of simplified camouflage net. (**a**) simplified local structure of camouflage Net. (**b**) simplification of the overall camouflage Net.

From the perspective of heat transfer, camouflage net essentially functions as a highly porous, anisotropic porous medium. Its interwoven fibers and the air within the pores collectively form a complex solid-gas coupled heat transfer system. When the scale of investigation far exceeds the diameter of a single fiber (typically on the micrometer scale), constructing a microscopic geometric model directly would incur enormous computational overhead and is unnecessary for engineering applications. Therefore, this paper employs a volume averaging approach, treating the complex fiber skeleton and pores as a continuous, homogeneous equivalent medium.

It should be noted that the accuracy of this simplified model depends on the specific application scenario. To quantify the error introduced by simplification, we define the characteristic scale ratio:

$$\lambda = \frac{L}{d} \tag{1}$$

Here, L represents the surface area of the camouflage network I am studying, while d represents the diameter of the holes on the surface of this network.

where L is the scale of interest on the target surface (e.g., the diameter of the area covered by the camouflage net), and d is the pore diameter of the camouflage net. When λ approaches 1 and local thermal equilibrium is achieved.Then, the approach we discussed in the previous text of treating the porous structure as an isotropic flat plate is reasonable. For the scale under study in this paper, the value of λ is much greater than 1. Therefore, the above simplified model is valid.

When $\lambda \approx 1$, assuming local thermal equilibrium, the temperature difference between the solid skeleton and the pore fluid can be neglected. Under these conditions, approximating the porous structure as a homogeneous plate is valid. Under the experimental conditions of this study, the value of λ is significantly greater than 1. Therefore, the simplified model adequately captures the overall trend of the surface temperature distribution.

In order to make the equivalent parameters more reasonable, I specially introduced three equivalent methods for porous media to determine the thermal properties of the model.

My first method is the volume averaging method. This method assumes that solids and air coexist in the porous medium and calculates the weighted average based on their respective volume fractions. The second method is the Maxwell-Owen model. I assume that the porous medium is composed of discrete phases and continuous phases, where the discrete phases are completely surrounded by the continuous phases. The third method is the series-parallel model. I assume that there is anisotropy in both the thickness direction and the plane direction, thus enabling the superposition calculation of the series and parallel models.

Although the volume averaging method can provide reasonable equivalent values for both density and specific heat capacity, at this point, the thermal conductivity will have a large error. This is because the thermal conductivity of the camouflage net fabric is much higher than that of air. So I chose to calculate the equivalent value of the thermal conductivity using the Maxwell-Owen model.

In this article, the thickness of the camouflage net fabric pieces I used was approximately 0.2 mm; however, at this thickness, it was difficult to generate high-quality grids using the grid generation software. To facilitate the creation of high-quality grids for subsequent simulation calculations, I increased the thickness of the geometric model to 10 mm. After increasing the thickness of the geometric model, the corresponding material parameters also needed to be equivalentized. The equivalentized material parameters can be calculated using the following equation.

$$\rho = \frac{m}{V} \tag{2}$$

ρ, m, V respectively represent the density, mass and volume of the model.

The equivalent specific heat capacity can be calculated using the following equation.

$$c = \frac{Q}{m * (\Delta t)} \tag{3}$$

c, Q, m and Δt respectively represent the specific heat capacity, the amount of heat absorbed, the mass, and the temperature change of the camouflage net.

The following equation can be used to calculate the equivalent thermal conductivity.

$$q = \lambda \frac{\Delta t}{\delta} \tag{4}$$

In the above equation, λ, Δt and δ represents the thermal conductivity, temperature change, and thickness of the camouflage net, respectively. As the thickness increases, the equivalent thermal conductivity increases accordingly.

2.2 Simplification of Heat Transfer Models

The heat exchange between the camouflage net, the target and the environment mainly includes heat conduction, self-radiation, solar radiation, sky background radiation, convective heat exchange with the air, and the mutual heat exchange between the camouflage net, the vehicle and the ground. Additionally, latent heat exchange caused by water evaporation needs to be considered. Figure 2 provides a schematic diagram of the heat exchange.

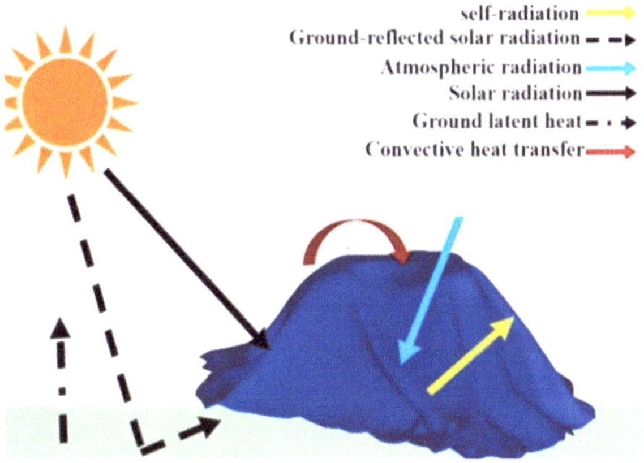

Fig. 2. Schematic Diagram of Camouflage Net Heat Exchange.

1. Direct solar radiation.

The direct solar radiation on the surface of the camouflage net can be calculated using the following equation.

$$q_{sund} = rI_{sc}P^{m} \cos i \tag{5}$$

q_{sund} represents direct solar radiation, and r is the correction factor. I_{sc} represents the solar constant, P indicates atmospheric transparency, m represents the mass of the atmosphere, and i represents the incident angle of sunlight on an inclined surface.

2. Solar scattered radiation.

The solar scattered radiation on the surface of the camouflage net can be calculated using the following equation.

$$q_{suns} = c_1 * (\sinh)^{c2} * \frac{1 + \cos\beta}{2} \tag{6}$$

Here, q_{suns} stands for the scattered solar radiation hitting the inclined surface, with c_1 and c_2 acting as empirical coefficients.

3. Ground reflection of solar radiation.

The ground reflection of solar radiation on the surface of the camouflage net can be calculated using the following equation.

$$q_{sunref} = \frac{1}{2}(q_{sund} + q_{suns})\alpha_{ground}\frac{1 + \cos\beta}{2} \tag{7}$$

Q_{sunref} represents the radiation reflected by the ground, while q_{sund} and q_{sunc} respectively denote the direct solar radiation and the scattered solar radiation from the sun. α_{ground} represents the reflectivity of the ground, and β is the inclination angle.

4. Atmospheric radiation.

Based on existing studies, for clear, cloudless skies, the longwave radiation from the sky background can be expressed as:

$$q_{sky} = \sigma T_a^4 \left(a_0 + b_0\sqrt{e_a}\right) = \varepsilon_a \sigma T_a^4 \tag{8}$$

In this expression, q_{sky} is the longwave radiation from the sky under clear conditions. a_0 and b_0 are empirical coefficients, e_a refers to the near-surface water vapor pressure (kPa), and ε_a stands for the sky background's equivalent emissivity.

5. Ground latent heat.

Because some ground surfaces lose moisture, processes like evaporation and condensation occur. As this happens, latent heat is exchanged with the surrounding air. The rate of this exchange, known as latent heat flux, depends on several factors—such as air and ground temperature, humidity, and wind speed. It can be calculated using the following equation:

$$LE = \rho_a h_{fg} C_D v w_s (q_a - q_c) \tag{9}$$

h_{fg} represents the latent heat of vaporization; w_s represents the surface moisture content; q_a represents humidity; q_c represents the saturated specific humidity at the ground level.

6. Convective heat transfer.

Due to the highly complex surface structure of the camouflage net after cutting, both the surface patterns and holes significantly affect convective heat transfer. A review of literature reveals no readily available empirical formulas for direct application. For the convective heat transfer coefficient in the above equation, we propose using the linear equations for windward and leeward surfaces proposed by J.A. Palyvos, which are applicable to solid surfaces.

Formula for windward convective heat transfer coefficient:

$$h_{net\_upy} = 7.4 + 4.0v_{conv} \tag{10}$$

Formula for calculating the convective heat transfer coefficient for downwind flow:

$$h_{net\_upb} = 4.2 + 3.5v_{conv} \tag{11}$$

where h_{net_upy} denotes the simplified windward convective heat transfer coefficient on the upper surface of the camouflage net structure; h_{net_upb} denotes the simplified leeward convective heat transfer coefficient on the upper surface of the camouflage net structure; v_{conv} represents the surface wind velocity of the camouflage net.

The apertures in the camouflage net permit sunlight and background radiation to penetrate to the target surface; this ratio is termed the transmittance. This paper treats a simplified flat-plate model as a translucent wall surface, employing FLUENT's translucent model to address the multiphysics coupling of transmission-radiation-convection.

The surface temperature of the camouflaged target is influenced by three factors: external environmental radiation (sun, sky, and surrounding background), internal and external convective heat transfer, and thermal conduction between different sections along with the material's thermal properties. The temperature distribution across the camouflage net, target, and ground must be determined through coupled solutions of internal heat conduction and internal/external thermal boundary conditions (Deng et al. 2024).

$$\frac{\partial T}{\partial t} = \nabla \cdot (\frac{\lambda}{\rho c_p} \nabla T) + \frac{q_V}{\rho c_p} \tag{12}$$

where T is the surface temperature of the camouflage net or standard body, t is time, ρ is the material density, λ is the thermal conductivity of the material, c_p is the specific heat capacity of the material, and q_V is the internal heat source term.

Since this study investigates only the surface temperature and infrared radiation characteristics of the target during steady-state cooling, the internal heat source term in the governing equation is set to zero. The governing equation can thus be rewritten as:

$$\frac{\partial T}{\partial t} = \nabla \cdot (\frac{\lambda}{\rho c_p} \nabla T) \tag{13}$$

The above establishes a temperature field solution model for the camouflage net and target. The solution process flowchart is shown in Fig. 3.

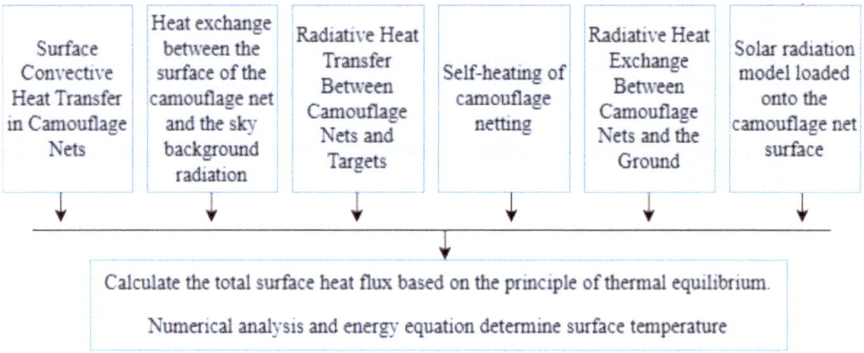

Fig. 3. Schematic Diagram of Temperature Field Solution Process.

3 Experimental and Temperature Field Model Validation

3.1 Field Testing

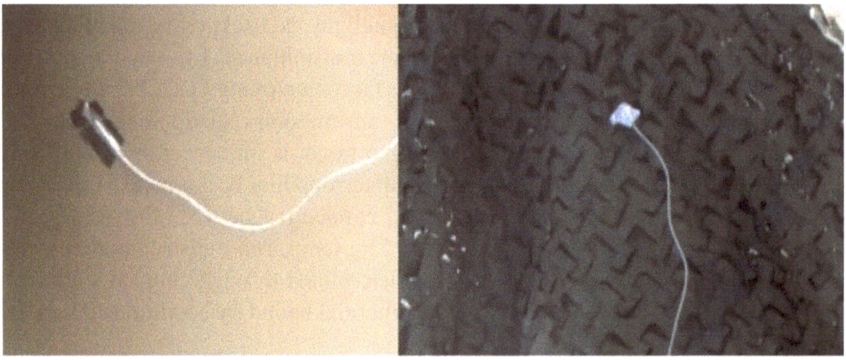

Fig. 4. Schematic diagram of thermocouple arrangement

To validate the accuracy of the simplified camouflage net model established above, a field test was conducted. Two identical square iron cavities and a camouflage net served as test targets. I covered one of the cavities with the camouflage net, making that cavity a target for camouflage; while I did nothing to the other cavity, leaving it as an uncamouflaged target. I set up a weather station to collect the meteorological data on the day of the experiment, and used the meteorological data I collected as the input conditions for the simulation. Thermocouples were employed to continuously measure and record 24-h temperature data on the target surfaces. These recorded temperature data served as parameters for validating the correctness of the simplified camouflage net model. During the test, thermocouple measurement contacts were affixed to the fabric surfaces of the square cavity and the camouflage net, respectively. The Fig. 4 shows the placement of thermocouples on the camouflage net and the cavity surface.

3.2 Model Validation

Comparative analysis was conducted between the simulated data from the temperature monitoring points of the simulation model and the measured data from the field test. The curves showing the changes over time for the test point temperature, the simulated point temperature, and the temperature difference are shown in the Fig. 5.

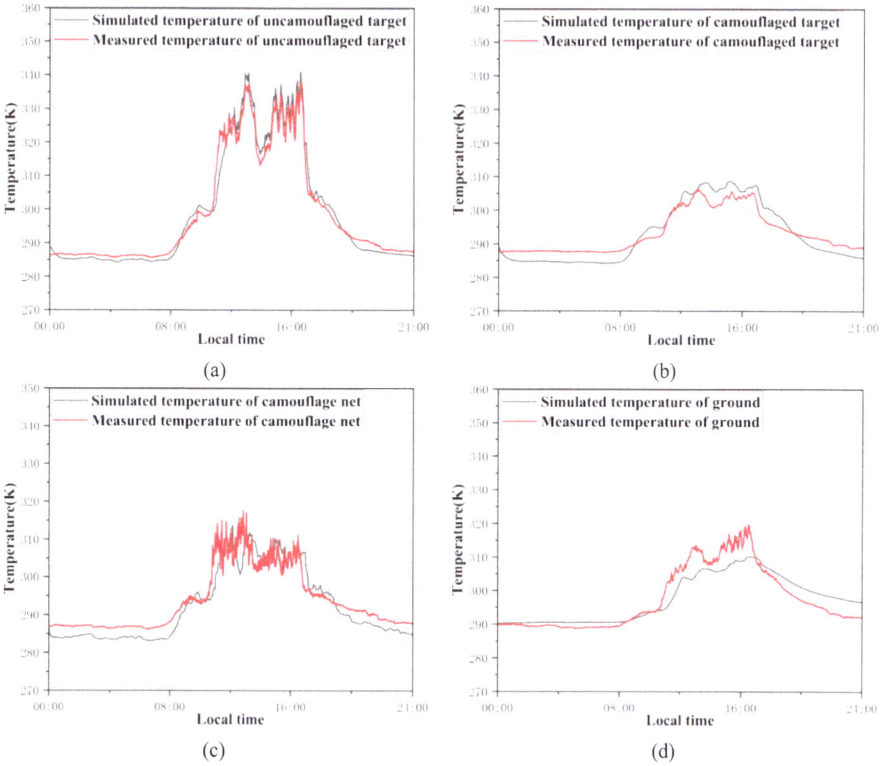

Fig. 5. Comparison chart of simulated temperature and measured temperature

As shown in the figure above, the measured temperature variation over time aligns well with the trend of surface temperature over time derived from computer simulation. The measured surface temperature of the camouflage net fluctuates significantly during daytime. This is primarily due to the thin fabric material of the camouflage net, which is heavily influenced by solar radiation during daylight hours, causing its surface temperature to fluctuate dramatically.

To investigate the influence of environmental wind speed on the surface temperature field of camouflage net, this study selected two time points—noon at 12:00 and midnight at 0:00—and conducted comparative analyses with wind speed gradients ranging from 0 to 7 m/s. The curves showing the variation of net surface temperature with wind speed are presented in Fig. 6.

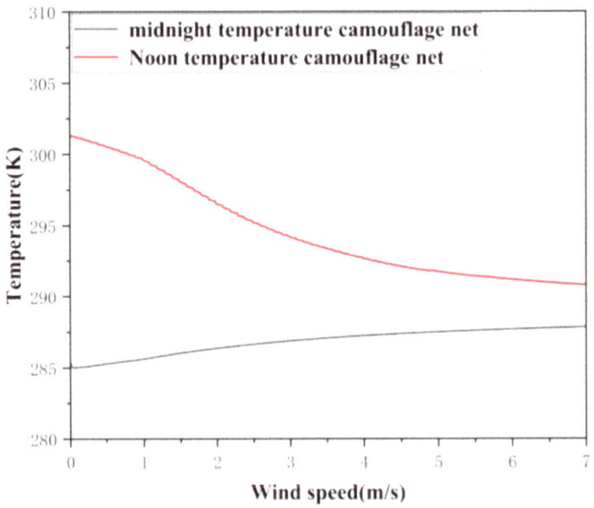

Fig. 6. Surface temperature of camouflage net varies with wind speed

Results indicate that wind speed significantly modulates the equilibrium temperature of camouflage net, with opposite effects during day and night: surface temperature increases with wind speed at night, while decreasing with increasing wind speed during the day. When wind speed rises from 0 m/s to 7 m/s, nighttime surface temperature increases by 2.45 K, while daytime temperature decreases by 10.53 K. Thus, wind speed significantly influences the surface temperature of camouflage net, with a more pronounced effect during the day. This is primarily because its porous structure enhances convective heat transfer to some extent.

4 Conclusions

Simplifying complex camouflage nets into flat canvas sheets of specific thickness and applying equivalent treatment to their physical parameters can significantly streamline computational workflows while maintaining computational accuracy. Treating the simplified model as a translucent wall and setting the corresponding transmittance in FLUENT effectively simulates the radiation transmission characteristics of the camouflage net. This resolves the multi-physics coupling issues involving radiation, transmission, and convection associated with its complex structure. The feasibility of this approach was validated through comparison with experimental data.

References

Akhmadiev, F.G., Gizzyatov, R.F.: Mathematical simulation of heat transfer through the multilayer filler structures of buildings. J. Eng. Phys. Thermophys. **95**(5), 1134–1144 (2022). https://doi.org/10.1007/s10891-022-02578-9

Bu, X., Bai, H.: Recent progress of bio-inspired camouflage materials: From visible to infrared range. Chem. Res. Chin. Univ. **39**(1), 19–29 (2023). https://doi.org/10.1007/s40242-022-2170-2

Deng, Q., Chen, H., Yang, C., Zhao, X., Cheng, W.L.: Study on transient heat transfer characteristics of heat flow under low frequency periodic thermal boundary conditions. Int. Commun. Heat Mass Transfer **159**, 108021 (2024). https://doi.org/10.1016/j.icheatmasstransfer.2024.108021

Gross, W., Queck, F., Schreiner, S., Mispelhorn, J., Küster, J., Middelmann, W., & Kneubühler, M. (2023). Experimental approach to camouflaged target detection and camouflage evaluation. In IGARSS 2023–2023 IEEE International Geoscience and Remote Sensing Symposium (pp. 2149–2152). IEEE. https://doi.org/10.1109/IGARSS52108.2023.10282243

Mikkelsen, A., Selj, G.K. (2020). Spectral reflectance and transmission properties of a multilayered camouflage net: Comparison with natural birch leaves and mathematical models. In Target and Background Signatures VI **11536**, pp. 65–75. SPIE. https://doi.org/10.1117/12.2573476

Zhang, Y., Bu, Z., Kong, H., Liu, W., Dong, M. (2022). Camouflage effect evaluation standard design. In Second International Conference on Optics and Image Processing (ICOIP 2022) **12328**, pp. 109–117. SPIE. https://doi.org/10.1117/12.2644308

Deformation Evolution Mechanism and Multi-Objective Collaborative Optimization of Large Crankshaft Machining

Yongsheng Zhao[1,2], Pengfei Gao[1,2], Xipeng Liang[1,2], and Jingjing Xu[1,2](✉)

[1] Beijing Key Laboratory of Design and Intelligent Machining Technology for High Precision Machine Tools, Beijing University of Technology, Beijing, China
xujj0908@bjut.edu.cn

[2] Institute of Advanced Manufacturing and Intelligent Technology, Beijing University of Technology, Beijing, China

Abstract. As a core component with weak stiffness in ship power systems, the large crankshaft undergoes dynamic coupling effects from multiple factors such as cutting force, centrifugal force, and clamping support during its turning process, making it difficult to accurately predict and control machining deformation. To address this question, this paper proposes a method for machining deformation prediction and process parameter optimization that integrates finite element simulation, empirical modeling, and multi-objective optimization. Taking the turning of the main shaft journal as the object, a three-dimensional cutting simulation model is established based on ABAQUS to obtain cutting force and deformation data under multi-factor coupling conditions. A high-precision deformation prediction model was constructed using an exponential empirical formula combined with nonlinear least squares method. The model's goodness of fit R^2 was 0.9343, and the predicted results were in good agreement with the simulation values. Based on this, the NSGA-II genetic algorithm is introduced to minimize machining deformation, minimize total cutting force, and maximize material removal rate as optimization objectives. The Pareto optimal solution set is obtained to achieve the coordination and unity between machining accuracy, efficiency, and cutting load. By combining simulation, prediction, and optimization techniques, the inherent relationship between cutting parameters and deformation patterns can be clearly presented. This method makes the selection of process parameters more scientific and efficient, and also provides a replicable technical solution for controlling machining deformation of low stiffness components such as large crankshafts.

Keywords: Large crankshaft · Processing deformation · Construction of cutting force model · Finite element simulation · Deformation prediction

1 Introduction

Large crankshafts are the core components of ship power plants, and in actual operation, they must withstand extremely strong mechanical loads and repeated fatigue effects (Zhang et al. 2026). They must adapt to extremely harsh stress environments (Li et al.

T. Mankovits and N. Trisovic (Eds.): MDS 2026, LNME 2026.
https://doi.org/10.1007/978-981-92-2440-1_31

2025), and during operation, they will be subjected to various alternating loads such as bending, twisting, and impact (Amouzgar et al. 2025; Baoyi et al. 2024). These types of parts not only have complex structures and use special materials, but also have strict standards for machining accuracy (Shen et al. 2022). The traditional process design approach is no longer suitable for the current data-driven and real-time changing production needs (Dong et al. 2022), and it is also difficult to clarify the relationship between processing parameters and part performance. In the production process of large-scale crankshafts, cutting is the core process. The specifications, wear status, and cutting parameters of the cutting tools not only directly affect the surface quality and machining accuracy of the crankshaft (Nogueira et al. 2022; Wu et al. 2022), but also change the residual stress distribution and overall deformation inside the parts. In recent years, computer technology and general finite element software have continued to mature (Gomes et al. 2018), and numerical simulation of cutting processes has become an important method for studying the deformation laws of crankshaft machining. Through cutting simulation, residual stress, surface deformation, and stress distribution during the machining process can be accurately predicted, providing theoretical basis for improving machining schemes, enhancing product quality, and saving production costs. Therefore, the systematic study of crankshaft cutting process has important application value for practical production.

2 Modeling Method for Cutting Force in Large Crankshaft Machining

Cutting force has a direct impact on machining deformation and cutting heat generation, and can also change the wear condition and service life of cutting tools, as well as the machining accuracy and surface quality of workpieces. In addition, cutting force is also an important reference for calculating cutting power, determining reasonable cutting parameters, designing cutting tools, machine tools, and fixtures. Considering the limitations of the theoretical model itself, this study adopts a simulation prediction method based on cutting force and deformation to analyze the machining process of key parts of large crankshafts.

2.1 Establishment of Finite Element Model for Cutting Force in Machining Process

A three-dimensional cutting simulation finite element model was established based on ABAQUS software. The establishment process of the three-dimensional cutting simulation model is shown in Fig. 1, which includes geometric model establishment, material parameter assignment, assembly of tool-workpiece model, mesh generation, boundary condition setting, contact property setting, and submitting for analysis and calculation. The simulation results are finally obtained, as shown in Figs. 2 and 3.

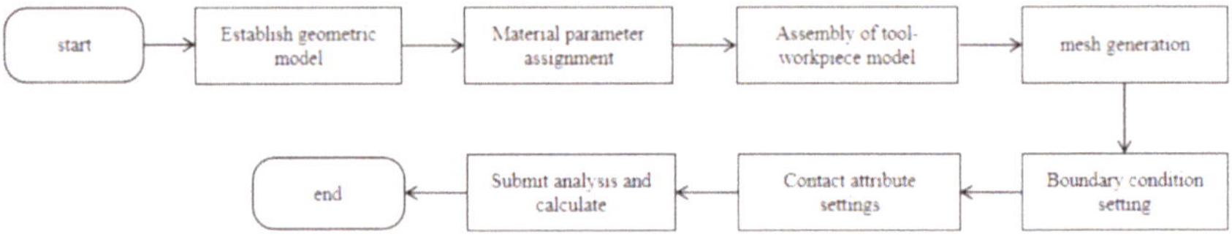

Fig. 1. Establishment process of 3D turning simulation model.

Fig. 2. Simulation result diagram.

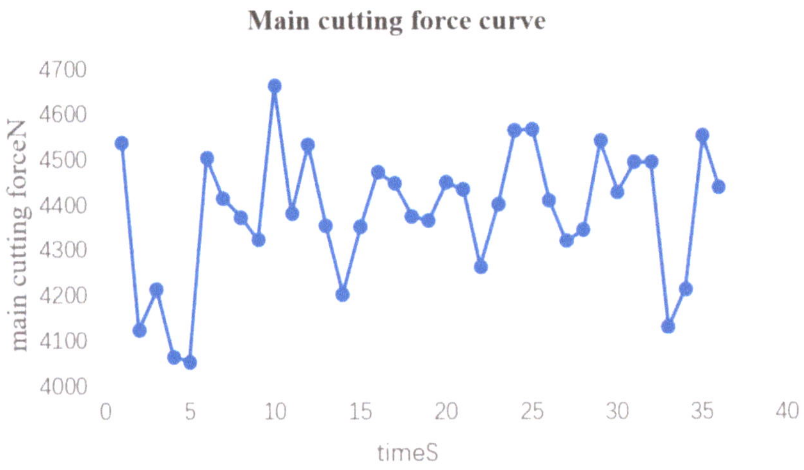

Fig. 3. Simulation result diagram.

2.2 Establishment of Cutting Force Prediction Model

The relationship between cutting force and cutting parameters is often described using an exponential empirical model, which is widely applied in industrial production and academic research due to its concise form and clear physical significance. The general expression for the turning force exponential empirical model is:

$$F_i = C \cdot v^a \cdot f^b \cdot a_p^c \cdot K \tag{1}$$

Among them, Fi represents the turning force, including the main cutting force Fc, feed resistance Ff, and back force Fp; v is the cutting speed; f is the feed rate; ap is the cutting depth; C is the model coefficient, which comprehensively reflects factors such as tool material and workpiece material; a, b, and c are the exponential parameters for cutting speed, feed rate, and cutting depth, respectively, indicating the influence of each parameter on the turning force; K is the correction factor. Taking the natural logarithm of this model can transform it into a linear form:

$$\ln F = \ln C + a \ln v + b \ln f + c \ln a_p + \ln K \tag{2}$$

Subsequently, the linear regression method can be employed to solve for the various parameters in the model.

$$\begin{aligned} F_p &= -\left(3.1817 \times 10^{-2}\right) \cdot V_c^{0.6211} \cdot a_p^{1.9489} \cdot f_r^{0.5773} \\ F_t &= 6.9249 \times 10^3 \cdot V_c^{-0.1658} \cdot a_p^{0.7053} \cdot f_r^{0.6739} \\ F_c &= 1.0568 \times 10^4 \cdot V_c^{-0.0039} \cdot a_p^{0.8921} \cdot f_r^{0.7281} \end{aligned} \tag{3}$$

This study used an exponential empirical formula to construct a deformation prediction model, rather than machine learning methods, mainly based on the following considerations: the parameters in the exponential form have clear physical meanings, making it easy to intuitively explain the influence of cutting parameters on cutting forces; At the same time, the model still has good fitting stability under small sample conditions, which can effectively avoid the problem of overfitting in machine learning models due to insufficient data. This modeling method balances the practicality of engineering applications with the reliability of modeling results.

3 Simulation Analysis of Machining Deformation of Large Crankshafts Due to Cutting Forces

The deformation during crankshaft machining is affected by the dynamic coupling of multiple factors such as cutting force, gravity, centrifugal force, and support reaction force. Single factor analysis is difficult to reveal its essential laws. This section uses the three center frame support as the clamping method to simulate extreme working conditions with the same direction of centrifugal force and cutting force under high rigidity conditions. It comprehensively analyzes the overall deformation of the crankshaft under the coupling effect of multiple factors, explores its deformation mechanism and optimization strategy, and provides theoretical basis and practical reference for high-precision machining.

Taking into account the coupling effects of multiple factors such as gravity, clamping conditions, rotational speed, and cutting force, the maximum deformation of the crankshaft follows a certain pattern with changes in cutting speed vc, cutting depth ap, and feed rate fr. According to the data in the table, the increase in cutting depth and feed rate significantly increases the cutting force, with the tangential force Fc showing the most significant increase (as in experiments 1 and 3, Fc increased from 3548.575N to 6134.650N). Correspondingly, the maximum deformation of the crankshaft increased from 7.272μm to 8.526μm. The influence of cutting speed exhibits nonlinear characteristics: when the cutting speed increases from 140 m/min to 160 m/min, the tangential force Fc decreases from 4830.671N to 4236.298N, but the maximum deformation of the crankshaft increases from 7.397μ m to 8.899μ m, which is mainly attributed to the superposition effect of centrifugal force enhancement; When the speed further increased to 180 m/min, Fc increased to 5271.352N and the deformation also increased to 10.77μm, further exacerbating the degree of deformation. The radial force Fp (negative value indicates pointing towards the center) and axial force Ff respectively affect the deformation shape of the crankshaft mid span and the stress distribution at the support point. Their effects show a significant superposition effect with changes in cutting parameters (for example, in experiments 7 and 9, when Fp is about -345N and Ff is about 1150N, the deformation after superposition reaches 10.77 μm). Table 1 lists the maximum deformation under different parameters.

Table 1. Maximum deformation in cutting simulation experiment.

Test Number	v (m/min)	a_p (mm)	f_r (mm/r)	radial forceF_p (N)	axial forceF_f (N)	tangential forceF_c (N)	maximum deformation μm
1	140	5	0.3	−319.820	1020.047	3548.575	7.272
2	140	5.5	0.4	−461.732	1393.017	4830.671	7.397
3	140	6	0.5	−600.180	1670.568	6134.650	8.526
4	160	5	0.4	−428.847	1246.057	4383.634	8.887
5	160	5.5	0.5	−543.656	1407.429	5458.003	8.783
6	160	6	0.3	−367.725	1204.351	4236.298	8.899
7	180	5	0.5	−494.316	1351.751	5145.252	10.77
8	180	5.5	0.3	−345.671	1148.951	3973.823	11.03
9	180	6	0.4	−504.726	1514.220	5271.352	10.77

In summary, an increase in cutting depth and feed rate directly exacerbates crankshaft deformation, while an increase in cutting speed can reduce cutting force to some extent, but centrifugal force increases, leading to a non-linear increase in deformation. Therefore, it is necessary to comprehensively optimize cutting parameters and combine them with high rigidity three center frame support to effectively control machining deformation and provide a mechanism basis for subsequent deformation prediction and optimization. Figure 4 shows the comprehensive deformation cloud map.

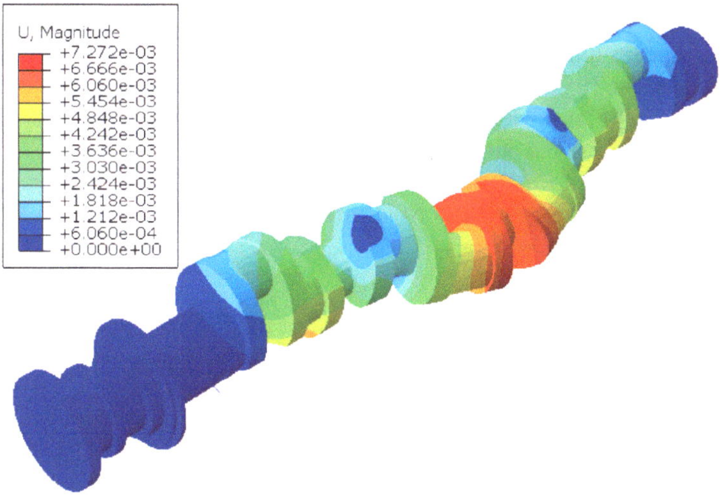

Fig. 4. Comprehensive impact.

4 Prediction and Optimization Methods for Deformation in Large Crankshaft Machining

This section takes the main journal turning as the research object and proposes a collaborative framework of "neural network prediction-multi-objective optimization". By integrating physical simulation and experimental data, a prediction model with high generalization ability is constructed, and an evolutionary algorithm is used to solve the Pareto front of process parameters, achieving multi-dimensional optimization of machining accuracy, efficiency, and cost. This method not only breaks through the trial-and-error limitations of traditional process development but also provides a transferable intelligent decision-making paradigm for machining with complex features.

4.1 Neural Network Prediction Method for Deformation in Crankshaft Main Journal Turning Process

In metal cutting research, empirical formulas in the form of power functions are commonly used to describe the influence of machining parameters on cutting outcomes, such as cutting force and machining deformation. This section establishes a mathematical model based on simulation experimental data, relating the maximum deformation D (μm) to cutting speed vc (m/min), cutting depth ap (mm), and feed rate fr (mm/r). The basic form of this model is presented below.

$$D = C \cdot v_c^x \cdot a_p^y \cdot f_r^z. \quad (4)$$

where C is the coefficient related to material and processing conditions, and x, y, and z are the influence indices of cutting speed, cutting depth, and feed rate, respectively. The four undetermined parameters in the model are accurately identified using experimental data through nonlinear least squares method.

Using MATLAB's lsqcurvefit function (Levenberg Marquardt algorithm) for nonlinear least squares fitting, iteratively minimizing the sum of squared residuals between predicted values and experimental data. The fitting steps are as follows:

① Define nonlinear model functions, input undetermined coefficients and independent variable matrices, and output predicted values;

② Using logarithmic linearization to obtain high-quality initial parameters, in order to improve convergence speed and avoid local optima, rewrite Eq. (1) in logarithmic form:

$$\ln(D) = \ln(C) + x \cdot \ln(v_c) + y \cdot \ln(a_p) + z \cdot \ln(f_r) \tag{5}$$

③ Call lsqcurvefit, pass in the model function, initial parameters, and experimental data, and automatically iterate optimization;

④ Write MATLAB code to solve and run it to obtain the model parameter identification results.Running the MATLAB code yields the following fitting results and analysis. The model parameters identified through the nonlinear least squares method are shown in Table 2.

Table 2. Fitting parameters for maximum deformation.

Parameter	Numerical value	Goodness of fitR^2
Coefficient$C$	0.0059	0.9343
Cutting speed index$x$	1.3800	
Cutting depth index$y$	0.2308	
Feed rate index$z$	0.0543	

Substituting the aforementioned parameters into the model, we obtain the final empirical formula for the maximum deformation:

$$D = 0.0059 \cdot v_c^{1.3800} \cdot a_p^{0.2308} \cdot f_r^{0.0543} \tag{6}$$

To evaluate the prediction accuracy of the model, we calculated the comparison between the model's predicted values and experimental values, and computed the goodness of fit, R^2. As shown in Table 3.

Table 3. Goodness of fit R^2 and prediction error.

Simulation experiment number	Simulated experimental value (μm)	Estimate (μm)	error
1	7.272	7.319	0.65%
2	7.397	7.600	2.74%
3	8.526	7.848	7.95%
4	8.887	8.938	0.58%

(continued)

Table 3. (*continued*)

Simulation experiment number	Simulated experimental value (μm)	Estimate (μm)	error
5	8.783	9.249	5.30%
6	8.899	9.178	3.14%
7	10.770	10.644	1.17%
8	11.030	10.583	4.05%
9	10.770	10.968	1.84%

The value of R^2 is 0.9343, which is very close to 1, indicating a good fitting effect. From the error analysis table and the visualization results in Fig. 5, it can be seen that the overall trend of the model's predicted values aligns well with the experimental values, and the relative errors for most points are relatively small.

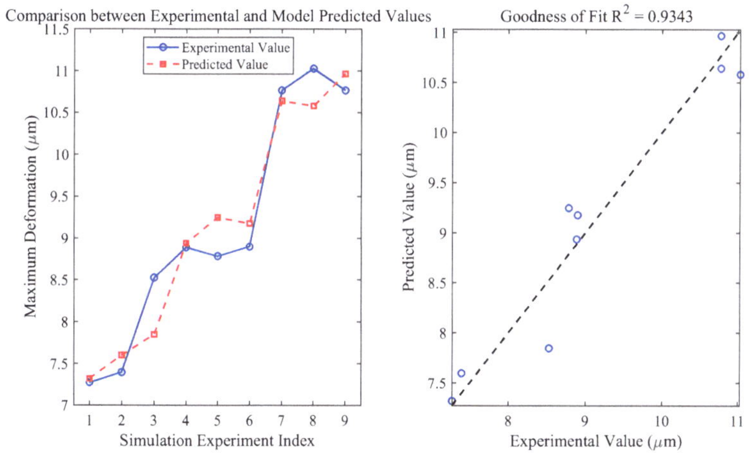

Fig. 5. Nonlinear fitting results and analysis.

4.2 Optimization Method for Turning Process Parameters Based on Minimizing Machining Deformation

A multi-objective optimization study was conducted on the three key parameters of cutting speed vc, cutting depth ap, and feed rate fr in turning machining, with the objectives of maximizing material removal rate Q, minimizing workpiece deformation D, and minimizing total cutting force F_total. The three conflicting objectives constitute a typical multi-objective optimization problem, which is solved using the NSGA-II algorithm to obtain the Pareto optimal solution set. This solution set can provide scientific decision-making basis for process parameter selection based on different production needs such as efficiency priority or accuracy priority, taking into account processing efficiency, quality, and equipment load.

The three-direction cutting force functions are respectively:

$$\begin{aligned} F_c &= 1517.2727 \cdot v_c^{0.0185} \cdot a_p^{0.9907} \cdot f_r^{0.6891} \\ F_f &= 401.8148 \times 10^3 \cdot v_c^{-0.0297} \cdot a_p^{1.0749} \cdot f_r^{0.5246} \\ F_p &= -239.5935 \cdot v_c^{-0.0316} \cdot a_p^{0.9433} \cdot f_r^{0.8853} \end{aligned} \tag{7}$$

The maximum deformation function is as follows:

$$D = 0.0059 \cdot v_c^{1.3800} \cdot a_p^{0.2308} \cdot f_r^{0.0543} \tag{8}$$

The objective function for material removal rate in turning processes is:

$$Q = 1000 \cdot v_c \cdot a_p \cdot f_r. \tag{9}$$

Consider the interval constraints (cutting parameter ranges) for each deformation variable:

$$\begin{cases} 120\text{m/min} \le v_c \le 180\text{m/min} \\ 5\text{mm} \le a_p \le 6\text{mm} \\ 0.3\text{mm/r} \le f_r \le 0.5\text{mm/r} \end{cases} \tag{10}$$

Using the gamultiobj solver in MATLAB, which is implemented based on the classic non-dominated sorting genetic algorithm (NSGA-II), we obtained the solution sets of the three objective functions under different combinations of cutting parameters by running the Matlab optimization algorithm of NSGA-II. The 3D Pareto front solution sets are shown in Fig. 6.

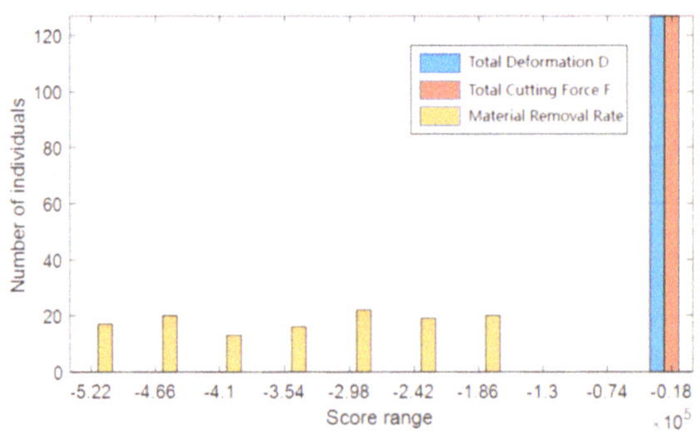

Fig. 6. Score histogram of Pareto optimal solution set.

Figure 6 is a score histogram showing the numerical distribution of the non dominated solution set on three objectives: deformation D, total cutting force F, and material removal rate Q. Figure 7 visually presents the conflict relationship between the three objectives: the surface point set is the optimal parameter combination. The high efficiency zone (high Q) corresponds to high D and high F; the high-precision/low loss zone (low D, low F) is accompanied by low Q, sacrificing efficiency; The balancing zone is located in the middle of the surface, achieving varying degrees of compromise among the three objectives.

In actual production, parameters can be optimized on the Pareto front based on processing objectives: the precision machining stage can focus on minimizing deformation, the rough machining stage pursues maximum material removal rate, and when the equipment load is sensitive, the total cutting force needs to be controlled. Multiple criteria decision-making methods such as TOPSIS can also be introduced to comprehensively rank the frontier solution set, providing quantitative basis for process parameter selection.

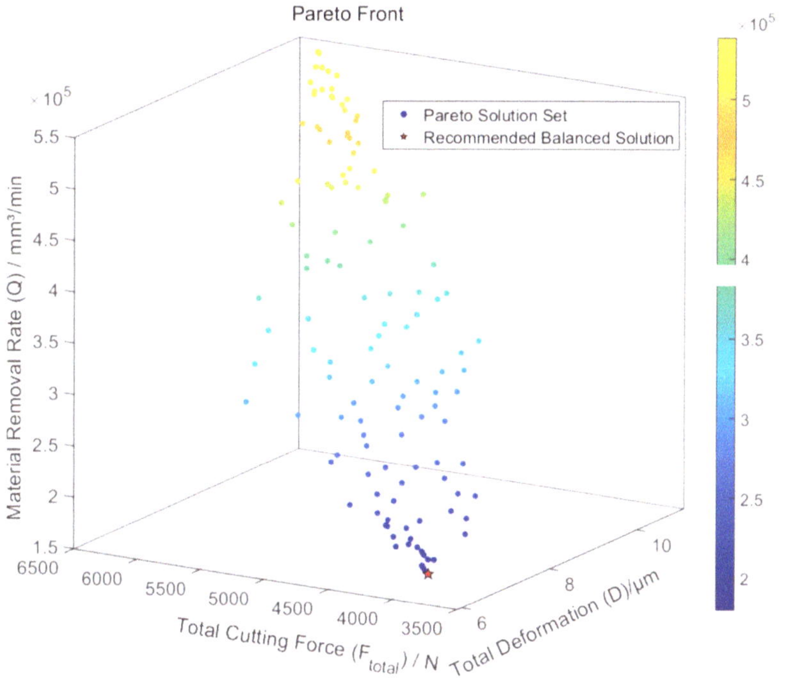

Fig. 7. 3D pareto frontier solution set.

5 Conclusion

This study focuses on the turning of large crankshaft main journals, using a combination of finite element simulation, nonlinear fitting, and NSGA-II algorithm to conduct research on machining deformation prediction and multi-objective optimization. Establish a multi factor coupled simulation model based on ABAQUS to obtain cutting force and deformation data; Constructing a high-precision deformation prediction model with a goodness of fit $R^2$of 0.9343; And with the goal of minimizing machining deformation and cutting force, and maximizing material removal rate, the Pareto front is solved.

The results show that the collaborative framework can reasonably reflect the deformation response law of machining, improve the efficiency of process parameter decision-making, and verify its applicability and promotion potential in deformation control of large crankshafts and similar weak stiffness structural components during machining.

References

Amouzgar, K., Wang, W., Eynian, M., Ng, A.H.C.: Smart process planning of crankshaft machining through multiple objectives optimization. Procedia CIRP **134**, 241–246 (2025). https://doi.org/10.1016/j.procir.2025.03.018

Dong, J., Zhao, S., Wang, Z., Wei, L., Bian, H., Liu, Y.: Failure mechanism tracing of the crankshaft for reciprocating high-pressure plunger pump. Eng. Fail. Anal. **141**, 106595 (2022). https://doi.org/10.1016/j.engfailanal.2022.106595

Gomes, J., Gaivota, N., Martins, R.F., Silva, P.P.: Failure analysis of crankshafts used in maritime V12 diesel engines. Eng. Fail. Anal. **92**, 466–479 (2018). https://doi.org/10.1016/j.engfailanal.2018.06.020

Li, X., Pang, Z., Li, J.: Analysis of the effect of crystal evolution on tool diffusion wear based on the change of cutting parameters. Sci. Rep. **15**(1), 8847 (2025). https://doi.org/10.1038/s41598-025-92946-1

Nogueira, M.L., Greis, N.P., Shah, R., Davies, M.A., Sizemore, N.E.: Machine learning classification of surface fracture in ultra-precision diamond turning using CSI intensity map images. J. Manuf. Syst. **64**, 657–667 (2022). https://doi.org/10.1016/j.jmsy.2022.04.011

Shen, N., et al.: Analysis of the relationship between flow characteristics of crankshaft core and fracture failures of S43CMnV steel. Eng. Fail. Anal. **141**, 106675 (2022). https://doi.org/10.1016/j.engfailanal.2022.106675

Wu, Z., Shen, J., Peng, Y., Wu, X.: Review on ultra-precision bonnet polishing technology. Int. J. Adv. Manuf. Technol. **121**(5–6), 2901–2921 (2022). https://doi.org/10.1007/s00170-022-09501-9

Xu, L. D., Xu, E. L., Li, L.: Industry 4.0: State of the art and future trends. Int. J. Prod. Res., **56**(8), 2941–2962 (2018). https://doi.org/10.1080/00207543.2018.1444806

Zhang, R., Yu, Y., Cao, B., Xie, L., Hu, L., Ma, B.: Research on the engine crankshaft stress inversion method based on vibration information. Eng. Fail. Anal. **184**, 110346 (2026). https://doi.org/10.1016/j.engfailanal.2025.110346

Zhu, B., Xiong, L., Chen, Y.: Evaluation of constitutive models used in orthogonal cutting simulation based on coupled Eulerian-Lagrangian formulation. Int. J. Adv. Manuf. Technol. **131**(1), 183–199 (2024). https://doi.org/10.1007/s00170-024-13104-x

A Study on Model Simplification Method for ANCF Based on an Improved Local Linearization Approach

Weifan Zhang and Yizhong Wu(✉)

School of Mechanical Science & Engineering, Huazhong University of Science and Technology, Wuhan, China
cad.wyz@hust.edu.cn

Abstract. The Absolute Nodal Coordinate Formulation (ANCF) is extensively utilized in the field of flexible multibody dynamics because it offers a constant mass matrix and inherently eliminates Coriolis forces. However, ANCF requires the computation of complex nonlinear elastic internal forces at each time step, which imposes a significant computational burden. To alleviate this burden, researchers have developed local linearization (LL) methods, but these suffer from error accumulation and rely on displacement-based update criteria that are inefficient for systems with large rigid-body motion. This paper proposes an improved local linearization (I-LL) method to address these issues. Two key enhancements are introduced: (1) the update criterion for the tangent stiffness matrix is modified from displacement-based to elastic strain-based, enabling more accurate and size-independent updates; (2) accurate elastic internal force calculations are inserted within the local linearization iteration cycle to mitigate error accumulation. The accuracy and effectiveness of the I-LL algorithm are demonstrated through two numerical examples.

Keywords: ANCF · Local linearization · Model simplification · Error accumulation

1 Introduction

In the past few years, the ANCF has become a widely adopted approach in flexible multibody dynamics simulations. By incorporating both global displacement vectors and slope vectors as generalized coordinates, ANCF naturally yields a constant mass matrix and eliminates Coriolis and centrifugal force terms from the equations of motion (Shabana 1997). To capture deformations, the formulation relies on the nonlinear Green–Lagrange strain measure, which leads to strongly nonlinear elastic force expressions (Shabana 1997). Unlike conventional finite element analysis (FEA) that rely on nodal displacements and rotational degrees of freedom (DOFs), ANCF avoids the need for coordinate transformations and is particularly well-suited for large-displacement and large-rotation problems (Shabana 2005, Dibold 2009, Bauchau 2014, Romero 2008). These distinctive features have motivated extensive research, with ANCF being successfully applied to

T. Mankovits and N. Trisovic (Eds.): MDS 2026, LNME 2026.
https://doi.org/10.1007/978-981-92-2440-1_32

the modeling of flexible robotic manipulators, deployable space structures, and other complex systems (Guo 2026, Xue 2026, Wang 2025).

Compared with traditional FEA, ANCF incorporates more generalized coordinates. This results in ANCF equations having much higher dimensionality than traditional finite element equations. In addition, as ANCF elements have progressed from rudimentary two-node beams to advanced elements including triangular, rectangular, tetrahedral, and hexahedral elements, the dimension of the dynamic equations and the required computational effort have grown substantially (Pappalardo 2017, Wang 2022, Kim 2018, Zhang 2018, Li 2024, Shabana 2023, Zhang 2022). Thus, improving the computational efficiency of ANCF dynamic equations remains a central challenge in the field (Otsuka 2022). The local linearization method has emerged as a prominent focus in recent research on ANCF model order reduction.

A local linearization method was first introduced by Tang et al (2019). In their approach, linearization is performed at a series of quasi-static equilibrium configurations, with the intervals determined by the required accuracy of the dynamic computation. Through the application of the Craig-Bampton method, a reduced-order model was constructed. This was achieved by projecting the incremental displacements of the locally linearized system onto the local modal basis associated with the quasi-static equilibrium configurations. The efficiency and accuracy of the proposed method were demonstrated through three numerical examples. Tian, Lan, and colleagues performed a local linearization of the dynamic equations to derive a constant tangent stiffness matrix (Tian 2020). Subsequently, they applied a free-interface component mode synthesis technique to construct a reduced-order model. In contrast to the Craig–Bampton method, this approach eliminates the need to retain interface coordinates, enabling a more significant reduction in the system's DOFs. The validity and effectiveness of the proposed method were demonstrated through numerical examples, including deployable solar array systems. In 2025, Tian et al. employed the Absolute Nodal Coordinate Formulation (ANCF) to model rigid-flexible-thermal coupled multibody systems (Tian 2025). Through the application of Taylor expansion, they locally linearized both the dynamic equations and the heat transfer equations, resulting in constant tangent stiffness and damping matrices. This approach was combined with dimensionality reduction to obtain simplified governing equations for numerical solution. Meanwhile, Yu et al. investigated nonproportionally damped viscoelastic flexible multibody systems using a complex modal synthesis method (Yu 2025). By performing local linearization around quasi-static equilibrium configurations, they also derived constant tangent stiffness and damping matrices. To maintain energy conservation and guarantee solution accuracy while utilizing a minimal set of retained modes, an energy-based criterion was developed for mode selection. The validity of this approach was confirmed through three numerical case studies.

However, in these studies employing the local linearization method, the simulation duration for flexible body motion is typically short. For simpler models, the simulation duration generally does not exceed 20 s, while for more complex models, it lasts only a few seconds (Tang 2019, Tian 2020, Tian 2025, Yu 2025). When the simulation time is extended, error accumulation from the local linearization method can severely affect the computational accuracy of the simplified models. Therefore, controlling error accumulation over longer simulation times and ensuring the computational accuracy of the simplified models hold significant research value.

2 Implementation of ANCF Planar Triangular Element

For meshing and modeling purposes, this study utilizes the Bézier-triangle-based ANCF planar element originally developed by Wang (Pappalardo 2017, Wang 2018, Wang 2020). The present section provides a concise description of the implementation procedure for this planar triangular element.

2.1 ANCF Triangular Element Coordinate Transformation

The location of any point P within a planar triangular element is characterized by the area coordinates $\boldsymbol{\xi} = (\xi_1, \xi_2, \xi_3)$. Defined as $\xi_i = A_i/A\ (i = 1, 2, 3)$, these coordinates rely on the total triangle area A and the areas A_i of the three sub-triangles created when point P is joined to each vertex.

From the above definition, it follows that the area coordinates are inherently subject to the following constraint:

$$\xi_1 + \xi_2 + \xi_3 = 1 \tag{1}$$

Fig. 1. (**a**) Schematic of area coordinates for a triangular element (**b**) Control point configuration for a bézier triangle

For a planar triangular element, the position vector of any point P can be written in the form:

$$\mathbf{r} = \mathbf{S}\mathbf{e} \tag{2}$$

Here, **S** represents the shape function matrix, and **e** is the element's nodal coordinate vector, expressed as:

$$\mathbf{e} = [\mathbf{r}^1\ \mathbf{r}^1_{\xi_2}\ \mathbf{r}^1_{\xi_3}\ \mathbf{r}^2\ \mathbf{r}^2_{\xi_3}\ \mathbf{r}^2_{\xi_1}\ \mathbf{r}^3\ \mathbf{r}^3_{\xi_1}\ \mathbf{r}^3_{\xi_2}]^T \tag{3}$$

where, $r^i, i = 1, 2, 3$ are the nodal position vectors, while $r^i_{\xi_j}, r^i_{\xi_k}, i, j, k = 1, 2, 3$ represent the area-gradient coordinates. The vector **e** contains only two gradient entries per node because the three area coordinates are not independent; since they are subject to a constraint (see Eq. (1)), the third gradient coordinate can be derived from the other two.

Within the Bézier triangle, the coordinates of point P are determined through interpolation of its control points:

$$\mathbf{r} = \sum_{k=1}^{10} g_k \mathbf{q}_k \tag{4}$$

In this context, $\mathbf{q}_k, k = 1, 2, \cdots 10$ represent the control points of the Bézier triangle, as illustrated in Fig. 1(b), while $g_k, k = 1, 2, \cdots 10$ correspond to the Bernstein basis functions. The transformation linking the coordinates of these control points to the element's nodal coordinates is established as follows:

$$\left.\begin{array}{l} \mathbf{r}^1 = \mathbf{q}_1, \mathbf{r}^1_{\xi_2} = 3(\mathbf{q}_4 - \mathbf{q}_1), \mathbf{r}^1_{\xi_3} = 3(\mathbf{q}_9 - \mathbf{q}_1) \\ \mathbf{r}^2 = \mathbf{q}_2, \mathbf{r}^2_{\xi_3} = 3(\mathbf{q}_6 - \mathbf{q}_2), \mathbf{r}^2_{\xi_1} = 3(\mathbf{q}_5 - \mathbf{q}_2) \\ \mathbf{r}^3 = \mathbf{q}_3, \mathbf{r}^3_{\xi_1} = 3(\mathbf{q}_8 - \mathbf{q}_3), \mathbf{r}^3_{\xi_2} = 3(\mathbf{q}_7 - \mathbf{q}_3) \end{array}\right\} \tag{5}$$

Combining Eq. (2) with Eq. (4), the shape function **S** expressed in terms of area coordinates can be obtained. Although the shape function **S** is formally concise, the discontinuity of its area gradient vectors at the nodes affects the assembly between elements. Wang (Pappalardo 2017, Wang 2018, Wang 2020) introduced a coordinate transformation technique to convert the element nodal coordinates from the form **e** (which incorporates area gradient coordinates) to the form **p** (which incorporates Cartesian gradient coordinates), thereby resolving the issue of discontinuity in the nodal gradient coordinates. The resulting nodal coordinates **p** are given by:

$$\left.\begin{array}{l} \mathbf{e} = \mathbf{T}\mathbf{p} \\ \mathbf{p} = [\mathbf{r}^1\ \mathbf{r}^1_x\ \mathbf{r}^1_y\ \mathbf{r}^2\ \mathbf{r}^2_x\ \mathbf{r}^2_y\ \mathbf{r}^3\ \mathbf{r}^3_x\ \mathbf{r}^3_y]^T \end{array}\right\} \tag{6}$$

The following expression relates the nodal coordinates **e** to the nodal coordinates **p**:

$$\left.\begin{array}{l}\mathbf{T}=\begin{bmatrix}\mathbf{T}^1 & 0 & 0\\ 0 & \mathbf{T}^2 & 0\\ 0 & 0 & \mathbf{T}^3\end{bmatrix}\\ \mathbf{T}^1=\begin{bmatrix}\mathbf{I} & 0 & 0\\ 0 & 3\mathbf{r}_{0,x}^{1T}(\mathbf{q}_0^4-\mathbf{q}_0^1)\mathbf{I} & 3\mathbf{r}_{0,y}^{1T}(\mathbf{q}_0^4-\mathbf{q}_0^1)\mathbf{I}\\ 0 & 3\mathbf{r}_{0,x}^{1T}(\mathbf{q}_0^9-\mathbf{q}_0^1)\mathbf{I} & 3\mathbf{r}_{0,y}^{1T}(\mathbf{q}_0^9-\mathbf{q}_0^1)\mathbf{I}\end{bmatrix}\\ \mathbf{T}^2=\begin{bmatrix}\mathbf{I} & 0 & 0\\ 0 & 3\mathbf{r}_{0,x}^{2T}(\mathbf{q}_0^4-\mathbf{q}_0^1)\mathbf{I} & 3\mathbf{r}_{0,y}^{2T}(\mathbf{q}_0^4-\mathbf{q}_0^1)\mathbf{I}\\ 0 & 3\mathbf{r}_{0,x}^{2T}(\mathbf{q}_0^9-\mathbf{q}_0^1)\mathbf{I} & 3\mathbf{r}_{0,y}^{2T}(\mathbf{q}_0^9-\mathbf{q}_0^1)\mathbf{I}\end{bmatrix}\\ \mathbf{T}^3=\begin{bmatrix}\mathbf{I} & 0 & 0\\ 0 & 3\mathbf{r}_{0,x}^{3T}(\mathbf{q}_0^8-\mathbf{q}_0^3)\mathbf{I} & 3\mathbf{r}_{0,y}^{3T}(\mathbf{q}_0^8-\mathbf{q}_0^3)\mathbf{I}\\ 0 & 3\mathbf{r}_{0,x}^{3T}(\mathbf{q}_0^7-\mathbf{q}_0^3)\mathbf{I} & 3\mathbf{r}_{0,y}^{3T}(\mathbf{q}_0^7-\mathbf{q}_0^3)\mathbf{I}\end{bmatrix}\end{array}\right\} \tag{7}$$

In this expression, $\mathbf{r}_{0,x}^i$ and $\mathbf{r}_{0,y}^i$ $(for\ i = 1, 2, 3)$ are the Cartesian gradient coordinates at each element node in the initial configuration; $\mathbf{q}_0^i (for\ i = 1, 2, \cdots 9)$ refers to the initial position vector of the $i - th$ control point; and **I** denotes the identity matrix.

2.2 ANCF Dynamic Equation

By applying the principle of virtual work, the governing equations of motion for the ANCF triangular planar element can be established as follows (Pappalardo 2017, Wang 2018, Wang 2020):

$$\mathbf{M}\ddot{\mathbf{p}} = \mathbf{Q_{ex}} - \mathbf{Q_{in}} \tag{8}$$

where **M** is the mass matrix:

$$\mathbf{M} = \mathbf{T}^T (h \int_{A_0} \rho \mathbf{S}^T \mathbf{S} dA) \mathbf{T} \tag{9}$$

In this expression, h stands for the thickness of the element, and A_0 represents the area of the element in its initial configuration. The term $\mathbf{Q_{ex}}$ corresponds to the external forces applied to the element, whereas $\mathbf{Q_{in}}$ refers to the internal elastic forces generated by deformation. The formulation for $\mathbf{Q_{in}}$ is given as follows:

$$\left.\begin{array}{l}\mathbf{Q}_{in} = h\mathbf{T}^T \int_{A_0} \left(\frac{\partial \boldsymbol{\varepsilon}}{\partial \mathbf{e}}\right)^T \mathbf{D}\varepsilon dA_0\\ \mathbf{D} = \frac{E}{1-\nu^2}\begin{bmatrix}1 & \nu & 0\\ \nu & 1 & 0\\ 0 & 0 & (1-\nu)/2\end{bmatrix}\end{array}\right\} \tag{10}$$

where, E represents Young's modulus, and v denotes Poisson's ratio. The Green-Lagrange strain tensor $\boldsymbol{\varepsilon}$ can be expressed as:

$$\left.\begin{aligned}&\boldsymbol{\varepsilon} = [\varepsilon_{xx}\,\varepsilon_{yy} 2\varepsilon_{xy}]^T\\&\varepsilon_{ij} = \frac{1}{2}(\mathbf{r}_i^T\mathbf{r}_j - \mathbf{r}_{i0}^T\mathbf{r}_{j0})\\&i, j = x, y\end{aligned}\right\} \tag{11}$$

Here, the subscripts i and j indicate partial differentiation with respect to the spatial coordinates, specifically $i, j = x, y$. The terms $\mathbf{r}_{i0}$ and $\mathbf{r}_{j0}$ denote the partial derivatives of the position vector in the initial configuration with respect to x and y, respectively.

3 Improved Local Linearization Method

Solving the dynamic equations of ANCF necessitates the evaluation of elastic internal forces at every discrete time step, creating a significant computational bottleneck. Reducing the computational burden associated with these forces is therefore a key factor in accelerating ANCF-based simulations. One common strategy employed to streamline this process is local linearization (Tang 2019, Tian 2020, Tian 2025, Yu 2025), which reduces the complexity of evaluating elastic forces. According to the formulation in Eq. (8), the governing dynamic equations of ANCF can be expressed as:

According to Eq. (8), we know the dynamic equation of ANCF is:

$$\mathbf{M}\ddot{\mathbf{p}} = \mathbf{Q_{ex}} - \mathbf{Q_{in}} \tag{12}$$

Performing a Taylor expansion of the elastic internal force $\mathbf{Q_{in}}$ about the configuration $\mathbf{p}(t)$ yields:

$$\begin{aligned}\mathbf{Q}_{in}(\mathbf{p}(t+t_n)) = \mathbf{Q}_{in}(\mathbf{p}(t)) + \frac{\partial \mathbf{Q}_{in}}{\partial \mathbf{p}}(\mathbf{p}(t)) \cdot \Delta\mathbf{p}_n + \cdots + \frac{\partial \mathbf{Q}_{in}^n}{n!\partial^n \mathbf{p}}(\mathbf{p}(t)) \cdot \left[\Delta\mathbf{p}_n\right]^n\\ + \mathbf{R}_n(\mathbf{p}(t))\end{aligned} \tag{13}$$

where, t_n denotes a brief time interval, with $\mathbf{p}(t + t_n) = \mathbf{p}(t) + \Delta\mathbf{p}_n$, where $\Delta\mathbf{p}_n$ is a small displacement increment. This interval t_n may span several time steps. Within the range defined by $\Delta\mathbf{p}_n$, higher-order terms in the Taylor series are omitted, keeping only the linear component.

Consequently, the elastic internal force is approximated as:

$$\left.\begin{aligned}&\mathbf{Q}_{in}(\mathbf{p}(t+t_n)) \approx \mathbf{Q}_{in}(\mathbf{p}(t)) + \frac{\partial \mathbf{Q}_{in}}{\partial \mathbf{p}}(\mathbf{p}(t)) \cdot \Delta\mathbf{p}_n\\&\mathbf{K_t}(\mathbf{p}(t)) = \frac{\partial \mathbf{Q}_{in}}{\partial \mathbf{p}}(\mathbf{p}(t))\end{aligned}\right\} \tag{14}$$

where, $\mathbf{K_t}(\mathbf{p}(t))$ is the tangent stiffness matrix of the system at the configuration $\mathbf{p}(t)$. The expression for $\frac{\partial \mathbf{Q}_{in}}{\partial \mathbf{p}}$ is as follows:

$$\left.\begin{aligned}&\frac{\partial \mathbf{Q}_{in}}{\partial \mathbf{p}} = \mathbf{T} \cdot \frac{\partial \mathbf{Q}_{in}}{\partial \mathbf{e}}\\&\frac{\partial \mathbf{Q}_{in}}{\partial \mathbf{e}} = h\int_{A_0} (\frac{\partial^2 \varepsilon_{xx}}{\partial \mathbf{e}^2}\sigma_{xx} + \frac{\partial^2 \varepsilon_{yy}}{\partial \mathbf{e}^2}\sigma_{yy} + 2\frac{\partial^2 \varepsilon_{xy}}{\partial \mathbf{e}^2}\sigma_{xy} + \left(\frac{\partial \boldsymbol{\varepsilon}}{\partial \mathbf{e}}\right)^T \mathbf{D}\frac{\partial \boldsymbol{\varepsilon}}{\partial \mathbf{e}})\mathrm{d}A_0\end{aligned}\right\} \tag{15}$$

Within the range defined by $\Delta \mathbf{p}_n$, the tangent stiffness matrix $\mathbf{K_t}(\mathbf{p}(t))$ remains unchanged, allowing the elastic internal force $\mathbf{Q}_{in}(\mathbf{p}(t + t_n))$ to be evaluated directly from this constant matrix. As a result, it is not necessary to recompute the stiffness matrix at every time step over the interval $\Delta \mathbf{p}_n$, which leads to a significant reduction in computational effort.

3.1 Improved Local Linearization Update Rule

In previous model order reduction using local linearization (Tang 2019), the update rule adopted was:

$$\left.\begin{aligned} &\max(\left\| \Delta \mathbf{p}_n^j \right\|_2) \le \delta, j = 1 \cdots n_e \\ &\delta = \frac{L}{1000} \end{aligned}\right\} \tag{16}$$

where, $\Delta \mathbf{p}_n^j$ represents the displacement of the j-th node within the current local linearization range, and n_e denotes the number of finite element nodes in the system. The value of δ is an empirical parameter. In previous studies, δ was determined based on the dimensions of the simulated flexible body, for example, $\delta = L/1000$, where L is the longer side length of the flexible body. However, this update rule has several shortcomings. First, when dealing with objects of different sizes, the selection of δ relies entirely on empirical judgment. When the flexible body itself is small, the chosen δ may become very small. For irregularly shaped objects, it is also difficult to determine δ using the above empirical formula.

Secondly, δ characterizes the displacement of a node within the local linearization time range. Node displacement may result from both rigid body motion and elastic deformation. However, as can be seen from Eqs. (10 and 11), the elastic internal forces of the flexible body are only related to its current configuration—that is, its deformation state—and not to its displacement. In simulation examples where the flexible body undergoes large-scale rigid body motion (e.g., an elastic sphere falling from a height and colliding with the ground), using displacement as the local linearization update criterion can lead to unnecessary updates during rigid body motion.

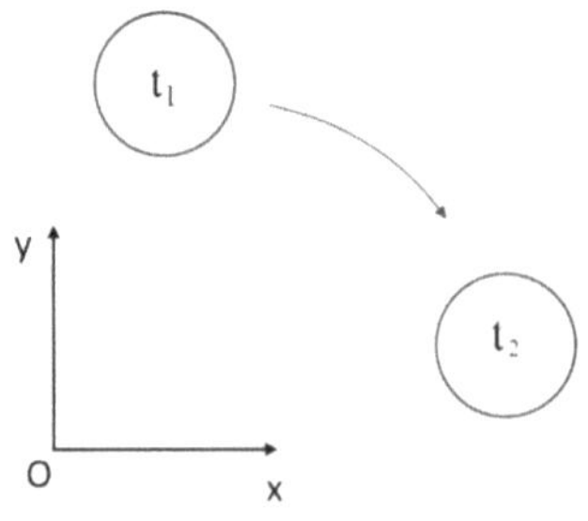

Fig. 2. Schematic diagram of the displacement of elastic sphere under the influence of gravity

For example, as shown in Fig. 2, a sphere moves under gravity from time t1 to time t2, disregarding air resistance. Since no external forces other than the uniform gravitational field act on it, the configuration of the elastic sphere remains unchanged from t1 to t2. If the update rule in Eq. (16) is adopted, several stiffness matrix updates would occur during the displacement from t1 to t2. However, because the configuration of the elastic sphere does not change during the displacement, its elastic internal forces remain constant, and thus, no updates to the tangent stiffness matrix are necessary. Therefore, using Eq. (16) as the update rule in scenarios involving large-scale displacement leads to unnecessary stiffness updates.

To avoid the aforementioned shortcomings, this paper proposes a new update rule. We adopt the Green–Lagrange strain as the criterion for determining whether to update the tangent stiffness matrix. Its expression is as follows:

$$\left.\begin{array}{l} \max(\left\|\Delta\boldsymbol{\varepsilon}_n^j\right\|_2) \le \delta, j = 1\cdots n_e \\ \Delta\boldsymbol{\varepsilon}_n^j = \boldsymbol{\varepsilon}_{t+t_n}^j - \boldsymbol{\varepsilon}_t^j \\ \delta = 0.001 \sim 0.005 \end{array}\right\} \tag{17}$$

where, $\Delta\boldsymbol{\varepsilon}_n^j = \boldsymbol{\varepsilon}_t^j - \boldsymbol{\varepsilon}_{t_n}^j$ represents the difference between the elastic strain $\boldsymbol{\varepsilon}_{t+t_n}^j$ at the current time step and the elastic strain $\boldsymbol{\varepsilon}_t^j$ at the moment when the tangent stiffness matrix was last updated, for the j-th node. We stipulate that the tangent stiffness matrix should be updated when the change in elastic strain exceeds 0.1% to 0.5% for any node of the flexible body. In ANCF, the Green–Lagrange strain used is numerically approximate to the Cauchy strain under small deformations. That is, we consider that the stiffness matrix needs to be updated when the flexible body undergoes elastic deformation equivalent to 0.1% to 0.5% of its own dimensions.

Adopting elastic strain as the update criterion effectively addresses the issues present when using displacement as the rule. Elastic strain is dimensionless, thus eliminating the need to consider the size of the flexible body when setting the update threshold. Furthermore, elastic strain is independent of displacement. Regardless of whether the flexible body undergoes large-scale displacement, elastic strain can accurately capture the magnitude of elastic deformation, enabling more effective updates of the tangent stiffness matrix.

3.2 Incorporating Full Elastic Force Calculation Within the Local Linearization Cycle

In previous model order reduction using local linearization, performing the Taylor expansion of the elastic internal force $\mathbf{Q_{in}}$ and retaining only the first-order linear term introduces an error on the order of $O((\Delta p)^2)$. Since the calculation of the elastic internal force for the subsequent step is based on the previously obtained force containing this error, error accumulation occurs. During prolonged simulations, this accumulated error leads to a mismatch between the elastic internal forces and the configuration of the flexible body, thereby compromising the computational accuracy of the simplified model. To extend the feasible simulation duration of the simplified model, we propose a method to mitigate error accumulation.

During the simulation, we periodically perform a full calculation of the elastic internal forces. This is not the elastic internal force obtained through simplified computation via local linearization, but rather the accurate elastic internal force that matches the current flexible body configuration, derived from Eqs. (10) and (11). Subsequent local linearization iterations are then based on this accurately obtained elastic internal force. This approach corrects the mismatch between the elastic internal forces and the flexible body configuration caused by prolonged local linearization calculations, ensuring that the next step of elastic internal force computation iterates from a correct foundation, thereby mitigating the error accumulation associated with long-term local linearization. In the two subsequent case studies, we calculate the accurate elastic internal force using Eq. (10) every 1000 time steps and the time step size of $1 \times 10^{-3} s$ was adopted for both examples.

The flowchart for the improved local linearization simplified computation process is shown in Fig. 3.

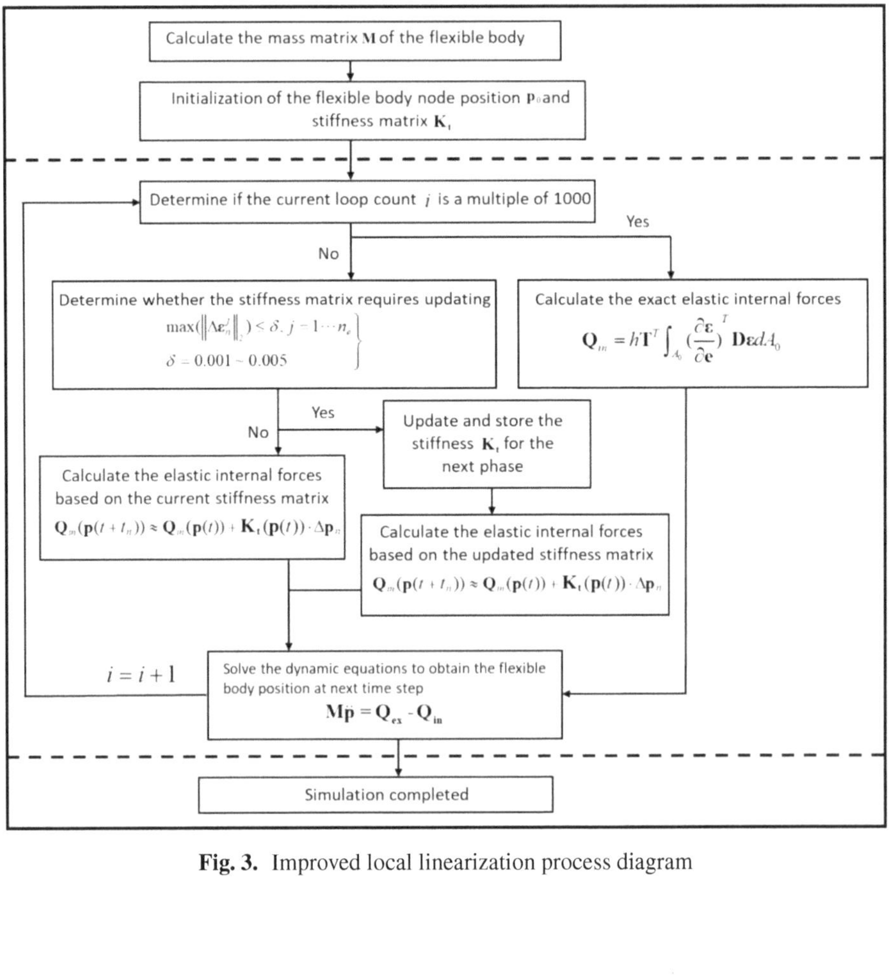

Fig. 3. Improved local linearization process diagram

4 Case Studies

In order to validate the precision of the enhanced local linearization algorithm for extended-duration simulations, two numerical case are investigated herein. The simulations were carried out using MATLAB 2024a on a workstation equipped with an Intel(R) Core(TM) Ultra 5 225H processor (1.70 GHz) and 32 GB of RAM.

4.1 Flexible Sphere Collision

To verify the accuracy and effectiveness of the improved local linearization algorithm in long-term simulations, a dynamics simulation of a two-dimensional flexible sphere falling under a uniform gravitational field and colliding with the ground was conducted. The initial state of the flexible sphere is shown in Fig. 4, and the relevant parameters for the dynamics simulation are listed in Table 1.

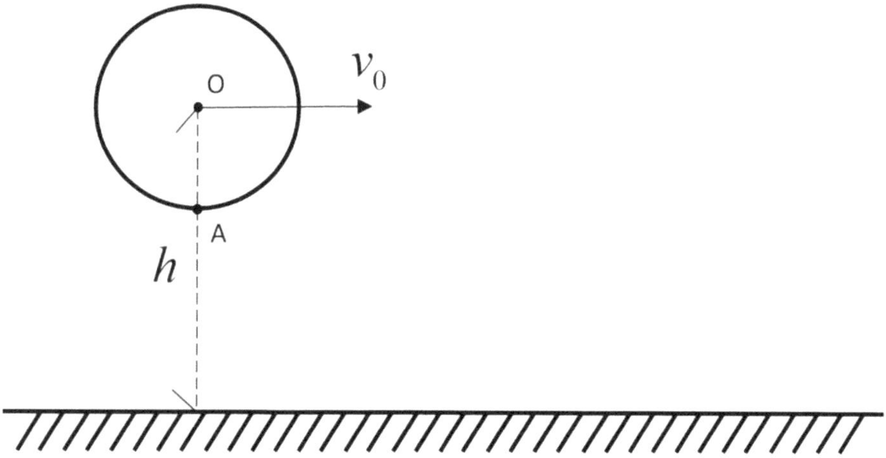

Fig. 4. Illustration of the initial configuration of the flexible sphere

Table 1. Structure parameters

Structural parameters	Value
Sphere radius R	$5m$
The distance from the center of the sphere to the ground h	$15m$
Young's modulus E	$9 \times 10^7 Pa$
Poisson's ratio v	0.3
Element thickness t	$0.01m$
Gravitational acceleration g	9.81

(continued)

Table 1. (*continued*)

Structural parameters	Value
Horizontal initial velocity v_0	$2m/s$
Ground friction coefficient μ	0.6

During the collision between the flexible sphere and the ground, significant instantaneous impact loads are generated, imposing high demands on the simulation accuracy of the simplified model. The spring–damper contact model and the Coulomb friction model were employed to simulate the collision and friction, and the dynamic response of the flexible sphere over 20 s was analyzed. Elastic internal force cloud map of the flexible sphere at its current position were generated every 2 s, as shown in Fig. 5(a). The trajectory of point A at the sphere center is illustrated in Fig. 5(b) and (c).

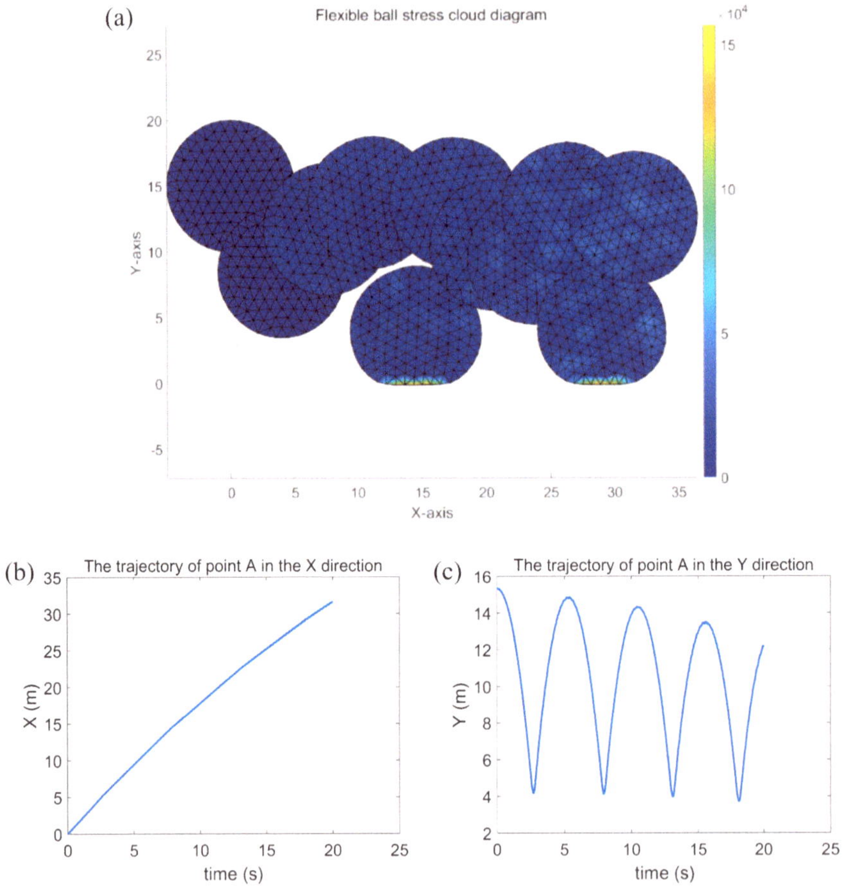

Fig. 5. (**a**) Stress cloud map of the flexible sphere (**b**) X-direction trajectory of point A (**c**) Y-direction trajectory of point A

Simplified computations were performed using both the traditional local linearization method(LL) and the improved local linearization method (I-LL). In the traditional local linearization method, the update criterion is $\max(\left\|\Delta\mathbf{p}_n^j\right\|2) \leq 0.001$; for the improved local linearization method proposed in this paper, the update criterion is $\max(\left\|\Delta\boldsymbol{\varepsilon}_n^j\right\|2) \leq 0.003$, and an accurate calculation of the elastic internal forces is performed every 1000 time steps. As shown in Eq. (17), we consider that the tangent stiffness matrix can be updated when the elastic strain difference is in $0.001 \sim 0.005$. The reason why this example uses $\max(\left\|\Delta\boldsymbol{\varepsilon}_n^j\right\|2) \leq 0.003$ is that the elastic strain is relatively large during the sphere collision, so the update criterion can be appropriately raised to achieve a balance between accuracy and computational speed. The computational results are shown in Fig. 6.

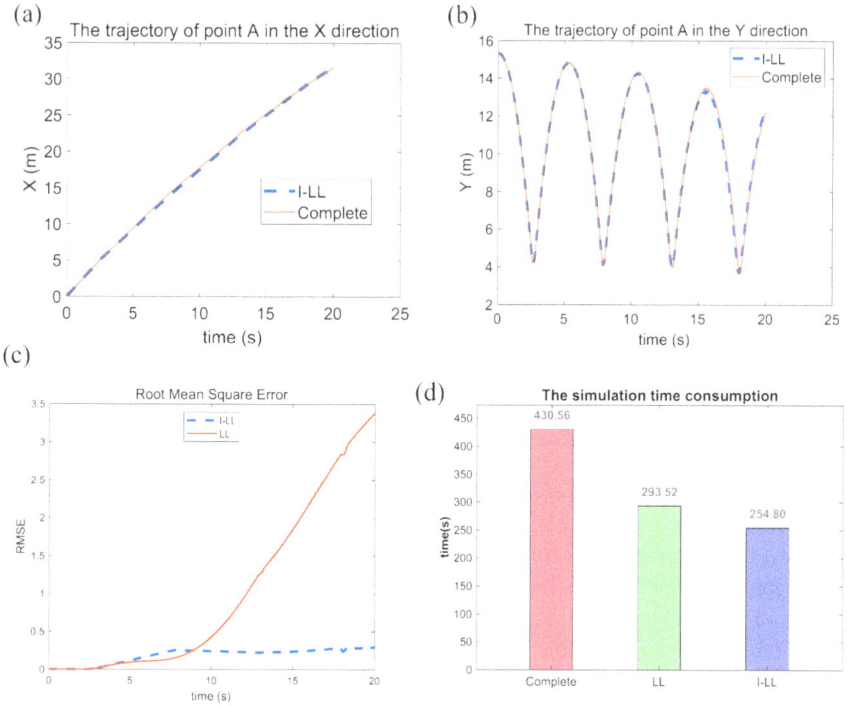

Fig. 6. (**a**) Comparative results for the X-direction trajectory of point A as predicted by the I-LL and complete model (**b**) Comparative results for the Y-direction trajectory of point A as predicted by the I-LL and complete model (**c**) Comparison of the RMSE between the LL model and the I-LL model (**d**) Comparison of the simulation time for the complete model, LL model, and I-LL model

Table 2. Comparison of the times of updates the tangent stiffness matrix in the complete model, the LL model, and the I-LL model

	Complete model	Local linearization	Improved Local linearization
Times of updates the tangent stiffness matrix	20000	10000	8764

As illustrated in Fig. 6(a) and (b), the trajectories obtained from the I-LL model agree well with those from the complete model. Figure 6(c) shows that the Root Mean Square Error (RMSE) between the I-LL model and the complete model is maintained below 0.3. In contrast, the traditional local linearization model maintains good accuracy only up to 10 s, after which the RMSE begins to increase rapidly. By the 20 s of simulation, the RMSE of the traditional local linearization model has expanded to ten times that of the I-LL model.

As illustrated in Fig. 6(d) and Table 2, the simulation time of the I-LL model is reduced by 40.82% compared to the complete model, and by 13.19% compared to the traditional local linearization model. The number of tangent stiffness matrix updates for I-LL is reduced by 12.36% relative to the traditional local linearization approach. These simulation results demonstrate that, compared with the traditional local linearization model, the I-LL model reduces simulation time while maintaining simulation accuracy. The strategy of inserting accurate elastic internal force calculations during the local linearization process effectively mitigates error accumulation, while the adoption of strain as the update criterion reduces the frequency of tangent stiffness matrix updates and more accurately reflects the deformation state of the flexible body.

4.2 Flexible Double Pendulum

To further assess the performance of the I-LL method, a dynamic simulation is carried out on a flexible double pendulum model. The pendulum is subjected to a constant vertical downward force applied at its free end. It comprises two flexible beams connected by a revolute joint at point A, with the left end of the first beam attached to a wall by another revolute joint. A schematic representation of the system is provided in Fig. 7, and the corresponding structural parameters are summarized in Table 3.

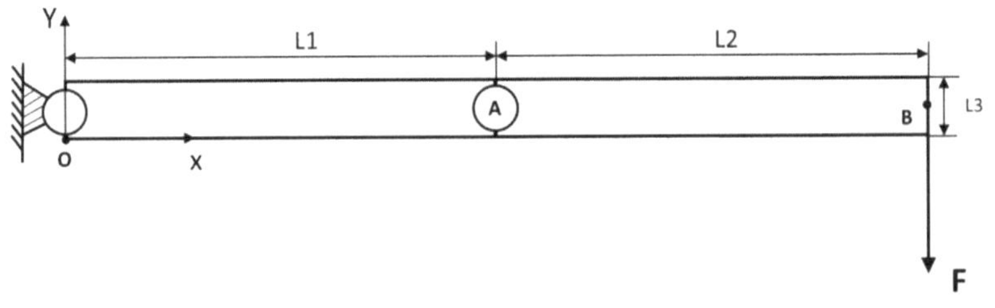

Fig. 7. Schematic diagram of the flexible double pendulum structure

Table 3. Structure parameters

Structural parameters	Value
$L1$ Beam length	$20m$
$L2$ Beam length	$20m$
Beam width h	$2m$
Young's modulus E	$6 \times 10^7 Pa$
Poisson's ratio v	0.3
Density ρ	$4800kg/m^3$
Element Thickness t	$0.01m$
Vertical downward Force **F**	$-3000N$

A 50-s dynamic simulation of the flexible double pendulum is conducted using the complete model, and the trajectory of point B at the end of the flexible beam is shown in Fig. 8.

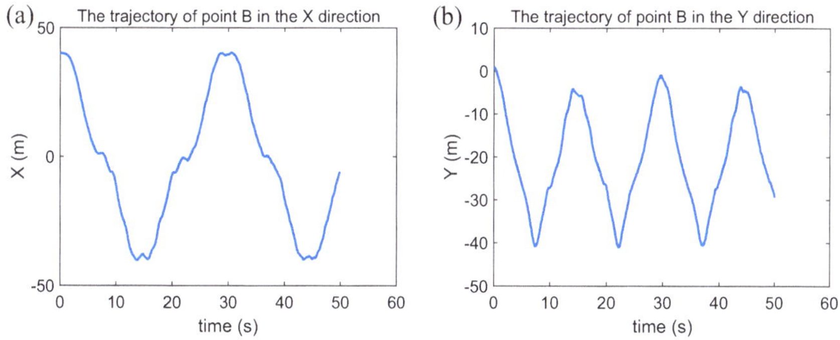

Fig. 8. (**a**) X-direction trajectory of point B (**b**) Y-direction trajectory of point B

We simulated the flexible double pendulum using both the traditional local linearization model and the improved local linearization model. In the traditional local linearization approach, the update criterion is $\max(\left\|\Delta \mathbf{p}_n^j\right\|2) \leq 0.001$; for the improved local linearization method proposed in this paper, the update criterion is $\max(\left\|\Delta \boldsymbol{\varepsilon}_n^j\right\|2) \leq 0.001$, and an accurate calculation of the elastic internal forces is performed every 1000 time steps. In the flexible double pendulum example, we use $\max(\left\|\Delta \boldsymbol{\varepsilon}_n^j\right\|2) \leq 0.001$ instead of 0.003 used in the sphere example. As the extent of elastic strain experienced by the flexible double pendulum throughout its motion cannot be predetermined, adopting a lower update threshold serves to maintain numerical precision in scenarios where the system behavior is not fully known. The computational results of the I-LL model are shown in Fig. 9.

Table 4. Frequency of tangent stiffness matrix updates in the complete model, LL model, and I-LL model

	Complete model		Local linearization		Improved local linearization	
Times of updates for the tangent stiffness matrix	Flexible beam L1	Flexible beam L2	Flexible beam L1	Flexible beam L2	Flexible beam L1	Flexible beam L2
	50000	50000	23796	24910	19760	24680

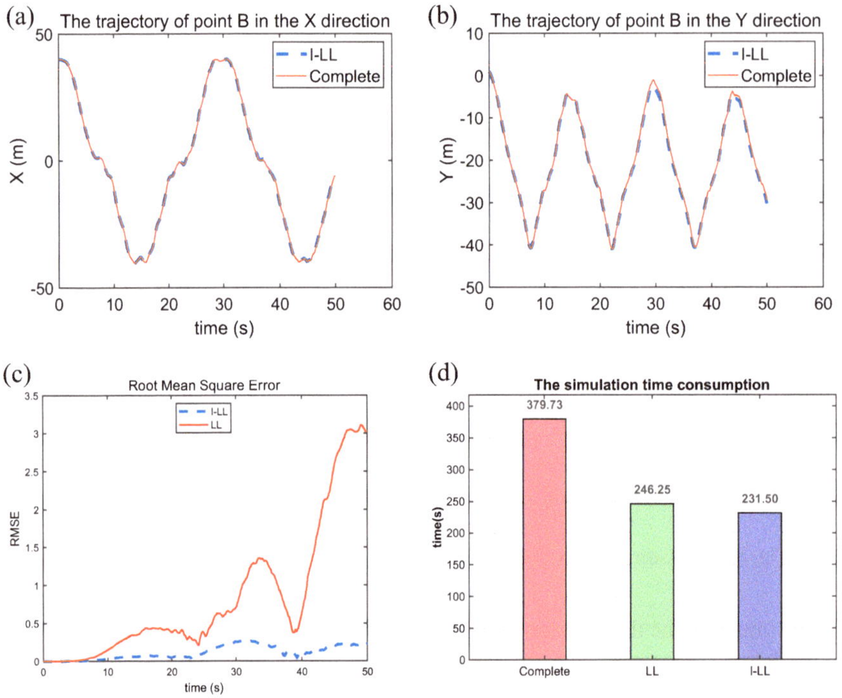

Fig. 9. (**a**) Comparative results for the X-direction trajectory of point B as predicted by the I-LL and complete model (**b**) Comparative results for the Y-direction trajectory of point B as predicted by the I-LL and complete model (**c**) Comparison of RMSE between LL model and I-LL model (**d**) Comparison of computing time for complete model, LL model, and I-LL model

As can be observed from Fig. 9(a) and (b), the trajectories obtained from the I-LL model are in good agreement with those from the complete model. Figure 9(c) shows that the RMSE between the I-LL model and the full model is maintained below 0.3. In contrast, the traditional local linearization model maintains good accuracy only up to 20 s, after which the RMSE begins to increase rapidly. By the 50 s of simulation, the

RMSE of the traditional local linearization model has expanded to ten times that of the I-LL model.

As can be seen from Fig. 9(d) and Table 4, the simulation time consumption of the I-LL model is reduced by 39.04% compared to the complete model, and by 5.9% compared to the traditional local linearization model. Due to the differing motion amplitudes of the two flexible pendulums, local linearization was performed separately for the L1 and L2 flexible beams in both the LL and I-LL models. Compared with the traditional LL model, the number of tangent stiffness matrix updates for the L1 beam in the I-LL model is reduced by 16.96%, and for the L2 beam by 0.92%. These simulation results demonstrate that, compared with the traditional LL model, the I-LL model reduces simulation time while maintaining simulation accuracy. Although the rigid body displacement in the motion of the flexible double pendulum model is relatively small, resulting in a marginal difference between using elastic strain and displacement as the update criterion in terms of simulation time reduction, the I-LL model remains highly effective in mitigating error accumulation.

5 Conclusion

To address the issues of error accumulation in traditional local linearization methods and the adaptability of the tangent stiffness matrix update criterion, this paper proposes an improved local linearization method (I-LL). The I-LL method enhances the conventional LL approach in two main aspects. First, to mitigate the rate of error accumulation, I-LL periodically inserts an accurate calculation of the elastic internal forces at specified time intervals and uses the obtained forces as the basis for the subsequent local linearization iteration. Second, the update criterion for the tangent stiffness matrix is modified from displacement-based to elastic strain-based. Elastic strain accurately captures the extent of deformation of the flexible body and provides a more effective criterion for determining whether the tangent stiffness matrix needs to be updated, particularly in scenarios involving large rigid body displacements.

The accuracy and effectiveness of the I-LL method are demonstrated through two numerical cases. Compared with the traditional local linearization method, I-LL exhibits significant advantages in mitigating error accumulation and maintaining long-term simulation accuracy. Furthermore, in both examples, I-LL requires fewer tangent stiffness matrix updates and less computational time than the conventional LL approach. This advantage is especially pronounced in the sphere–ground collision example, which involves substantial rigid body motion. In future work, we will further explore the application of the I-LL method to more complex examples for long-duration, high-precision simulations, and develop more effective simplified computation strategies for elastic internal forces to further alleviate error accumulation.

Acknowledgments. This work was supported by the National Key Research and Development Program of China (Grant 2024YFB3310201).

References

Bauchau, O.A., Han, S., Mikkola, A., Matikainen, M.K.: Comparison of the absolute nodal coordinate and geometrically exact formulations for beams. Multibody Sys.Dyn. **32**, 67–85 (2014). https://doi.org/10.1007/s11044-013-9374-7

Dibold, M., Gerstmayr, J., Irschik, H.: A detailed comparison of the absolute nodal coordinate and the floating frame of reference formulation in deformable multibody systems. J. Comput. Nonlinear Dyn. **4**(2), 071008 (2009). https://doi.org/10.1115/1.3079825

Guo, Y.Q., Li, L., Zhang, D.G., Liao, W.H.: Thermo-electro-mechanical modeling and analysis of flexible system containing dielectric elastomers. Int. J. Mech. Sci. **311**, 111150 (2026). https://doi.org/10.1016/j.ijmecsci.2026.111150

Kim, E., Cho, M.: Design of a planar multibody dynamic system with ANCF beam elements based on an element-wise stiffness evaluation procedure. Struct. Multidiscip. Optim. **58**(3), 1095–1107 (2018). https://doi.org/10.1007/s00158-018-1954-y

Li, K., Yu, Z., Lan, P., Tian, Q.L., Liu, N.L.: ALE-ANCF circular cross-section beam element and its application on the dynamic analysis of cable-driven mechanism. Multibody Sys.Dyn. **60**(2), 417–446 (2024). https://doi.org/10.1007/s11044-023-09929-1

Otsuka, K., Makihara, K., Sugiyama, H.: Recent advances in the absolute nodal coordinate formulation: Literature review from 2012 to 2020. J. Comput. Nonlinear Dyn. **17**(8), 080803 (2022). https://doi.org/10.1115/1.4054113

Pappalardo, C.M., Wang, T., Shabana, A.A.: Development of ANCF tetrahedral finite elements for the nonlinear dynamics of flexible structures. Nonlinear Dyn. **89**(4), 2905–2932 (2017). https://doi.org/10.1007/s11071-017-3635-6

Romero, I.: A comparison of finite elements for nonlinear beams: The absolute nodal coordinate and geometrically exact formulations. Multibody Sys.Dyn. **20**, 51–68 (2008). https://doi.org/10.1007/s11044-008-9105-7

Shabana, A.A.: Definition of the slopes and the finite element absolute nodal coordinate formulation. Multibody Sys.Dyn. **1**(3), 339–348 (1997). https://doi.org/10.1023/A:1009740800463

Shabana, A.A.: Flexible multibody dynamics: Review of past and recent developments. Multibody Sys.Dyn. **1**(2), 189–222 (1997). https://doi.org/10.1023/A:1009773505418

Shabana, A.A.: Dynamics of multibody systems (3rd ed.). Cambridge University Press (2005). https://doi.org/10.1017/CBO9780511610523

Shabana, A.A.: An overview of the ANCF approach, justifications for its use, implementation issues, and future research directions. Multibody Sys.Dyn. **58**(3), 433–477 (2023). https://doi.org/10.1007/s11044-023-09890-z

Tang, Y.X., Hu, H.Y., Tian, Q.: Model order reduction based on successively local linearization for flexible multibody dynamics. Int. J. Numer. Meth. Eng. **118**(1), 159–180 (2019). https://doi.org/10.1002/nme.6011

Tian, Q., Pan, C., Liu, Z., Chen, X.: Model order reduction for rigid–flexible–thermal coupled viscoelastic multibody system via the modal truncation with complex global modes. Actuators **14**(10), 479 (2025). https://doi.org/10.3390/act14100479

Tian, Q.L., Lan, P., Yu, Z.Q.: Model-order reduction of flexible multibody dynamics via free-interface component mode synthesis method. J. Comput. Nonlinear Dyn. **15**(10), 101008 (2020). https://doi.org/10.1115/1.4047868

Wang, T.: Two new triangular thin plate/shell elements based on the absolute nodal coordinate formulation. Nonlinear Dyn. **99**(4), 2707–2725 (2020). https://doi.org/10.1007/s11071-019-05448-x

Wang, T., Mikkola, A., Matikainen, M.K.: An overview of higher-order beam elements based on the absolute nodal coordinate formulation. J. Comput. Nonlinear Dyn. **17**(9), 091001 (2022). https://doi.org/10.1115/1.4054348

Wang, T., Tinsley, B., Patel, M.D., Shabana, A.A.: Nonlinear dynamic analysis of parabolic leaf springs using ANCF geometry and data acquisition. Nonlinear Dyn. **93**(3), 2487–2515 (2018). https://doi.org/10.1007/s11071-018-4338-3

Wang, W., et al.: Dynamics modeling and control method for non-cooperative target capture with a space netted pocket system. Actuators **14**(7), 358 (2025). https://doi.org/10.3390/act14070358

Xue, Y.C., Li, L., Zhang, D.G., Liao, W.H., Guo, X., Guo, Y. B.: Modeling and analysis of underwater oscillation of a flexible biomimetic caudal fin driven by MFC based on ANCF. Thin-Walled Structures, **218**(Part A), 113970 (2026). https://doi.org/10.1016/j.tws.2025.113970

Yu, Z.Q., Liu, Z., Wang, Y., Tian, Q.L.: Complex modal synthesis method for viscoelastic flexible multibody system described by ANCF. J. Comput. Nonlinear Dyn. **20**(3), 031002 (2025). https://doi.org/10.1115/1.4067522

Zhang, Y., Wei, C., Zhao, Y., Tan, C.L., Liu, Y.J.: Adaptive ANCF method and its application in planar flexible cables. Acta Mech. Sin. **34**(1), 199–213 (2018). https://doi.org/10.1007/s10409-017-0721-4

Zhang, Z.W., Ren, W., Zhou, W.J.: Research status and prospect of plate elements in absolute nodal coordinate formulation. Proceedings of the Institution of Mechanical Engineers, Part K: Journal of Multi-body Dynamics **236**(3), 357–367 (2022). https://doi.org/10.1177/14644193221098866

Fluid–Structure Interaction Analysis and Multi-Objective Optimization Design of Rectangular Microchannels

Zhaoyang Li[1], Shengzi Lu[1], Pengan Zhu[1], Huasheng Hu[1], Ganlu Zeng[1], Cong Deng[1], and Huikun Cai[2](✉)

[1] Guangdong Institute of Special Equipment Inspection and Research, Foshan 528251, China
[2] Pen-Tung Sah Institute of Micro-Nano Science and Technology, Xiamen University, Xiamen 361000, China
caihuikun@xmu.edu.cn

Abstract. To tackle the degradation of heat transfer performance in microscale flows, an active regulation strategy based on structural parameter optimization is proposed in this study. Taking a microchannel heat exchanger as the research object, a parametric model is established by integrating multiple design variables, such as characteristic dimensions, channel height, and inlet flow rate. Subsequently, a multi-objective optimization framework is constructed, with the objectives of controlling average temperature, minimizing total mass, and maximizing heat transfer efficiency, while subject to the constraints of inlet flow rate and overall system pressure drop. By coupling computational fluid dynamics (CFD) simulations with intelligent optimization algorithms, the thermo–fluid–structure interaction mechanisms under the synergistic effects of various structural parameters are systematically investigated. The results demonstrate that a channel aspect ratio of 5:1 achieves superior comprehensive thermo-hydraulic performance, realizing the simultaneous improvement of heat transfer efficiency, suppression of pressure drop losses, and lightweight design of the system. This research provides theoretical guidance and a design paradigm for the thermal management of high-power-density microdevices.

Keyword: Heat exchanger · Optimal design · Multi-objective

1 Introduction

With the increasing demand for higher energy utilization efficiency and high heat-flux dissipation, the design and optimization of high-performance heat exchangers have become a major research focus in the field of engineering thermophysics. Enhancing heat transfer structures, optimizing flow channels, and improving internal geometric configurations can significantly improve heat transfer efficiency while reducing flow resistance. In recent years, related studies have mainly focused on fin optimization, flow enhancement structures, and multi-objective optimization methods.

T. Mankovits and N. Trisovic (Eds.): MDS 2026, LNME 2026.
https://doi.org/10.1007/978-981-92-2440-1_33

In terms of fin structure optimization, extended surfaces can effectively increase the heat transfer area. (Hajabdollahi 2022) employed a multi-objective optimization approach and found that the newly designed constructal fins improved efficiency by 7% while reducing the total heat transfer area by 9.47% compared with conventional fins. (Yang 2025) applied topology optimization to the fin design of rectangular parallel-plate heat exchangers. The topology-optimized fins effectively enhanced heat transfer, increasing the Nusselt number by 5.77%–8.23%, while the performance evaluation criterion reached 1.71–2.12.

Regarding heat transfer enhancement structures, (Jaiswal 2023) optimized truncated-prism ribs in a rectangular channel using a genetic algorithm, achieving a 168.6% improvement in overall thermal performance compared with the reference design. (Negi 2024) conducted numerical simulations showing that optimized rib geometries in threaded tubes could induce secondary flows, increasing the heat transfer rate by approximately 17% with only minor changes in pressure drop. (Wang 2025) introduced vortex generators into plate-fin heat exchangers and optimized their arrangement, resulting in a 25.3% increase in the overall thermal performance factor.

In terms of flow distribution and advanced optimization methods, (Li 2025), (Jing 2023), and Yicong optimized microchannel distribution structures, fin-tube heat exchanger circuits for air-conditioning systems, and chevron corrugated plate parameters, respectively, achieving coordinated improvements in heat transfer and flow performance. Furthermore, (Petrovic 2023) and (Wang 2024) applied topology optimization and response surface methodology to heat exchanger design, providing efficient pathways for structural optimization.

Although substantial progress has been made in heat exchanger structural optimization and heat transfer enhancement, studies on the coupled effects of flow, heat transfer, and structural deformation in microscale rectangular channels remain relatively limited. In microchannels, variations in pressure and temperature induced by fluid flow may cause structural deformation, which in turn influences the flow and heat transfer processes. Therefore, fluid–solid coupling analysis combined with multi-objective optimization is essential for improving the comprehensive performance of microchannel heat exchangers.

Based on this background, the present study investigates a rectangular microchannel, establishes a fluid–solid coupling numerical model, and applies multi-objective optimization to key structural parameters. The objective is to achieve coordinated optimization between heat transfer performance and flow resistance, thereby providing theoretical guidance and technical support for the design of high-performance microchannel heat exchangers.

2 Parametric Structural Design of Microchannel Models

Based on the geometric configuration of the rectangular heat sink illustrated in Fig. 1 (overall dimensions: $L \times W \times H = 8\ mm \times 8\ mm \times 2\ mm$), a parameterized flow channel model was constructed with copper as the substrate material. The total flow cross-sectional area S of the heat sink is governed by the coupled effects of the following key parameters.

$$S = N \times b \times h \tag{1}$$

In Eq. (1):

N denotes the number of flow channels, representing the quantity of parallel flow paths; b denotes the single-channel width (as labeled in the figure), defined as the spacing between adjacent fins;

h denotes the channel height (as labeled in the figure), corresponding to the flow dimension in the vertical direction between fins.

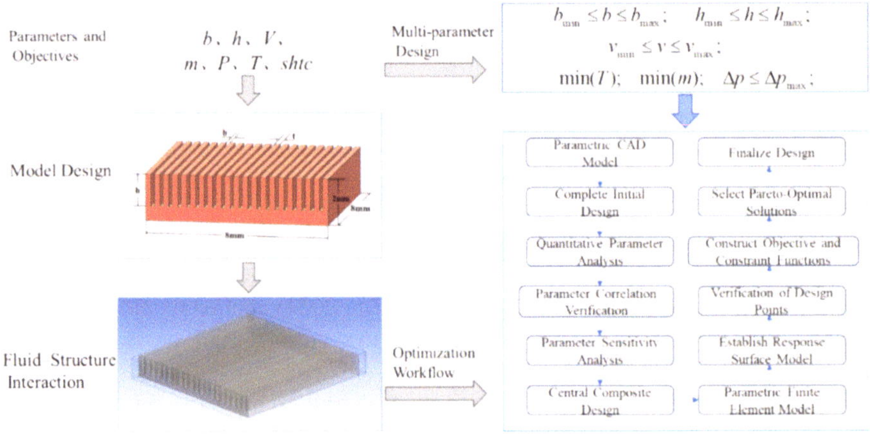

Fig. 1. Conceptual framework of microscale heat transfer enhancement.

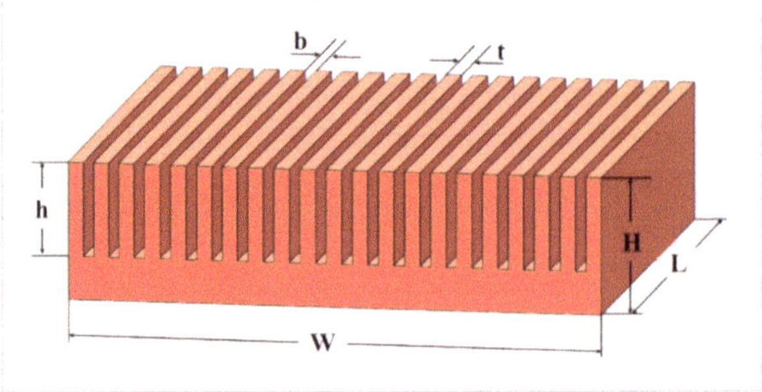

Fig. 2. Three-dimensional schematic of the microchannel heat exchanger.

As shown in Fig. 2, the total height of the heat sink, $H = 2\text{mm}$, consists of the substrate thickness and the channel height h, with the constraint $h < H$ to ensure structural integrity. By tuning the matching relationship among N, b, and h, an optimal balance can be achieved between thermal resistance, pressure drop, and heat dissipation efficiency.

3 Numerical Analysis of the Heat Exchanger

To ensure the reliability of the multi-objective optimization results, a rigorous grid independence verification was performed on the computational model prior to parameter optimization, using the Fluent module in ANSYS Workbench 2024R2. Helium was employed as the working fluid, with an inlet temperature of 300 K, while a heat source of 12.8 W was applied at the bottom of the heat exchanger.

Monitoring parameters: mass-weighted average temperature at each channel outlet and pressure drop. Convergence criteria: residuals reduced to 10 − 6 for the energy equation and 10 − 5 for the momentum equations, with monitored parameter fluctuations within 0.1% sustained over 100 iterations.

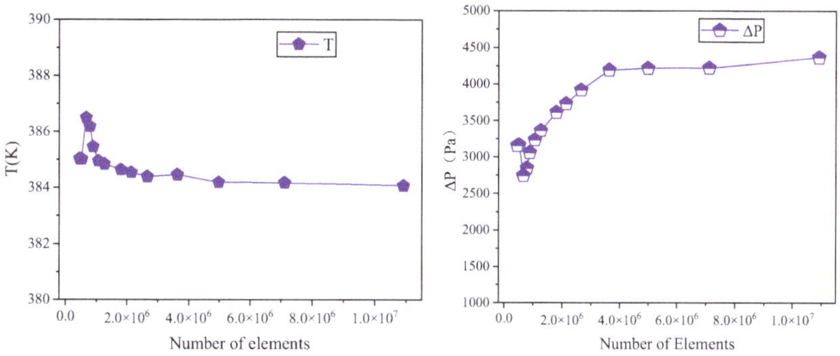

Fig. 3. Grid independence verification curves.

From Fig. 3, it can be seen that when the grid number reaches 3.6 million, the fluctuations of the mass-weighted outlet temperature Tout and average pressure drop ΔP become negligible. Refining the mesh to 10 million cells yields a relative difference in Tout of less than 0.1% and ΔP fluctuations below 1%, confirming grid independence of the temperature field and stability of flow resistance prediction. Therefore, a grid size of 3.6 million was adopted as the computational baseline, reducing single-case simulation time by approximately 64% compared to the 10 million cell model, while maintaining outlet thermal parameter accuracy within 2% and greatly improving multi-condition simulation efficiency.

3.1 Quantitative Analysis of Input Parameters

3.1.1 Analysis of Different Channel Numbers

Based on previous studies of the heat exchanger, the total flow channel cross-sectional area ($N \times b \times h$, where N is the number of channels) is approximately half of the total channel cross-sectional area ($W \times H$). This relationship enables the determination of the number of channels N. The inter-channel wall thickness t and the substrate thickness (H-h) can then be determined accordingly.

As shown in the pressure contour of Fig. 4, the inlet pressure at a channel number of 5 is only about 6% of that at 50 channels. This indicates that with fewer channels, the larger cross-sectional area of each channel reduces fluid velocity, thereby significantly lowering flow resistance and inlet pressure requirements. In contrast, as the channel number increases, the narrower passages intensify wall friction and viscous dissipation, causing a sharp rise in inlet pressure. Although a larger channel number expands the effective heat transfer area, it simultaneously leads to substantially higher flow resistance and pumping power consumption, reflecting the fundamental trade-off between pressure drop and heat transfer enhancement in the design of microscale heat exchangers.

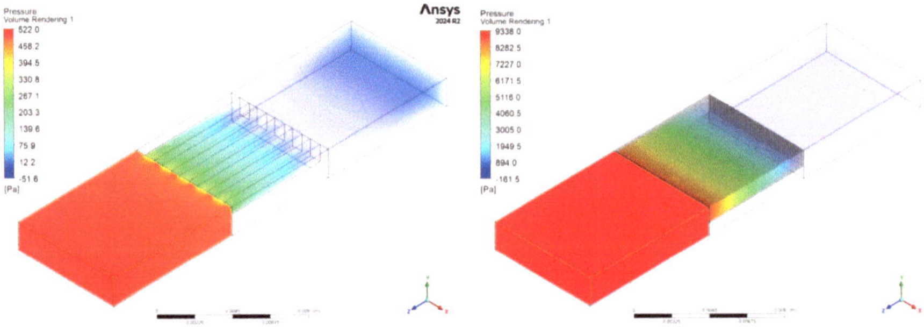

Fig. 4. Pressure distribution.

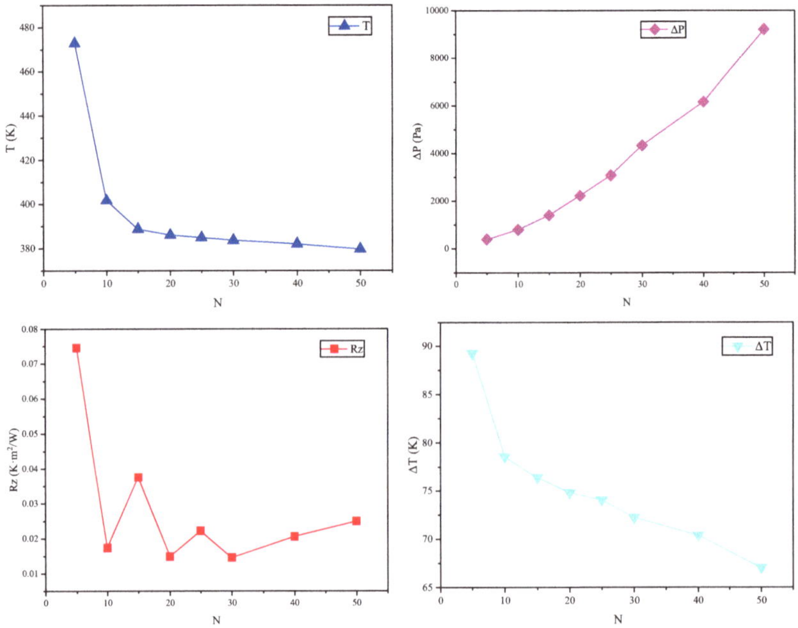

Fig. 5. Effect of channel number N on relevant parameters.

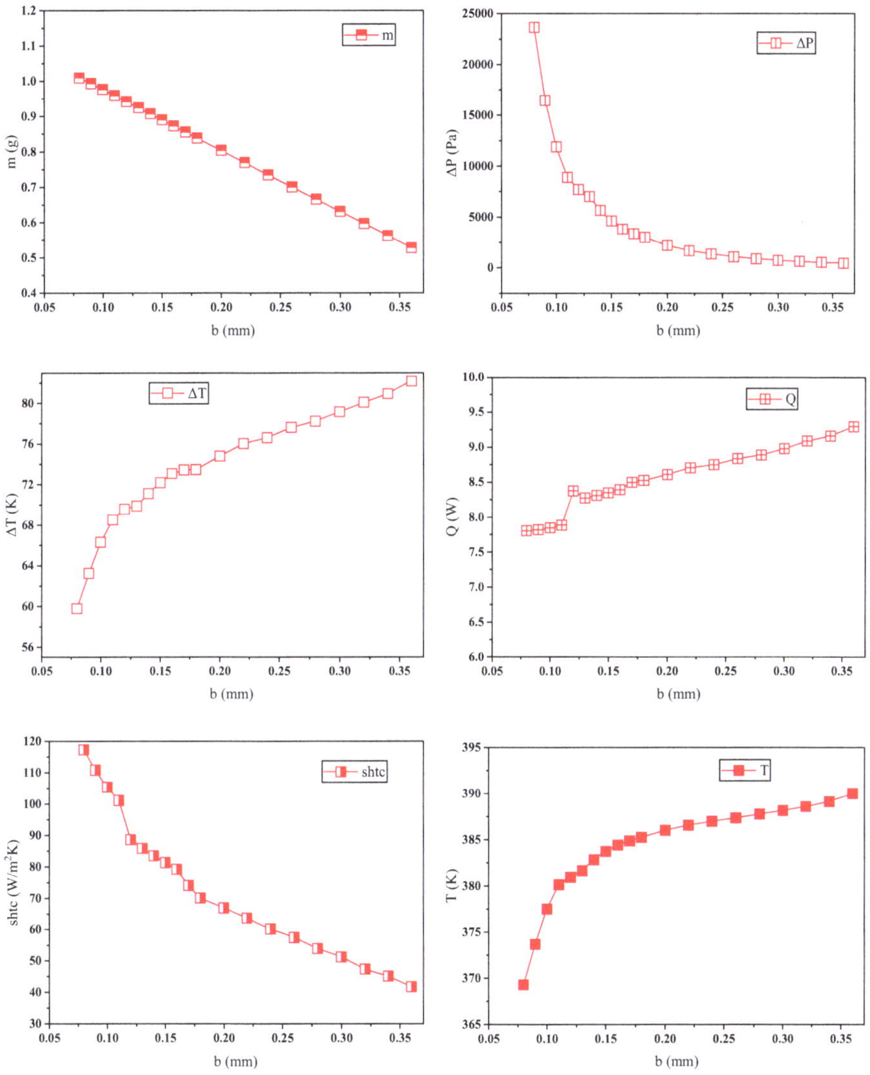

Fig. 6. Effect of channel width *b* on relevant parameters.

As shown in Fig. 5, under the condition of constant channel height and fixed total cross-sectional area, increasing the number of channels leads to a gradual reduction in the width of each channel. Numerical results indicate that when the number of channels increases from 5 to 50, the overall solid temperature decreases, the average pressure drop significantly increases, and the inlet–outlet temperature difference gradually decreases. Further comparative analysis of heat transfer performance reveals that the comprehensive heat transfer coefficient reaches its maximum at 20 channels, while the corresponding thermal resistance remains relatively low. Considering the coupling effects between heat transfer enhancement and flow resistance, as well as engineering economy, 20 channels are ultimately selected as the optimal configuration for subsequent investigation.

3.1.2 Analysis of Different Channel Widths

In this section, the inlet cross-sectional dimensions of the heat exchanger correspond to its overall width *W*and height *H*. The channel height was initially set to $h = 1.2$mm, and the inlet cross-sectional velocity was set to 10.42m/s. Simulations were then conducted for different channel widths *b* ranging from 0.08mm to 0.32mm. The fitted results are presented as the curve shown in Fig. 6.

As shown in Fig. 6, with the channel width *b* increasing from 0.08 mm to 0.37 mm, the total mass of the heat exchanger decreases gradually from 1.01 g to 0.60 g, the mass-weighted average pressure drop at the channel inlet decreases from 13,700 Pa to 590 Pa, while the temperature difference between the channel outlet and inlet increases from 66.4 K to 81 K. The convective heat transfer area of the channels expands with increasing *b*, whereas the fluid velocity decreases, resulting in a convective heat transfer coefficient that first increases and then decreases. The reduction in the fin conduction area causes the thermal resistance to exhibit a decreasing–then–increasing trend, thereby affecting the overall heat transfer performance. As a result, the solid temperature increases with *b*.

Notably, within the range of $b = 0.08$ mm to 0.11 mm, significant fluctuations in the relevant variables are observed, indicating pronounced microscale effects that may involve changes in heat transfer mechanisms and flow regimes. In the parameter optimization of the heat exchanger, the effect of channel width on heat transfer performance should be comprehensively considered, balancing solid mass and inlet–outlet pressure drop to determine the optimal channel width *b* for efficient thermal management.

3.1.3 Analysis of Different Channel Heights

The channel width *b* was initially set to 0.16 mm and the inlet velocity *V* to 10.42 m/s. Simulations were conducted for different channel heights *h* ranging from 0.7 mm to 1.7 mm. The fitted results are presented as the curve shown in Fig. 7.

The results shown in Fig. 7 indicate that as the channel height *h* increases from 0.7 mm to 1.7 mm, the solid mass of the heat exchanger gradually decreases from 0.99 g to 0.76 g. The fluid velocity correspondingly decreases, leading to a reduction in the convective heat transfer coefficient from 94.5 to 50 W/(m^2·K), while the thermal resistance increases from 0.01 K/W to 0.02 K/W. As the channel cross-sectional area expands, the inlet–outlet pressure drop decreases from 6,157 Pa to 2,186 Pa, and the heat transfer rate of the heat exchanger decreases from 7.95 W to 7.59 W, whereas solid temperature rises from 382.9 K to 385.3 K.

These findings demonstrate that the channel height *h* has a significant impact on heat transfer performance. Its variation not only affects the fluid flow state and heat transfer efficiency but also determines the mass, pressure drop, and temperature distribution of the heat exchanger.

3.1.4 Analysis of Different Inlet Velocities

The channel height was initially set to $h = 1.4$ mm and the channel width to $b = 0.16$ mm. Simulations were conducted for different inlet velocities V ranging from 4 m/s to 50 m/s. The fitted results are presented as the curve shown in Fig. 8.

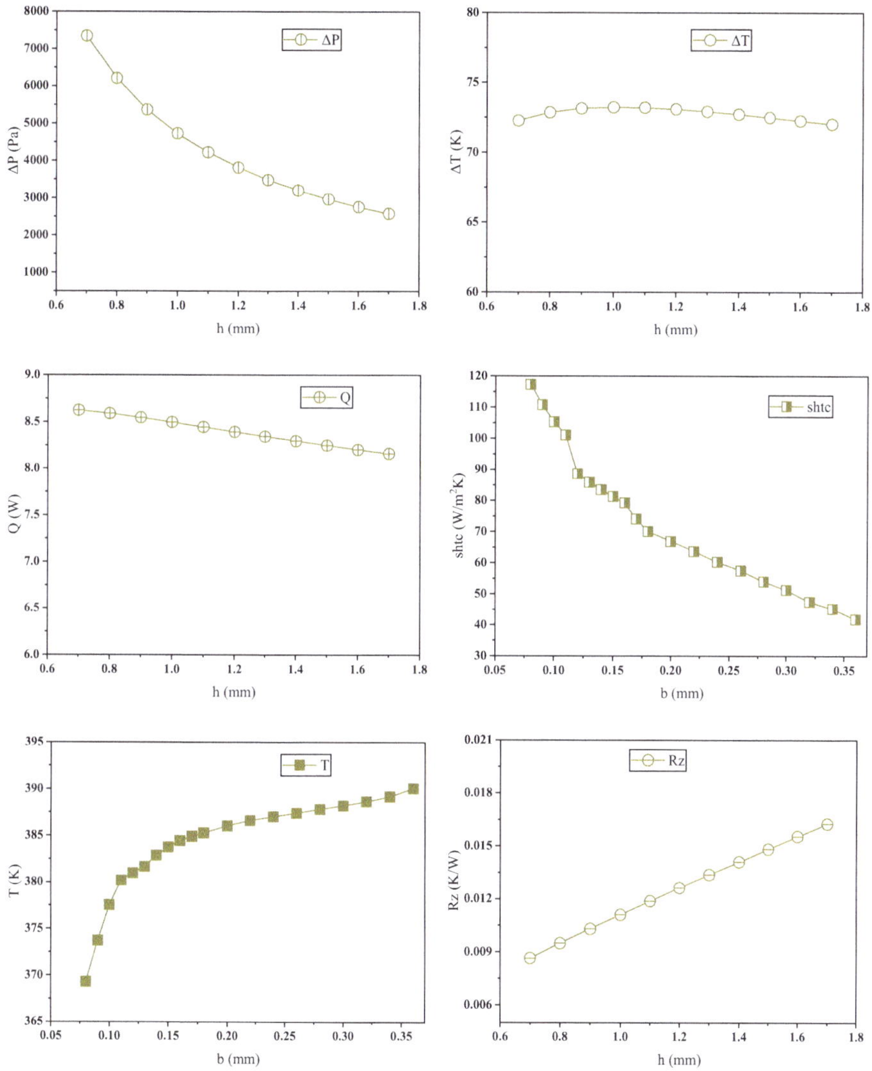

Fig. 7. Effect of channel height h on relevant parameters.

As shown in Fig. 8, with the inlet velocity increasing from 4 m/s to 50 m/s, the solid mass of the heat exchanger remains constant at 0.83 g, while the convective heat transfer coefficient significantly increases from 34.2 W/(m^2·K) to 116.1 W/(m^2·K), and the thermal resistance gradually decreases from 0.03 K/W to 0.008 K/W. The inlet–outlet pressure drop increases with velocity, rising from 1,392 Pa to 14,470 Pa. The heat dissipation rate of the heat exchanger initially increases from 6.3 W to a peak of 8.38 W, then slightly decreases to 7.7 W, indicating the existence of an optimal velocity range for maximizing heat transfer performance. The solid temperature decreases steadily from 530 K to 313 K with increasing velocity, while the fluid inlet–outlet temperature difference reduces from 161.5 K to 13.3 K, reflecting enhanced heat transfer at higher velocities.

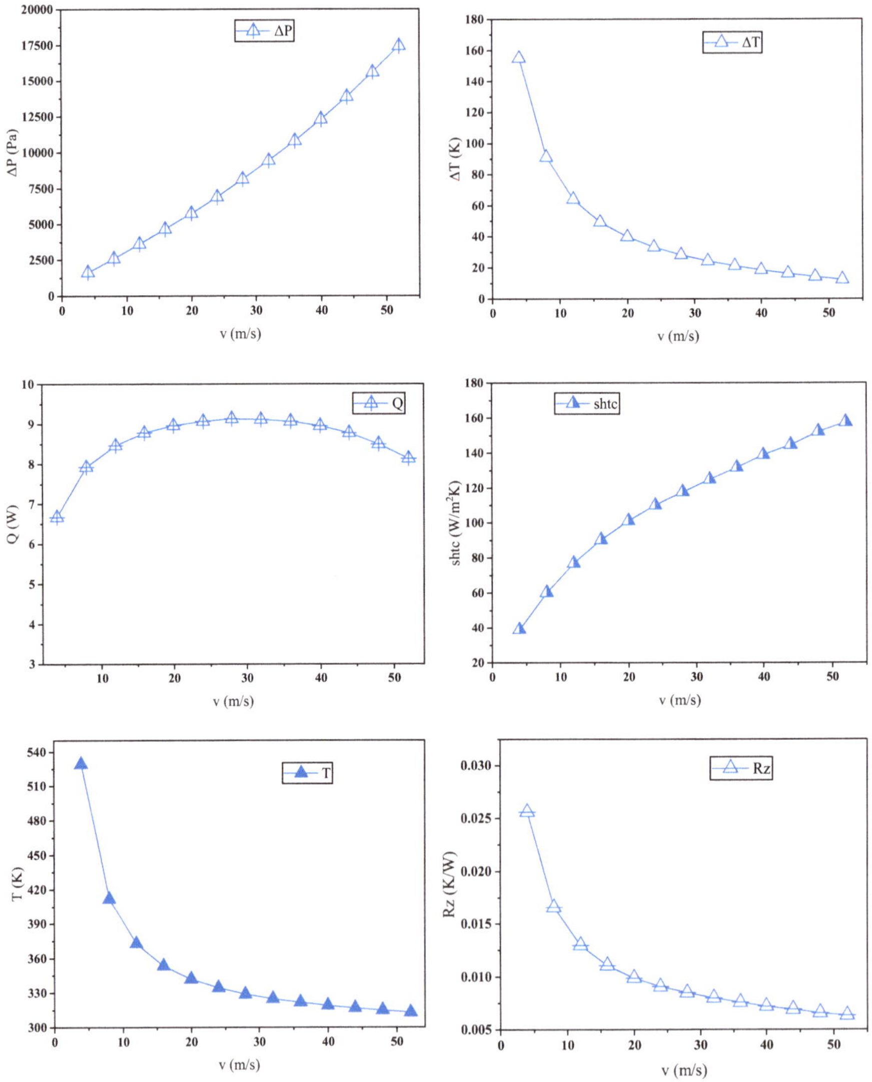

Fig. 8. Effect of inlet velocity *V* on relevant parameters.

In summary, the inlet velocity V has a significant impact on the heat transfer and flow characteristics of the heat exchanger. Its variation determines the convective heat transfer efficiency, channel pressure drop, and temperature distribution. Therefore, in the design and optimization of heat exchangers, it is necessary to balance heat transfer performance, flow resistance, and system energy consumption to determine the optimal inlet velocity for efficient thermal management.

3.2 Quantitative Analysis of Input Parameters

A sensitivity analysis, also referred to as parameter correlation analysis, was conducted in Fluent to investigate the relationships between input parameters—channel width (b), channel height (h), and inlet cross-sectional velocity (V)—and output parameters including solid mass (m), average solid temperature (T), convective heat transfer coefficient ($shtc$), inlet–outlet temperature difference (ΔT), average inlet–outlet pressure drop (ΔP), heat transfer rate per unit time (Q), and thermal resistance (Rz).

To ensure the accuracy and reliability of the analysis, the optimization ranges for the parameters were defined as follows: b from 0.08 mm to 0.32 mm, h from 0.3 mm to 1.9 mm, and V from 4 m/s to 50 m/s.

Regarding sample point selection, it is generally recommended that the number of samples be at least ten times the number of input parameters to ensure statistical significance. However, excessive sampling increases computational cost. Therefore, this study employed 150 sample points, balancing analysis accuracy with computational efficiency. The samples were generated using an automated method to ensure uniform distribution and representativeness in the parameter space.

For correlation analysis, the Spearman method was chosen. Spearman's rank correlation coefficient is primarily used to examine the monotonic relationship between two parameters, i.e., whether one parameter consistently increases or decreases with another. Compared to Pearson's method, Spearman's approach does not rely on a linear relationship between variables, making it more suitable for analyzing nonlinear or non-normally distributed data.

The geometric parameters of microchannels and the inlet velocity exhibit significant correlations with system performance metrics, revealing the following trends:

The solid mass (m) shows a negative correlation with channel width (b) and channel height (h), indicating that increasing channel dimensions reduces the solid mass. The average solid temperature (T) and the inlet–outlet temperature difference (ΔT) are both negatively correlated with the inlet cross-sectional velocity (V), suggesting that increasing velocity lowers the average solid temperature and reduces the temperature difference along the channel.

The convective heat transfer coefficient ($shtc$) decreases with increasing channel height (h) but significantly increases with higher inlet velocity (V), demonstrating the enhancing effect of velocity on heat transfer efficiency. The average pressure drop (ΔP) is negatively correlated with both channel width (b) and height (h), but positively correlated with inlet velocity (V), indicating that enlarging channel dimensions reduces pressure losses while increasing velocity amplifies flow resistance.

The heat transfer rate per unit time (Q) is positively correlated with channel dimensions (b, h), implying that enlarging the channel cross-section enhances heat exchange capacity. The thermal resistance (Rz) increases with channel height (h) but decreases with increasing inlet velocity (V), reflecting the suppressive effect of velocity on thermal resistance.

Overall, channel width (b), channel height (h), and inlet velocity (V) have decisive impacts on both thermodynamic characteristics (such as heat transfer rate and thermal resistance) and fluid dynamic performance (such as pressure drop and temperature difference) of the system.

4 Numerical Analysis of the Heat Exchanger Establishment of Response Surface Optimization Model

4.1 Construction of the Response Surface Model

A Design of Experiments (DOE) was implemented based on the Central Composite Design (CCD) method, targeting three input parameters: channel width (b), channel height (h), and inlet velocity (V), and seven output responses: convective heat transfer coefficient, pressure drop, thermal resistance, average solid temperature, solid mass, heat transfer rate, and inlet–outlet temperature difference. A total of 25 design points were generated to form the experimental sample set. Expanding the range of input parameters enables a broader exploration of feasible objectives, thereby providing greater design flexibility for multi-objective optimization. The design ranges were set as follows: channel width b from 0.07 mm to 3.7 mm, channel height h from 0.3 mm to 1.9 mm, and inlet cross-sectional velocity V from 4 m/s to 50 m/s.

Simulation data obtained from these design points were used to develop a surrogate model using the Kriging method. Kriging, based on Gaussian process regression, effectively captures nonlinear relationships while providing prediction confidence intervals, making it well-suited for data with non-uniform distributions or high local sensitivities. In ANSYS, the Kriging model is constructed by optimizing kernel hyperparameters (such as the Gaussian kernel parameter θ) and global trend terms (e.g., constant or linear regression), achieving high-fidelity fits. The resulting response surfaces can be integrated with tools such as Fluent for multi-objective optimization (e.g., minimizing thermal resistance and pressure drop) or parameter sensitivity analysis, thus replacing computationally expensive high-fidelity simulations and significantly improving the efficiency and reliability of complex fluid system design optimization. The response surface is illustrated in Fig. 9.

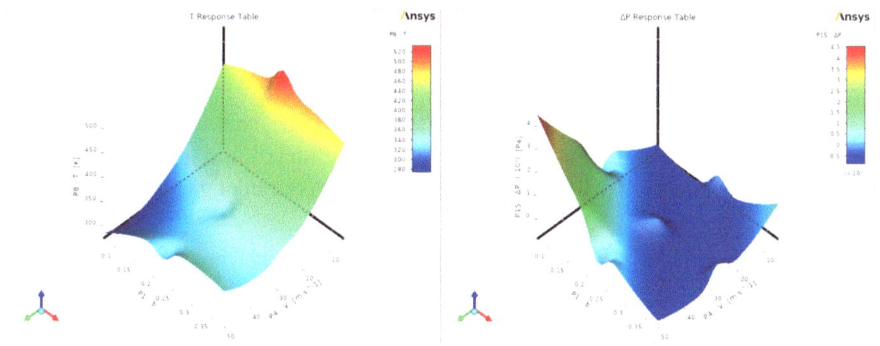

Fig. 9. Response surface.

4.2 Multi-Objective Optimization Design

In this section, parameter optimization is performed on the basis of the response surface. The optimization framework is constructed with the inlet velocity and the average pressure drop between the inlet and outlet as constraints, while minimizing the solid mass and maximizing the heat transfer rate as the objective functions.

(1) In this study, the design range of the channel width b is set from 0.07 mm to 3.7 mm, and the channel height h is varied from 0.3 mm to 1.9 mm. The inlet gas temperature is fixed at 300 K, and the inlet velocity is maintained at 28 m/s under a constant inlet mass flow rate condition.

Constraints: The inlet velocity is prescribed as $V = 28\text{m/s}$, with the pressure drop constrained within $100\text{Pa} \leq \Delta P \leq 101{,}325\text{Pa}$.

Objectives: To maximize the total heat transfer rate Q while maintaining the exchanger mass $m < 1\text{g}$. The optimization is carried out using the Multi-Objective Genetic Algorithm (MOGA), whose primary aim is to optimize multiple conflicting objective functions simultaneously and to identify a set of trade-off solutions, i.e., the Pareto-optimal front. In this work, the design points satisfying the prescribed inlet velocity and pressure drop constraints are obtained. Table 1 presents the three optimal candidate points.

Table 1. Optimized candidate points

Parameter name	h(mm)	b(mm)	Q (W)	m(g)	T(K)	ΔP(Pa)
Optimal candidate 1	1.833	0.338	8.85	0.26	333.9	772
Optimal candidate 2	1.896	0.218	8.41	0.56	330.9	2305
Optimal candidate 3	1.084	0.218	8.40	0.81	333.5	4686

The candidate points are ranked in descending order of heat transfer rate Q, as summarized in Table 1. Among the three candidates, the maximum heat transfer rate reaches 8.85 W, while the minimum is 8.23 W. The heat exchanger mass ranges from a maximum of 0.96 g to a minimum of 0.26 g. The solid temperature varies between a maximum of 333.9 K and a minimum of 330.9 K, whereas the pressure drop spans from as low as 772 Pa to as high as 10,041 Pa. The aspect ratio of channel height to width is approximately 5:1.

(2) The optimization was further conducted with the channel width b varied within the range of 0.07 mm to 3.7 mm, channel height h within 0.3 mm to 1.9 mm, and inlet cross-sectional velocity V within 4 m/s to 50 m/s, with the inlet gas temperature fixed at 300 K.

Constraints: $100\text{Pa} \le \Delta P \le 101{,}325$ Pa.

Objective: Maximize the heat transfer rate Q while maintaining the heat exchanger mass $m < 1$ g. The converged candidate points obtained via the genetic algorithm are summarized in Table 2.

Table 2. Optimization Candidate Points

Parameter name	h(mm)	b(mm)	V(m/s)	Q(W)	m(g)	T(K)	ΔP(Pa)
Optimal candidate 1	1.221	0.260	48.9	8.73	0.69	322.9	5045
Optimal candidate 2	1.257	0.311	12.0	8.16	0.59	376.9	675
Optimal candidate 3	1.429	0.246	5.5	7.08	0.64	466.0	573

It can be clearly observed that in multi-objective optimization, not all objectives need to be simultaneously optimized. Instead, priorities should be ranked according to actual operating conditions. By establishing target hierarchies and selecting appropriate parameter values, the most suitable microscale channel heat exchanger configuration can be identified for fabrication.

5 Conclusion

This study numerically investigated microscale constant heat-flux heat transfer in microchannel heat exchangers, clarifying the coupled effects of geometric parameters and inlet conditions on thermo–hydraulic performance. The results reveal a clear trade-off between heat transfer enhancement and flow resistance with increasing channel number, fin spacing, and channel height. Inlet velocity strongly promotes convective heat transfer and reduces solid temperature, with an optimal velocity range identified for maximum heat dissipation. Geometric optimization indicates that an aspect ratio of 5:1 achieves the best overall thermal–flow performance. These findings provide valuable guidance for the structural design and performance optimization of microscale heat exchangers.

Acknowledgments. The presented work was supported by the National Natural Science Foundation of China (Grant No. 52076183), Science and Technology Planning Project of Fujian Province (Grant No. 2023T3009).

References

Hajabdollahi, H., Shafiey Dehaj, M., Masoumpour, B., Ataeizadeh, M.: Optimal design analysis of a tubular heat exchanger network with extended surfaces using multi-objective constructal optimization. Frontiers in Energy **16**(5), 862–875 (2022). https://doi.org/10.1007/s11708-022-0839-3

Jaiswal, A.K., Afzal, A.: Design optimization of prismatic rib turbulators in a rectangular channel based on multi-objective criterion. Int. J. Therm. Sci. **185**, 108091 (2023). https://doi.org/10.1016/j.ijthermalsci.2022.108091

Jing, N., Xia, Y., Ding, Q., Liu, Y.: Simulation and optimization study on the performance of fin-and-tube heat exchanger. Sustainability **15**(15), 11587 (2023). https://doi.org/10.3390/su151511587

Li, Y., Sun, C., Liu, W., Zhang, H.: Structural parameter design of welded plate heat exchanger based on multi-objective optimization algorithm. Int. Commun. Heat Mass Transfer **146**, 106900 (2023). https://doi.org/10.1016/j.icheatmasstransfer.2023.106900

Li, B., Hou, J., Xu, K., Chen, L.: Optimal designs for flow uniformity at inlet of microchannel flat tube heat exchanger. Appl. Therm. Eng. **226**, 120300 (2023). https://doi.org/10.1016/j.applthermaleng.2023.120300

Negi, V.P.S., Pandey, V.K., Ranganayakulu, C.: Optimization of rifled tube by parametric changes using CFD for goodness factor enhancement. Heat Transfer **53**(2), 688–706 (2024). https://doi.org/10.1002/htj.22970

Petrovic, M., Fukui, K., Kominami, K.: Numerical and experimental performance investigation of a heat exchanger designed using topologically optimized fins. Appl. Therm. Eng. **218**, 119232 (2023). https://doi.org/10.1016/j.applthermaleng.2023.119232

Wang, Z., Feng, Y., Yang, Y., Zhao, J.: Multi-objective optimization of rectangular cooling channel design using design of experiments (DOE). Appl. Therm. Eng. **242**, 122507 (2024). https://doi.org/10.1016/j.applthermaleng.2023.122507

Wang, Y., Wang, J., Zhang, W., Li, X.: Performance optimization of plate-fin heat exchanger with curved vortex generators based on response surface method. Int. J. Therm. Sci. **210**, 109659 (2025). https://doi.org/10.1016/j.ijthermalsci.2024.109659

Yang, C., Wang, D., Ali, S., Kim, S.: Topology optimization of rectangular parallel plate heat exchanger unit. Int. J. Therm. Sci. **208**, 109503 (2025). https://doi.org/10.1016/j.ijthermalsci.2024.109503

Simulation Study on the Influence of Food Viscosity on Single-Screw Pump Performance

Mingshuo Wang[1], Rongyu Ge[1](✉), and Haiwen Xiao[2]

[1] Shandong Provincial Key Laboratory of Sensor Technology and High Precision Weighing Instruments, College of Mechanical Engineering, University of Jinan, Jinan 250000, Shandong, China
me_gery@ujn.edu.cn
[2] Shandong Labor Vocational and Technical College, Jinan 250000, Shandong, China

Abstract. Single-screw pumps are extensively employed in food processing due to their continuous and stable conveyance process, minimal shear action, and strong adaptability to high-viscosity media. However, the significant variations in viscosity among different food materials and their systematic influence on single-screw pump performance remain to be thoroughly investigated. This study focuses on food-grade single-screw pumps, examining the impact of food material viscosity changes on pump performance parameters. An internal flow field computational model for the single-screw pump was established. Comparative analyses were conducted on pressure distribution, torque analysis, and theoretical versus actual displacement under various food medium viscosities to investigate the influence of food viscosity variations on the pump's overall conveying performance. The findings indicate that as the viscosity of the food medium changes, the flow state and performance parameters within the pump exhibit distinct differences. Appropriately matching food viscosity with pump structural parameters can enhance conveying stability and efficiency.

Keywords: Single screw pump · Fluent · Food viscosity · Volumetric efficiency

1 Introduction

Screw pumps are positive displacement pumps that convey media through progressive compression achieved by the interaction between stator and rotor. Compared to dynamic pumps such as centrifugal pumps, they offer smoother flow characteristics and superior self-priming capability (Gamboa et al. 2003). Based on the mating configuration of rotor and stator, they are categorised into rubber-stator screw pumps with interference fits and all-metal single-screw pumps with clearance fits. This paper focuses on the all-metal single-screw pump as its subject of study (Müller Kouhi et al. 2021). With heightened hygiene requirements, optimising the design and manufacturing processes of sanitary-grade screw pumps as critical conveyance equipment is paramount (Liu et al. 2025). Therefore, systematically investigating the influence of food viscosity on single-screw pump performance and studying the internal flow field and its variations can provide a basis for parameter optimisation, thereby enhancing the efficiency and service life of single-screw pumps (Rodrigues et al. 2024).

T. Mankovits and N. Trisovic (Eds.): MDS 2026, LNME 2026.
https://doi.org/10.1007/978-981-92-2440-1_34

This study investigates the effects of varying food viscosity on the flow characteristics, pressure distribution, and internal flow efficiency of food-grade single-screw pumps. Employing simulation techniques within the Fluent software platform, the research incorporates food material models across different viscosity grades. By comparing performance parameter variations under diverse viscosity conditions, the study elucidates the underlying mechanisms through which food viscosity influences pump operation. These findings provide theoretical guidance for optimising single-screw pump performance within the food industry.

2 Simulation Method

2.1 Three-Dimensional Modelling

The food-grade single-screw pump selected for this study employs an all-metal single-screw design with a 1:2 head profile for both inner and outer rotors. The model parameters are detailed in Table 1. A clearance fit of 1 mm is maintained between the rotor and stator. As the pump's maximum flow rate is unaffected by the number of stages, a two-stage screw pump configuration was utilised in the simulation to reduce computational time. The two-stage pump was modelled and assembled using modelling software, with the three-dimensional model depicted in Fig. 1.

Table 1. Single screw pump model parameters.

Parameter	Outer diameter(mm)	Pitch(mm)	Lead(mm)	Eccentricity(mm)
Rotor	38	95	190	7.4
Stator	76	190	190	7.4

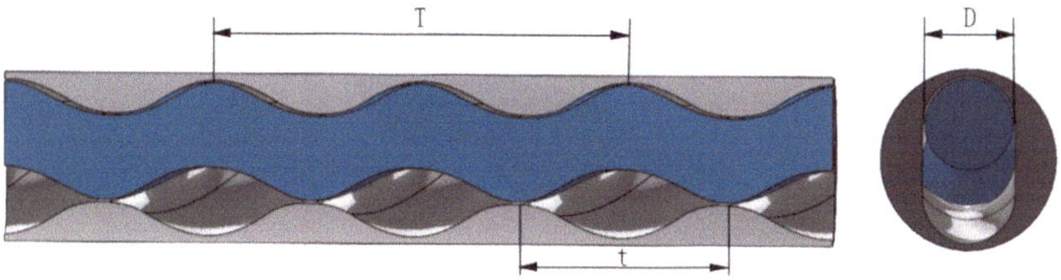

Fig. 1. Three-dimensional model of a single-screw pump

2.2 A Subsection Sample

Since food-grade single-screw pumps employ a clearance fit between the stator and rotor, the gap between them tends to represent the minimum wall thickness in the fluid domain. However, this area is prone to strong shear forces. To better analyze fluid transport and stress distribution within the chamber, the following assumptions are made: (1) The fluid is a non-Newtonian fluid and is incompressible. (2) The analysis does not account for temperature or wear-induced non-uniform deformation of the stator and rotor. (3) The chamber is filled with a single fluid consisting of food material. (4) Since inertial forces are far smaller than viscous and shear forces, the effects of inertial forces are not considered.

Based on the above assumptions, the gap between the stator and rotor was selected for filling. The rotor within the filled region was extracted using the Boolean function. The resulting fluid domain, consisting of the gap between the rotor and stator as well as the inlet and outlet end faces, is shown in Fig. 2.

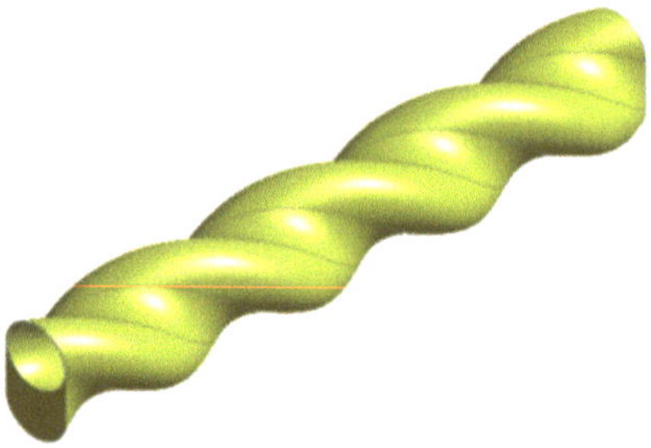

Fig. 2. Fluid domain for single-screw pumps

Due to the narrow gap between the stator and rotor and the complex flow patterns, the mesh generator in ANSYS software was used to refine the mesh in the thin regions of the single-screw pump and to generate a mesh for the fluid domain. To obtain more accurate numerical simulation results, a structured mesh was selected in this study to balance computational efficiency with accuracy. The final mesh comprised 377,450 elements, and the fluid domain mesh is shown in Fig. 3.

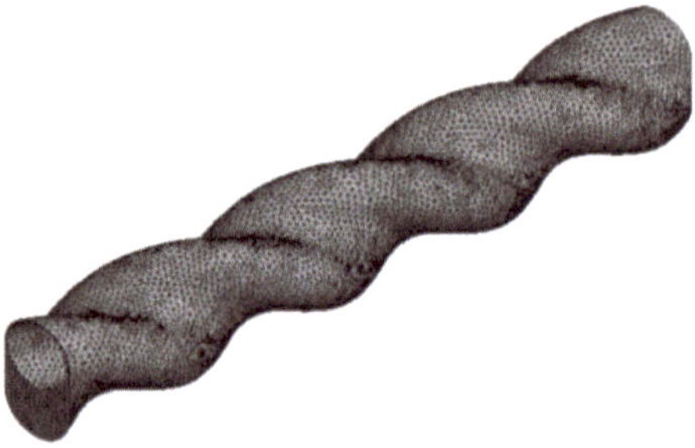

Fig. 3. Mesh partitioning for single-screw pumps

2.3 Boundary Conditions and Solution Methods

Following the completion of mesh generation, boundary conditions were established. Given the single-stage pressurisation function of the single-screw pump, boundary conditions were set as pressure inlet and pressure outlet. The single-stage pressurisation value was 0.35 MPa, with inlet pressure set at 2 MPa and outlet pressure at 2.7 MPa.

The screw rotation speed was 300 r/min, while the stator was defined as a stationary wall. The k-epsilon turbulence model is selected. The viscosity of the food material is set to 1350 kg/cm^3. A global standard initialisation is performed with 800 iterations to analyse the impact of varying viscosities on the pump. Figure 4 illustrates the boundary conditions for the fluid simulation calculation.

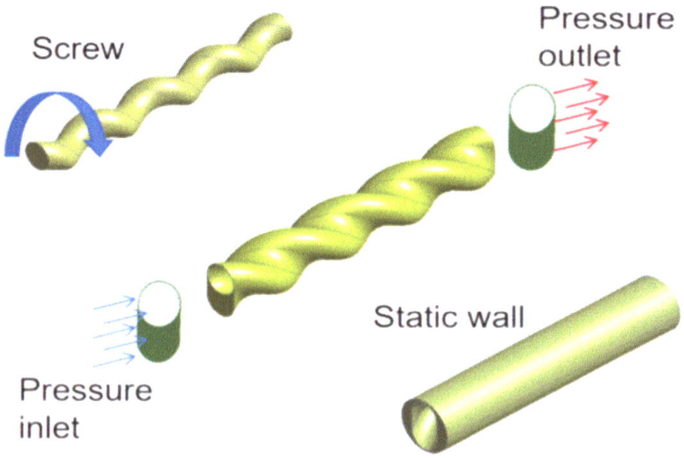

Fig. 4. Fluid domain boundary conditions

3 Simulation Results

3.1 CFD Analysis of Pressure Contour Maps at Different Viscosities

To analyse the mechanisms by which varying food viscosity conditions influence their flow characteristics and performance, the following section will investigate the flow field of a single-screw pump under different viscosity food medium conveyance conditions. Density arises from the gravitational effects on matter, whereas viscosity stems from molecular forces; the two possess no intrinsic correlation. Consequently, during simulation, the density value is set as a constant, while viscosity is configured to vary within a defined range (Zhang, 2018).

As shown in Fig. 5, the pressure contour plots for operating conditions with viscosities of 0.5 Pa·s, 1 Pa·s, 1.5 Pa·s, 2 Pa·s, 3 Pa·s, and 4 Pa·s are depicted. The pressure distribution within the internal flow field remains consistent across varying viscosities, with pressure progressively increasing from inlet to outlet. Pressure distribution within individual pump chambers is uniform, whereas abrupt pressure changes occur in the stator meshing zones between chambers. As the viscosity of the conveyed medium increases, the overall pressure level within the single-screw pump rises, with the axial pressure gradient becoming more pronounced. High-pressure zones gradually concentrate towards the outlet end, indicating that viscous resistance dominates the flow behaviour of high-viscosity fluids within the pump. This places heightened demands on the pump's energy consumption and structural load-bearing capacity.

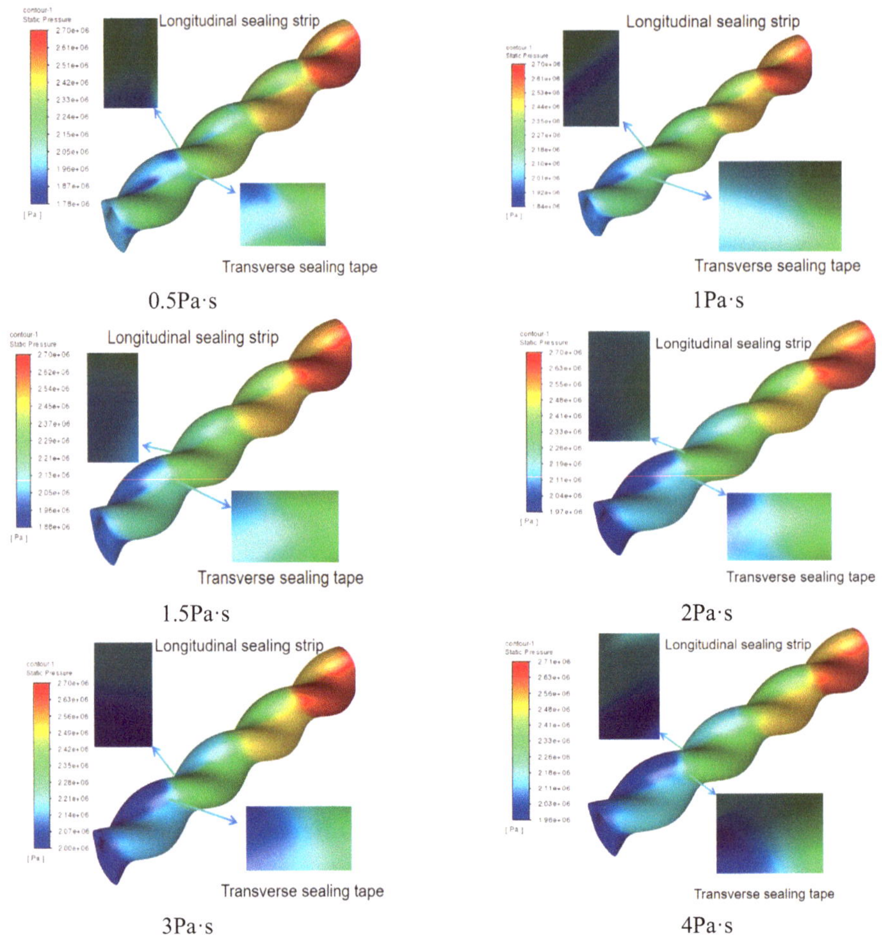

Fig. 5. Pressure contour map for different viscosities

3.2 Analysis of Viscosity Effects on Torque and Volumetric Efficiency

To investigate the variation in pump torque under different medium viscosities, this study employs Fluent software for numerical simulation and monitoring analysis. During the calculation process, a torque report was defined, with the wall selected as the monitoring region. Torque monitoring was established with the rotation axis as the torque direction, and torque values were recorded in real time after steady-state convergence; Subsequently, calculations were performed for multiple operating conditions with medium viscosities ranging from 0.5 Pa·s to 4 Pa·s, and the final torque data for each condition were extracted; finally, the viscosity and corresponding torque values were imported into Origin for fitting, generating a curve showing the variation of pump torque with medium viscosity, as shown in Fig. 6, which provides a basis for analyzing the effect of viscosity on pump performance.

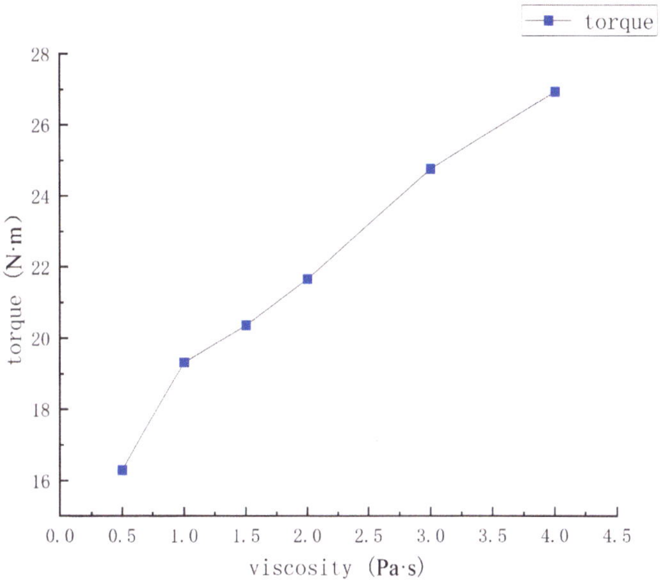

Fig. 6. Torque curves curves at different viscosities

This curve illustrates how pump torque varies with fluid viscosity, showing that torque is directly proportional to viscosity. The primary reason for this is that an increase in viscosity increases the fluid's viscous resistance; to meet engineering requirements for stable flow, the pump's output torque must increase accordingly as viscous resistance rises.

To determine the effect of different viscosities on the performance of a single-screw pump, it is necessary to analyze the relationship between its actual displacement and theoretical displacement. The theoretical displacement of a single-screw pump is determined by its structural parameters and can be derived from its geometric delivery characteristics:

$$Q_0 = \frac{4enDT\ 60}{10^6} \tag{1}$$

$$V = \frac{Q}{Q_0} \times 100\% \tag{2}$$

In the formula, Q_0 represents the theoretical displacement; e denotes the rotor eccentricity; D signifies the rotor diameter; T indicates the stator pitch; Q denotes the actual displacement; n represents the rotor rotational speed; ηv denotes the volumetric efficiency.

Based on the data in the above table, the theoretical flow rate of this single-screw pump is calculated to be 92.4 m^3/d. To investigate the effect of different medium viscosities on pump performance, the actual flow rate was determined by defining a flow rate report, imported into Origin for fitting, and used to generate a volumetric efficiency curve, as shown in Fig. 7.

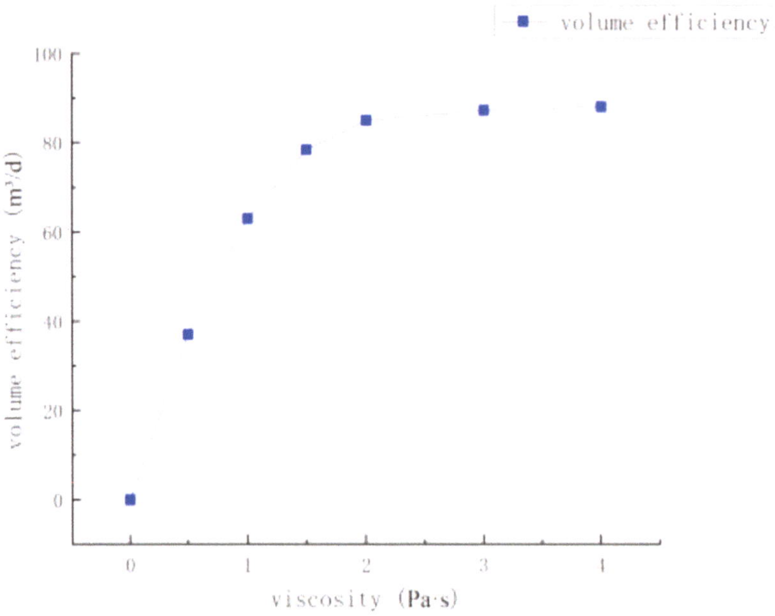

Fig. 7. Volumetric efficiency curves at different viscosities

The volumetric efficiency curve in Fig. 7 exhibits a nonlinear growth pattern. The actual displacement of the single-screw pump increases rapidly as the viscosity of the medium rises, but then tends to stabilize and grow more slowly as viscosity continues to increase. In the low-viscosity range of 0–1 Pa·s, the rate of displacement increase is significant, and the actual displacement is notably lower than the theoretical displacement of 92.4 m^3/d. This results in lower volumetric efficiency in the low-viscosity range, indicating that leakage and backflow are the primary factors affecting single-screw pumps. Under low-viscosity conditions, food products exhibit high fluidity, leading to severe leakage and backflow through the gap between the screw and the stator, which reduces the pump's effective delivery capacity. As viscosity continues to increase, the rise in displacement rate slows, and the actual displacement begins to approach the theoretical displacement. This indicates that as viscosity increases, backflow through the gap between the rotor and stator is suppressed, forming a sealed conveying zone within the pump chamber, allowing for stable transport. The volumetric efficiency approaches 1, indicating that under high-viscosity conditions, the performance of the single-screw pump is primarily influenced by viscous resistance and mechanical losses.

4 Conclusion

(1) Under low-viscosity conditions, the actual displacement deviates significantly from the theoretical displacement, indicating that single-screw pumps have limited adaptability to low-viscosity fluids. In food production and transportation, methods such as lowering temperature can be employed to increase the viscosity of food fluids, thereby enhancing production efficiency.

(2) An optimal viscosity operating range exists for food production and transportation. At ambient temperatures, this single-screw pump performs best within the medium-viscosity range around 2 Pa·s. At this viscosity, the actual displacement approaches the theoretical value, volumetric efficiency is high and stable, pressure variation is minimal, and torque requirements are low. The pump operates more smoothly under these conditions, with its structural characteristics well-matched to the fluid properties.

(3) The effect of high viscosity on displacement enhancement diminishes. Whilst leakage is further reduced during high-viscosity production, increased fluid viscous resistance and energy loss make it difficult to further improve displacement and volumetric efficiency. Moreover, under high-viscosity conditions, the pump's output torque must overcome resistance to perform work, intensifying friction between the rotor and stator and reducing service life. In practical production, viscosity can be reduced by increasing temperature.

Acknowledgements. This research was jointly funded by the Shandong Provincial SME Innovation Capacity Enhancement Programme (Project Title: Research on Key Technologies and Equipment for Municipal Pipeline Water-Jet Vacuum Excavation; Project Number: 2024TSGC0629) and the Shandong Provincial Science and Technology Enhancement Project (Project Title: R&D and Industrialisation of Key Technologies for High-Efficiency Food-Grade Screw Pumps; Project Number: 2025TSGCCZZB0587).

References

Gamboa, J., Oliveira, J.L., Teixeira, A.F., Rosa, A.J.: Experimental study of a progressing cavity pump handling viscous oil and air-liquid mixtures. SPE J. **8**(2), 188–196 (2003). https://doi.org/10.2118/84137-PA

Liu, X., Wu, Z., Wang, J.: Design and manufacturing process optimization of sanitary screw pump. In: Proceedings of the Academic Symposium on Artificial Intelligence and Economic Engineering Development (II), pp. 644–647 (2025). https://doi.org/10.26914/c.cnkihy.2025.013708

Müller, J., Kouhi, Y., Leonow, S., Mönnigmann, M.: Modeling pressure pulsation and backflow in progressing cavity pumps with deformable stator. J. Petrol. Sci. Eng. **203**, 108402 (2021). https://doi.org/10.1016/j.petrol.2021.108402

Rodrigues, R., Lima, J.A., Paladino, E.: Experimental and CFD modeling of a progressive cavity pump (PCP) using overset unstructured mesh part 1: single-phase flow. Geo. Sci. Eng. **234**, 212602 (2024). https://doi.org/10.1016/j.geoen.2023.212602

Zhang, W.: Flow field analysis and structural improvement of screw pump (Master's thesis). Southwest Petroleum University (2018). https://kns.cnki.net/kcms2/article/abstract?v=iB5Z0i9DW_xTKzFeQZEnWOuoVwouJ5Y-WZhkrYTNCOb9S0aKkSCFAgFthC1wqOWf-8rDxTKRMlopDN3HGSJ1eR_keR6Sf1N7gLjnUkg405T60ndKgeEKLH1AiNmCQ_cGRyWEDGFJZpsQbVKD7_-0g-ou4mSWlbkW09hBhHx6uDsPzb_1sJGy8w==&uniplatform=NZKPT&language=CHS

Finite Element Simulation Study of Human Brain Safety Threshold Under Medium-Speed Collision Load

Junhui Chang[1], Chao Wan[2,3](✉), Hongpeng Yu[4], and Jian Zhao[1](✉)

[1] School of Technology, Beijing Forestry University, Beijing 100083, China
zhaojian1987@bjfu.edu.cn

[2] School of Aerospace Engineering, Beijing Institute of Technology, Beijing 100081, China
chaowan@bit.edu.cn

[3] Tangshan Research Institute, Beijing Institute of Technology, Tangshan 063000, China

[4] Shenyang Institute of Automation, Chinese Academy of Sciences, Shenyang 110169, China

Abstract. A high-fidelity finite element model of the human head was developed to investigate brain injury under medium-speed impact and to quantify impact conditions related to injury thresholds. The model was reconstructed from MRI data, and the coupled Eulerian-Lagrangian method was used to model the fluid-structure interaction of cerebrospinal fluid. A six-degree-of-freedom stiffness constraint was introduced for the atlantoaxial joint. After validation against the Nahum experiment, intracranial responses under frontal, occipital, vertex, and temporal impacts were compared. A three-factor, three-level response surface method was then applied to the most hazardous impact direction. The model showed good agreement with experimental results. The results showed that occipital impact indicated the greatest injury risk and that impact velocity was the dominant factor affecting injury response. Under $m = 4$ kg and $S = 208$ cm^2, the critical velocities for 6 kPa von Mises stress, and 18% maximum principal strain were about 1.75 m/s and 4.0 m/s, respectively.

Keywords: Head impact biomechanics · Finite element model · Coupled eulerian-lagrangian method · Response surface methodology · Safety threshold

1 Introduction

Traumatic brain injury (TBI) is a common and severe injury in traffic accidents, contact sports, and falls (Maas 2022). Its occurrence depends not only on external loading, but also on load transfer among the skull, cerebrospinal fluid (CSF), and brain tissue, together with pressure wave propagation and local shear deformation (Ji 2022, Yang 2022). In particular, medium-speed impact is a representative loading condition because it may not cause severe skull fracture, but can still induce clear intracranial pressure and shear response. Meanwhile, impact mass, velocity, and contact area have coupled effects on injury thresholds. Therefore, studying intracranial mechanical response during head impact is important for understanding TBI mechanisms and improving safety assessment.

T. Mankovits and N. Trisovic (Eds.): MDS 2026, LNME 2026.
https://doi.org/10.1007/978-981-92-2440-1_35

Traditional head safety evaluation mainly uses global indices such as the head injury criterion (HIC). However, these indices cannot directly describe the actual loading state of brain tissue. With the development of computational biomechanics, research has shifted from global head response to tissue-level prediction. Intracranial pressure, maximum principal strain (MPS), and von Mises stress are widely used because they better represent different injury mechanisms (Clark 2020, Sbriglio 2025).

Finite element head models provide an effective framework for linking external loading with intracranial response and injury outcome. In this study, a high-fidelity finite element model of the human head was developed from MRI data, with the coupled Eulerian-Lagrangian (CEL) method adopted to represent CSF behavior and physiologically relevant neck constraints introduced at the head-neck junction. After validation against the Nahum experiment, frontal, occipital, vertex, and temporal impacts were compared to identify the most hazardous impact location, and a three-factor, three-level response surface analysis was further conducted for the critical direction to quantify the relationships between injury metrics and impact parameters.

2 Materials and Methods

2.1 Head Finite Element Model

A high-fidelity head finite element model was constructed from MRI data and included the skull, CSF, gray matter, white matter, cerebellum, and brainstem. The anatomical structures were first segmented in Mimics and then imported into ABAQUS for meshing with four-node tetrahedral elements. In the final model, the total number of elements reached 1,031,006, including 770,566 elements in the brain.

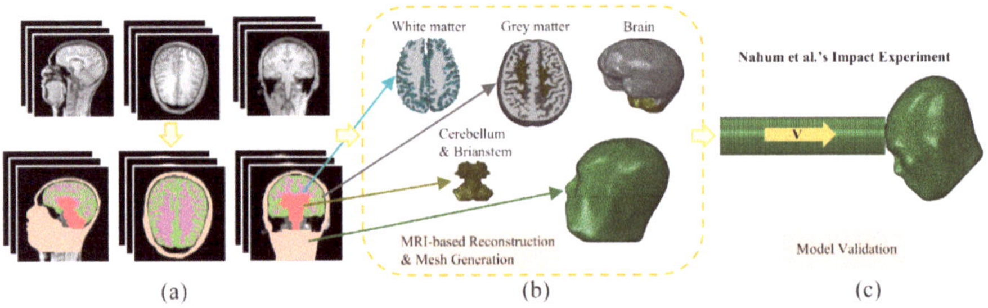

Fig. 1. Human head FEM and illustration of model validation.

2.2 Material Properties and Boundary Conditions

Brain tissues were modeled as isotropic linear viscoelastic materials with time-dependent shear behavior described by a Prony series (Tse 2014).

$$G(t) = G_\infty + (G_0 - G_\infty)e^{-\beta t} \tag{1}$$

where G_∞ is the long-term shear modulus, G_0 is the initial shear modulus, and β is the decay coefficient. The skull was defined as a linear elastic material ($E = 8$GPa, $v = 0.22$, $\rho = 2.0$ g/cm^3) (Horgan 2003). CSF was treated as a nearly incompressible Newtonian fluid ($\rho = 1.040$ g/cm^3, $\mu = 0.001$ Pa·s) governed by the Mie-Grüneisen equation of state. The CEL method was used to simulate the fluid-structure interaction between CSF and brain tissue (Na 2020).

$$p = \frac{\rho_0 c_0 \eta}{(1 - s\eta)^2}\left(1 - \frac{\Gamma_0 \eta}{2}\right) + \Gamma_0 \rho_0 E_m \quad (2)$$

where ρ_0 is the reference density, $\eta = 1 - \rho/\rho_0$ is the nominal volumetric compressive strain, and E_m is the internal energy. Under the present medium impact conditions, a simplified linear compressible response controlled by $c_0 = 1500$ m/s was adopted. Table 1 lists the relevant parameters of brain tissue.

Table 1. Material properties of human brain tissues in the head model.

Tissue	Density (g/cm^3)	Poisson's ratio	G_0(kPa)	G_∞(kPa)	β(s^{-1})
Cerebral gray matter	1.040	0.4996	34	6.4	700
Cerebral white matter	1.040	0.4996	41	7.8	700
Cerebellum	1.060	0.4996	10	2	80
Brainstem	1.060	0.4996	22.5	4.5	80

To represent neck support, a six-degree-of-freedom stiffness constraint was introduced at the head-neck junction based on the mechanical characteristics of the atlantoaxial joint (Panjabi 2001). The definitions of stiffness for each degree of freedom of the cervical joint are shown in Table 2.

Table 2. Stiffness for each degree of freedom of the atlantoaxial joint in the head model.

Type		Stiffness (N/mm or N·mm/rad)
Translational	Lateral shear	281.8
	Anterior-posterior shear	115.5
	Posterior-anterior shear	323.4
	Compression	2501.75
	Tension	900.9
Rotational	Flexion	4658.2
	Extension	4735.2
	Lateral bending	17629.5
	Axial rotation	2017.5

2.3 Model Validation

The model was validated against the classical Nahum cadaver head impact experiment (Nahum 1977). As shown in Fig. 1c, a 5.59 kg impactor struck the forehead at 9.94 m/s, and the simulated contact force and intracranial pressure responses were compared with the reported experimental results.

2.4 Impact Cases Under Different Directions

As shown in Fig. 2, four typical impact cases: frontal, occipital, vertex, and temporal impacts were designed. All cases used the same impactor to compare the effect of impact location on intracranial response. The impactor ($m = 5.50$ kg, $\rho = 3.99$ g/cm^3, $E = 2 \times 10^7$ kPa, $v = 0.3$) impacted at 3.30 m/s. A front elastic pad ($t = 50$ mm, $\rho = 0.08$ g/cm^3, $E = 3.15 \times 10^4$ kPa) was attached to the front of the impactor.

Fig. 2. Four typical impact conditions: **a** Frontal impact; **b** Occipital impact; **c** Vertex impact; **d** Temporal impact.

2.5 Parametric Analysis for the Hazardous Direction

A three-factor, three-level response surface design was used. The factors were impactor mass, impact velocity, and contact area, with levels of m = 2/4/6 kg, v = 0.10/0.75/1.40 m/s, S = 16/208/400 cm^2, respectively. Peak von Mises stress and peak MPS of the brain were selected as response variables, and quadratic regression models were established for threshold analysis.

3 Results

3.1 Results of Validation

The verification results are shown in Fig. 3. The peak pressure error was within 5%, and the deviation in the main peak time was within 0.3 ms.

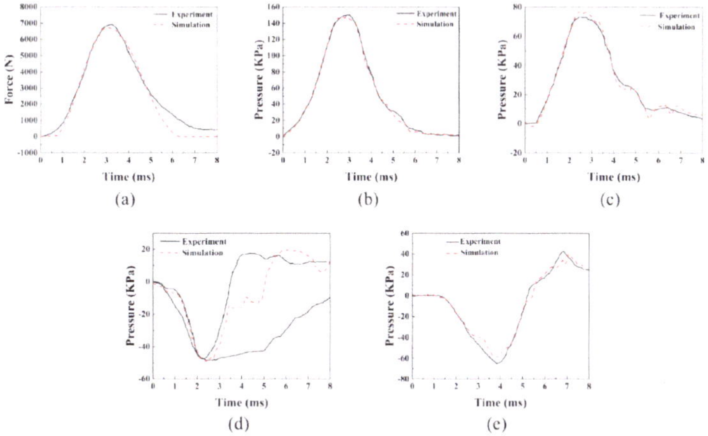

Fig. 3. Comparison of contact force and ICP at four locations between our simulation and reported experiment results: **a** Contact force; **b** Impact side pressure; **c** Parietal pressure; **d** Occipital pressure; **e** Posterior pressure.

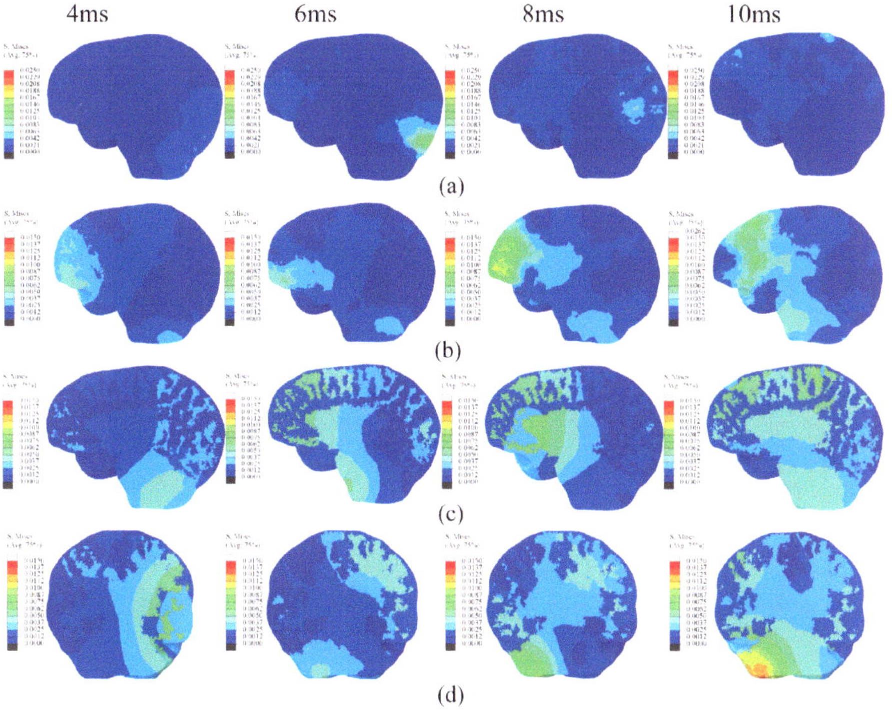

Fig. 4. The von Mises stress distribution of the brain under four impact scenarios: **a** Frontal impact; **b** Occipital impact; **c** Vertex impact; **d** Temporal impact.

The model reproduced the main features of load input, pressure wave propagation, and contrecoup response, confirming its suitability for subsequent impact analysis.

3.2 Brain Responses Under Different Impact Directions

Six milliseconds after the impact, in the four impact cases, occipital impact produced the highest peak von Mises stress in the whole brain, reaching 12.42 kPa. Frontal impact ranked second, while vertex and temporal impacts showed relatively lower responses.

As shown in Figs. 4 and 5, both coup and contrecoup regions exhibited obvious loading in all cases, but the spatial distribution of high-risk regions differed clearly with impact location.

3.3 Parametric Analysis Under the Hazardous Direction

The response surface results showed in Table 3, indicating that both von Mises stress and MPS increased mainly with impact velocity. The effect of mass was secondary but still significant, whereas the effect of contact area was relatively weak. This indicates that impact velocity is the dominant factor controlling brain injury response under medium-speed occipital impact.

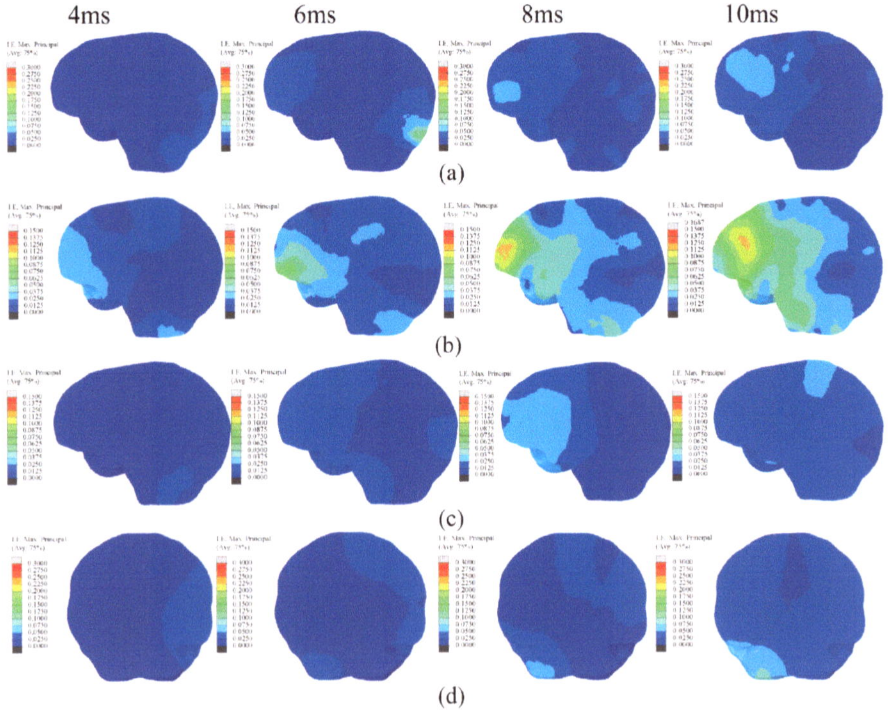

Fig. 5. The MPS distribution of the brain under four impact scenarios: **a** Frontal impact; **b** Occipital impact; **c** Vertex impact; **d** Temporal impact.

Table 3. Simulation experiment results.

Scheme number	Impact mass m(kg)	Impact velocity v(m/s)	Impact body end face area S(cm^2)	von Mises stress(kPa)	MPS(%)
1	2	0.10	208	1.413	2.264
2	6	0.10	208	1.436	2.349
3	2	1.40	208	3.072	3.816
4	6	1.40	208	5.108	6.158
5	2	0.75	16	1.405	2.326
6	6	0.75	16	2.648	3.205
7	2	0.75	400	1.467	2.157
8	6	0.75	400	2.624	3.200
9	4	0.10	16	1.433	2.322
10	4	1.40	16	5.350	6.240
11	4	0.10	400	1.536	2.315
12	4	1.40	400	4.424	5.395
13	4	0.75	208	2.314	2.669
14	4	0.75	208	2.287	2.655
15	4	0.75	208	2.100	2.467

Based on the experimental data in Tables 4 and 5, quadratic polynomial fitting was performed on the peak von Mises stress (kPa) and peak MPS (%), respectively, and the regression equations R_1 and R_2 were obtained as follows:

$$\begin{aligned} R_1 = 0.265 + 0.626\,m - 1.756v - 0.00004S + 0.387\,m \cdot v - 0.00006\,m \cdot S \\ - 0.002v \cdot S - 0.078\,m^2 + 1.980v^2 + 3.131 \times 10^{-6}S^2 \end{aligned} \quad (3)$$

$$\begin{aligned} R_2 = 2.318 + 0.220\,m - 3.263v - 0.003S + 0.434\,m \cdot v + 0.0001\,m \cdot S \\ - 0.002v \cdot S - 0.037\,m^2 + 2.835v^2 + 7.409 \times 10^{-6}S^2 \end{aligned} \quad (4)$$

Table 4. Mises stress ANOVA table.

Source	Sum of squares	df	Mean squares	F-value	P-value	
Model	25.39	9	2.82	34.75	0.0006	Significant
A-m	2.49	1	2.49	30.61	0.0026	
B-v	18.41	1	18.41	226.76	<0.0001	
C-S	0.0770	1	0.0770	0.9487	0.3748	
AB	1.01	1	1.01	12.48	0.0167	

(*continued*)

Table 4. (*continued*)

Source	Sum of squares	df	Mean squares	F-value	P-value	
AC	0.0018	1	0.0018	0.0228	0.8859	
BC	0.2647	1	0.2647	3.26	0.1308	
A^2	0.3619	1	0.3619	4.46	0.0885	
B^2	2.58	1	2.58	31.83	0.0024	
C^2	0.0492	1	0.0492	0.6058	0.4716	
Residual	0.4059	5	0.0812	–	–	
Lack of fit	0.3788	3	0.1263	9.30	0.0987	Not significant
Pure error	0.0272	2	0.0136	–	–	
Cor total	25.79	14	–	–	–	

Table 5. MPS ANOVA table.

Source	Sum of squares	df	Mean squares	F-value	P-value	
Model	28.70	9	3.19	35.46	0.0005	Significant
A-*m*	2.36	1	2.36	26.28	0.0037	
B-*v*	19.09	1	19.09	212.26	< 0.0001	
C-*S*	0.1316	1	0.1316	1.46	0.2805	
AB	1.27	1	1.27	14.16	0.0131	
AC	0.0067	1	0.0067	0.0748	0.7955	
BC	0.1756	1	0.1756	1.95	0.2212	
A^2	0.0810	1	0.0810	0.9006	0.3862	
B^2	5.30	1	5.30	58.90	0.0006	
C^2	0.2754	1	0.2754	3.06	0.1405	
Residual	0.4498	5	0.0900	–	–	
Lack of fit	0.4243	3	0.1414	11.12	0.0837	Not significant
Pure error	0.0254	2	0.0127	–	–	
Cor total	29.15	14	–	–	–	

Figure 6 further shows that the overall variation patterns of the two types of response surfaces are basically consistent: the response surface rises significantly with increasing speed; in the high-speed region, the differences between different quality levels further widen; while along the end-face area direction, the change of the response surface is relatively gradual. This result is consistent with the conclusions of the analysis of variance.

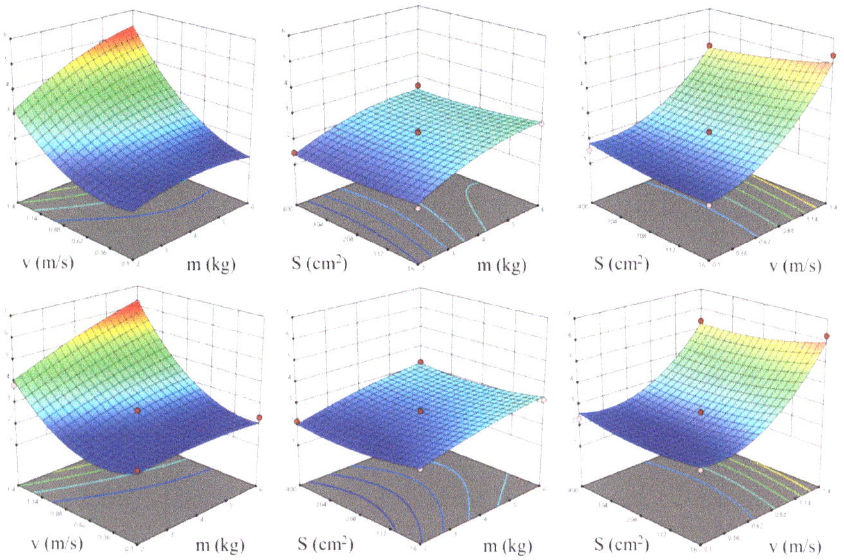

Fig. 6. Response surface analysis of von Mises stress and MPS.

3.4 Threshold-Based Safety Assessment

With 6 kPa von Mises stress and 18% MPS taken as reference thresholds, the regression equations were used to determine the corresponding impact conditions. For the case of $m = 4$ kg and $S = 208$ cm^2, the predicted critical velocities were about 1.75 m/s for 6 kPa von Mises stress and 4.0 m/s for 18% MPS. The validation simulations gave values of 6.27 kPa and 18.21%, which were close to the predicted thresholds.

4 Discussion

Among the four typical medium-velocity impact directions, occipital impact produced the highest stress level and the widest high-risk region, suggesting that it is the most dangerous of the four typical impact locations considered in this study.

The clear contrecoup response may come from several factors. Pressure travels quickly through the nearly incompressible CSF. It then reflects from the inner skull surface. At the same time, brain motion lags because of inertia. These effects raise tensile and shear loading on the side opposite the impact.

The parametric study also shows a clear trend. Velocity has the strongest effect on injury response. Mass matters too, but less. Contact area mainly changes local loading and has only a small effect on the whole-brain peak response. These findings are useful for injury evaluation and protective design.

5 Conclusions

A high-fidelity human head finite element model was developed from MRI data, and it was developed and validated for medium-speed impact analysis. The CEL method was used to model CSF at the skull-brain interface, and a six-degree-of-freedom stiffness constraint was introduced to represent neck support. The model reproduced the Nahum frontal impact experiment well.

Among the four typical impact directions, occipital impact produced the highest brain stress and the largest high-risk region, making it the most hazardous case among the four loading scenarios.

For occipital impact, velocity was the dominant factor affecting injury response, followed by mass, while contact area had a weaker effect. Under $m = 4$ kg and $S = 208$ cm^2, the critical velocities for 6 kPa von Mises stress, and 18% MPS were respectively about 1.75 m/s and 4.0 m/s.

Acknowledgments. This work was supported by the National Key Research and Development Program (2023YFB4704600).

References

Clark, J.M. et al.: Proposed injury thresholds for concussion in equestrian sports. J. Sci. Med. Sport. **23**(3), 222–236 (2020). https://doi.org/10.1016/j.jsams.2019.10.006

Horgan, T.J., Gilchrist, M.D.: The creation of three-dimensional finite element models for simulating head impact biomechanics. Int. J. Crashworthiness. **8**(4), 353–366 (2003). https://doi.org/10.1533/ijcr.2003.0243

Ji, S. et al.: Use of brain biomechanical models for monitoring impact exposure in contact sports. Ann. Biomed. Eng. **50**(11), 1389–1408 (2022). https://doi.org/10.1007/s10439-022-02999-w

Maas, A.I.R., Menon, D.K., Manley, G.T.: Traumatic brain injury: Progress and challenges in prevention, clinical care, and research. Lancet Neurol. **21**(12), 1004–1060 (2022). https://doi.org/10.1016/S1474-4422(22)00309-X

Na, M., Beavers, T.J., Chandra, A., Bentil, S.A.: Simulation of brain response to noncontact impacts using coupled Eulerian–Lagrangian method. J. Biomech. Eng. **142**(5), 051011 (2020). https://doi.org/10.1115/1.4045047

Nahum, A., Smith, R., Ward, C.: Intracranial pressure dynamics during head impact. In: Proceedings of the 21st Stapp Car Crash Conference. New Orleans, LA (1977) https://doi.org/10.4271/770922

Panjabi, M.M. et al.: Mechanical properties of the human cervical spine as shown by three-dimensional load-displacement curves. Spine. **26**(24), 2692–2700 (2001). https://doi.org/10.1097/00007632-200112150-00012

Sbriglio, C., Ptak, M., Kwiatkowski, A.: Advances in computational modelling of head injury biomechanics – a comprehensive review. Arch. Comput. Methods Eng. **32**(1), 1–43 (2025). https://doi.org/10.1007/s11831-025-10463-w

Tse, K.M., Tan, L.B., Lee, S.J., Lim, S.P., Lee, H.P.: Development and validation of two subject-specific finite element models of human head against three cadaveric experiments. Int. J. Numer. Methods Biomed. Eng. **30**(3), 397–415 (2014). https://doi.org/10.1002/cnm.2609

Yang, S., Tang, J., Nie, B., Zhou, Q.: Assessment of brain injury characterization and influence of modeling approaches. Sci. Rep. **12**(1), 13597 (2022). https://doi.org/10.1038/s41598-022-16713-2

The Transient Response Mechanism and Application of Strain Gauges under Laser Excitation

Zhichao Liu, Xiangjun Bi(✉), Yangxv Li, and Caihua Zhou

Dalian University of Technology, Dalian 116024, China
bixj@dlut.edu.cn

Abstract. The demand for the development of large-scale load-bearing structures within the aerospace industry has been experiencing continuous and significant growth. At present, the conventional methodology for networking the massive arrays of strain gauges and data channels associated with these large structures relies predominantly on manual coding coupled with manual verification. The inherent inefficiency of this traditional approach renders it inadequate for satisfying the stringent requirements of contemporary, intensive structural assessment and validation procedures. In this paper, the transient response characteristics of strain gauge under laser excitation are studied by simulation, and the surrogate model of laser parameters to strain gauge response is constructed. Finally, a fast correlation method based on laser excitation-strain response is proposed. Ultimately, the method is verified by shell test.

Keywords: Laser excitation · Transient response characteristics · A fast correlation method

1 Introduction

As aerospace technology rapidly advances, large load-bearing structures, such as rocket shells and aircraft wings, face very strict mechanical evaluation requirements during both development and ground testing (Azad et al. 2024). In current traditional electrical measurement tests involving massive arrays of sensors, the networking information of strain gauges and data channels is acquired through manual pasting, coding, and the node-by-node verification of information. This method is not only time-consuming and labor-intensive but also highly susceptible to human error. Its low efficiency has become a critical bottleneck restricting the intensive assessment and rapid validation of modern aerospace equipment.

To enhance networking efficiency, non-contact excitation and automated addressing technologies have increasingly become research hotspots in the field of Structural Health Monitoring (SHM) (Scarselli and Nicassio 2025). Among these, utilizing the laser excitation-strain response as a correlation pathway has demonstrated immense application potential. However, the thermomechanical coupling process of strain gauges under laser excitation is highly complex, and their transient response characteristics

T. Mankovits and N. Trisovic (Eds.): MDS 2026, LNME 2026.
https://doi.org/10.1007/978-981-92-2440-1_36

are subject to profound non-linear influences from laser parameters and environmental thermal drift (Lou et al. 2021). Although traditional high-fidelity Finite Element Method (FEM) analysis can accurately reveal the underlying physical mechanisms, the excessive time required for a single solution cannot meet the demands for real-time, rapid identification efficiency at engineering sites. Furthermore, conventional decision algorithms in existing studies, which rely solely on a single static threshold, are highly prone to misjudgment in complex test environments due to interference from mechanical impacts or low-frequency environmental thermal noise.

To address the aforementioned engineering challenge of low automated networking efficiency in large-scale load-bearing structures, this paper investigates the transient response characteristics of strain gauges under laser excitation through advanced simulation and proposes a novel rapid correlation method based on the laser excitation-strain response paradigm. This study extracts typical physical features under excitation and introduces Gaussian Process Regression (GPR) to construct a high-precision surrogate model, effectively replacing time-consuming FEM calculations (Dai and Chen 2026). Subsequently, based on this surrogate model, the mathematical mapping of laser parameters is systematically elucidated, and excitation constraints that balance testing efficiency with sensor safety are clearly delineated. Ultimately, this innovative method is systematically validated through a networking test on a large, thin-walled cylindrical shell, providing a highly efficient and stable networking approach to significantly elevate the operational efficiency of large-scale electrical measurement tests.

2 Methodology

The spatial energy distribution of a real laser beam follows a Gaussian profile, characterized by exceptionally high energy at the center that decays exponentially towards the edges. To accurately replicate the response mechanism of strain gauges subjected to laser excitation, a Gaussian non-uniform heat source is selected in this section. To ensure that the thermal energy inputted into the model matches the beam profile while conserving total power, a mathematical equivalent derivation of the Gaussian heat flux density was conducted in this study (Zhang et al. 2026).

Let P represent the overall laser power and R the effective spot radius. The attenuation function of the local heat flux density q(r) as a function of radius r, is defined in Eq. (1):

$$q(r) = q_{\max} \cdot \exp\left(-2\frac{r^2}{R^2}\right) \tag{1}$$

Where q_{max} denotes the peak heat flux density at the exact center of the laser spot (r = 0). Energy conservation requires the total laser power P to equal the two-dimensional integral of the heat flux density across the entire 2D plane, as expressed in Eq. (2):

$$P = \int_0^{\infty} q_{\max} \cdot \exp\left(-2\frac{r^2}{R^2}\right) \cdot 2\pi r dr \tag{2}$$

Solving the integral in Eq. (2) yields the following relationship:

$$P = q_{\max} \cdot \left(\frac{\pi R^2}{2}\right) \tag{3}$$

From this, it can be inversely deduced that the central peak heat flux density of the Gaussian beam is exactly double the value expected from a perfectly flat laser profile, which is given by Eq. (4):

$$q_{\max} = \left(\frac{2P}{\pi R^2}\right) \tag{4}$$

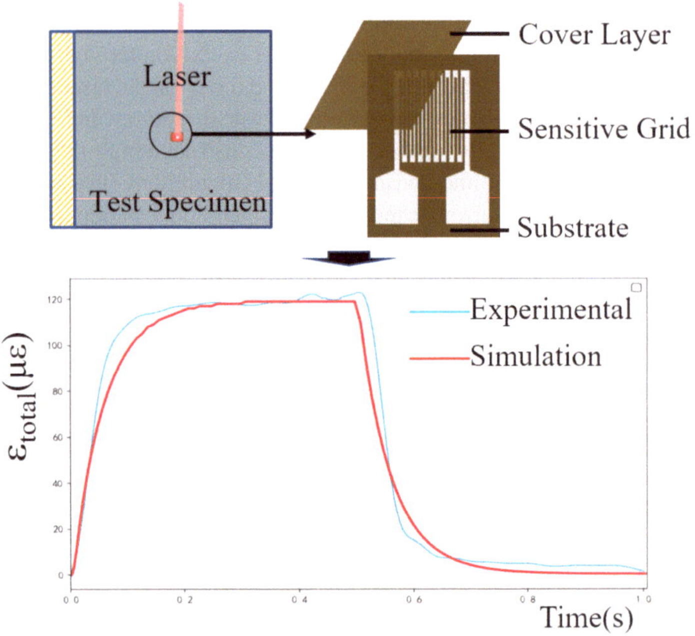

Fig. 1. Strain-time curve of the strain gauge under laser excitation.

Under transient thermal shock, to prevent mesh distortion and computational divergence caused by severe local expansion, we used simplified contact definitions between layers. For the interfaces among the cover layer, sensing grid, and substrate, given the ultra-thin nature of the sensing grid and substrate, heat transfers almost instantly, so we applied a Tie constraint to rigidly bind them together. Furthermore, cohesive contact properties are introduced between the substrate and the stainless steel surface. By assigning a reasonable normal stiffness K_{nn} and moderately reduced tangential/shear stiffnesses K_{ss} and K_{tt}, the shear lag effect of the adhesive layer is accurately reproduced.

In this study, a three-dimensional solid model comprising the cover layer, sensing grid, and substrate was established using the finite element simulation software ABAQUS (Zhao et al. 2023). Coupled with the previously mentioned laser model, the thermomechanical coupling process was simulated with a laser power of 120 mW, a spot radius of 1 mm, and an excitation duration of 0–0.5 s. The resulting simulation and experimental curves are illustrated in the Fig. 1.

As Fig. 1 shows, the ABAQUS results closely track the experimental data, successfully replicating the three-stage core characteristics of the strain gauge response under laser excitation: "rapid surge — a steady climb to stabilization — sudden drop and residual strain". However, there are reasonable physical deviations in certain local features. To elucidate the underlying mechanisms, we break down these three stages below:

(1) Rapid surge stage (0–0.1 s): Initially, the two curves nearly overlap. Subsequently, both the strain value and the initial temperature rise slope of the experimental curve exceed those of the simulation. The simulation treats the laser as an ideal Gaussian heat source, which contributes to this difference. In the actual physical environment, however, factors such as the focal shift induced by the thermal micro-deformation of the laser optics and the microscopic roughness of the sensor surface significantly enhance the actual photothermal absorption rate, thereby triggering a more severe transient thermal shock.
(2) Slow heating and steady-state stage (0.1–0.5 s): The peak strain reached in the experiment (approximately 123 $\mu\varepsilon$) is slightly higher than the simulated predicted value (approximately 120 $\mu\varepsilon$), presenting an error of about 2.4% and accompanied by minor fluctuations. The core mechanism underlying this discrepancy is that the simulation model employs a "Tie" constraint (i.e., zero thermal contact resistance), whereas the actual strain gauge contains an adhesive layer. This layer leads to localized heat accumulation in the sensing grid region. Coupled with the viscoelastic creep of the adhesive resulting from localized heating and ambient lab noise, the steady-state strain gradually climbs as the system dynamically adjusts its temperature.
(3) Sudden drop and thermal residual stage (0.5–1.0 s): Upon laser cut-off at 0.5 s, both curves drop precipitously. However, the experimental curve recovers slower and retains minor residual strain. The simulation assumes ideal convective heat transfer, whereas actual air temperature fluctuations and delayed heat dissipation.

Overall, these experimental comparisons confirm that the finite element model is reliable while highlighting the superimposed effects of actual laser hot spots, adhesive heat accumulation, and environmental disturbances. Because individual FEM runs take too much time, the following section constructs a high-precision surrogate model that rapidly captures how transient features evolve under varying excitation parameters for real-time identification.

Based on the established high-fidelity simulation model, we constructed a surrogate model using Gaussian Process Regression (GPR) to capture the typical response characteristics of the strain gauge. (Yu et al. 2025).

Taking comprehensive account of engineering factors like hardware limits, preventing gauge burnout, and keeping the testing fast, three core parameters were selected and their effective design spaces were defined: laser power P (10–150 mW, to ensure a pronounced response while avoiding thermal damage), spot radius R (0.5–1.5 mm, to balance positioning tolerance and a distinct response), and excitation duration t (0.1–2.0 s, to guarantee rapid identification). We applied Latin Hypercube Sampling (LHS) across the designated ranges for laser power, spot radius, and duration to generate a representative sample matrix of 150 sets with high spatial uniformity, which avoids the "curse of dimensionality" while keeping computational costs low.

Based on the high-fidelity computational results of the aforementioned samples, the following two physical features are extracted as the output metrics for the surrogate model:

The peak strain ε_p: Characterizes the maximum macroscopic effect of the local thermal shock.

Initial temperature rise slope k_i: Reflects the instantaneous thermal shock intensity induced by the surface layer's laser absorption during the initial 0.05 s.

The resulting response surfaces, which map laser parameters to strain responses, are illustrated in Fig. 2. The determination coefficient (R^2) of the GPR model is greater than 0.92, showing good global generalization ability and local fitting accuracy. Looking at Fig. 2, both the peak strain and initial temperature rise slope increase sharply as the spot radius decreases and laser power increases.

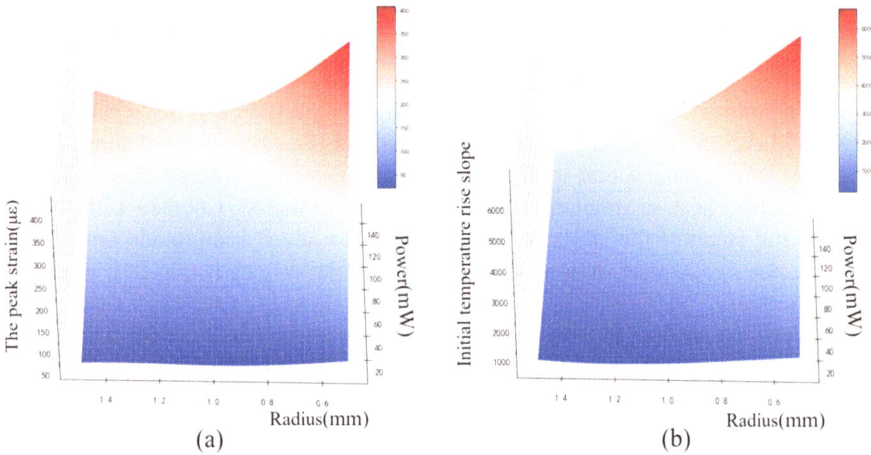

Fig. 2. Response surfaces of the response characteristics: **a** Response surface of the peak strain; **b** Response surface of initial temperature rise slope.

Likewise, based on how the peak strain and initial slope react to the laser settings, it is necessary to secure a pronounced response during the experimental excitation while preventing thermal degradation from extreme laser intensity or an overly small spot size. To prevent this, we restricted the laser settings to: power of 50–130 mW, excitation time of 0.2–0.5 s, and the spot radius controlled within 1 mm.

3 Results

Before deploying these parameters on the actual Φ1000 × 600 mm cylindrical shell, we must account for real-world lab conditions:

(1) Surface finish impacts: Real-world aerospace structures often have rust-preventive coatings or complex curved geometric features. Because the laser directly acts on the cover layer of the strain gauge and subsequently conducts to the sensitive grid,

surface reflectivity has minimal impact on how the cover layer initially absorbs the laser energy. Although the coating thickness and structural curvature can significantly alter the contact thermal resistance between the strain gauge substrate and the specimen—impeding or accelerating heat dissipation and causing uncontrollable fluctuations in the steady-state peak strain—the transient response feature extracted in this paper, namely the "initial temperature rise slope," occurs within the first 0.05 s of laser irradiation. At this stage, heat conduction has not yet deeply penetrated the substrate interface. Relying on this initial heating rate provides the system with exceptionally strong anti-interference capabilities, allowing it to ignore underlying surface conditions such as surface finish or shell curvature.

(2) Influence of spatial standoff distance: During actual system deployment, the response characteristics are highly sensitive to the spatial standoff distance. However, the defined constraints for spot radius and power naturally restrict the robotic arm's working distance (in the subsequent tests of this study, the distance between the laser and the strain gauge was strictly controlled within the range of 100–300 mm). Fundamentally, the influence mechanism of spatial distance is equivalent to the coupled effect of surface heat flux and spot radius on the strain gauge.

(3) Environmental noise and suboptimal conditions: Engineering sites inevitably experience slow thermal drift caused by ambient light, background machinery vibrations, and sudden impacts (which typically do not exceed 10 $\mu\varepsilon$). By combining a sliding window approach with a conservative amplitude threshold, we created a time-domain filter that isolates the interference of conventional environmental fluctuations from a mathematical logic perspective. Furthermore, in extended field deployments, these gauges might experience suboptimal conditions such as localized degradation (aging of the adhesive layer or minor thermal resistance changes caused by sensitive grid excitation), which can directly distort the slope at the onset of heating. In response, this study strictly limits the duration of laser excitation to minimize strain gauge damage while ensuring the emergence of identifiable response features. To accommodate these non-ideal factors, a dynamic tolerance coefficient α is utilized in the identification criterion to provide a certain margin (since the strain gauges in the subsequent tests of this study were brand new, α was empirically set to 0.95). Nevertheless, future work will still need to establish detailed error statistical models for such long-term fatigue and degradation states, and further evaluate the identification confidence intervals of the system under extreme working conditions.

The dynamic relative amplitude $\Delta\varepsilon_i$ must exceed the environmental thermal drift, i.e., $\Delta\varepsilon_i > \varepsilon_{threshold}$.

The response slope must be greater than a specific lower tolerance limit of the predicted slope derived from the surrogate model under identical excitation parameters, i.e., $k_{actual,\, i} \geq \alpha \cdot k_{predict}$.

To verify the stability and efficiency of the proposed rapid correlation method based on the transient response characteristics under laser excitation, a proof-of-concept validation was conducted using a smooth, thin-walled cylindrical shell with end faces measuring $\Phi 1000 \times 600$ mm. The specific experimental process and the corresponding results are detailed in Fig. 3.

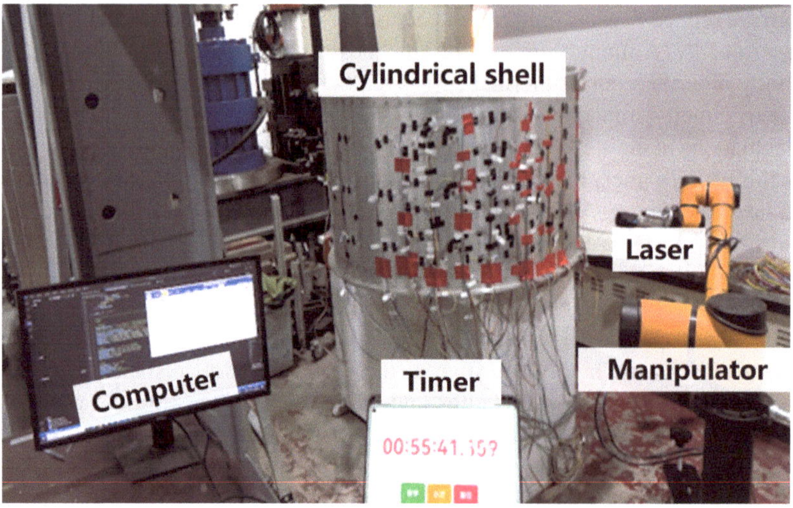

Fig. 3. Test site.

A total of 200 strain gauges were distributed on the outer wall of the test specimen, all of which were connected to the data acquisition channels. Throughout the test, the identification criterion program for the response features ran continuously on the computer. A robotic arm systematically traversed the locations of the strain gauges, applying laser excitation to each individual sensor. The program automatically recorded the corresponding responding channels, ultimately acquiring 200 sets of strain gauge-channel correlation data with an absolute accuracy rate of 100%. These results compellingly demonstrate the stability of the identification criteria and validate the feasibility of the proposed rapid correlation method based on laser excitation transient response characteristics, highlighting its substantial engineering value.

In the field of automated sensor networking and structural health monitoring (SHM), frontier addressing technologies such as electrical time-domain reflectometry (TDR) and acoustic excitation currently exist. However, for the preliminary system setup of electrical measurement tests on large-scale aerospace structures, the laser excitation method proposed in this paper exhibits unique advantages. TDR technology primarily relies on electrical pulse reflection time for localization to acquire the one-dimensional electrical length of cables, failing to directly determine the three-dimensional physical spatial coordinates of sensors on the structural surface. Although acoustic excitation technology can be utilized for wide-area localization, it is highly susceptible to the interference of acoustic wave reflection, scattering, and multipath effects within complex thin-walled structures, resulting in low spatial resolution. In contrast, the laser excitation method proposed herein utilizes a robotic arm to guide the laser to accurately traverse strain gauges with known coordinates, achieving true "point-to-point" high-resolution addressing. This method can accurately establish the absolute physical mapping relationship between "3D spatial coordinates" and "data acquisition channels." Admittedly, this method relies on unobstructed line-of-sight operations, requires relatively high initial hardware costs (e.g., robotic arms), and still necessitates manual pre-mapping and input of strain gauge coordinates prior to testing.

To address the aforementioned hardware costs and operational complexity, the ultimate engineering goal of this research is to develop a highly reusable, intelligent electrical measurement networking system. The future technical roadmap is not limited to traversing sensors with known coordinates but involves a deep integration of machine vision technology: without the need to pre-input the location information of strain gauges, an industrial camera mounted on the end-effector of the robotic arm will be utilized to perform real-time global scanning and identify strain gauge features; subsequently, the host computer software will execute dynamic path planning, guiding the robotic arm to automatically align and apply laser excitation; ultimately, the correlation information between the actual spatial coordinates of the strain gauges and the acquisition channels will be synchronously obtained. This networking system, integrating "visual recognition, dynamic excitation, and response matching," will completely eliminate the reliance on manual pre-mapping. It can be repeatedly applied to strength or fatigue electrical measurement tests across various models of large load-bearing structures, demonstrating immense engineering application value and industrialization potential.

4 Conclusion

Based on the thermal response mechanism, this paper establishes a high-fidelity simulation model of strain gauges subjected to laser excitation. This model reveals the underlying patterns of strain under transient conditions, which serves as the foundation for extracting core response features. Subsequently, utilizing these extracted features, a high-precision surrogate model is constructed via Gaussian Process Regression (GPR). We then use this model to determine the operating constraints necessary for securing stable responses from laser-excited strain gauges. Ultimately, an identification criterion predicated on these response features is formulated and successfully implemented in the electrical measurement networking experiments of large-scale load-bearing structures, demonstrating significant practical value. Ultimately, the targeted laser excitation mechanism and robust identification criteria validated in this study provide the groundwork to develop a fully automated, vision-integrated networking system without the need for prior coordinate inputs.

Declaration of Generative AI and AI-Assisted Technologies in the Writing Process. During the preparation of this work, the author used AI tools to improve the readability of certain overly literal translations. After using this tool, the author reviewed and edited the content as needed and takes full responsibility for the content of the publication.

References

Azad, M.M., Kim, S., Cheon, Y.B., Kim, H.S.: Intelligent structural health monitoring of composite structures using machine learning, deep learning, and transfer learning: a review. Adv. Compos. Mater. **33**(2), 162–188 (2024). https://doi.org/10.1080/09243046.2023.2215474

Dai, Y.R., Chen, Z.G.: Multi-output Gaussian process regression for rapid multi-nutrient prediction in soil using near-infrared spectroscopy. Agriculture. **16**(4), 485 (2026). https://doi.org/10.3390/agriculture16040485

Lou, C. et al.: Polymer-coated quartz tuning fork for enhancing the sensitivity of laser-induced thermoelastic spectroscopy. Opt. Express. **29**, 12195–12205 (2021). https://doi.org/10.1364/OE.421356

Scarselli, G., Nicassio, F.: Machine learning for structural health monitoring of aerospace structures: a review. Sensors. **25**(19), 6136 (2025). https://doi.org/10.3390/s25196136

Yu, Y.J. et al.: Intelligent quadrant-cavity-integrated FBG strain sensor: a multifunctional composite tube for real-time monitoring of vertical deformation in slope geohazard early warning systems. Opt. Express. **33**, 46655–46677 (2025). https://doi.org/10.1364/OE.576065

Zhang, H., Wang, S., Wang, J., Yan, Z.: A comparative study and introduction of a new heat source model for the macro-scale numerical simulation of selective laser melting technology. Materials. **19**(3), 480 (2026). https://doi.org/10.3390/ma19030480

Zhao, Y., Zhang, F., Ai, Y., Tian, J., Wang, Z.: Structural uncertainty analysis of high-temperature strain gauge based on Monte Carlo stochastic finite element method. Sensors. **23**(20), 8647 (2023). https://doi.org/10.3390/s23208647

Analysis and Calculation of Flow-Induced Vibration for Serpentine Finned-Tube Heat Transfer Equipment

Shan He, Haonan Xu(✉), and Shuijin Wu

China institute of atomic energy, Beijing, China
zhirou120@163.com

Abstract. In integral fast reactors (IFRs), serpentine finned-tube heat exchangers serve as critical heat transfer components in the decay heat removal system (DHRS), transferring thermal energy from the accident residual heat to ambient air. These serpentine finned tubes function as the critical boundary interface between the sodium coolant and the atmospheric environment, and their structural integrity is paramount to the safe operation of the entire reactor. The heat exchanger tubes in sodium-air heat exchangers are subjected to tube bundle vibration induced by external cross-flow gas impingement, potentially leading to wear, fatigue failure, and even rupture of the serpentine finned tubes. To ensure reactor safety and reliability, flow-induced vibration (FIV) analysis of the heat transfer equipment must be conducted during the design phase. Flow-induced vibration in tube bundles is primarily governed by the flow medium and velocity, tube geometry, and support structure configuration. Currently, three design codes—GB/151, ASME, and TEMA—are available for analyzing flow-induced vibration in tube bundles operating in gaseous environments. However, the computational methods in these codes were developed primarily for conventional working fluids such as water inside heat exchanger tubes, whereas the serpentine finned tubes in this application contain liquid metal sodium. Consequently, the applicability of these codes to sodium-air heat exchangers requires further evaluation. This study employs both design code methodologies and numerical simulation approaches to conduct flow-induced vibration analyses of turbulent buffeting and Kármán vortex shedding for the sodium-air heat exchanger in an integral fast reactor. The vibration frequencies of serpentine finned tubes, external flow velocities, and tube bundle vibration responses are determined to ensure that flow-induced vibration instability will not occur in the sodium-air heat exchanger.

Keywords: Serpentine finned tube · Flow-induced vibration (FIV) · Fluidelastic instability · Turbulent buffeting

1 Introduction

The IFR utilizes an integrated serpentine finned tube heat exchanger structure as the key component of the sodium-air heat exchanger, enabling a more compact equipment arrangement, reducing the number of welds, and improving equipment safety and reliability. While achieving the same heat transfer capacity within a smaller spatial envelope,

T. Mankovits and N. Trisovic (Eds.): MDS 2026, LNME 2026.
https://doi.org/10.1007/978-981-92-2440-1_37

the corresponding flow field velocity increases and the tube bundle length extends. As a critical component, the integrity of the sodium-air heat exchanger is essential for the safe operation of the reactor. The tube side of the serpentine finned tube sodium-air heat exchanger contains liquid metal sodium, supported by hanging loop supports, which differs significantly from conventional finned tube heat exchangers in terms of tube-side medium and structural configuration. Under transverse external air flow flushing, flow-induced vibration may occur, potentially leading to wear, fatigue failure, or even rupture of heat transfer tubes (Kim et al. 2016). To ensure safety and reliability, flow-induced vibration analysis of the heat exchanger is required. The main flow-induced vibration mechanisms studied for the sodium-air heat exchanger include: Kármán vortex shedding, turbulent buffeting, fluidelastic instability, and acoustic vibration (Nie 2013). Kim et al. conducted structural and vibration analysis of serpentine finned tube heat exchangers, focusing on the effects of flow velocity on vibration characteristics and the relationship between fin spacing and vibration (Rao et al. 1987). The calculated values of tube bundle vibration frequency and vibration response directly affect the predicted frequencies and amplitudes of Kármán vortex shedding and turbulent buffeting (Jo et al. 2008; Pettigrew et al. 2003); therefore, research on calculation methods for flow field velocity of serpentine finned tubes is necessary. Pettigrew systematically reviewed calculations related to flow-induced vibration, laying a solid theoretical foundation (Pettigrew et al. 1989; Xu et al. 2019). Xu et al. investigated calculation methods for flow field velocity in heat exchangers incorporating ASME design criteria (Xu et al. 2015). Based on the aforementioned research, this paper presents a flow-induced vibration analysis of the integrated serpentine finned tube sodium-air heat exchanger, focusing on calculating and analyzing the vibration frequency and response of serpentine finned tubes, as well as analyzing and calculating the flow field characteristics, to establish a comprehensive calculation methodology applicable to flow-induced vibration analysis of serpentine finned tubes (Kim et al. 2004).

2 Equipment Structure

The sodium-air heat exchanger comprises an external shell, upper and lower headers, finned heat transfer tubes, a flow distributor, support structures, and inlet/outlet dampers. The upper and lower headers are cylindrical vessels serving as tube-side headers for the bundle. A flow distribution structure is installed within the lower shell section to ensure uniform air flow distribution (the finned tube geometry is shown in Fig. 1. Both ends of the serpentine finned tubes are connected to the flow distributors, while intermediate sections are constrained by perforated support plates. The design employs integrally formed serpentine finned tubes to minimize weld joints. Sodium flows downward inside the tubes (from top to bottom), while air flows upward through the tube bundle gaps (from bottom to top) in a counter-flow arrangement.

3 Modal Analysis

The Modal and Pre-Stress Modal modules in the finite element analysis software ANSYS Workbench were employed to conduct modal analysis on an integrally formed serpentine finned tube. Due to the complex surface geometry of the finned tube, a numerical

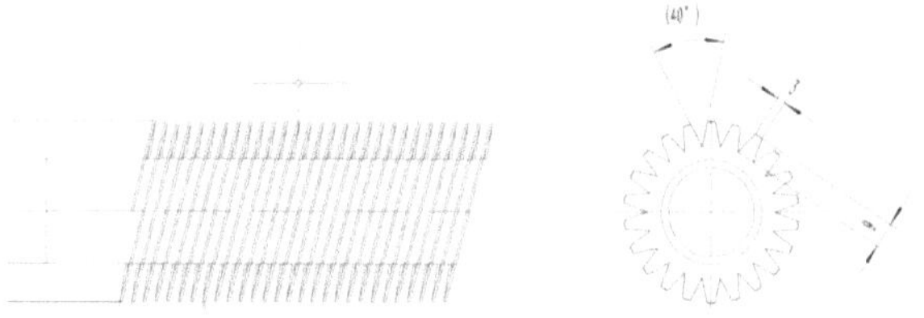

Fig. 1. Structure of finned tube.

simulation approach was adopted wherein the added mass of the fins was calculated and subsequently applied to a smooth tube of equivalent diameter. The actual computational model consisted of a bare tube with bends at both ends, with its cross-sectional geometric parameters and material properties specified in Table 1. The density of the heat exchange tube was simplified and calculated using an equivalent density, defined as the sum of the average density of the sodium medium inside the heat exchange tube and the equivalent average density of the external fins.

The density of the finned tube was determined by equivalently incorporating both the liquid sodium inside the tube and the external fins into the base tube. The calculation formula is as follows:

$$\rho = \left[\rho_3\frac{\pi}{4}\left(D^2 - d^2\right) + \rho_2\frac{\pi}{4}d^2\right]/\left[\frac{\pi}{4}\left(D^2 - d^2\right)\right]$$

Where $\rho 2$ represents the density of liquid sodium, D denotes the outer diameter of the pipe at the corresponding location, and d denotes the inner diameter of the pipe at the corresponding location. Meanwhile, $\rho 3$ is calculated using the following formula:

$$\rho_3 = \frac{\rho_1(V_1 + V_2)}{V_1}$$

Where V_1 represents the volume of the base tube per unit length, V_2 represents the volume of the fin portion attached to the corresponding section of the base tube, $\rho 1$ and represents the density of the base tube.

Table 1. Geometric and material parameters of serpentine finned tube.

Parameter	Unit	Value
Elastic modulus	GPa	160
Equivalent density	Kg/m3	19478
Tube outer diameter	mm	38
Linear thermal expansion coefficient	K^{-1}	$2.02*10^{-5}$

The serpentine finned heat transfer tube has an outer diameter of 38 mm with five finned spans, and a Poisson's ratio of 0.3. The tube features non-uniform span lengths, with the structural configuration of span lengths shown in Fig. 2. The tube density is 873.43 kg/m^3; Young's modulus is 1.93×10^{11} Pa; Poisson's ratio is 0.3; and water density is 983 kg/m^3. The material density is 7930 kg/m^3. Fluid parameters: the flow field media are air and liquid metal sodium with a density $\rho = 845$ kg/m^3.

The analysis employs a pre-stress modal method to calculate the natural frequencies of various heat exchanger tubes. First, the end faces of the heat transfer tube are constrained to restrict displacement in the y and z directions. Constraints are applied at the 2nd, 3rd, and 4th contact regions between the finned tube sections and the support structure to restrict x and y direction displacements, while the z direction (aligned with the straight tube section) remains free. Additionally, considering a 30 mm displacement difference due to thermal expansion between the header and the heat transfer tube support structure, a 30 mm displacement along the x direction is applied to the header outlet pipe, with corresponding temperature fields applied at the end sections, as shown in Fig. 2. Finally, the tube-side pressure of 1 MPa is applied to the inner wall surface. The pre-stress field solved with the aforementioned boundary conditions is imported into the modal solver to obtain a natural frequency of 11.885 Hz for the serpentine finned tube bundle, as shown in Fig. 3.

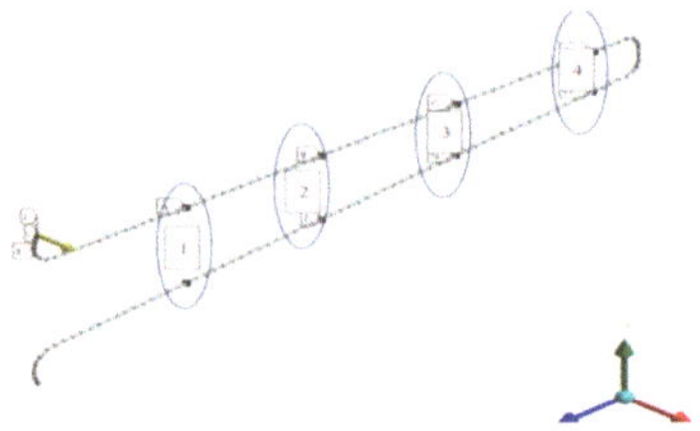

Fig. 2. Support structure layout of serpentine finned-tube bundle.

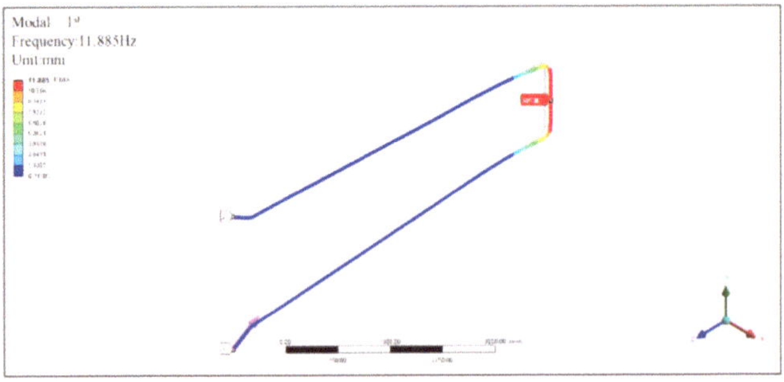

Fig. 3. Fundamental modal shape calculation of serpentine finned-tube bundle.

4 Flow-Induced Vibration Analysis and Calculation

4.1 Flow Field Analysis

Flow-induced vibration (FIV) analysis is performed based on the obtained modal characteristics, incorporating flow field computation for the sodium-air heat exchanger. With air as the external medium and liquid metal sodium as the internal working fluid, equivalent tube mass methods are employed to evaluate FIV in integrally formed serpentine finned tubes, encompassing critical parameters such as Kármán vortex shedding and turbulent buffeting [11].

The critical velocity V_{cn} for the n -th mode fluid-elastic instability of the tube bundle (with the serpentine finned tube equivalently treated as a smooth tube) can be determined from the critical flow velocity calculation formula provided in the design code GB/T 151.

$$V_{cn} = \beta f_n \left(\frac{2\pi \xi_n m_t}{\rho_0} \right)^{1/2}$$

Where β denotes the Connors constant, fn represents the natural frequency of the n-th mode, ξn represents the damping coefficient, mt denotes the equivalent mass per unit length of the pipe, and $\rho 0$ represents the average fluid density surrounding the pipe. By substituting the calculated first-order natural frequency into the aforementioned formula, and selecting and determining the numerical values of the key parameters in accordance with the design standard, the calculated critical velocity is determined to be 6.19196 m/s.

By specifying the inlet flow boundary conditions and conducting actual geometric modeling, the external flow field was simulated to obtain the flow characteristics within the serpentine finned tube bundle region [12]. The computational results indicate that the maximum inter-tube clearance velocity reaches 15.2 m/s in the axial flow direction. However, for flow-induced vibration analysis, the transverse (cross-flow) velocity is the primary parameter of concern. The cross-flow velocity in the tube bundle transverse region is 4.57 m/s (as shown in Fig. 4), which was compared against the critical velocity calculated in accordance with the design code GB/T 151. This maximum flow velocity serves as one of the key criteria for subsequent vibration response analysis and for determining whether the integrally formed serpentine finned tube will experience flow-induced vibration (ASME 2015).

4.2 Flow-Induced Vibration Analysis

The flow-induced vibration (FIV) analysis of serpentine finned tubes is performed using the obtained natural frequencies and flow field velocities as input parameters. Analyses of Kármán vortex shedding and turbulent buffeting are conducted through numerical simulation and theoretical calculation to determine tube bundle vibration responses (as shown in Fig. 5). The FIV characteristics are subsequently evaluated against the assessment criteria specified in relevant design codes (Fig. 6).

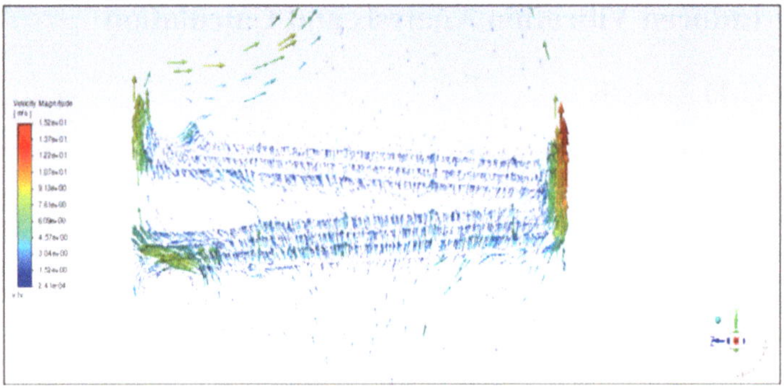

Fig. 4. Flow field streamlines of serpentine finned-tube bundle

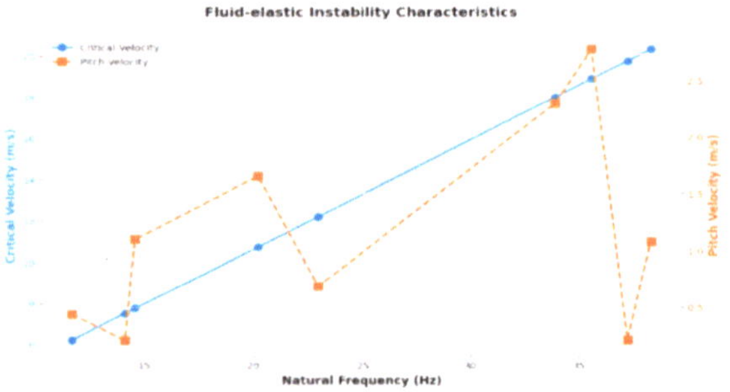

Fig. 5. Critical velocity for fluidelastic instability as a function of natural frequency

5 Conclusions

Please note that the first paragraph of a section or subsection is not indented. The first paragraphs that follows a table, figure, equation etc. does not have an indent, either.

The natural frequency of the heat exchanger tubes was calculated under current support constraints, yielding a value of 11.885 Hz, which is relatively low.

Analysis of the flow field velocity revealed that the maximum transverse gap velocity in the tube bundle clearance flow field under the current design parameters is 4.57 m/s. When compared with the critical velocity of 6.19196 m/s calculated by substituting the first-order natural frequency into the formula, the results demonstrate that the tube bundle does not undergo flow-induced vibration under cross-flow conditions.

Vibration responses of the serpentine finned-tube bundles were computed for multiple modal configurations. Results reveal significantly higher vibration amplitudes at locations remote from the support constraints. Additional supports are recommended at high-amplitude response locations to mitigate tube bundle vibration.

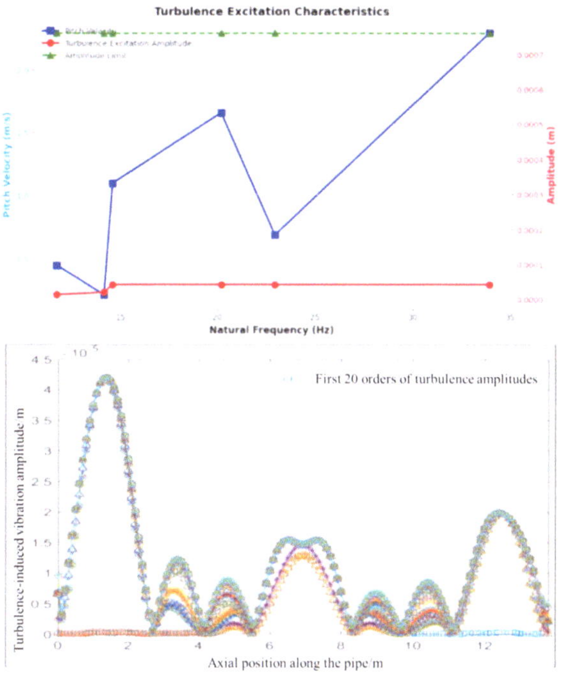

Fig. 6. Turbulent buffeting amplitude response

References

American Society of Mechanical Engineers: ASME N-1300 Standard for Operation and Maintenance of Nuclear Power Plants. ASME, New York (2007)

American Society of Mechanical Engineers: ASME BPVC Section III: Rules for Construction of Nuclear Facility Components. ASME, New York (2015)

Jo, J.C., Jhung, M.J.: Flow-induced vibration and fretting-wear predictions of steam generator helical tubes. Nucl. Eng. Des. **238**(4), 890–903 (2008)

Kaviri, A., Hosseini, S.M., Torabi, M.: Failure analysis and structural improvement for cracked circular finned tube. Eng. Fail. Anal. **92**, 95–106 (2018)

Kim, S.H., Ahn, B.H., Ha, J.M., et al.: Structural and vibration analysis considering the flow velocity of the heat exchanger. Int. J. Precis. Eng. Manuf. **17**(6), 725–732 (2016)

Kim, W.T., Kim, H.S., Lee, Y.S., et al.: Vibration analysis of pipe with elbow subject to internal pressure and temperature. In: Proceedings of the KSNVE Annual Spring Conference, pp. 223–226 (2004)

Nie, Q.D.: Flow-Induced Vibration of Shell-and-Tube Heat Exchangers. Beijing: China Petrochemical Press, 2013. Note: Duplicate of Ref. [4]—to be removed

Pettigrew, M.J., Taylor, C.E.: Vibration analysis of shell-and-tube heat exchangers: an overview—part 1: flow, damping, fluidelastic instability. J. Fluids Struct. **18**(5), 469–483 (2003)

Pettigrew, M.J., Taylor, C.E., Fisher, N.J., et al.: Flow-induced vibration: recent findings and open questions. Nucl. Eng. Des. **185**(2–3), 249–276 (1998)

Rao, S.M., Gupta, G.D., Eisinger, F.L., et al.: Computer modeling of vibration and wear in multi-span tubes with clearances at tube supports. In: Proceedings of the International Conference on Flow Induced Vibrations (1987)

Standardization Administration of China: GB/T 151–2014 Heat Exchangers. Standards Press of China, Beijing (2015)

Tubular Exchanger Manufacturers Association: TEMA Standards, 7th edn. TEMA, New York (2007)

Xu, S.G., Wang, C., Wang, W.Q.: Failure analysis of stress corrosion cracking in heat exchanger tubes during start-up operation. Eng. Fail. Anal. **51**, 1–8 (2015)

Xu, W., Li, L.Q., Xu, B.: Study on pulsating flow-induced tube bundle vibration in air coolers. Light. Ind. Mach. **37**(1), 36–42 (2019)

Multi Objective Topology Optimization of Gearbox Based on Combination Weighting and Variable Density

Quanwen He[1,2](✉), Xiaohua Qiu[1,3], and Zhijiang Xie[1,4]

[1] School of Mechanical Engineering, Southwest Jiaotong University, Chengdu 610031, China
1150850056@qq.com

[2] School of intelligent manufacturing and Information Engineering, Sichuan Technology & Business College, Dujiangyan 611830, China

[3] Wuxi Hongyuan Elastic Equipment Co., Ltd., Wuxi, Jiangsu 214000, China

[4] Chengdu Lingding Technology Co., Ltd., Chengdu 610052, China

Abstract. To address the issue of weight allocation in multi-objective topology optimization of planetary gearboxes relying on subjective experience, this paper proposes a combined weighting method integrating the Analytic Hierarchy Process (AHP) and Entropy Weight Method (EWM) to scientifically determine the optimization weights for static stiffness and multiple natural frequencies. Based on this, the Solid Isotropic Material with Penalization (SIMP) method is employed to perform multi-objective topology optimization on a cast aluminum gearbox housing with volume fraction as the constraint. The optimized structure achieves a 16.4% reduction in mass, while the maximum stress and deformation decrease by 24.4% and 26.8%, respectively. The first six natural frequencies remain stable (variation rate: 0.03% ~ 1.02%). This approach significantly enhances the static stiffness and lightweight performance of the housing while ensuring dynamic characteristics.

Keywords: Structural topology optimization · Gearbox · Combined weighting method · Variable density method

1 Introduction

The planetary gearbox is the core component of the transmission system, and its housing structure is crucial to the system performance. Topology optimization technology has been widely applied, achieving significant weight reduction and performance improvement in components such as racing gearbox housings (Zeng et al. 2022), automotive torsional dampers (Yan et al. 2026), and aluminum die cast alloy connecting rods (Nouri 2026).

However, the performance optimization of planetary gearboxes faces special challenges, as their load uniformity and dynamic characteristics are highly dependent on the distribution of box stiffness (Du et al. 2025), which requires balancing multiple objectives

T. Mankovits and N. Trisovic (Eds.): MDS 2026, LNME 2026.
https://doi.org/10.1007/978-981-92-2440-1_38

such as static stiffness and multiple natural frequencies during optimization. In traditional multi-objective topology optimization, the weight allocation of each objective often relies on subjective experience and lacks objective basis.

To overcome the shortcomings of subjective and objective weighting methods, this paper proposes a combined weighting method that combines Analytic Hierarchy Process (AHP) and Entropy Weight Method (EWM). This method introduces game theory to seek the optimal equilibrium solution of subjective and objective weights, thereby scientifically determining the comprehensive weights of each optimization objective.

2 Structural Model

The main body of the gearbox is a cylindrical shell (Fig. 1), surrounded by evenly distributed installation lugs with through holes, forming the main connection and fixing interface. The center of the structure is a large stepped concentric circular hole, used for precise support of the sun gear or main shaft bearings; Five planetary wheel axle mounting holes evenly distributed in a circular array around the central hole are used to position the planetary gear system (Table 1).

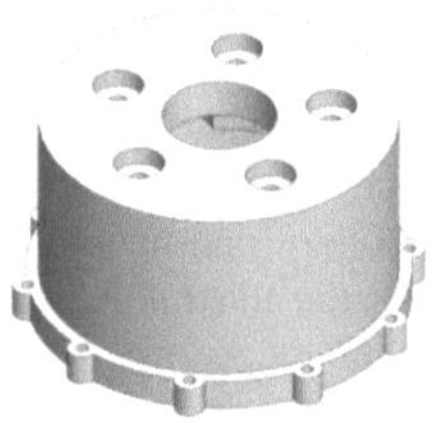

Fig. 1. Three dimensional structural model of gearbox.

Table 1. Gearbox operating parameters.

Main parameters	Numerical value
Input speed	3500
Output torque	55
Input power	0.8
Transmission ratio	24
Material	Cast aluminum

3 Static Analysis of Gearbox

3.1 Static Analysis of Gearbox

Based on finite element analysis, the maximum stress (109.29 MPa) of the gearbox prototype under rated operating conditions is located at the root of the rounded corner of the bottom installation boss, and there is also high stress at the connection of the middle reinforcement rib; The maximum displacement (0.56 μm) occurs at the edge of the top center hole, and the stiffness of the gearbox meets the design requirements (Figs. 2 and 3). The strength and stiffness design of the gearbox are sufficient, and there is a lot of room for optimization design.

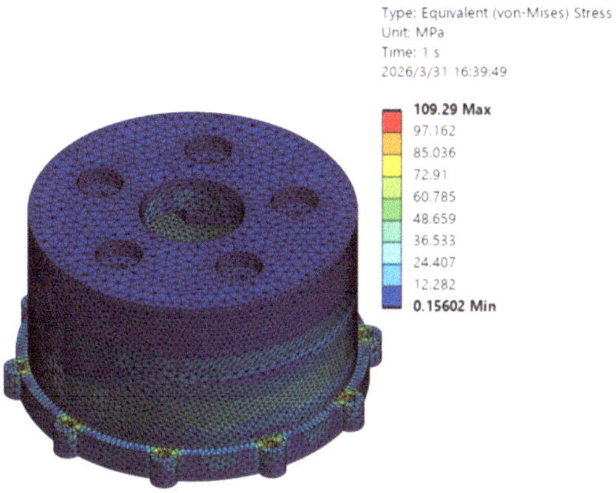

Fig. 2. Stress cloud map before optimization.

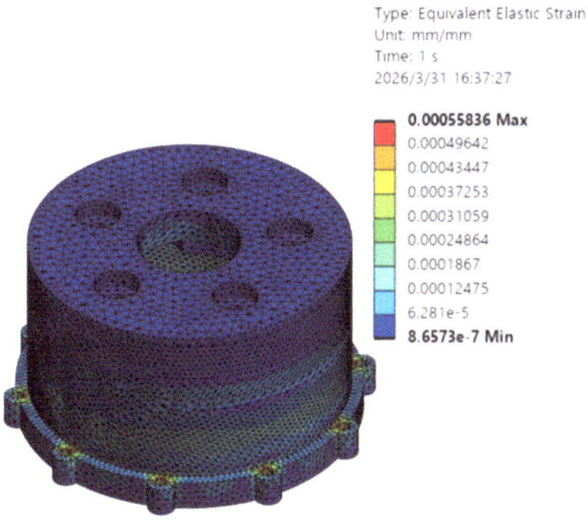

Fig. 3. Displacement cloud map before optimization.

3.2 Modal Analysis of Gearbox

To evaluate the dynamic characteristics of the gearbox and avoid resonance risks, a constrained modal analysis was conducted on the pre optimized box body, obtaining its first six natural frequencies and corresponding vibration modes (Table 2). The low order modes of the box body mainly exhibit locally dominant vibrations (Abouseda et al. 2025; Mabrouk et al. 2024).

Table 2. Natural frequency and vibration mode of the box.

Order	Frequency	Mode shape
1st order	6358 Hz	Lateral swing along the X-axis
2st order	6398 Hz	Lateral swing along the Y-axis
3st order	6398 Hz	Torsional vibration around the Z-axis
4st order	6884 Hz	Local vibration
5st order	6886 Hz	The overall bending of the side wall of the box is the main factor
6st order	7237 Hz	Composite vibration of local structure and overall wall panel

4 Combination Weighting Method for Weight Allocation

The common methods for weighting indicators mainly include subjective weighting and objective weighting. Subjective weighting methods (such as Analytic Hierarchy Process, AHP) can reflect experts' profound understanding of the static and dynamic performance of gearboxes, but are easily influenced by their experience and preferences; Objective weighting methods (such as entropy weighting method, EWM) are strictly based on the degree of variation of the data itself, but may overlook important factors in engineering practice.

To overcome the limitations of a single weighting method and seek the optimal equilibrium solution for subjective and objective weights, this paper introduces game theory and combines AHP subjective weighting with entropy weighting to construct a combined weighting model, thereby scientifically determining the weights of optimization indicators.

4.1 Subjective Weight Determination (AHP Method)

The Analytic Hierarchy Process (AHP) decomposes complex decision-making problems into levels such as objectives, criteria, and solutions by establishing a hierarchical structure model, and conducts qualitative and quantitative analysis on this basis. This article takes the optimal static and dynamic performance of the gearbox as the objective layer, and six parameters including the 5th order frequency, 1st to 4th order natural frequencies, and stiffness as the criterion layer. The 1–9 scale method is used to compare each indicator at the same level pairwise, and the scale values represent the comparison results of the relative importance of the two indicators. The judgment matrix form is as follows:

$$A = \left(a_{ij}\right)_{n \times n} = \begin{bmatrix} 1 & a_{12} & \cdots & a_{1n} \\ a_{21} & 1 & \cdots & a_{2n} \\ \vdots & \vdots & \ddots & \vdots \\ a_{n1} & a_{n2} & \cdots & 1 \end{bmatrix} \tag{1}$$

a_{ij} denotes the comparison value of the importance of the i indicator relative to the j indicator, and satisfies $a_{ji} = 1/a_{ij}$.

Construct a 6th order judgment matrix:

$$A = a_{ij} = \begin{bmatrix} 1 & 9 & 5 & 3 & 2 & 1/2 \\ 1/9 & 1 & 1/5 & 1/7 & 1/8 & 1/9 \\ 1/3 & 5 & 1 & 1/3 & 1/4 & 1/6 \\ 1/2 & 7 & 3 & 1 & 1/2 & 1/5 \\ 1/2 & 8 & 4 & 2 & 1 & 1/4 \\ 2 & 9 & 6 & 5 & 4 & 1 \end{bmatrix} \quad (2)$$

Use the root mean square method to calculate the eigenvectors, and perform normalization and consistency checks. The calculated subjective weight vector is:

$$W_s = [0.276\ 0.020\ 0.060\ 0.106\ 0.150\ 0.388]^T$$

By conducting consistency checks, it is determined that the matrix meets the consistency requirements.

4.2 Objective Weight Determination (Entropy Weight Method)

Entropy weighting method is an objective weighting method based on the concept of information entropy. The greater the degree of variation of the indicator, the smaller the information entropy, the greater the amount of information provided by the indicator, the greater its role in evaluation, and the greater its weight. Six design parameters form the raw data matrix:

$$X = (x_{ij})_{m \times n} \quad (3)$$

x_{ij} represents the value of the i alternative on the j criterion.

To eliminate the differences in dimensions and orders of magnitude of different indicators, it is necessary to standardize the original data. For benefit indicators (frequency, stiffness), the larger the value, the better. The following formula is used:

$$r_{ij} = \frac{x_{ij} - \min(x_j)}{\max(x_j) - \min(x_j)} \quad (4)$$

For cost based indicators, the smaller the value, the better, then:

$$r_{ij} = \frac{\max(x_j) - x_{ij}}{\max(x_j) - \min(x_j)} \quad (5)$$

After standardization, the matrix is obtained:

$$R = (r_{ij})_{m \times n}, 0 \le r_{ij} \le 1 \quad (6)$$

The coefficient of difference reflects the degree of differentiation of each indicator in the evaluation. The larger the coefficient of difference, the greater the importance of the indicator in evaluation. The coefficient of difference is defined as:

$$g_j = 1 - e_j \tag{7}$$

Obtain objective weights using entropy weighting method:

$$W_{OJ} = \frac{g_j}{\sum_{k=1}^{6} gk} \tag{8}$$

$$W_o = \begin{bmatrix} 0.271\ 0.0237\ 0.191\ 0.206\ 0.080\ 0.004 \end{bmatrix}^T$$

4.3 Combination Weighting Based on Game Theory

Consider the AHP weight Ws and EWM weight Wo as two sides in a game, and seek the comprehensive weight W* that balances the two by optimizing the combination coefficients. Let the comprehensive weight be:

$$W = aW_s + \beta W_o \tag{9}$$

α and β are linear combination coefficients, and $\alpha, \beta > 0$

The goal of the game theory approach is to find the optimal α and β that minimize the deviation between the comprehensive weight W and W_s, W_o. The model is:

$$\min\left(\left\|\alpha W_s^T + \beta W_o^T - W_s\right\|_2 + \left\|\alpha W_s^T + \beta W_o^T - W_o\right\|_2\right) \tag{10}$$

After obtaining the solution for α and β, normalization is performed to derive the comprehensive weight W^* (Table 3).

Table 3. Comparison of weight allocation results among various weighting methods.

Optimization metrics	AHP	EWM	Combined weight
1st Order	0.020	0.024	0.022
2st Order	0.060	0.191	0.112
3st Order	0.106	0.206	0.144
4st Order	0.150	0.080	0.121
5st Order	0.276	0.271	0.274
Stiffness	0.388	0.004	0.247

4.4 Sensitivity Analysis of the Combined Weighting Scheme

To evaluate the robustness of the AHP-EWM combined weighting method proposed in this paper against subjective judgment uncertainty, bounded perturbations were applied to the key comparison values with the greatest impact on the overall weights in the constructed judgment matrix to observe changes in the final combined weights.

Select the two indicators with the highest weights in the criterion layer: static stiffness and the 5th natural frequency. The comparison value $a_{ij} = 2$ indicates that "static stiffness" is ranked between "equally important" and "slightly more important" relative to "the 5th natural frequency".

Disturb the comparison values by ± 1, set $a'_{ij}=1$ (equally important) and $a''_{ij}=3$ (slightly important) respectively. Recalculate the AHP weight vectors W_s' and W_s'', and solve for the new combination weights W'^{*} and W'^{*} based on the game theory model. Compare the results (Table 4).

Table 4. Weight changes under sensitivity analysis.

Optimization metrics	Original combination weight	Combination weight after disturbance	Rate of change	Combination weight after disturbance	Rate of change
1st Order	0.022	0.023	+4.5%	0.021	−4.5%
2st Order	0.112	0.114	+1.8%	0.110	−1.8%
3st Order	0.144	0.147	+2.1%	0.141	−2.1%
4st Order	0.121	0.123	+1.7%	0.119	−1.7%
5st Order	0.274	0.263	−4.0%	0.284	+3.6%
Stiffness	0.247	0.230	−6.9%	0.262	+6.1%

Analysis shows that even with a significant disturbance of ± 1 level on the most important subjective judgments, the absolute value of the maximum change rate of the weights of each indicator combination is less than 7%. Therefore, the combination weighting model constructed based on game theory can effectively absorb reasonable fluctuations in subjective judgments, and the resulting weight allocation scheme has good stability and robustness.

5 Topology Optimization of Gearbox Structure

5.1 Topology Optimization of Box Structure

To synergistically improve the dynamic and static performance of the gearbox, multiple optimization sub objectives need to be integrated into a solvable single objective function . This article uses the compromise programming method to construct this comprehensive objective function (Huang et al. 2025):

$$\min F(\rho) = \Big[\omega_1(1-\mu\lambda_1(\rho))^2 + \omega 2(1-\mu\lambda_2(\rho))^2 + \omega_3(1-\mu\lambda_3(\rho))^2 + \omega_4(1-\mu\lambda_4(\rho))^2 + \omega_5(1-\mu\lambda_5(\rho))^2 + \omega_C(1-\mu\lambda_C(\rho))^2\Big]^{\frac{1}{2}} \quad (11)$$

ω_1, ..., ω_5, ω_C are the weighting coefficients for sub-objective; λ_1, ..., λ_5 are the corresponding first five natural frequencies.

The gearbox housing was topologically optimized using the SIMP variable-density method (Ma et al. 2023), with the optimization criterion method employed as the solver. The maximum iteration steps were set to 200, the minimum relative density to 0.001, and the convergence tolerance to 0.15%, thereby seeking the optimal material distribution within the design space (Fig. 4).

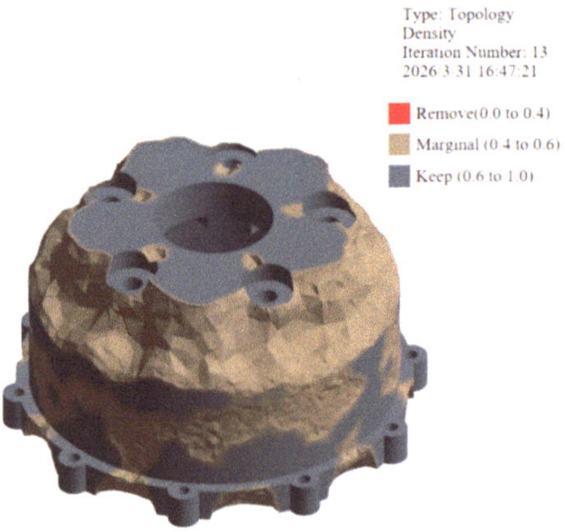

Fig. 4. Topology optimization cloud map.

Geometric reconstruction and feature redesign were carried out based on casting process constraints, following the following key manufacturing principles:

Minimum wall thickness control. To ensure the filling and solidification quality of cast aluminum, the minimum wall thickness in all areas of the reconstructed model is controlled to be above 4 mm, meeting the process requirements of typical sand casting for this material.

Add draft angle. To facilitate demolding, a 1.5° draft angle has been added to all feature surfaces that may form cavities or protrusions.

Feature geometry simplification. Transform the transition region formed in topology optimization into rounded corners that are easier to process and have a more reasonable stress distribution.

Through the above processing, the final box model (Fig. 5) not only inherits the lightweight configuration optimized by topology.

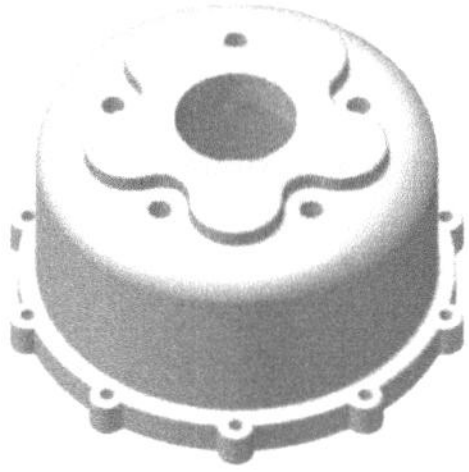

Fig. 5. Optimized gearbox model.

5.2 Comparative Analysis After Optimization

Comparison of static and modal performance. Reduce quality by 16.4% and achieve significant lightweighting goals; The static performance has been synchronously improved, with a maximum stress reduction of 24.4%, a maximum deformation reduction of 26.8%, and a significant increase in structural static stiffness. The rate of change of the first six natural frequencies ranges from 0.03% to 1.02%, all of which remain stable or slightly increase, proving that optimization did not cause harmful modal frequency drift and avoided resonance risk.

Verification of transient response under dynamic meshing load. Under dynamic excitation, the maximum equivalent stress of the box is 94.7 MPa, which occurs in the rounded corner area of the bottom installation protrusion, consistent with the static stress concentration position. The maximum dynamic stress is about 14.7% higher than the static maximum stress (82.58 MPa), which is due to the superposition effect of dynamic loads, but its value is still much lower than the 240 MPa of cast aluminum material (Table 5).

Table 5. Comparison of performance parameters before and after optimization.

Optimization metrics	Before optimization	Optimized	Rate of change	Effect
Mass (kg)	1.63	1.36	−16.4%	Significant lightweighting
Max. stress (MPa)	109.29	82.58	−24.4%	intensity enhancement
Max. deformation (um)	0.56	0.41	−26.8%	stiffness enhancement
1st Order	6358	6360	+0.03%	Maintain stability
2st Order	6398	6403	+0.08%	Maintain stability
3st Order	6398	6452	+0.84%	Stable and rising

(continued)

Table 5. (*continued*)

Optimization metrics	Before optimization	Optimized	Rate of change	Effect
4st Order	6884	6886	+0.03%	Maintain stability
5st Order	6886	6956	+1.02%	Stable and rising
6st Order	7237	7245	+0.11%	Maintain stability

6 Conclusions

By using game theory to seek the optimal balance of subjective and objective weights, and combining it with variable density topology optimization methods, the subjective problem of weight allocation in multi-objective topology optimization of gearboxes is solved, achieving structural topology optimization of gearboxes. Under the constraint of a volume fraction of 0.7, the optimization scheme achieved significant weight reduction, with a 16.4% decrease in mass and a 24.4% and 26.8% decrease in maximum stress and deformation, respectively. Additionally, the first six natural frequencies remained stable (varying by 0.03% to 1.02%). This method effectively balances engineering experience and data objectivity, significantly improving static stiffness and achieving lightweight goals.

References

Abouseda, A.I., Doruk, R., Emin, A., Guede, J.M.L.: Experimental investigation of energy efficiency, SOC estimation, and real-time speed control of a 2.2 kW BLDC motor with planetary gearbox under variable load conditions. Energies. **19**(1), 36 (2025). https://doi.org/10.3390/en19010036

Du, X. et al.: Study on plastic overload fracture of planet gears in wheel-side reducer based on dynamic simulation and fracture analysis. J. Fail. Anal. Prev. **26**(1), 1–10 (2025). https://doi.org/10.1007/s11668-025-02345-x

Huang, H., Wang, Z., Yang, Z., Liu, H., Lu, Z.: Multi-objective topology optimization of aero-engine accessory gearbox case based on compromise programming method. SCIENCE CHINA Technol. Sci. **68**(7), 1720601 (2025). https://doi.org/10.1007/s11431-024-2915-3

Ma, W., Lu, Y., Wang, P., Wang, Y., Wang, J.: Double optimization design of the formula racing car frame based on the variable density method and the joint variable method. Appl. Sci. **13**(18) (2023). https://doi.org/10.3390/app131810155

Mabrouk, M., Gao, P., Yan, K., Xie, Y., Yan, Q.: A comparative analysis of multi-scale and Rayleigh approaches in capturing eigenfrequencies and mode shape evaluation in planetary gear transmission systems of medium and heavy trucks. Machines. **12**(1) (2024). https://doi.org/10.3390/machines12010048

Nouri, H.: Multi-objective topology optimization of aluminium die casting alloy connecting rod using the bacterial foraging algorithm. Int. J. Interact. Des. Manuf. **1–19** (2026). https://doi.org/10.1007/s12008-026-02506-8

Yan, Z., Wang, S., Yao, M., Zhang, N., Liu, S.: Structural topology and parameter optimization of torsional dampers used in parallel hybrid electric vehicles based on the working conditions. J. Vib. Control. **32**(5–6), 1219–1231 (2026). https://doi.org/10.1177/10775463241312949

Zeng, Z., Zhang, Y., Tang, D.: Lightweight design of gearbox housing of Baja racing car based on topology optimization. J. Phys. Conf. Ser. **2235**(1), 012084 (2022). https://doi.org/10.1088/1742-6596/2235/1/012084

Defining Clinically Relevant Polyethylene Wear Debris Thresholds in Total Knee Arthroplasties: Implications for Osteolysis Risk Assessment

Gusztáv Fekete(✉)

Department of Materials Science and Mechanical Engineering, Audi Hungaria Faculty of Engineering, Széchenyi István University, Egyetem tér 1, 9026 Győr, Hungary
fekete.gusztav@sze.hu

Abstract. This paper, in one hand, investigates the clinically tolerable, and therefore admissible, volume of ultra-high-molecular-weight polyethylene (UHMWPE) wear in total knee arthroplasties (TKAs) through an extensive literature review. On the other hand, the paper evaluates whether the existing numerical wear models can reliably predict TKA wear volumes and whether the estimated results for a given TKA geometry are credible enough to classify the implant's osteolysis risk as low, medium, or high. Based on the findings of the first part, three clinically meaningful linear wear rate thresholds have been established from current orthopaedic and tribological literature: below 0.05 mm/year (low osteolysis risk), 0.05–0.1 mm/year (medium risk), and above 0.1 mm/year (high risk). By the use of typical tibial insert contact areas, a lower and upper level have been determined (12.5–37.5 mm^3/year or million cycle), which range covers most TKA designs. From the outcomes of the second part, the Archard-based wear formulations, implemented in Finite Element Models (FEM) and Multibody Dynamics models (MBD), showed good agreement with experimental knee simulator data under ISO 14243-1 loading, confirming that numerical models validity and the investigated TKAs reliability for long-term use. This summary demonstrated that (a) the available FEM and MBD models coupled with different wear equations can reliably predict TKA wear volumes (b) the less sophisticated MBD model is also capable to estimate global wear propagation, therefore to decide the implant's osteolysis risk.

Keywords: Archard equation · Volumetric wear · Osteolysis · Threshold

1 Introduction

Ultra-high-molecular-weight polyethylene (UHMWPE) wear particles in total knee arthroplasties (TKAs) have been detected as possible long-term source of inflammatory reaction, which are inflicted by the appearing macrophages (Taki et al. 2005). These macrophages engulf and digest pathogens, such as cancer cells, microbes, or foreign substances such as UHMWPE wear debris since these particles do not act as healthy body cells on their surface (Hallab et al. 2012; Ingham and Fisher 2005). This sustained inflammation can be the cause of accelerated localized bone degradation near the

T. Mankovits and N. Trisovic (Eds.): MDS 2026, LNME 2026.
https://doi.org/10.1007/978-981-92-2440-1_39

implant, which can lead to aseptic loosening and eventually to implant failure (Green et al. 2000; Ormsby et al. 2019).

Since the reduction of the biological reaction is the key question in this matter, it can be achieved either by reducing wear particle generation or applying a pharmacological modulation on the material of the tibial insert to lower the macrophage-driven inflammatory cascade.

With regard to the reduction of wear particle formation, there are several possibilities from the use of highly cross-linked UHMWPE (Randall et al. 2025; Thoen et al. 2025) to the optimized implant design, which include lower conformity (Brockett et al. 2017), patient-specific design (Koh et al. 2019) and novel fixation methods to reduce micro motion and backside wear (Choudhari et al. 2024).

In the matter of pharmacological modulation, most studies focus on either metformin as an anti-inflammatory agent (Yan et al. 2018; Rocha et al. 2010), which could limit the osteolysis (progressive loss of bone tissue) induced by implant debris, or on the mesenchymal stem cell use, which has been proven to be particularly effective in attenuating the UHMWPE-induced osteolysis (Ding et al. 2025).

Nevertheless, from a mechanical engineering perspective, it is essential to quantify the amount of wear debris which does not yet mean a harmful amount for the human body, and also to know how can be the amount of wear determined with regard to different TKAs.

Therefore, this paper aims to provide answers to two scientific questions. First, based on the available relevant literature, it will deduce how much UHMWPE wear debris mean (a) low risk of osteolysis or virtually no inflammation and/or biological reaction; (b) mild risk of osteolysis; (c) high risk of osteolysis and directly leading to aseptic loosening.

Second, the paper reviews numerical wear models, which are capable of accurately quantifying wear propagation in the knee joint. The result of this review will help determining whether a given TKA configuration is likely to generate wear debris at levels, which are associated with mild or high osteolysis risk.

2 Materials and Methods

2.1 Conversion Method of Different Wear Rates

In the literature, let it be orthopaedic or engineering, wear rate is often quantified as the amount of wear debris volume (mm^3) produced either per million cycle (mc) or per year. The one million cycle equals roughly a year is a convention used by researchers and testing labs. However, another commonly used rate is the linear wear rate (mm/year), which can be converted to mm^3/year (or mm^3/mc), as follows:

$$Volumetric\ wear \left(\frac{mm^3}{year}\right) \approx Linear\ wear \left(\frac{mm}{year}\right) \cdot \left(A_{eff.med.} + A_{eff.lat.}\right) \quad (1)$$

Where $A_{eff.med.}$ and $A_{eff.lat.}$ represent the effective contact area (or worn area) of the tibial insert. To use a conversion, the contact area must be known for the applied TKA. Typical ranges are show in Table 1.

It must be noted that the contact area depend on the motion (Zou et al. 2023), and the angle of flexion as well (Castellarin et al. 2023), however an estimation can be given if these values are considered.

Table 1. TKA designs with the range of contact area (Zou et al. 2023)

Implant design	Contact area (mm^2)
MB CR (Mobile Bearing Cruciate Retaining)	500–550
MB CRU (Mobile Bearing Cruciate Retaining Ultracongruent)	700–750
FB PS (Fixed Bearing Posterior Stabilized)	350–400
FB CCK (Fixed Bearing Constrained Condylar Knee)	250–300

2.2 Threshold of Wear Debris Volume

With regard to the tolerable amount of wear debris, it must be mentioned that quantitative thresholds are mostly defined in the hip literature, however the biological response to wear debris is similar across all joints (Couto et al. 2020).

Therefore, the following summary (see Table 2) has been deduced with regard to the relevant literature. The data in the table indicate that linear wear rates below 0.05 mm/year mean low-risk, as they are consistently associated with a minimal incidence of osteolysis.

Linear wear rates in the range of 0.05–0.1 mm/year represent a medium risk zone, where the amount of the cumulative particles may become sufficient to trigger macrophage-induced bone loss over long-term.

Finally, linear wear rates exceeding 0.1 mm/year are widely recognized as a high-risk threshold for osteolysis. Above this amount of wear debris, active biological reaction and higher probability of periprosthetic bone loss can be expected. These thresholds provide a clinically meaningful framework for interpreting wear performance in contemporary TKA designs.

Table 2. Summary of polyethylene linear wear rate versus osteolysis risk

Risk category	Linear wear rate (mm/year)
Low risk of osteolysis (Lachiewicz et al. 2016; Dumbleton et al. 2002)	< 0.05
Medium risk of osteolysis (Wilkinson et al. 2005; Elke and Rieker 2018)	≈ 0.05–0.1
High risk of osteolysis (Zhu et al. 2001)	> 0.1

By having these thresholds, it is possible to convert linear wear rates, from Table 1, to estimate the tolerable amount of wear volume, where the risk of osteolysis is low. In addition, it is worthy to plot the the maximal contact area range of the different TKAs versus the expected wear rate per million cycle (or per year) to investigate possible trends (see Fig. 1). Each prosthesis type (FB CCK, FB PS, etc.) was associated with its contact surface (shown as a blue column) and the corresponding wear volume, which indicates low risk of osteolysis (shown as an orange circle).

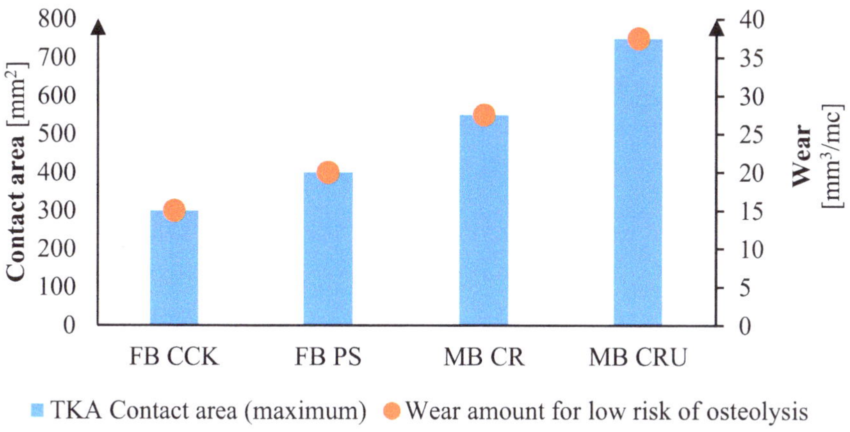

Fig. 1. Contact area of different TKAs with their allowable wear rate

Let us use Eq. (1) to calculate the minimal amount of wear for low risk of osteolysis (in case FB CCK TKA):

$$Minimal\ volumetric\ wear \approx 0.05 \cdot 250 = 12.5 \left(\frac{mm^3}{year}\ or\ \frac{mm^3}{mc} \right) \tag{2}$$

While the maximum amount of wear for low risk of osteolysis (in case MB CRU):

$$Maximal\ volumetric\ wear \approx 0.05 \cdot 750 = 37.5 \left(\frac{mm^3}{year}\ or\ \frac{mm^3}{mc} \right) \tag{3}$$

Based on these results, it can be analyzed how the available wear models, coupled with finite element method or multibody dynamic method, can estimate wear debris volume, and whether the estimated values fall into the low risk zone of osteolysis.

2.3 Wear Models for Predicting Volumetric Wear

Several wear models are available in the relevant literature (see Table 3), which models are mostly based on the Archard and Hirst equation (Archard and Hirst 1956).

Where dV is the infinitesimal wear volume (mm^3), k is the wear factor (mm^3/Nm), A is the contact area (mm^2), p is the surface pressure (N/mm^2), ds is the infinitesimal slide length [m], C is the dimensionless wear coefficient (−), CS is the cross-shear ratio [−], E is the elastic modulus of the tibial insert (N/mm^2), μ is the coefficient of friction (−), S/R is the slide-roll ratio (−), v_{slide} is the relative sliding velocity between the femoral and the tibial component (m/s), dt is the infinitesimal time (s) and F_N is the applied normal force (N).

Most above-mentioned wear models (Shu et al. 2001) were integrated into Finite Element Models (FEM), while the model of Fekete (Fekete 2026) was integrated into a Multibody Dynamic (MBD) model to simulate wear propagation under force controlled ISO 14243–1 (ISO 14243-1 2009) gait standards, in order that the numerical results could

Table 3. Summary of the relevant wear models in the literature

Author	Equation	Main strength
Archard and Hirst 1956	$dV = k \cdot A \cdot p \cdot ds$	Simple baseline model
Shu et al. 2001	$dV = A \cdot C(CS, p, E) \cdot ds$	Captures CS and material stiffness effects
Zhang et al. 2017	$dV = A \cdot C(CS) \cdot ds$	Captures CS effect
Abdelgaied et al. 2011	$dV = A \cdot C(CS) \cdot ds$	Captures CS effect
Turell et al. 2003	$dV = k \cdot A \cdot CS \cdot F_N \cdot ds$	Links CS directly with applied load
Koh et al. 2020	$dV = k \cdot A \cdot p \cdot \sqrt{(1 + 3\mu^2)} \cdot ds$	Incorporates frictional effect
Fekete 2026	$dV = k \cdot \mu \cdot F_N \cdot S/R \cdot v_{slide} \cdot dt$	Incorporates frictional-load-, and kinematic effects

be compared to wear data found in the literature. Regarding the material properties, it is worth noting that the cited authors used UHMWPE GUR 1050 or GUR 1030 for the tibial insert in their simulations.

3 Results and Discussion

As a result of the comparison, it could be stated that the available wear models, combined with FEM or MBDM models, obtained results which correlated well with the experimental ones in case of standardized gait motion (see Fig. 2). Based on these synthesized findings several key conclusions can be drawn.

First, although FEM models are considered the most adequate numerical mean to determine wear propagation since they can integrate contact mechanics (material nonlinearities, cross-shear ratio, stress redistribution during gait cycles, etc.), MBD models, if they are augmented with other wear influencing parameters (Fekete 2026), can also provide suitable and accurate information about global wear propagation.

Second, MBD models have the advantages that they can be quickly built and they require low computation cost during simulation. In case of analysing new TKA geometries whether they are suitable for long-term wear propagation, MBD modelling can be a quick and fairly adequate way.

The results also show that the less sophisticated MBD model is also capable to estimate global wear propagation, therefore to decide the implant's osteolysis risk.

It must be also mentioned that presented results suggest an association between increasing tibiofemoral contact area and higher allowable polyethylene wear volumes (Fig. 1) before reaching a threshold associated with low osteolysis risk. It is well established that mobile-bearing TKA configurations, particularly cruciate-retaining and unconstrained variants, exhibit substantially larger maximum contact areas compared with fixed-bearing designs, coinciding with a greater tolerance for volumetric wear. However, it has also been lately exposed that constrain level (on rotation, anterior-posterior movement, etc.) due to different configuration may have minimal effect on abrasive wear propagation (Fekete 2025). This suggest that contact area cannot solely influence TKA wear.

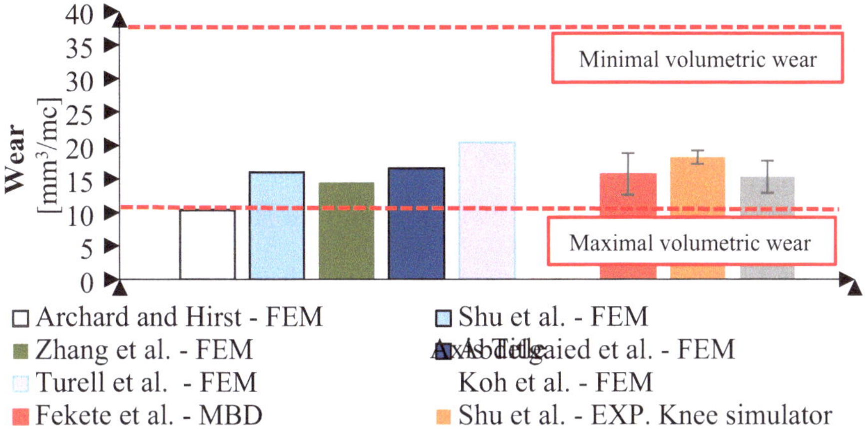

Fig. 2. Numerical and experimental volumetric wear from different sources

Conclusively, the methodological novelty of this work lies in the proposed clinically valuable osteolysis thresholds, which is based on the amount of polyethylene wear determined numerically. By converting linear wear thresholds into volumetric wear thresholds using realistic tibial contact areas, the study enables direct, risk-based interpretation of FEM and MBD wear predictions. In addition, it demonstrates that an augmented MBD model can reliably classify osteolysis risk, providing a computationally efficient alternative for early-stage TKA evaluation.

4 Conclusions

This paper established clinically significant UHMWPE wear-rate thresholds for TKAs and demonstrated their utility in estimating osteolysis risk. Based on the orthopaedic and tribological literature, three linear wear rate categories were identified: below 0.05 mm/year, 0.05–0.1 mm/year, and above 0.1 mm/year, which correspond to low, medium, and high biological risk, respectively. When converted to volumetric wear using typical tibial insert contact areas, these thresholds provide a practical framework for interpreting wear performance in contemporary TKA designs. The review of numerical wear models showed that Archard-based formulations, implemented in both FEM and MBD environments, yield volumetric wear predictions that correlate well with standardized ISO gait data. Although FEM remains the most comprehensive approach for modelling contact mechanics, augmented MBD models can offer a computationally efficient alternative capable of estimating global wear propagation. Together, these findings support the use of numerical wear modelling as a credible tool for preclinical osteolysis-risk assessment in TKA development.

Acknowledgments. The Author wishes to thank the kind support of the Department of Materials Science and Mechanical Engineering, Audi Hungaria Faculty of Engineering, Széchenyi István University and the Ningbo University.

References

Abdelgaied, A., Liu, F., Brockett, C., Jennings, L., Fisher, J., Jin, Z.: J. Biomech. **44**(6), 1108–1116 (2011). https://doi.org/10.1016/j.jbiomech.2011.01.027

Archard, J.F., Hirst, W.: Proc R Soc A. **236**(1206), 397–410 (1956). https://doi.org/10.1098/rspa.1956.0144

Brockett, C.L., Carbone, S., Fisher, J., Jennings, L.M.: Proc. Inst. Mech. Eng. H. **232**(2), 127–134 (2017). https://doi.org/10.1177/0954411917746433

Castellarin, G., Bori, E., Rapallo, L., Pianigiani, S., Innocenti, B.: Art. **5**(1), 3 (2023). https://doi.org/10.1186/s42836-022-00157-0

Choudhari, A. et al.: Applications of Biotribology in Biomedical Systems, pp. 345–432 (2024). https://doi.org/10.1007/978-3-031-58327-8_13

Couto, M. et al.: Front. Mater. **7**, 274 (2020). https://doi.org/10.3389/fmats.2020.00274

Ding, K. et al.: J. Inflamm. Res. **18**, 15717–15735 (2025). https://doi.org/10.2147/JIR.S557244

Dumbleton, J.H., Manley, M.T., Edidin, A.A.: J. Arthroplast. **17**(5), 649–661 (2002). https://doi.org/10.1054/arth.2002.33664

Elke, R., Rieker, C.B.: Proc. Inst. Mech. Eng. H. **232**(8), 753–758 (2018). https://doi.org/10.1177/0954411918784982

Fekete G.: J Mech Med Biol. (2025). doi:10.1142/S0219519425400561.

Fekete, G.: J. Tribol. **148**(5), 051110 (2026). https://doi.org/10.1115/1.4070693

Green, T.R., Fisher, J., Matthews, J.B., Stone, M.H., Ingham, E.: J. Biomed. Mater. Res. **53**(5), 490–497 (2000). https://doi.org/10.1002/1097-4636(200009)53:5<490::AID-JBM7>3.0.CO;2-7

Hallab, N.J., McAllister, K., Brady, M., Jarman-Smith, M.: J Biomed Mater Res B Appl Biomater. **100**(2), 480–492 (2012). https://doi.org/10.1002/jbm.b.31974

Ingham, E., Fisher, J.: Biomaterials. **26**(11), 1271–1286 (2005). https://doi.org/10.1016/j.biomaterials.2004.04.035

ISO 14243-1. Geneva: International Organization for Standardization (2009)

Koh, Y.G., Jung, K.H., Hong, H.T., Kim, K.M., Kang, K.T.: J. Clin. Med. **8**(11), 2023 (2019). https://doi.org/10.3390/jcm8112023

Koh, Y.-G., Park, K.-M., Lee, H.-Y., Park, J.-H., Kang, K.-T.: J. Orthop. Surg. Res. **15**(1), 24 (2020). https://doi.org/10.1186/s13018-020-1548-4

Lachiewicz, P.F., Soileau, E.S., Martell, J.M.: Clin. Orthop. Relat. Res. **474**(2), 365–371 (2016). https://doi.org/10.1007/s11999-015-4319-5

Ormsby, R.T. et al.: Acta Biomater. **87**, 296–306 (2019). https://doi.org/10.1016/j.actbio.2019.01.047

Randall, Z.D., Mologne, M.S., Gaziano, D.J., Clohisy, J.C., Bendich, I.: J. Arthroplast. **S0883-5403**(25), 01563–01563 (2025). https://doi.org/10.1016/j.arth.2025.12.019

Rocha, M.F.G., Mansur, A.A.P., Martins, C.P.S., Barbosa-Stancioli, E.F., Mansur, H.S.: Open Biomed. Eng J. **4**, 107–112 (2010). https://doi.org/10.2174/1874120701004010107

Shu, L., Hashimoto, S., Sugita, N.: Ann. Biomed. Eng. **49**(1), 322–333 (2021). https://doi.org/10.1007/s10439-020-02555-4

Taki, N. et al.: J. Orthop. Res. **23**(2), 376–383 (2005). https://doi.org/10.1016/j.orthres.2004.08.023

Thoen, P.S., Lindalen, E., Nordsletten, L., Röhrl, S.M.: Bone Joint J. **107-B**(11), 1182–1188 (2025). https://doi.org/10.1302/0301-620X.107B11.BJJ-2025-0273.R1

Turell, M., Wang, A., Bellare, A.: Wear. **255**(7–12), 1034–1039 (2003). https://doi.org/10.1016/S0043-1648(03)00357-0
Wilkinson, J.M., Hamer, A.J., Stockley, I., Eastell, R.: J. Orthop. Res. **23**(3), 520–525 (2005). https://doi.org/10.1016/j.orthres.2004.11.005
Yan, Z. et al.: Mol. Med. **24**(1), 20 (2018). https://doi.org/10.1186/s10020-018-0013-x
Zhang, J., Chen, Z., Wang, L., Li, D., Jin, Z.: Tribol. Int. **109**, 382–389 (2017). https://doi.org/10.1016/j.triboint.2016.10.050
Zhu, Y.H., Chiu, K.Y., Tang, W.M.: J. Orthop. Surg. (Hong Kong). **9**(1), 91–99 (2001). https://doi.org/10.1177/230949900100900117
Zou, D. et al.: Bioengineering (Basel). **10**(3), 290 (2023). https://doi.org/10.3390/bioengineering10030290

Author Index

T. Mankovits and N. Trisovic (Eds.): MDS 2026, LNME 2026.
https://doi.org/10.1007/978-981-92-2440-1

MIX
Papier aus verantwortungsvollen Quellen
Paper from responsible sources
FSC® C105338

If you have any concerns about our products,
you can contact us on
ProductSafety@springernature.com

In case Publisher is established outside the EU,
the EU authorized representative is:
Springer Nature Customer Service Center GmbH
Europaplatz 3, 69115 Heidelberg, Germany

Printed by Libri Plureos GmbH
in Hamburg, Germany